The Soviet Atomic Project

How the Soviet Union Obtained the Atomic Bomb

The Soviet Atomic Project

How the Soviet Union Obtained the Atomic Bomb

Lee G. Pondrom

University of Wisconsin-Madison, USA

NEW JERSEY · LONDON · SINGAPORE · BEIJING · SHANGHAI · HONG KONG · TAIPEI · CHENNAI · TOKYO

Published by

World Scientific Publishing Co. Pte. Ltd.

5 Toh Tuck Link, Singapore 596224

USA office: 27 Warren Street, Suite 401-402, Hackensack, NJ 07601

UK office: 57 Shelton Street, Covent Garden, London WC2H 9HE

Library of Congress Cataloging-in-Publication Data

Names: Pondrom, Lee G., 1933–   author.

Title: The Soviet atomic project : how the Soviet Union obtained the atomic bomb /
  Lee G. Pondrom, University of Wisconsin-Madison, USA.

Description: Singapore ; Hackensack, NJ : World Scientific Publishing Co. Pte. Ltd., [2018] |
  Includes bibliographical references and index.

Identifiers: LCCN 2018018462| ISBN 9789813235557 (hardcover ; alk. paper) |
  ISBN 9813235551 (hardcover ; alk. paper)

Subjects: LCSH: Atomic bomb--Soviet Union--History. | Nuclear physics--Research--
  Soviet Union--History. | Scientists--Soviet Union. | Engineers--Soviet Union. |
  Nuclear industry--Soviet Union--History--20th century. |
  Nuclear weapons--Soviet Union--20th century.

Classification: LCC QC789.2.S65 P66 2018 | DDC 355.8/25119094709045--dc23

LC record available at https://lccn.loc.gov/2018018462

British Library Cataloguing-in-Publication Data

A catalogue record for this book is available from the British Library.

First published 2018 (Hardcover)
Reprinted 2020 (in paperback edition)
ISBN 978-981-122-137-8 (pbk)

Copyright © 2018 by World Scientific Publishing Co. Pte. Ltd.

For any available supplementary material, please visit
https://www.worldscientific.com/worldscibooks/10.1142/10865#t=suppl

Acknowledgements

Many people helped with the preparation of this book. Olga Pobedinskaya was my Russian native speaker who made sure that my translations were correct. Judith Kozminski of Media Solutions at UW Madison was my chief formatter and publications expert, who guaranteed that the document was 'camera ready.'

Several Russian émigré colleagues at the University of Wisconsin took an interest in the book, and helped on occasion with some difficult technical passages in Russian. I would like to thank Vladimir Mirnov, Natalya Perkins, and Andrei Chubukov. Perkins and Chubukov are now at the University of Minnesota, where they join another group of physics emigres that often spend summers in Aspen. Colorado. I had useful discussions in Aspen with Arkady Vainshtein of the University of Minnesota.

Kerry Kresse, the Physics Department librarian, was very helpful in obtaining sources, particularly through interlibrary loan. I used the collections at UW Memorial Library extensively.

Tanya Buckingham and Alex Nelson of the UW Madison Cartography Department prepared my maps. Brandy Wilcox of the UW Madison German Department helped me with translation of von Ardenne's memoir.

I am grateful to the Department of Physics of UW Madison for its support during the preparation of the manuscript.

My two NKVD master spies, Pavel Sudoplatov and Aleksandr Feklisov, are no longer alive, but they wrote very interesting memoirs that I read before I started working on this book. They served as inspirations.

Valery Burov on the scientific staff at JINR Dubna has given valuable aid in supplying me with bound copies of the HISAP96 Proceedings that furnished the basis for this story, and in continuing encouragement and support. Elena Rusakovich at Dubna has arranged for my many visits to Russia, and obtained access for me to the Kurchatov Institute. Anatoly Shikov, Raisa Kuznetsova, and Ekaterina Kolesnikova of the Kurchatov Institute, Moscow, and Artem Bekzhanov of JINR Dubna were my companions and guides through the Institute grounds in 2014. Artem took my picture in front of the F-1 reactor – Figure 7.3.

I corresponded by email with several authors, some of whom responded to my queries. Pavel Oleinikov of Wesleyan University gave me useful information and encouragement, and Alex Wellerstein of Stevens Institute did me a huge favor by posting on a web site the Russian files that I refer to as "Atmonyy_proekt_SSSR" in over 100 footnotes in this book. The files are transcriptions of de-classified official Soviet documents and correspondence pertaining to the Soviet Atomic Project prepared by L. D. Ryabev of the Laboratory VNIIEF Sarov, referred to in this book as Arzamas-16/KB-11. Much of my detailed information came from this source.

I should thank Sergey Brin and Larry Page for inventing GOOGLE, without which this book would have been impossible. Russian GOOGLE is alive and well, and I use it daily.

Last but not least I am grateful to my wife Cyrena for her forbearance and support over the five years it took to do this.

Preface

I was trained as a physicist, not an historian. I am a member of the 'next generation' of physicists, those who entered the profession in the 1960's, and profited from the federal support of science that was a result of the efforts of our predecessors. I was fortunate to be a graduate student at the University of Chicago in the 1950's, when the Manhattan Project was only ten years old, and there was plenty of aura from its achievements. This book is partly the result of this early exposure. Manhattan Project lore was hard wired as a part of my education.

I have long had a certain fascination for the Soviet Union. It was a detached fascination – I never had a desire to live under its umbrella – but I found the monolith to be interesting, and I have visited there many times. My graduate school required two foreign languages, and mine were German and French. I actually knew some German, and that came in handy in reading the memoirs of German physicists who worked in the USSR after WWII, but my French was, and still is, pitiful. However, I passed the language exam. This is one reason why foreign languages are no longer required for a graduate degree in physics – the exams did not really test language competence. Another reason is that almost everyone now speaks English.

I started learning Russian on active duty as a lieutenant in the US Air Force after graduate school. A Russian émigré on the air base offered language classes in our spare time. That was a long time ago, but I have kept it up, and now read it fairly well.

That is the third component in the composition of this book – a knowledge of the Manhattan Project, an interest in the Soviet Union, and an ability to read Russian. The fourth component is an empathy with the Soviet physicists. The cold war placed many hurdles in the path of scientific exchange, but did not cut it off. Mutual interest and enthusiasm for physics research resulted in international conferences that were well attended by both sides, and there were, and still are, East-West exchange agreements for visitors to carry out experiments at reciprocal national laboratories. Soviet physicists have worked in Western Europe and the United States, and western physicists have done experiments in the Soviet Union and the Russian Federation.

Thus it was that around 1970 a group of scientists, technicians, and their families from the Joint Institute for Nuclear Research (JINR) Dubna, USSR, arrived in the western suburbs of Chicago to collaborate on one of the first experiments at Fermi National Accelerator Laboratory, where the author also worked. They remained at Fermilab for several years, and I became acquainted with the Dubna staff.

My colleagues at Dubna are second generation, as am I. This book is about their parents, those who pioneered physics research in the Soviet Union, and who successfully delivered nuclear weapons to the Motherland. The end product is a terrible weapon, but along the path to its achievement are many experiments and calculations that are physics problems that could be done in the peaceful environment of any research university or laboratory. Both the Manhattan Project and the Soviet Atomic Project applied a curtain of secrecy to university physics department research, but continued to carry on with it. The story of Leonid Nemenov and parts of a cyclotron in blockaded Leningrad during WWII, told in Chapter 7, is experimental physics in unusual circumstances, completely unrelated to atomic bombs.

One consequence of my physics background is a natural curiosity regarding how things work. In following the development of nuclear weapons I ran across many unfamiliar processes and techniques that I attempted to understand to the best of my ability. The results of my efforts are displayed in the Appendices to this book, written at the level of an undergraduate physics text. The reader not interested in technical information may skip most of these, but I would even so recommend Appendix G on encryption and decryption, for some insight into the secret world of spy communications, and Appendix H, where some of the discrepancies among accounts of Soviet espionage are explored. Neither appendix has any differential equations.

I hope the reader finds the tale of successful pursuit of nuclear energy by scientists and technicians working in Joseph Stalin's Soviet Union to be as absorbing as I have.

To my wife, Cyrena

WESTERN SOVIET UNION
500 km
Baltic Sea
Black Sea
Caspian Sea
TURKEY
CHINA
Tallinn
Leningrad
Riga
Vilnius
Virebsk
Minsk
Mogilev
Smolensk
Vyazma
Moscow
Rostov
Serpukhov
Tula
Gomel
Bryansk
Orel
Kiev
Kursk
Voronezh
Kharkov
Odessa
Sevastopol
Sochi
Sukhumi
Tbilisi
Stalingrad
Gorky
Kazan
Sverdlovsk
Novosibirsk
Chelyabinsk
Magnitogorsk
Semipalatinsk
Baikonur
Tashkent
45 N
45 E
30 E
60 E

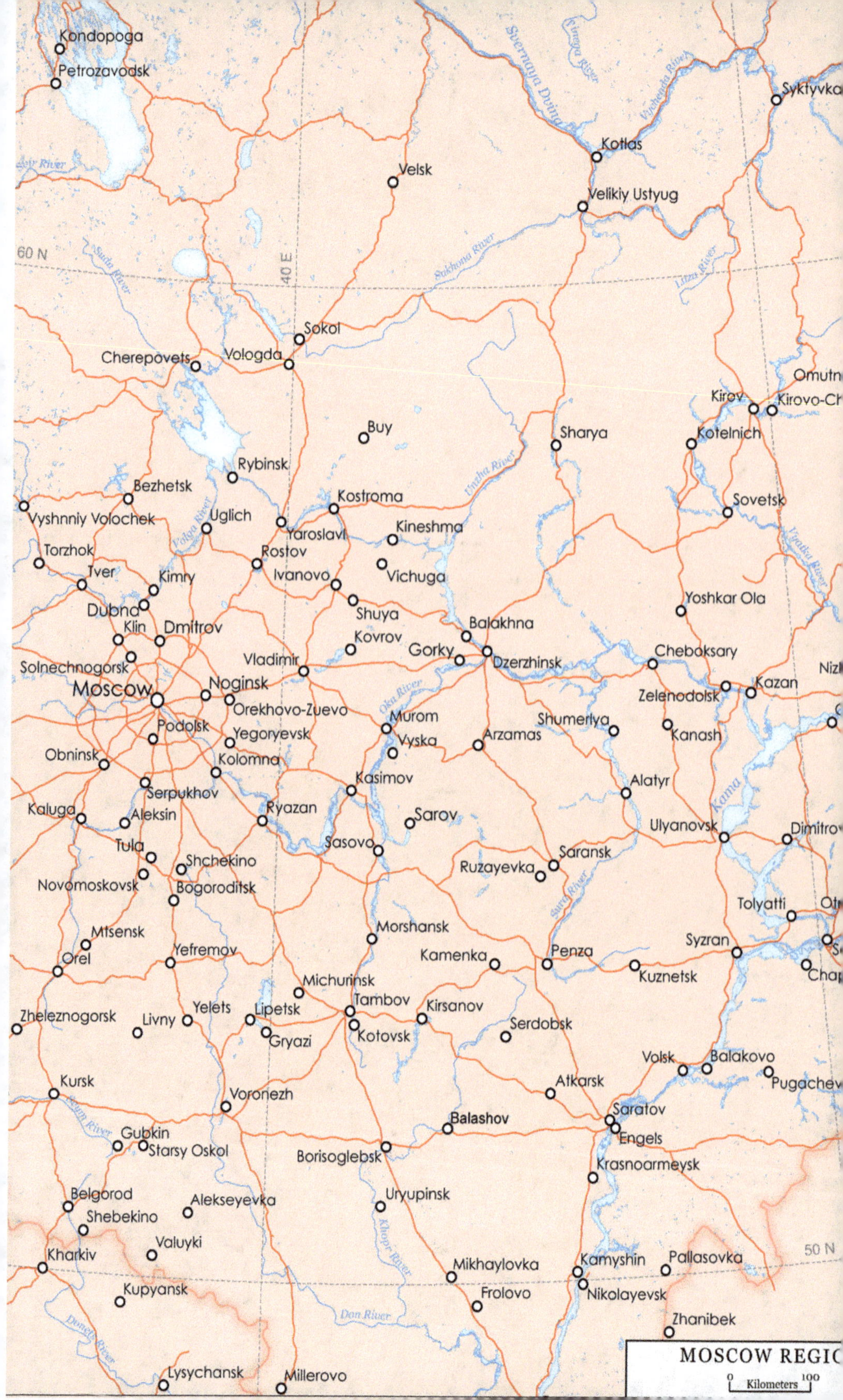

Kondopoga
Petrozavodsk
Syktyvka
Kotlas
Velsk
Velikiy Ustyug
60 N
40 E
Sokol
Cherepovets
Vologda
Omutn
Kirov
Kirovo-Ch
Buy
Sharya
Kotelnich
Rybinsk
Kostroma
Sovetsk
Bezhetsk
Vyshnniy Volochek
Uglich
Kineshma
Torzhok
Yaroslavl
Rostov
Vichuga
Yoshkar Ola
Tver
Kimry
Ivanovo
Dubna
Shuya
Balakhna
Klin
Dmitrov
Kovrov
Gorky
Cheboksary
Solnechnogorsk
Vladimir
Dzerzhinsk
Kazan
Niz
Moscow
Noginsk
Zelenodolsk
Orekhovo-Zuevo
Murom
Shumerlya
Podolsk
Yegoryevsk
Vyska
Arzamas
Kanash
Obninsk
Kolomna
Kasimov
Alatyr
Kaluga
Serpukhov
Ryazan
Sarov
Ulyanovsk
Dimitrov
Aleksin
Sasovo
Saransk
Tula
Shchekino
Ruzayevka
Novomoskovsk
Bogoroditsk
Tolyatti
Mtsensk
Morshansk
Ot
Yefremov
Kamenka
Syzran
Orel
Penza
Michurinsk
Kuznetsk
Chap
Zheleznogorsk
Yelets
Lipetsk
Tambov
Kirsanov
Livny
Kotovsk
Serdobsk
Gryazi
Volsk
Balakovo
Kursk
Atkarsk
Pugachev
Voronezh
Saratov
Gubkin
Balashov
Engels
Starsy Oskol
Borisoglebsk
Krasnoarmeysk
Belgorod
Alekseyevka
Uryupinsk
50 N
Shebekino
Kamyshin
Pallasovka
Kharkiv
Valuyki
Mikhaylovka
Nikolayevsk
Kupyansk
Frolovo
Zhanibek
Lysychansk
Millerovo
MOSCOW REGIO
0   Kilometers   100

INDUSTRIAL
URALS
Central Area
0   Kilometers   50
Tura River
Tura River
Nizhhyaya Tura
Verkhnyaya Tura
Krasnouralsk
Verkhnyaya Salda
Nizhny Tagil
Alapayevsk
Sverdlovsk
Chusovaya River
Nevyansk
Verkhny Tagil
Artemovskiy
Verkh-Neyvinsky
Asbest
Pervouralsk
Sverdlovsk
Polevskoy
Kamensk Uralskiy
Shadrinsk
Verkhniy Ufaley
Lake Irtyash
Lake Kizyltash
Techa River
Kyshtym
Ozersk
Chelyabinsk
Zlatoust
Chelyabinsk
55 N
Bakal
Miass
Belaya River
60 E
Yemanzhelinsk

10 E
POLA
Oder River
Warta
Berlin
Potsdam
Hannover
Braunschweig
Magdeburg
Cottbus
Gottingen
Kassel
Weser River
Leipzig
Dresden
Erfurt
Jena
Gera
Chemnitz
Lib
GERMANY
Aue
Annaberg
Usti Nad Laber
Johanngeorgenstadt
Erzgebirge Mountains
Elbe River
Hof
Jachymov
Coburg
Prag
50 N
Main River
CZECHOSLOVA
Wurzburg
Pizen
Vltava River
Nornberg
Furth
Regensburg
Ceske Budejovice
Ingolstadt
Passau
Danube River
Ulm
Augsburg
Linz
Danube River
Inn River
Munich
AUSTRIA
Rosenheim
Salzburg
EAST GERMAN
CZECHOSLOVA
Kilometers

Contents

List of Chapter Figures

List of Appendices Figures

Introduction

The weather was chilly and rainy in Moscow on Sunday, June 24, 1945. At 10 am Red Square was filled with victorious Soviet Armed Forces on parade. Marshal Rokossovsky was parade master. Marshal Zhukov gave the keynote address after riding around on his favorite white horse. Joseph Stalin and his comrades were on the review stand atop Lenin's tomb. Thousands of troops paraded past, together with tanks, mobilized guns, and every imaginable war transport. Captured Nazi banners and standards were piled in a heap, a symbol of the Soviet victory.[1]

The war with Germany, which began on June 22, 1941, almost exactly four years earlier, was over. The German army had occupied all of the western Soviet Union, and their advance and retreat caused much devastation. The farthest eastern penetration of German forces was at Stalingrad, on the Volga River. The territory under German occupation as far east as Stalingrad can be seen in Map #1. Here, in November, 1942, the north and south arms of Soviet Operation Uranus met near Kalach, encircling the German sixth army in the city of Stalingrad and its suburbs. Commanded by Field Marshal von Paulus, the sixth army contained about 200,000 men. They held out in the trap for three months, eventually surrendering to Soviet forces on January 31, 1943. The remaining German troops were marched into captivity. A few thousand prisoners were repatriated to Germany after the war. The loss at Stalingrad began two

---

[1] Soviet victory parade www.youtube.com/watch?v=DPEWh8yMXg0. See also Pravda, 25 June 1945, #151, p 1.

years of hard fighting in retreat for the German army. The outcome of the Soviet-German war was no longer in doubt.

Red Army soldiers who participated in Operation Uranus experienced the happiest days of the war – happier than the invasion of Germany in 1945. The Red Army had political commissars (Nikita Khrushchev was one)[2], but life in the trenches was relatively free of political suppression. The Bolshevik government under Stalin had to relax its rules in order to sustain the Red Army's efforts. Officers' epaulettes, which were eliminated in the Civil War of 1918-1920, reappeared in 1942, a small gesture towards military rank and privilege. In the absence of epaulettes, rank was displayed on the collars of military jackets.

During the war, many citizens of the Soviet Union considered the political repression of the 1930's to be a thing of the past, and they hoped that after the war their lives would be easier. This sentiment was certainly present on June 24, 1945 at the armed forces parade. After all, the Soviet Union won the war, and secured eastern Europe as a buffer zone between its territory and that of Germany. Germany itself was divided into East and West, and considerably weaker as a result. The partition of eastern Europe was ratified at the Yalta Conference in February, 1945, although the territory occupied by the Red Army at that time left little room for the US and British allies to maneuver.

The Red Army was all powerful, arguably the most feared fighting force in the world in 1945. This was one of Stalin's problems. What to do with a large army, and a substantial number of general officers, several with rank of marshal, and some immensely popular in the country, when economic recovery was desperately needed?

Three weeks later, on the other side of the Atlantic Ocean in the New Mexico desert, the first atomic bomb exploded. The Trinity Test near the town of Alamogordo, NM, took place in the early hours of Monday morning, July 16, 1945. It was the first successful detonation of an

---

[2] Sergei Khrushev, Trilogia ob Otse, Reformator, Vremya, Moskva, 2010.

implosion type Plutonium bomb. An essentially exact copy of this bomb was subsequently tested by Soviet scientists at Semipalatinsk in Kazakhstan four years later, on August 29, 1949.

The goal of this book is to describe the process in the Soviet Union from the end of the war in 1945 through the detonation of the first Soviet bomb in 1949. To do this, one must look on both sides of the Atlantic before, during, and after World War II. The Soviet Union was an ally of the US and British forces against Nazi Germany, but this alliance did not extend to intentional sharing of classified data on the Uranium problem. The Soviet Union was however viewed sympathetically by a group of US citizens and European émigrés, who thought that the experiment with communism was a noble one for the human race. There was also a feeling among some atomic scientists that sharing of weapons secrets would lead to a better balance of power after the war, and hence world peace. As a result, the Soviet spy network in the US and Britain had easier access to classified information, which was to prove crucial in the Soviet development of atomic energy.

Much has been written about the huge scope of the Manhattan project, its construction of laboratories in many separate places, the problems encountered and solved, the outflow of large amounts of money, and the interaction of the personalities involved in the decisions.[3]

The Soviet project has been covered by at least one history book in English.[4] This book will use extensively the proceedings of a symposium held at Dubna, a laboratory town near Moscow, in May, 1996,[5] and the memoirs of two Soviet master spies, Aleksandr Feklisov[6] and Pavel

---

[3] Richard Rhodes, "The Making of the Atomic Bomb," 25th Anniversary edition, Simon and Schuster paperbacks, NY 2012. See also Leslie M. Groves, "Now It Can Be Told", Harper, NY, 1962.

[4] David Holloway, "Stalin and the Bomb," Yale University Press, New Haven, 1994.

[5] History of the Soviet Atomic Project (HISAP 96), 3 Volumes, Izdat, Moskva, 1999.

[6] Aleksandr Feklisov, "Kennedi i Sovetskaya Agentura," Eksmo Algoritm, Moskva, 2011.

Sudaplatov[7] Like the Manhattan project on which it was based, the Soviet project built several laboratory sites and testing grounds throughout the Soviet Union, and spent a very large amount of the nation's gross national product. The top Soviet scientists were as capable as anyone. However, living conditions in the Soviet Union in the 1940's for workers in construction, in the factories, and in the uranium mines were appalling. The national sacrifice required to accomplish the mission of building an atomic bomb prolonged the wartime restrictions and suffering for the population.

---

[7]   Pavel Sudaplatov, "Razvedka i Kreml," TOO Geya, Moskva, 1996.

Chapter 1

Wartime Soviet Industry

1.1. The American Challenge

The Manhattan project was carried out in the continental United States in wartime. While daily life in the US in the 1940's was restricted by shortages, ration cards, and meatless Tuesdays, and the economy was on a wartime footing, the country was not invaded by enemy forces. The Manhattan project cost $2 billion in then year dollars, 95% of which was spent on construction.[1] To accomplish its mission in a timely fashion required vast industrial capability, electric power, and modern infrastructure. General Groves, the overall head of the US project, has said that the bomb would never have been built in peacetime.[2] It simply required too many resources. The industry created was comparable to the automobile industry in the US at that time.[3]

Soviet physicists in 1945 knew that an atomic bomb would work, which left no uncertainty in the achievability of their goal. Nevertheless, in order to reach that goal in only four years after the end of WWII – the

---

[1] Leslie Groves, "Now It Can Be Told," p 360. To put the cost in perspective, a WWII Jeep cost $400, a P-47 combat fighter plane cost $10,000, and a British Spitfire cost £4500. So to convert to 2013, multiply by about 50. The official inflation figure is more like a factor of 10 across the domestic economy. US federal spending in 1943 was $78 billion, which was 44% of the GDP; FY 2012 numbers are $3.5 trillion and 23% respectively. (White House Office of Management and Budget). According to these numbers the GDP increased a factor of 85 in 70 years. Spreading the cost over 3 years, the Manhattan Project was about 0.1% of the US federal budget.

[2] Groves, *ibid.*, p 413.

[3] Rhodes, *ibid.*, p 605 quoting Bertrand Goldschmidt, a chemist who worked with Glen Seaborg.

5

successful test of an atomic bomb – the Soviet Union had to create industries for isotope separation and plutonium production comparable to that of the United States. This chapter describes how the infrastructure needed to accomplish this task came into being in the Soviet Union under very difficult circumstances before and during the war.

1.2. Soviet Industrialization Before the War

First it is necessary to briefly describe the situation before the war started. The Soviet Union began a concentrated effort to industrialize the country and catch up with the West soon after the end of the civil war in the 1920's. Reform of agriculture was necessary in parallel with industrialization, because peasant farming was not efficient enough to feed a growing number of city dwellers and factory workers. The construction of Magnitogorsk was one of the initiatives of Stalin's first 5-year plan. A site was selected at the southeast side of the Ural Mountains, on the Ural River in the Chelyabinsk Region. Iron ore was available nearby, although coal had to be delivered by railroad. The plan was to build from scratch the world's largest steel mill, located well into the interior of the country, and safe from possible invaders. The US Steel plant in Gary, Indiana was used as a model for both the plant and the town where the workers lived.

Construction began early in 1929, and the railroad reached the construction site in June of that year. The construction workers came from regional farmers displaced by collectivization, imported workers from the western Soviet Union, and prisoners. Living conditions were initially very primitive. Workers and families lived in tents and makeshift dormitories in a climate which had severe winters. Later on, in the 1930's, Magnitogorsk attracted Western volunteers interested in the communist experiment. The first steel was produced in 1932, and the mill was up to full capacity by 1937. Half of the steel for tanks in WWII was produced at Magnitogorsk.[4] The region also had hydroelectric power, and coal mines.

---

[4]   www.macalester.edu/courses/geog61/aritz/magnitogorsk.html

Chelyabinsk is a town to the north of Magnitogorsk, over 100 years older, and an industrial center itself. It had served as a focus for work on the trans-Siberian railroad during the reign of Tsar Nicholas II. It too enjoyed industrial upgrades from the Soviet government in the 1930's, including a tractor plant and a metallurgical plant, which formed the base for manufacturing tanks for the Red Army in WWII. North of Chelyabinsk lies Ekaterinburg, which was called Sverdlovsk during Soviet times. This town is famous for the assassination of the Tsar and his family by Bolsheviks in 1918. Sverdlovsk furnished new homes for displaced industry from the western Soviet Union.

"The Urals," wrote president of the USSR Academy of Sciences V.L. Komarov, "is the richest country for iron, non-ferrous metals, light metals, fuel and chemical resources. This region, stretching parallel to the front lines, at a distance of one to two thousand kilometers, is a powerful line of economic fortifications, the richest deposits, factories, and electric power stations, created during the course of the five year plans."[5] This region was to play an important role in the Soviet atomic project.

1.3. German Invasion Causes Eastern Migration

In 1941 significant industrial and agricultural resources existed in the western Soviet Union. There were large factories in the Moscow and Leningrad regions, productive agriculture in western Ukraine, and factories in Kiev and Kharkov. The German invasion threatened to overrun all of this territory. Moscow, Leningrad, and Kharkov were also important centers for experimental and theoretical work in nuclear physics before WWII. Evacuations disrupted research as well as industrial production. One of the best known Soviet theoretical physicists, Lev D. Landau, was at Kharkov before the war. A clever nuclear experimentalist, Fritz Houtermans, was also there. Later Landau was at Moscow, as was Peter

---

[5] Istoria Velikoi Otechenvennoi Voiny Sovetskogo Soyuza (IVOVSS), 5 volumes, Moskva, 1963. p 141. The official Soviet WWII history, while hardly unbiased, is still a valuable source of information. A typescript English translation exists.

Kapitza, another internationally known physicist, who had spent 13 years in Cambridge, England, working with Ernest Rutherford.

Industries, arts, women and children, foreign dignitaries, and Vladimir Lenin were all evacuated from Moscow. Wait a minute, you say, Lenin died in 1924, how could he catch a train out of Moscow in 1941? Lenin was, and is preserved in his mausoleum on Red Square. Fear that the Germans would capture Moscow and Lenin in his tomb caused the Soviet government to move the body to Tyumen, in Siberia, where it stayed until March, 1945.[6]

Leningrad was host to Georgy Flerov, Igor Kurchatov, and Yuli Khariton, all of whom we shall learn much more later, as they played key roles in the Soviet Atomic Project. In the 1930's the institutes at Kharkov and Leningrad each had charged particle accelerators for the study of nuclear reactions. In October, 1938, the Academy of Sciences of the USSR formally established a commission for atomic nuclei. Kharkov had a 1.2 MeV electrostatic accelerator to make neutrons by deuteron stripping. Construction of a cyclotron was started there, but not completed before the war. There was a working cyclotron for neutron production at the Radium Institute in Leningrad, and an electrostatic accelerator at the Leningrad Physico-Technical Institute.[7] The USSR physics laboratories before the war had capability and equipment for the study of nuclear reactions on par with other leading laboratories in the world.[8] The Leningrad Physico-Technical Institute began construction on another cyclotron in 1939, which was completed after the war in Moscow. During the Leningrad blockade the Physico-Technical Institute went to Kazan.[9]

The evacuation of the city of Leningrad began in July, 1941, with the packaging of priceless works of art from the Hermitage museum, to be

---

[6] Andrew Nagorski, "The Greatest Battle", Simon and Schuster, NY, 2007. p 54. The author describes the recollections of the team that had the responsibility of caring for the body during that period.

[7] HISAP'96, *ibid.*, Vol 1, p 23.

[8] HISAP'96, *ibid.*, Vol 2, p 32.

[9] HISAP'96, *ibid.*, Vol 1, p 41.

shipped by train to Sverdlovsk.[10] Women and children were also evacuated to the east beginning in July, but the population of the city did not drop that much, because of an influx of refugees from surrounding territories.[11] Evacuation of factories making war supplies was problematic, because the products were needed immediately at the front, which was just outside the city. After the city was blockaded, some factory equipment was actually shipped out by air![12] Some laboratory equipment was similarly evacuated.

The Donets Basin, or Donbass, in eastern Ukraine, was and still is a heavily industrialized region. By the autumn of 1941, some three months after the invasion on June 22, Leningrad and Kiev had been surrounded, and Moscow itself was in grave danger. Shortly after the invasion Stalin proclaimed a 'scorched earth' policy, in which no useful equipment or resources should be left for the Germans. When possible, resources were to be moved east rather than destroyed in place. Thus the great movement of factories, machine tools, and all kinds of supplies from west to east ahead of the invading armies began.

Boris L. Vannikov, Peoples Commissar for Armaments from January, 1939 was arrested on June 7, 1941 for 'exceeding his authority,' and was sitting in prison when the Germans invaded. He was released on July 25, 1941, and immediately put to work on the evacuation of Soviet industry. This was a complete surprise to him, as he expected the war to be fought on German soil rather than on territory of the USSR.[13] His boss was Lavrenty P. Beria, Minister of Internal Affairs, and Stalin's right hand man. Vannikov was arrested again in December, 1941, and released the following January, in an episode told in Chapter 9. Beria as Minister of Internal Affairs had control over the entire country, industry, economy, prison camps, free labor - all resources except the Red Army. Beria and Vannikov played crucial roles in the Soviet Atomic Project after WWII, and we shall meet them again later.

---

[10] Harrison Salisbury, The 900 Days, Harper and Rowe, NY, 1969, p 149.

[11] Harrison Salisbury, *ibid.*, p 206.

[12] Harrison Salisbury, *ibid.*, p 277.

[13] HISAP'96, *ibid.*, Vol 2, p 189. Article on Vannikov by G.A. Sosnin.

On July 5, 1941, orders were given to evacuate the Leningrad Kirov Plant and the Kharkov Tractor Plant to the Chelyabinsk region. On August 7, the Moscow factories "Hammer and Sickle" and "Electrostal" were ordered to relocate in the Urals.[14] "Electrostal," when restored to Moscow after the war, will play a role in the Soviet Atomic Project.

Evacuation receiving points were created in Sverdlovsk, Gorky, Kuybyshev, Chelyabinsk, Novosibirsk, Magnitogorsk, Tashkent, and other cities. Wood was used for construction, and a large building could be finished in 15-20 days.

"During the winter, in one of the outskirts of Sverdlovsk, where they were assembling the plant "Bolshevik," which had been shipped from Kiev, it was possible to observe the following scene. Underneath the pine trees, in which hung electric lamps, the machine tools were working. On a spot nearby, welders cut steel, showering the snow with golden sparks. At the flaming furnace, the blacksmiths forged the burning metal. Thus was born the now famous Ural Order of Red Labor Banner Chemical Machinery Building Plant."[15]

In order to finance the shift of the economy to a war footing, military expenses for the second half of 1941 were increased by 20.6 billion rubles, and civilian branches of the national economy were reduced by a similar amount.[16] Ordinary civilian goods disappeared. If your teapot broke in 1941, you would have to live without it for a while.

Evacuation of electric power involved special challenges, since nothing worked if the power was cut. As a result, the power stations were kept running until the last possible moment. But they too were dismantled and shipped east, where the new factories needed the extra power. What could not be moved was destroyed.[17]

---

[14] IVOVSS, *ibid.*, Vol 2, p 142.

[15] IVOVSS, *ibid.*, Vol 2, p 150.

[16] IVOVSS, *ibid.*, Vol 2, p 142. referat.znate.ru/text/index-32255.html gives the 1940 USSR national economy expenses of 174 billion rubles, so this is 12% of the total. The 1945 expenses were 298.6 billion rubles, reflecting the war costs. For the USSR the GDP and the government expenses were the same.

[17] IVOVSS, *ibid.*, Vol 2, p 145.

To quote Alexander Werth:

"This transplantation of industry in the second half of 1941 and the beginning of 1942 and its rehousing in the east, must rank among the most stupendous organizational and human achievements of the Soviet Union during the war."[18]

One example described by Werth[19] is the armor plate mill in Mariupol, a city in southern Ukraine on the Azov Sea, which was moved to Magnitogorsk. Armament plants from Leningrad, Moscow, and Tula were all sent east together with their workers, administrative staff, and families. This process was repeated over and over. Through December, 1941, a total of 994 thousand railway wagon-loads filled with factory equipment and workers with their families, was moved from west to east.[20] The maintenance of the Soviet rail system in the western part of the country where there was continuous bombardment by enemy aircraft, was especially challenging.

Of course, once moved, the factories had to be reassembled and started up again in production. The following story was translated by the author as a realistic description of conditions for moving factories during the German invasion of 1941.

Izvestia 1941 #300 page 3

On the Steppes

The snow here is dry, sharp like sand. The wind blows and drives the snow. It is difficult to breathe in the wind, the icy air affects breathing, and people go around bent over from the wind. Collars are pulled up against the wind. When the snow falls in your eyes, your eyelashes become grey

---

18 Alexander Werth, "Russia at War," E.P. Dutton, NY, 1964. P 213. The author was a war correspondent for the London Sunday Times based in Moscow throughout WWII.
19 Werth, *ibid.*, p 214.
20 IVOVSS, *ibid.*, Vol 2, p 169.

and stick out like needles. Sign boards rattle in the wind, like thunder; street lamps bounce around, their light interrupted by snowdrifts.

But this is just a Russian snowstorm, reminding you of when you found yourself out in the countryside in the wintertime.

The Steppe is covered with fallen snow like a featherbed. On the road, an automobile hurries along in a snowy cloud; sometimes a truck is visible in the rear. A driver in a fur coat stands near the truck. He turns his head, adorned with icicles, and his eyebrows thick and white with snow. He yells something, waves his hand in its large mitten, and the truck stops. The truck bounces over a ditch, and past the rushing automobiles, the tires crushing the snow, and this line of trucks moves on towards somewhere in the steppe.

And there through the leaden smoke appears an outline of some sort of building.

At first the outline resembles an oil refinery or a crane; the car approaches and one is able to see the powerful outlines of construction. In the steppe covered with icicles stands a structure, solid like stone, the ground scattered with scrap, the interior is filled with people working with cables, tubes, and conducting wires. The wooden frames rise up nearby the enormous bare walls, not yet covered with a roof. The people carry carpets and rugs from the truck, and drag them up the stairs to a brightly lit, fully constructed building. Below at the entrance stands a fog of frozen breath; on the stairway people in padded jackets and fur hats with earmuffs walk up and down. Floor mats are stretched in the third floor hall, smelling like ship's rope, along a line of spittoons. A man wearing high felt boots with cold fingers fixes signboards on the doors with names like 'chief engineer' and 'party commissar.'

That reminds you of places like Magnitka and Avtostroi, years of the first five year plan, and the great courage and concentration of the first construction projects.

We know well the familiar view of construction, beginning with iron beams and wooden poles, all covered with snow, climbing out of frozen red clay which has been heated by campfires. We also know the people. They are the same people, with wind scarred faces and red sleepless eyes. They have a fearless ability to work, to overcome all obstacles, and to complete the job. Their hands are strong, their fingernails like iron – hands which do not fear any task. Frozen metal will stick to your fingers and tear off skin. It is not possible to describe this feeling, but if you experience it once, you will never forget it.

These are the same people who worked on the first five year plan, but much has changed since then. The first five year plan was in peacetime, and now construction proceeds during the war.

The people came here with their own factories, built peacefully and planning for the future. They came in special trains with their equipment, wives and children. They loaded the machinery on flatcars, and loaded the boxcars where they would ride with bundles and pillows. The trains moved slowly, hindered by military transport, and the children cried. Wiry old ladies sat in the corner of the boxcar like wooden statues. People disembarked at stations along the way, and had a chance to warm up, read the news, and obtain provisions.

Then the train pushed further on. Someone ran after the train holding a jug of milk. Someone else jumped on the car, waving a pot of hot kasha. The train gathered speed, and on the open flatcars, next to machinery covered with tarps, day and night, in rain and snow, guards stood watch to protect the equipment.

And then the trains arrived at their destination.

However, the new shops for the machinery were not finished. Tall walls stood on the steppe, without roofs or windows. It was necessary to move the equipment as quickly as possible, because it was sitting on the flatcars, instead of where it would be useful. So they dragged the heavy machinery on rollers into the shop with cables by hand, yelling 'One –two

–Pull!' They tore the skin on their hands, but kept on pulling until they set up the machines on the floor of the roofless shop.

A roof and windows were installed on the shop building. Inside there were mounting blocks for the machines. One could not tell who was the engineer, or the lathe operator, or the machinist, because they were all working at the same thing – installing the heavy equipment, pulling heavy, icy pieces of metal. Morozov, the party secretary of one of the shops, constantly stood at the head of the unloading brigade. Senior master Malyuta assumed the management of the shop, and while unloading the equipment began installation on the mounting blocks, not only directing the work, but doing some of the hardest jobs himself. Lathe operator Grevtsov was team leader of one of the best brigades for disassembly and reassembly. Women worked to clean up rust on the steel surfaces. There was no electricity in the factory, so they worked 'as in the village,' from daylight to dusk, cleaning the metal which had rusted on the trip, and the metal warmed up and shined under their hands. The people silently and passionately gathered together the parts of their factory, which had to be assembled to restore it to life.

There were no accommodations for the workers families near the factory, so they were sent to collective farms in the district. This was not good, but it was not possible to do otherwise, and the children, wrapped in mufflers, were again placed on trains or carts, kissed, and sent on their way. Sleeping quarters were side by side bunks in barracks or dormitories. It was a difficult situation, but the people, accustomed to a comfortable life, endured the hardships silently and patiently, jumping to their feet at dawn and going to the factory. In the crowded temporary canteen, with moisture from the ceiling falling on the tables, they had to eat so fast that they burned their mouths on the hot food. But people saw that big shining kettles were being set up in a new canteen for the factory, that new houses and barracks were under construction, that shops were sprouting up at an unbelievably fast pace, that now, during the war, large construction projects were taking place, and they silently bore all hardships, pointing to the new buildings standing on the snowy steppe. They said with pride, with force, and with excitement: "Here will be a factory as impressive as the one we left at home."

Trainloads of equipment continued to arrive. The workers pulled the icy machinery into the plant and mounted it in place. Soon there stood smooth and shiny silent machines, and a bit later, in the same shop, the machines sprang to life, whistling and throwing sparks, turning metal. Actual work had begun.

The temperature inside the shop was -15 degrees C. (+5 degrees F).

Campfires and braziers boiled water at the workbenches. The factory communist leader, a short man with curly white hair, went around the shop without either a hat or a coat. That was the way he was used to going around shops earlier. Everybody shouted at him that he was going to catch the death of cold, but this quiet man with thick glasses, with a nicely pressed suit and necktie, inspired a certain tranquility and conviction, that everything was in order, that the factory was just so.

At a new place, in new, unbelievably difficult conditions, people learned how to work in a new way. Duties have changed, times allowed for projects and for overcoming obstacles have shortened. No one lost their head over the difficulties, over this trying new situation. And the machines one by one started to work, the factory, moved from its original place, still not fully up to speed, did fulfill its monthly plan, still working, still distributing, like a boss, its orders for the new year, 1942.

The products of the factory were coming alive. Shiny propellers and aircraft motors seemed like huge birds, endowed with powerful and intelligent force. And we saw what the deep interior of our country had become.

These are front lines, where people work for the war effort and for victory, where every worker is subject to severe conditions, requiring strength, fearlessness, and endurance.

We should not allow anyone to slander these people. Let us try therefore to tell about these people, who during wartime were in the rear. They conducted themselves like heroes. They experienced much and endured much more than would seem possible for human strength.

And together with those who fight, together with those who went to war and suffered wounds, they did everything asked of them necessary to win the war and achieve victory.

Tatyana TECC

There probably is no better way to describe the conditions than to quote an unknown local reporter, even though Izvestia was a tightly controlled journal, printing only what the Kremlin wanted its readers to see.

The transfer of manufacturing from west to east naturally caused a disruption in production. The results were acute shortages of supplies for the Red Army in 1941-42, which contributed to the conquest of territory by the German army.

Supplies of grain and livestock were also moved east when possible, but agricultural capability was not so easy to move. Western Ukraine was a substantial source of food, which was not available to the Soviet Union from the fall of 1941 until its liberation in 1944. This left the Volga country, the Urals, Western Siberia, and Kazakhstan as the Soviet Union's food base for the greater part of the war.[21] Food supplies remained severely restricted for the period of the war, and indeed for several years afterwards.

By mid 1942 the industrial capability in the east had stabilized, but further problems awaited the Soviet government, as the Germans moved south, and captured the Donets Basin, or Donbass, which contained

---

[21] IVOVSS, *ibid.*, Vol 2, p 163.

substantial industrial and mining infrastructure. Some factories from the Donbass were moved to Novokuznetsk, near Novosibirsk in Siberia.[22]

The German forces, commanded by Field Marshal von Manstein, were aiming for the oil fields of Baku, on the Caspian Sea, which events at Stalingrad prevented them from reaching. The Volga at Stalingrad and the Don at Voronezh were the deepest eastern penetrations of the German army. After the war, the French leader Charles de Gaulle visited Stalingrad, and had this conversation with Alexander Werth:

De Gaulle: "Ah, Stalingrad! C'est tout de meme un peuple formidable, un tres grand peuple."
Werth: "Ah, oui, les Russes…"
DeGaulle: "Mais non. Je ne parle pas des Russes, je parle des Allemands. Tout de meme, avoir pousse jusque la!"[23]

Industrial capability in the East, particularly in the Chelyabinsk region where Magnitogorsk is located, began during the prewar five year plans, but was much increased by the relocation of factories in advance of the invading German armies.

1.4. Restoration in Recaptured Territory

After Stalingrad, the troop movement was slowly but surely in the other direction, reclaiming territory lost in 1941-42. Industrial capability in the West was restored, but on the whole with new equipment. Factories moved to the East stayed put.

---

[22] Arkady Vainshtein, private communication. Professor Vainshtein was born in Novokuznetsk in 1942 of parents evacuated from Donbass with their factory. His father was an engineer.

[23] Werth, *ibid.*, p 930. "Ah Stalingrad! All the same, they are a formidable people, a very great people." "Ah yes, the Russians…" "No, I'm not talking about the Russians, I mean the Germans. They got this far!"

This is not a history of WWII, but some time line is necessary to explain how the West was liberated and re-industrialized.

Moscow was never captured, and was out of danger after the Germans were stopped on the western outskirts of the city in the fall of 1941. There are still monuments in Moscow's north-west suburbs commemorating the end of the German advance. There were evacuations from the city – the embassies and many government agencies were moved to Kuibyshev, now Samara, on the Volga between Kazan and Volgograd in the southeastern part of European Russia.[24] Stalin's daughter Svetlana was 15 years old in 1941, and was part of the evacuation.[25] She returned to Moscow from the summer retreat in Sochi in September, to find the city under bombardment, and widespread damage. Soon she was packed off to Kuibyshev, together with various relatives of Stalin, and Stalin's library, but not Stalin himself. Svetlana's mother had died in 1932, but Svetlana's grandparents still lived with her, and moved to Kuibyshev. She remained there for a year, attending school in the ninth grade. She noted that the inhabitants of Kuibyshev had been substantially inconvenienced by the arrival of dignitaries from Moscow. Factories and laboratories were also moved east, mostly to Kazan. Stalin remained in the capital for the entire war, and since he had complete control of everything all of the time, sufficient government remained in Moscow. Indeed, the US Ambassador was also in Moscow in the fall of 1941.[26] In any event, everything was moved from Kuibyshev back to Moscow in the summer of 1943.[27]

Life in Moscow during the war was grim, but the higher levels of government and the diplomatic corps functioned more or less normally. Harriman was US ambassador, and a confidant of President Roosevelt, so no ordinary citizen, but his memoir gives some insight into conditions,

---

[24] en.wikipedia.org/wiki/Samara, Russia

[25] Svetlana Alliluyeva, "Twenty Letters to a Friend", Avon Books, New York, 1967, p 175.

[26] Avrell Harriman and Elie Abel, "Special Envoy to Churchill and Stalin", Random House, NY. 1975.

[27] en.wikipedia.org, *ibid*

from initial contact in the fall of 1941 to setup Lend-Lease,[28] to later in the war in 1943-44.[29] It is particularly interesting for its description of many interactions with Stalin over the course of the war. Like other observers, Harriman was struck by Stalin's memory, and his attention to minute detail on any subject. In discussions of Lend-Lease, Stalin emphasized the need for 2.5 to 3 ton trucks, saying that the bridges in the Soviet Union could not support heavier loads, but that motorized transport was of prime importance.[30]

To quote Harrison Salisbury:

"All the big Soviet cities went on a ration card system July 18, 1941. Leningrad's ration was the same as the rest of the country-800 grams of bread a day for workers, 600 for employees, 400 for dependents and children. The meat ration was 2200 grams a month for workers, dropping to 600 for dependents and children. There were ample rations of cereals, fats, and sugar."[31]

Leningrad was surrounded by the initial thrust of the German invasion, in the fall of 1941, and not liberated until January, 1944.[32] The city itself was not captured, but its location on the Gulf of Finland allowed the German troops to virtually cut it off from any supplies from Soviet territories by land. In winter, an ice road over Lake Ladoga gave some relief by truck. The winter of 1941-42, just after the blockade, was particularly harsh. Food supplies for the population dropped to almost zero, and many people died of starvation.[33]

In the spring of 1943 the German army still occupied Ukraine, Beloruss, and most of European Russia. Although the Wermacht was in retreat, the war did not end. Germany hoped to achieve a stalemate with

---

[28] Harriman and Abel, *ibid.*, p 98.

[29] Harriman and Abel, *ibid.*, p 299.

[30] Harriman and Abel, *ibid.*, p 90. The trucks supplied by the US were mostly Studebakers and Dodges.

[31] Harrison Salisbury, *ibid.*, p 205. 1000 grams equals 2.2 lbs.

[32] Harrison Salisbury, *ibid.*, p 473.

[33] Harrison Salisbury, *ibid.*, p 391ff.

the USSR. Kharkov in eastern Ukraine changed hands four times![34] First the Germans captured the city in the summer of 1942. Then the Red Army counterattacked in February, 1943, only to be driven out again in March. The city was liberated for the last time in August, 1943, after the battle of Kursk.[35]

It is not clear how much of the Kharkov Physico-Technical Institute remained in the city during the occupations, but Fritz Houtermans visited the Institute in 1942 as part of a German Air Ministry commission sent to Kharkov and Kiev.[36] How this came about is an interesting, and probably not atypical, story of people mixed up in WWII.

When we left Fritz, he was working at Kharkov. Landau regarded him highly, and they became friends. Fritz was an accomplished physicist. In Göttingen in the late 1920's he and George Gamow made important contributions to the early development of nuclear physics. He is credited with the original suggestion of nuclear reactions as the source of energy in stars. The development of this idea eventually won a Nobel Prize for Hans Bethe.

Houtermans, a German national, was arrested as a German spy by the NKVD[37] in 1937. Several other physicists at Kharkov were also arrested, including the first director I.B. Obreimov, and the second director A.I. Leipunsky.[38] Even Landau was arrested in 1938.[39] Houtermans' wife Charlotte was a German physicist too. After his arrest, she managed to leave the USSR through Latvia with their two children. They eventually made it to the USA, and she tried repeatedly to learn of his fate, with little success. Fritz, in prison, was not informed of his family's whereabouts. He was tortured, and confessed to non-existent crimes. An exchange of

---

[34] Melvin, *ibid.*, p 342ff.
[35] Melvin, *ibid.*, p 385.
[36] Iosif Khriplovich, "The Eventful Life of Fritz Houtermans", Physics Today Vol **45**, p 29 (July, 1992).
[37] Narodny Komissariat Vnutrenikh Del', the Soviet Secret Police, later called the KGB.
[38] ru.wikipedia.org/Kharkovskii_fiziko-tekhnicheskii_institut
[39] Maya Bessarab, "Lev Landau", Oktopus, Moskva, 2007. p 72.

political prisoners took place between the USSR and Germany after the Molotov-Ribbentrop pact, and Houtermans was extradited back to Germany in 1940. Now in the hands of the Gestapo, he went from the frying pan into the fire. He was well known by German physicists, and Max von Laue of X-ray fame rescued him from prison. He was then able to work in a private laboratory run by Manfred von Ardenne in Lichterfelde, near Berlin. We shall hear of Herr von Ardenne again in the course of our story, because he volunteered to work in the USSR after the war on the Soviet Atomic Project. While at Lichterfelde, Houtermans worked on the theory of chain reactions, a subject of interest to the German atomic energy project. It is a consequence of this, and also of his prior association with the Physico-Technical Institute, that he went back to Kharkov under German occupation. Some Soviet physicists, including Peter Kapitza, accused Houtermans of treason after this episode, making it difficult for him to return to the Soviet Union. Research in Germany after the war was strictly limited, so he went to Switzerland, and spent the post war years until his death in 1966 in Berne. What he actually accomplished during his wartime visit is not known.

Alexander Werth flew into Kharkov in February, 1943, when the city was first liberated by the Red Army.[40] The population of the city was 900,000 before the war, increasing to 1,200,000 from refugees fleeing the invasion of Ukraine. Later, in October, 1941, the evacuation of Kharkov began – industrial plants, workers, and their families. Some 700,000 were left when the Germans came, but only half as many were there in 1943 when Werth visited, the other half having departed dead or alive. Nevertheless, a substantial population remained in the city to cope with the German occupation as best they could. The occupation government was there only for the exploitation of the human and natural resources of the captured territory. No effort was made to encourage Ukrainian independence ambitions, although an independent spirit was certainly alive in Ukraine at that time. Independence had its chance with the dissolution of the Soviet Union in 1991, and Ukraine is now an

---

[40] Werth, *ibid.*, p 604ff. The author does not say how he got permission to fly into Kharkov under military fighter escort, with the Germans some 60 miles away.

independent country.[41] A similar sentiment was latent in Belorussia, with the same results during the occupation, and ultimate independence in 1991. But Nazi Germany made no gesture of friendship and support. As a result, although many people in the occupied territories of the USSR were less than enthusiastic in their allegiance to Moscow, they soon decided that the Germans were much worse.

The return of the Red Army was not an occasion for unanimous rejoicing in Kharkov.[42] The NKVD was busy rounding up various agents and collaborators, either real or imagined. This became standard procedure in liberated territories in the western Soviet Union, and was pursued by SMERSH, an organization created explicitly to ferret out foreign spies and collaborators as the Red Army drove westward.[43] Even today, Russian citizens who lived in occupied territory are required to state this fact. Suspicion of possible corruption of Communist ideology from contact with the west went to the extreme of imprisoning repatriated Red Army soldiers who had been POW's in Germany.

The German army returned to Kharkov for the second time on March 15, 1943.[44]

Immediately after the war Laboratory No. 1 of the USSR Academy of Sciences was organized within the Kharkov Institute to work on Uranium enrichment and isotope separation for the Soviet Atomic Project.[45]

The railroad network in the occupied territories went through an interesting period.[46] As already mentioned, it played a crucial role before German occupation in the evacuation of Soviet industrial capability. Under German control, the same lines were necessary to move up supplies

---

[41] The Russian Federation is attempting to re-take eastern Ukraine as I write this.

[42] Werth, *ibid.*, p 616.

[43] Headed by Victor Abakumov, one of Beria's lieutenants. A.C. Tereschenko, "Abakumov," Tsentr "Akva-Term," Moskva, 2012.

[44] Werth, *ibid.*, p 618.

[45] ru.wikipedia.org/Kharkovskii_fiziko-technicheskii_institut.

[46] IVOVSS, *ibid.*, Vol 3, p 190.

and reinforcements, and to transport wounded and prisoners to the rear. Each side wanted to deny the other the use of this transportation network, through air attacks and partisan activities in the rear. Main lines were bombed, bridges were destroyed, and infrastructure disrupted on a daily basis. During 1943 the German air force made 7000 raids on Soviet railroads, about 20 attacks a day.[47]

A technical difficulty existed, beyond the continual bombardments. The rail gauge was not the same in Russia as in Germany and the rest of Western Europe.[48] The Germans knew that, of course, and hoped to capture enough railroad equipment to use Russian gauge rolling stock on Russian gauge tracks. That didn't work out, especially as regards locomotives, which the Soviets managed to remove from harm's way. So the Germans had to move one rail of the track 85 mm closer to the other rail in order to operate their trains. This they did, using prison labor to tear up the old rail and drive new spikes into existing cross ties for hundreds of kilometers of track.[49]

Then along came the Red Army, liberating the occupied territory. In order to use the railroads, they had to move the rail back out! The same problem occurred again when the USSR invaded European territory in 1944.[50] Werth describes a rail junction at Abganerovo, in the region West of Stalingrad, where the railroad was already re-tracked in January, 1943, soon after the Stalingrad battle, and Russian gauge rolling stock was running smoothly.[51] Repairs to portions of track destroyed by bombs were done from work trains carrying 'rail joints,' which are lengths of assembled track – two rails and cross ties – which could be installed in

---

[47] IVOVSS, *ibid.*, Vol 3, p 196.

[48] The standard gauge, used in the USA and most of Western Europe, is 1435 mm. The Russian gauge is 1520 mm, 85 mm, or about 3.3 inches, wider, and well outside tolerance for rail spacing (±10 mm). So the wheels do not fit from one system to the other.

[49] forum.axishistory.com/viewtopic.php?f=66&t=2304 gives a series of chat boxes about this phenomenon, without any outside references, but interesting nonetheless.

[50] Moving the rail out on a smaller gauge track could be problematic due to the width of the cross ties.

[51] Werth, *ibid.*, p 519.

place.[52] Any work involving railroad construction or maintenance is heavy industry, which when carried out in wartime conditions is especially taxing. Nevertheless, during 1943 the lines that had been restored in liberated territory of the USSR carried one fourth of the nation's total freight.[53]

New lines had to be built through marshy areas to restore railroad service to Leningrad after the blockade was lifted in January, 1944. The front lines were still not far away, and continued shelling disrupted rail operations for several more months.

There were plenty of problems with the Soviet rail system not directly affected by the German occupation. To quote IVOVSS:

"The relocation of a large number of enterprises to the east, the rapid growth in military production in the Volga area, the Urals, and Siberia, and the Red Army's offensive operations had placed a sharply increased burden on the railroads…Operations were also complicated by the fact that a portion of the rolling stock had been lost…Thus transportation had been unable to keep up with the growth in industry."[54]

A solution to the rail transport problem became a prime goal of the Soviet government. While troop movements had the highest priority, moving needed raw materials – coal, iron ore, etc – to the defense industries was next, with shipping finished war materials to the front not far behind. Non-essential passenger traffic did not exist. Half of the capital construction for railroads in 1943 – 1.5 billion rubles – was work behind the front lines, in the Urals and western Siberia, to improve the east-west flow.[55] Railroad workers were placed on military status in order to assure efficient operation of the system.

---

[52] forum.axishistory.com, *ibid.*
[53] IVOVSS, *ibid.*, Vol 3, p 199.
[54] IVOVSS, *ibid.*, Vol 3, p 191.
[55] IVOVSS, *ibid.*, Vol 3, p 192. 1.5 billion rubles was about 1% of the economy.

The resulting strengthened rail network in the east was to have a significant impact on the location of facilities for the Soviet Atomic Project.

Industrial war production increased dramatically in 1943. While plants in and near Moscow were operational, for the most part this increase was due to improved facilities in the East, rather than restoration of factories in the newly liberated West. For example, Soviet industry in the Volga and Ural regions increased the average monthly output of combat aircraft from 1800 in 1942 to 2500 in 1943. Total production of the aviation industry in 1943 was 35,000 planes.[56] This illustrates the capability of Soviet industry in the East for sophisticated aircraft manufacturing. Tank production showed similar gains. Assembly line methods were introduced to speed up production and decrease the number of man hours required for each finished product. A substantial fraction of the factory labor force was composed of women. Almost half of all workers in ferrous metallurgy and foundries were women.[57]

An important goal for the Soviet Union in 1943 was an increase in agricultural production. The loss of Ukraine, Belorussia, the Baltic States, and the richest grain areas of the Russian Republic – the Don, Stavropol, and Kuban – in 1941-42 decreased the amount of farmland by a factor of two.[58] To make matters worse, the manufacture of farm equipment had ceased to increase war production. As in the case of heavy industry, most of the gains in agriculture came from efforts in the rear areas, which were the chief suppliers of food.[59] The production of horse drawn farm equipment was increased in 1943 to make up for the lack of tractors.

Spring planting in 1943 in the liberated territories was hampered by shortage of seed, farm equipment, livestock, and manpower, to say nothing of the devastation caused by the combat operations and the German

---

56 IVOVSS, *ibid.*, Vol 3, p 167.
57 IVOVSS, *ibid.*, Vol 3, p 179.
58 IVOVSS, *ibid.*, Vol 3, p 182.
59 IVOVSS, *ibid.*, Vol 3, p 183.

occupation. In the entire country, the sown area totaled 232 million acres, an increase of 7% over 1942.[60] Despite the increase in area and the extra efforts, total agricultural production in 1943 remained approximately the same as in 1942, due partly to unfavorable weather.[61]

One might ask what role Lend-Lease played in the wartime economy of the USSR? It is hardly mentioned in IVOVSS, if at all. From 1941 to mid 1944 the USA sent to the Soviet Union about 7 million tons of supplies, including 2 million tons of food, and substantial amounts of raw materials – gasoline, aluminum, copper, zinc, steel, etc.[62] Some supplies came in through Vladivostok in the far east, a surprising amount by air via Great Falls, Montana to Siberia,[63] and some through the Black Sea before/after German occupation, but the bulk of the material arrived in Archangel via Arctic convoy.[64] This frozen route from Britain across the Barents Sea, loaded with mines and U boats, was crewed by British Navy and merchant seamen. Seventy-seven convoys, with 1400 merchant ships, made the trip during WWII, with a loss of 104 ships sunk. The war equipment used at Stalingrad in 1942 was all homemade, but later on the Lend-Lease did have a positive effect, particularly the trucks, which helped move the Red Army westward.[65] When hostilities in Europe ended in May, 1945, President Truman promptly canceled Lend-Lease, even ordering ships at sea to turn around and come home.[66]

The people in the wartime Soviet Union had to survive under very harsh conditions. Their standard of living was very low. Long working

---

[60] IVOVSS, *ibid.*, Vol 3, p 186.

[61] IVOVSS, *ibid.*, Vol 3, p 188.

[62] Werth, *ibid.*, p 625.

[63] George Racey Jordan, "From Major Jordan's Diaries," Harcourt Brace, New York, 1952. An account of airborne Lend-Lease.

[64] Michael G. Walling, "Forgotten Sacrifice," Osprey Publishing, Oxford, England, 2012.

[65] Lend-lease to the USSR administered by Harry Hopkins, totaled $9.5 billion over four years of war. George Racey Jordan, "Major Jordan's Diaries," *ibid.*, Chapters 8 and 9. See also Joint Committee on Atomic Energy, "Soviet Atomic Espionage," US Govt printing office, 1951, p 184ff. We return to this in Chapter 8.

[66] Bert Cochran, "Harry Truman and the Crisis Presidency," Funk and Wagnalls, New York, 1973, p 149.

hours, minimal rations, and stark living conditions were the norm for the home front and the front lines. Prisoners of war, political prisoners, and common criminals formed a substantial part of the labor force. There were prisoner battalions in the Red Army. There was the substantial territory of European Russia that was occupied by the Germans. Vast industrial capability was created in the East. Very few Soviet citizens could travel to the west in peacetime, and exposure to western ideas was regarded as very dangerous, stigmatizing those who were unfortunate enough to be in occupied territory, or to be captured by the invaders and sent to camps in the West. All of these conditions form the background for the Soviet Atomic Project, which began in earnest in 1945. A totalitarian dictatorship was able to mobilize national resources at will without regard to the quality of life of its citizens, in order to accomplish goals which it deemed necessary. Throughout the existence of the Soviet Union, the goal of ultimately achieving communism in one country was sufficient justification for sacrifice. The Soviet Atomic Project was one more requirement for this goal.

Chapter 2

Development of Nuclear Physics Before the Discovery of Fission

2.1. Physics in the Early 1900's

The discipline known as nuclear physics has existed for less than 100 years. Naturally occurring radioactivity, which is a nuclear phenomenon, was discovered by accident by Henri Becquerel in 1896. He wondered whether certain minerals exposed to sunlight, which emit visible light called fluorescence, might also emit the newly discovered penetrating X-rays. Accordingly, he exposed a sample of potassium uranyl sulfate sitting on a wrapped photographic plate to sunlight. Sure enough, the plate was darkened, showing penetrating radiation. But being a careful experimenter, he did the obvious control; he placed the uranium compound on a new wrapped photographic plate in a desk drawer. It turned dark too! The uranium compound was emitting some sort of penetrating radiation all by itself, independent of sunlight. Becquerel had discovered natural radioactivity.

Ernest Rutherford first classified the emitted radiation as α, β, and γ rays, based on their ability to penetrate matter and ionize air.[1] He determined that α particles are doubly ionized helium nuclei. He did this by allowing α particles to acquire electrons in an evacuated glass tube, exciting the atoms to radiate, and obtaining the helium spectral lines.[2] Beta

---

[1] Paul Tipler and Ralph Llewellyn," Modern Physics," 5th edition, W.H. Freeman, NY, 2008, p 477. There are many treatments of the development of this interesting subject. This text has been used by the author at the University of Wisconsin.

[2] Helium gas extracted from natural gas comes from α emission in the earth.

rays were identified as electrons, and γ rays as very shortwave length electromagnetic radiation, shorter than X-rays. The source of this energy was a mystery. No ordinary thermodynamic manipulation, heating or cooling, or mechanical pressure or squeezing had any effect whatsoever on the radiation intensity. You could dig the stuff out of the ground, bring it into the kitchen, and watch it emit rays.

Albert Einstein, in his famous 1905 paper on special relativity, observed that the formula relating mass and energy, $E = mc^2$, might have something to do with explaining the source of energy in radioactivity. Good idea.

Geiger and Marsden, working in Rutherford's laboratory, scattered 5.5 MeV α particles emitted by ^{214}Bi nuclei off a gold foil 1 μm thick.[3] They expected all of the α's to pass through unscathed, and were surprised to find a measurable fraction coming back at them, as if reflected by a mirror. This was the first scattering experiment, in which a particle beam, in this case α particles, is used to probe the structure of a target (gold atoms).[4] Rutherford in a classic paper derived the formula for the scattering of an α particle from a heavy positively charged object (the gold nucleus). The data showed that the radius of the nucleus was very much smaller than the known approximate size of the gold atom.[5] Thus the modern atomic picture was born - a cloud of electrons buzzing around a very much smaller positively charged nucleus. Atomic size is determined by the electrons, and atomic weight by the nucleus.

---

[3] MeV is an energy unit, standing for one million electron volts. The electron volt is a very useful unit in atomic physics. It is the energy gained by an electron falling through a potential of one volt. 1 μm = 0.000001 meters.

[4] H. Geiger and E. Marsden, Philosophical Magazine (**6**), 25, 605 (1913).

[5] Atoms are all roughly the same size, about 0.00015 μm. Atomic size can be estimated from the number of atoms in a gram molecular weight (Avogadro's number), and the dimensions of a crystal of one gram molecular weight. Think of a cube with equal spacing between atoms. Nuclear radii are 0.00001 times atomic radii, and monotonically increase with increasing mass number.

2.2. Bohr's Atomic Model

Based on this atomic picture, Niels Bohr that same year, 1913, proposed a model of the hydrogen atom which correctly predicted all of the hydrogen optical spectrum.[6] In the Bohr model, the frequency of the emitted light is proportional to the difference in energy between two sharp energy levels of the bound electron in the hydrogen atom. The proportionality constant is the reciprocal of Planck's constant. This universal constant of nature appears in all formulas relating quantum phenomena. The Bohr model correctly predicted the optical spectra of hydrogen, singly ionized helium, doubly ionized lithium, and all complex atoms with only one bound electron. It could not treat multielectron atoms. However, by a gift of fortune, it did a credible job of accounting for atomic X-ray lines, which come from deep inside the atom, closest to the nucleus.

The next big step in understanding atomic physics came in 1926, with the publication by Erwin Schrödinger of a sequence of seminal papers on quantum mechanics. Schrödinger's work placed quantum phenomena on a rigorous mathematical footing. Schrödinger's differential equation has generated more than 100,000 papers, and is successful in describing any phenomenon presented to it so far. It derives the Bohr formula for hydrogen from first principles, and can be applied to helium and more complex atoms, to molecules and crystals, even biology.

2.3. Structure of the Nucleus

The simplest nucleus is the proton, which binds with one electron to form the hydrogen atom. Heavier atoms make up the periodic table of chemistry. The atomic number equals the number of electrons, which is balanced by an equal number of protons in the nucleus. Coulomb repulsion of the positively charged protons would blow the nucleus apart, so a stronger short range force must act to overcome it. The atomic weight is greater than the atomic number times the mass of hydrogen, so something

---

[6]   N. Bohr, Philosophical Magazine (**6**), 26, 1 (1913).

contributes to the mass of the nucleus besides the protons. This question was answered in 1932 when James Chadwick discovered the neutron, a neutral particle with approximately the proton mass and radius – a hydrogen atom, except much smaller, with no electron in residence. Nuclei are now believed to be made up of protons and neutrons. The atomic number Z = number of protons. The nucleus is composed of Z protons and N neutrons, with the number of nucleons (protons or neutrons) A = N + Z. A is also called the mass number. For light nuclei Z and N are equal. As Z increases, N becomes larger than Z. A lead nucleus has 82 protons and 126 neutrons. The nuclear force between two protons is the same as between two neutrons, or a neutron and a proton, so the extra neutrons in a heavy nucleus help cancel the coulomb repulsion. Using $E = mc^2$ all of the masses can be written in MeV, or energy units. Then the mass energy of an atom is equal to the sum of Z times (the electron mass + the proton mass) plus N times the neutron mass minus the binding energy.[7] What is the binding energy, and why is it subtracted? Einstein's formula relating energy and mass tells us that a bound system weighs less than the sum of its parts, and that this deficiency is the binding energy.[8] The average binding energy per nucleon in a complex nucleus, about 6 MeV, is the energy which must be added to the nucleus to extract a nucleon (neutron or proton). In a complex atom, the binding energies of the electrons can usually be neglected in calculating the energy available for a nuclear transition. The electron mass usually (but not always) cancels out.

2.4. Nuclear Binding Energies

The binding energy per nucleon as a function of mass number A is shown in Figure 2.1. In the figure, the binding energy is positive, so larger numbers mean more tightly bound. The curve peaks at Fe (iron). Climbing up the curve to the left towards iron from large A represents the energy released by fission, while the steep climb to the right from small A towards iron is the energy release by fusion. The fission process is the source of power generation in nuclear power reactors, and the atomic bomb. The

---

[7] Electron mass = 0.511 MeV, proton mass = 938 MeV, neutron mass = 939.7 MeV.

[8] Quarks do not satisfy this rule. But quarks are outside our purview.

fusion process is the hypothetical generation of power by fusion, and the hydrogen bomb. Fission product nuclei tend to be around mass number A=110. At the present time the fusion analog of the nuclear power reactor, the fusion power generator, does not exist. Its absence during the development of the hydrogen bomb took away a useful experimental tool, which the nuclear power reactor furnished for the atomic bomb. Note that the common parlance 'atomic bomb,' which we use just like everyone else, is really a misnomer, because the source of energy is nuclear, not atomic. But the term is part of the language, and we continue to use it.

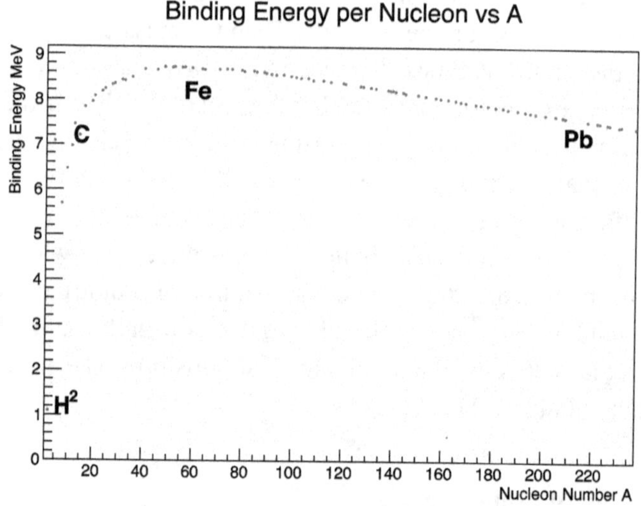

Figure 2.1. Binding energy per nucleon in MeV as a function of mass number A. The peak at iron means that this region is most tightly bound, and that energy is released by moving towards iron from either direction. Fission of ^{235}U gives two elements in the final state of unequal mass number, peaking around A = 90 and A = 140.

Chemical energies come from atomic forces, which are in the eV (electron volt) range. Nuclear binding energies are measured in MeV, 1,000,000 times larger. It is the variation of this binding energy with atomic number that enables the nucleus to be a source of energy for electric power generation or nuclear weapons far exceeding chemical reactions.

2.5.　Chemical Elements and the Periodic Table

As mentioned above, the mass number A gives the atomic weight, and is equal to the sum of the number of protons, Z, and the number of neutrons, N, viz $A = Z + N$. The atomic number Z is also the number of electrons. Electrons are negatively charged, and protons are positively charged, and the magnitudes of their charges are exactly equal to very high precision, so that an atom with all of its electrons is electrically neutral. Chemical properties are determined by the valence electrons, and depend on the atomic number. This classification gives the periodic table, where elements with similar chemical properties are grouped in vertical columns. Thus, the noble gases He, Ne, Ar, Xe, Rn are all in the right hand column of the periodic table, and are not chemically reactive.[9] The alkali metals Li, Na, K, Rb, Cs, Fr occupy the left hand column of the table, and are strongly chemically reactive. Atoms with the same Z but different values of A, and hence different numbers of neutrons in the nucleus, are called isotopes. Isotopes of an element cannot be chemically separated from one another. Most elements have more than one stable isotope, from $Z=1$ (hydrogen) to $Z=82$ (lead). For instance, hydrogen and helium each have two stable isotopes, ^{1}H, ordinary hydrogen, and ^{2}H, deuterium, and ^{3}He, and ^{4}He. ^{4}He is the common isotope; its nucleus is the α particle. It is a curious fact that there are no stable nuclei with $A=5$. Nuclei with Z greater than 82 are all radioactive, and hence unstable, although their half-lives vary over a wide range of values, from fractions of a second to millions of years.[10] Radioactive decays follow an exponential law – the number of nuclei decaying at any time is proportional to the number present. Half of the number of nuclei present will have decayed in one half-life. A senile nucleus lives just as long as one freshly made.

---

[9]　Students in chemistry were taught that the noble gases did not form compounds at all until 1962. See Neil Bartlett, Proceedings of the Chemical Society of London (**6**), 218, 1962. Xenon hexafluoride is an early compound, although many are now known.

[10]　A Table of isotopes of all of the elements, with atomic weights, half-lives, and decay products is given in Tipler and Llewelyn, *ibid*, Appendix A.

2.6. Alpha Decay

We started this discussion with the discovery of radioactivity, which is a naturally occurring phenomenon of all of the elements heavier than lead. Gamma rays, the short wavelength electromagnetic radiation emitted by radioactive elements, are analogous to atomic spectral lines. They come from transitions between energy levels of the nucleus. Beta decay, the emission of energetic electrons, is a complex subject that we probably won't need in this book, so we will not discuss it further. Alpha decay is the simplest to understand. The emission of an α particle will occur if the atomic weights of the parent and daughter nucleus allow it. Symbolically, $M(Z,A) \rightarrow M(Z-2,A-4) + M(2,4)$, where the last item is the helium atom. In the Table of Isotopes, the masses are atomic masses, which include the electrons, but the number of electrons on the right hand side of the reaction (Z) equals the number on the left, so they cancel out. Masses are given in atomic mass units, or amu, defined by convention as $^{12}C = 12.000000$. One amu = 931.5 MeV, less than the average mass of the proton and neutron by the binding energy per nucleon in ^{12}C.

With all this in hand, let us calculate the kinetic energy of the α particle from the decay of $^{214}Bi \rightarrow {}^{210}Tl + \alpha$:

1. $M(Z=83,A=214) = 213.998692$ parent mass
2. $M(Z=81,A=210) = 209.990057$ daughter mass
3. $M(Z=2,A=4) = 4.002602$ ^{4}He mass

1 -2 -3 = .006033 x 931.5 = 5.6 MeV α particle kinetic energy.[11]

These are the α particles used by Geiger and Marsden in their famous experiment.

---

[11] These numbers are from Tipler and Llewellyn, *ibid.*, Appendix A. Atomic masses are experimental numbers, and are subject to adjustment, so not all tables agree exactly. The above calculation gives the total kinetic energy available, but the heavy recoil nucleus absorbs only about 2%, so most of it goes to the α particle.

Naturally occurring α and β radioactivity is a form of nuclear transmutation. One element changes into another. In this case bismuth becomes thallium. ^{238}U, which plays a central role in our story, has a half-life of 4.4 billion years, about the same as the age of the earth, so half of it is still here. It is the parent of many daughters in radioactive decay chains studied by the Curies – Polonium, Radium, etc.[12] Alpha particles can initiate nuclear reactions too. For instance, a common neutron source is a mixture of radium and beryllium in powder form. Radium emits α particles, and ^{4}He + ^{9}Be $\rightarrow$ ^{12}C + n. Note that the numbers all add up: (Z=2,A=4)+(Z=4,A=9)=(Z=6,A=12)+(Z=0,A=1), the last item being the neutron. The conservation of electric charge is a cornerstone of electromagnetic theory, and is built in to all nuclear reactions. The energy liberated in the reaction can be obtained from the various atomic masses plus the incident α particle kinetic energy.

2.7. Nuclear Reactions Caused by Neutrons

After the discovery of the neutron in 1932 it was realized that the bombardment of nuclei with neutrons might cause artificial radioactivity, because their electric neutrality facilitated penetration into the nucleus, where an extra neutron might be disturbing. Fermi's group in Rome studied these processes systematically, and discovered that the nuclear reaction rates were substantially increased when the neutrons were slowed down in a moderator like paraffin.[13] Neutrons from a Radium-Beryllium or similar source have kinetic energies of several MeV. Paraffin contains a lot of hydrogen, which is made of protons, with almost the same mass as the neutrons. This means that the neutrons can transfer energy efficiently by colliding with the protons – just like billiard balls – resulting in a rapid loss of neutron kinetic energy. Neutrons which suffer many collisions in a moderator can be slowed down to thermal energy, which is a fraction of 1 eV. That slow neutrons are more efficient at nuclear activation is an experimental fact, explained by quantum mechanics.

---

[12] Nobel Address of Marie Curie, 1911.

[13] A summary account of Fermi's neutron work in Rome is given by Alberto De Gregorio, arxiv.org/pdf/physics/0501140.pdf

2.8. Charged Particle Accelerators

It was soon realized that in order to probe the nucleus in a more controlled way it was necessary to replace radioactive sources with beams of charged particles. The invention of particle accelerators greatly enhanced the study of nuclear reactions and transformations. Two types of accelerator appeared almost simultaneously. In 1931 Robert J. Van de Graaf at Princeton University in the USA created a high voltage on a metal sphere supported by an insulated column by spraying charge on a moving belt inside the column. The charge was transferred from the belt to the inside of the conducting sphere. A theorem of electrostatics guarantees that limitless charge can be transferred to the sphere in this way. The charge all resides on the outside of the sphere, where high voltage discharges in the air and breakdown along the support structure determined the achievable peak voltage, which equaled 1 million volts. Some engineering work on the design was necessary before a charged particle beam could be obtained in an environment useful for experimental studies of nuclear structure. At almost the same time, 1932, Cockcroft and Walton at the Cavendish Laboratory in Cambridge made a charged particle electrostatic accelerator using a direct high voltage, initially 0.8 million volts, to accelerate protons. The voltage was obtained by a multiplier circuit, now called a Cockcroft Walton circuit, in which capacitors are charged in parallel, and then hooked up in series, using high voltage switches. Today the moving belt design is still in use, and high voltage technology has progressed so that voltages of 10 million volts or more are possible. Cockroft Walton circuits are also still in use as sources for high energy accelerators. They are generally operated at lower voltages, but they have the advantage of no moving parts. A further description of electrostatic accelerators is presented in Appendix D.

Also in 1932, at the University of California, Berkeley, Ernest Lawrence started accelerating protons in his new cyclotron to energies in the 1 MeV range without the use of such high voltages.[14] The cyclotron

---

[14] Nuel Pharr Davis, "Lawrence and Oppenheimer," Fawcett World Library, NY, 1968, p 40.

principle, based on the (non-relativistic) equation that gives the rotation frequency of a charged particle in a magnetic field ($2\pi\times$the cyclotron frequency) as independent of the particle velocity, viz. $\omega = eB/m$, where ω is the angular frequency, e is the elementary charge, B is the magnetic field strength, and m is the particle mass.[15] This formula allows one to accelerate protons in increments of relatively low voltage – 10,000 volts per kick for example - applying the voltage alternating at frequency ω over and over again as the particle spirals out of a constant magnetic field. So in the 1930's nuclear physics had two kinds of particle accelerators: high voltage on a dome, measured directly; and high voltage acquired cumulatively in a cyclotron. Cyclotrons gave higher energy beams, but electrostatic accelerators offered more precise energy control, often useful in studying nuclear structure. More information regarding cyclotron design is given in Appendix D.

In 1937 Jean Frederic Joliot-Curie, with his wife Irene Joliot-Curie, built the first cyclotron in Europe at the College de France. The Joliot-Curies had earlier discovered artificially induced radioactivity by bombarding boron, aluminum, and magnesium with alpha particles from a radioactive source.[16] Frederic joined the French socialist party in 1934. During the German occupation in WWII Frederic was active in the French resistance, and in politics. He helped form the French communist party after the liberation of France. Throughout his professional life Frederic maintained close ties with physics in the USSR.

George Gamow,[17] a Soviet physicist who later emigrated to the US, was the first to apply quantum mechanics to the nucleus in his classic derivation of half-lives of α decays as a function of α kinetic energy, daughter nucleus atomic number, and radius. While simple arithmetic gives the energy, the half-life requires quantum mechanics. This is a common feature of nuclear physics. Energy considerations are relatively

---

[15] For protons, e $= 1.6\times10^{-19}$ coulombs, m $= 1.67\times10^{-27}$ kg, so the frequency f $= 15.2$ Mhz for a 1 Tesla magnetic field. This is the radio frequency of the electric field.

[16] www.nobelprize.org/nobel_prizes/chemistry/laureates/1935

[17] The correct transliteration of Gamow's name is Gamov. He adopted the German transliteration, and did not change it when he moved to the US.

simple, while calculations of decay rates and other dynamic quantities are more complicated. More complicated calculations are moved to the Appendices. Some discussion of Gamow's model is presented in Appendix A.

Let us summarize our outline of nuclear physics before the discovery of the fission of Uranium. Naturally occurring radioactivity is a nuclear phenomenon. The energy liberated is derived from Einstein's famous formula $E=mc,^2$ and the masses of the atoms involved. Atoms are classified by their atomic number Z, the number of electrons outside the nucleus, equal to the number of protons inside the nucleus; and their atomic mass number A= Z+N, N being the number of neutrons. Chemistry is determined by Z. Atoms with the same Z but different A are called isotopes. Separation of isotopes is a tough problem, as we shall see. The α decay kinetic energy can be analyzed by simple arithmetic, given the mass numbers for the parent atom, the daughter atom, and Helium. Neutrons can be used to create artificial radioactivity. In many cases the slower the neutron the better.

2.9. Nuclear Physics in the USA Before WWII

Western Europe and Great Britain were the leaders in pre-war development of nuclear physics, carried out at university laboratories and funded by private grants. Private industry in Europe paid for targeted research aimed at improving the product line. Support for research in the USA had similar sources. The US National Academy of Sciences, or any branch of the Federal Government, did not play an important role in financial support of research in the 1930's. Large private and state universities managed to fund modest research activities. The author is familiar with the history of the University of Wisconsin, Madison Physics Department, which may serve as an example of university research in nuclear physics in the US in the 1930's. The quantum theory was introduced at Wisconsin by John H. Van Vleck, who learned it from E.C. Kemble at Harvard. In those days, a few true believers taught the new physics to the masses.

Raymond G. Herb was a young experimental physicist at Wisconsin who had a knack for building electrostatic accelerators. In 1935 Herb constructed a working electrostatic accelerator, which produced a proton beam of 2.6 MeV kinetic energy for nuclear physics experiments. Herb's machine was of the Van de Graaf type, with a number of technical improvements, which increased performance.[18] Two electrostatic accelerators built by Herb were moved from Wisconsin to Los Alamos during the Manhattan Project for use as neutron sources for nuclear reactions relevant to the atomic bomb. Much of his prewar research was supported by private funds through the Wisconsin Alumni Research Foundation. Gregory Breit and Eugene Wigner were both at Wisconsin in the mid 1930's. To quote Wigner: "In 1936 the University of Wisconsin was rapidly becoming a center of nuclear physics research."[19]

Other universities had similar programs. There was a cyclotron at Harvard, and electrostatic accelerators at other sites in the Midwest. And of course, there was the Radiation Laboratory at the University of California, Berkeley, where Ernest Lawrence was building cyclotrons. Adoption of the new quantum theory into the curriculum of a physics department in the United States was likely to lead to a vibrant research program in modern physics.

There were very productive laboratories supported by the large corporations, but nuclear physics was not considered a field with any practical applications, so it was not favored by industry. However, they did maintain a high level of research in physics and chemistry in the US. General Electric, where Irving Langmuir studied plasmas, the DuPont Chemical Company, which worked on chemistry for consumer goods, public health, and agriculture, and AT&T Bell Telephone Laboratory,

---

[18] Raymond G. Herb, Biographical Memoirs, National Academy of Sciences, 1996.

[19] Andrew Szanton, "Recollections of Eugene P. Wigner," Plenum Press, NY, 1992, p 176.

which over the course of seven decades won seven Nobel Prizes[20] are all examples of industrial support of research in the 1930's. Bell Labs supported pure as well as targeted research, with great success all around.

2.10. Nuclear Physics in the USSR Before WWII

There was in principle no private enterprise in the USSR, so everything had to be government sponsored. The Bolsheviks realized early on that strong scientific infrastructure would enhance the industrialization of the country – one of the primary objectives of the five year plans. On the other hand, scientists, like artists and writers, are members of the intelligentsia, and hence inherent threats to the Communist Manifesto. It is interesting that the national laboratory structure to which we are currently accustomed in the United States began before WWII in the Soviet Union under the auspices of the Soviet Academy of Sciences. Graduate teaching at universities in the Soviet Union became separated from the research institutes. Many of the distinguished Soviet physicists worked full time at the laboratories without formal university appointments. Graduate students carried out laboratory experiments 'on loan' from a university. This model still obtains today. The laboratories also had strong theory groups, which trained students in theoretical physics, a field in which the Soviets excelled. So after a few years of course work at say Moscow State University, the aspiring graduate student had to depart for further training somewhere else. This split tended to isolate the university faculties from the research frontiers, and led to some curious disputes.

Soviet scientists had contacts with Western Europe, just as did the US physicists. Thus L.D. Landau from the USSR and J. Robert Oppenheimer from the USA, two prominent theoretical physicists, both traveled in Western Europe, were stimulated by progress in understanding the quantum mechanics, and returned home to educate the next generation in

---

[20] Clinton Davisson won the first in 1937 for the discovery of the wave nature of the electron, one of the cornerstones of quantum mechanics. Willard Boyle and George Smith won the most recent in 2009 the invention of charge coupled devices, used in digital cameras.

these important developments in modern physics. Yuli Khariton, who had important responsibilities in the Soviet Atomic Project, also spent two years at Rutherford's laboratory, and received his doctor's degree from Cambridge.

The Soviet Union always severely restricted foreign travel by its citizens, fearing that exposure to western decadence would somehow corrupt pure socialist thought. Landau seemed immune to these restrictions. He was a star student at the Leningrad physico-technical Institute, where he became a doctoral candidate in 1927 at the age of 19. An international physics conference was held in Moscow in 1928. The director of the Leningrad institute, Academician A.F. Ioffe, who we will meet again later on, was successful in inviting many famous physicists from Western Europe, including Niels Bohr, Paul Dirac, Leon Brillouin, and others.[21]

Landau as a graduate student gave three papers at the conference on the new field of quantum electrodynamics, the application of quantum mechanics to Maxwell's equations of electrodynamics. The talks were given in German. Landau spoke reasonable German with a Russian accent.[22] These talks made such a good impression that Landau was invited to visit the research centers in Germany, Denmark, and England. After the conference Landau received permission from the Peoples Commissariat of Education to visit the West.[23] This all seems perfectly ordinary, except we should not forget the government he was dealing with. He took a crash course to learn some English before the trip.

Landau's first stop was Berlin, where he met with Einstein, and, according to Landau's own account, he tried unsuccessfully to convince Einstein of the logical consistency of the quantum theory. In particular

---

[21] Maiya Bessarab, *ibid.*, p 19ff. Bessarab describes Landau's first trip to the famous laboratories of Western Europe. We quote freely from her text.

[22] German was the language of physics, as is English today. Many Soviet physicists published papers in German in Zeitschrift fur physik and other German journals.

[23] E.M. Lifshitz, "Life of L.D. Landau", Uspekhi fizicheskikh nauk, **97**, 169 (1969). English version Soviet Physics Uspekhi, 12, 135 (1969).

they talked about the uncertainty principle, about which Einstein and Bohr had a running debate over 20 years or so – Bohr in favor, and Einstein opposed. Einstein would propose a thought experiment that contained seemingly logical contradictions, which Bohr would then straighten out to demonstrate another victory for the quantum theory. Landau was no more successful in convincing Einstein than was Bohr.

The next stop was Göttingen, where Landau met with Max Born. Two years earlier Robert Oppenheimer, after a difficult year at the Cavendish Laboratory in Cambridge,[24] was a graduate student at Göttingen, and he and Max Born wrote a famous paper, which is still relevant today, applying quantum theory to diatomic molecules.[25] Landau and Oppenheimer did not meet as graduate students in Göttingen, but each traveled to Germany, a foreign country, to learn new physics. Werner Heisenberg of uncertainty principle fame was also a student of Max Born, so Göttingen was one of the hot spots for the development of quantum mechanics.[26]

Another hot spot was Copenhagen, where Landau went next, in April, 1930. Niels Bohr, who proposed his radical model of the hydrogen atom based on Rutherford's picture of the atom in 1913, had become a giant in modern physics, with a theory school in Copenhagen where the very best and brightest came together. Bohr greeted his new visitor from Leningrad as follows: "How good it is that you have come to visit us. You will teach us many things." Landau was very flattered, but later learned that Bohr greeted everyone that way. Bohr's institute in Copenhagen was at that time a continuously operating training ground for theoretical physicists. Bohr led seminars every day, in which there was focused but polite discussion of ideas and approaches to the solution of various physics problems. It was

---

[24] Kai Bird and Martin Sherwin, "American Prometheus," Alfred A. Knopf, NY, 2005. p 52.

[25] Nuel Pharr Davis, *ibid.*, p 21. The paper in German is: M. Born and J.R. Oppenheimer, Annalen der Physik 389 (**20**); 457-484 (1927).

[26] Born wrote an optics text which the author was obliged to try to study in German in graduate school in the 1950's. I did my best to learn the material somewhere else, but occasionally I had to tackle Born's formulas in German script.

a time of rapid development of modern physics. The common language of the Bohr Institute in the 1930's was German.[27] The application of quantum mechanics to a wide range of problems from atomic and nuclear structures to the properties of materials was an ever-present challenge. Bohr's presence assured politeness.

Different parts of the world have different customs for debates in seminars. American seminars tend to be low key. The speaker presents new material in outline form, and the audience listens politely, unless there is something obviously wrong. There is a question period after the talk, which can be spirited, but is usually restrained. The Russian custom, which Landau would follow if he were not inhibited by the famous Dane, is quite the opposite. Presentations, especially theoretical talks, are very detailed. Every line of a calculation is presented on the slide. The audience freely interrupts the speaker in mid-sentence to point out some apparent inconsistency or dumb mistake. A general lively discussion among the audience may follow, with the speaker left to his or her own devices on the stage. Sometimes the speaker finds it impossible to continue the presentation.

Landau spent about a month in Copenhagen, attending the seminars and working on the development of quantum electrodynamics. He wrote a paper on the subject with Rudolf Peierls.[28] Peierls was a German born British physicist who was to play an important role in nuclear weapons development. His wife Eugenia was Russian, also a physicist. They met at a conference in Odessa in 1930. And Peierls was later to hire Klaus Fuchs in 1940.[29] Peierls and his family were never suspected of any espionage for the Soviet Union, but like so many players in our story, he had Russian connections.

---

[27] Oral History of George Gamow, www.aip.org/history/ohilist/4325.html

[28] L. Landau and R. Peierls, Zeitschrift fur Physik **62**, 188 (1930). Most early papers on quantum electrodynamics are hard to read, but this one looks modern. The field was greatly simplified by Feynman and others after WWII, so that calculations are less cumbersome.

[29] www-history.mcs.st-and.ac.uk/Biographies/Peierls.html

Landau's travels were funded in part by the Rockefeller Foundation.[30] The fund was administered by Paul Ehrenfest, an Austrian and Dutch physicist, for the purpose of allowing young Soviet physicists to continue their education at the best universities of Europe. Ehrenfest was married to a mathematician born in Ukraine, giving him a link to the Soviet Union too. His wife was named Tatiana. They met in St Petersburg in 1910, before WWI and the Russian Revolution.[31] One of his good friends was Abraham Ioffe, Landau's mentor. Ehrenfest went to Leiden, but he wanted to return to the Soviet Union after the Revolution. He applied for an appointment at Kharkov, without success. Landau eventually obtained the Kharkov position. In 1933 Ehrenfest committed suicide. He is one of the tragic figures of 20th century physics.

2.11. Peter Kapitza Goes to Cambridge

Peter Kapitza was also in Cambridge, working with Ernest Rutherford at the Cavendish Laboratory. Kapitza was 14 years older than Landau, and already enjoyed an international reputation in physics research. He also was educated in A.F. Ioffe's Institute in Leningrad (Petrograd at that time), the same place where Landau studied. Ioffe was a very effective advocate for physics research in Russia after the Bolshevik revolution, and nurtured many of the country's rising stars.[32] Among other things he is credited with convincing the Russian Academy of Sciences to set up a commission for renewing scientific relations with foreign countries in the early 1920's, and in securing foreign currency to carry out the mandate. He also is credited with the establishment of the Physico-Technical Institute at Kharkov. He argued that the industrialization of the nation required active research centers in the physical sciences.[33] In the early 1930's Kharkov was the capital of Ukraine. Thus the new institute was called the Ukranian

---

[30] Maiya Bessarab, *ibid.*, p 24.

[31] Dirk Van Delft, "Paul Ehrenfest's final years," Physics Today **67**, 1 (2014) p 41.

[32] D. Schoenberg, F.R.S., Piotr Leonidovich Kapitza, Biographical Memoirs of Fellows of the Royal Society, 1984. The correct transliteration would be Kapitsa. Both forms are used in English. The Nobel Prize used Kapitsa. We follow the Royal Society.

[33] Delo UFTI, cripo.com.ua (online description of the history of UFTI).

Physico-Technical Institute, or UFTI in Russian. Ioffe's own institute was named similarly: the Leningrad Physico-Technical Institute, or LFTI. Each of these institutes was managed by the Ministry of Heavy Industry instead of the Soviet Academy of Sciences. The Ministry of Heavy Industry was not necessarily more enlightened, but at least it was different from the Academy of Sciences, which actually gave Kharkov and Leningrad more flexibility to develop research equipment.

Kapitza profited from Ioffe's efforts to encourage foreign travel for scientists, as did Landau later on. Obtaining foreign visas was difficult for Kapitza for two reasons: the Soviet government was suspicious of foreign travel; and in the early 1920's most of Western Europe had no diplomatic relations with the USSR.

Neither Kapitza nor Landau was born of peasant or 'worker' origin. Kapitza's father was a general in the Tsar's army engineers, and Landau's father was a civilian petroleum engineer. Each mother was educated, and took an interest in her precocious son. Landau's mother was a medical doctor. Kapitza's mother was an important figure in the literary world of St. Petersburg.[34] That is not to say that great achievements cannot flow from humble beginnings, but in these instances the beginnings were conspicuously upper class. Kapitza's wife Anna nee Krylova was the daughter of Alexei Krylov, a Russian admiral and also a famous mathematician.[35]

Ioffe managed to get to Berlin in February, 1921, and tried to get a visa for Kapitza. Germany was reluctant, but England was more accommodating, and in May, 1921, Kapitza got an English visa.[36] Kapitza and Ioffe visited Rutherford in Cambridge on July 12, 1921. The first meeting between Kapitza and Rutherford is such a good story that it is told

---

[34] D. Schoenberg, *ibid.*, p 327. (pagination refers to all biographies for 1984)

[35] George Gamow, "My World Line, an Informal Autobiography," Viking Press, NY 1970, p 80.

[36] D. Schoenberg, *ibid.*, p 330.

by Maiya Bessarab in Russian in her biography of Landau, and by Schoenberg in English in his biography of Kapitza. It goes like this:

Kapitza: "May I join the staff of your Cavendish Laboratory?"
Rutherford: "I'm sorry, but that is not possible, as we have too many people already."
Kapitza: "What accuracy do you usually achieve in your experiments?" (a seemingly irrelevant question)
Rutherford: "Oh, perhaps 2% to 3%." (There were about 30 researchers on the staff, so 3% accuracy is one person).
Kapitza: "Then my joining the staff is within experimental error. You won't notice me."[37]

Of course, Kapitza got the job. He was supposed to stay a few months, but apart from summer visits home in the Soviet Union, he remained in Cambridge for 13 years. He had been there nine years when Landau came to visit.

Landau was a theorist, skilled in mathematics and with excellent physical intuition. In his later career, back in the USSR he demonstrated his uncanny ability to solve any problem in physics in his own way. This is the reason that his papers contain few references to other work.[38]

Theoretical physics is a field with a small number of super powers, and a lot of other folks. It has been said that the gap between the very top and the next tier down is enormous.[39] Suppose you sit down to read a theoretical article. For most of them you might say, "If I worked just a little harder I could do that." But then there are a few papers which you would not be able to do in a million years. That is the gap. Landau himself was well aware of this phenomenon, and constructed a logarithmic scale to evaluate theoretical competence.[40] Logarithms have big gaps. The

---

[37] This use of statistical error is a characteristic of experimental physics. If Rutherford had said 1%, Kapitza would have been out of luck.

[38] E.M. Lifshitz, *ibid.*

[39] L.W Alvarez, autobiography.

[40] E.M. Lifshitz, *ibid.*

common logarithm of 10 is one, and of 100 is two, so you get the idea. Any creative human endeavor allows a range of skill. Great theoretical physics is similar to great art or literature or mathematics, with the extra condition that the end results have to correspond to nature. "The miracle of the appropriateness of the language of mathematics for the formulation of the laws of physics is a wonderful gift which we neither understand nor deserve."[41]

Kapitza was a skilled experimental physicist, inventor and engineer. He began at Cavendish working on problems of interest to Rutherford – how α particles lose energy passing through gases. Some of these measurements required magnetic fields stronger than were available at the laboratory, so he designed a new high current generator for a pulsed magnetic field, achieving fields over 10 Tesla for several milliseconds. One result of this work was a PhD degree, awarded by Cambridge University in 1923.[42] Kapitza turned his attention to studies of other phenomena affected by a very strong magnetic field, approaching 100 Tesla.[43] To approach such a high pulsed field, Kapitza designed a big dynamo, which was built by Metropolitan-Vickers, and paid for by a grant of £8000 from the Department of Scientific and Industrial Research. It was possible to get some government money to support research even in 1925. It was also the departure from 'string and sealing wax' experiments to big science at the Cavendish. It took a few years to get the whole thing working, and there was a problem with magnet coils burning up. Once he got things going, he began research in material science, measuring the electrical resistance of metals in high magnetic fields.[44] Kapitza was elected as a foreign member of the Royal Society in 1929. He continued research in material science by developing methods to liquefy hydrogen

---

[41] Eugene Wigner, "Unreasonable Effectiveness of Mathematics in the Natural Sciences," Comm. In Pure and Applied Mathematics, Vol 13, no. 1 (Feb. 1960) John Wiley & Sons, Inc., NY.

[42] D. Schoenberg, *ibid.*, p 334.

[43] This is a colossal magnetic field. One hundred tesla multi-shot magnets have only recently been built. See www.lanl.gov.

[44] D. Schoenberg, *ibid.*, p 338. Pulsed fields were below 100 tesla.

for low temperature work. He was starting in this direction when Landau came to visit.

2.12. Landau Visits Cambridge

In early May, 1930, Bohr invited Landau to accompany him to England. Bohr was to give the Faraday Lectures at Cambridge.[45] Apparently there were no problems with the necessary formalities, and soon Landau was in Cambridge talking to Paul Dirac about quantization of electron orbits in metals – the 'Landau levels.'[46]

Soon after Landau's arrival in Cambridge, an old friend from Leningrad, George Gamow, invited him to go on a motorcycle trip to Scotland. Landau rode in the side car. Gamow, Ioffe, and Peierls are shown in Odessa in 1930 in Figure 2.2. Gamow was also visiting the Cavendish at that time, and had purchased a second hand BSA motorcycle to get around town.[47] Landau enjoyed the trip immensely. Kapitza also enjoyed riding a motorcycle, but had perhaps outgrown this mode of transportation by the time Landau came to Cambridge. He had since graduated to automobiles, and owned a Vauxhall, because full blueprints of the devise were available, and he could take it apart.[48] Later on, back in the Soviet Union, Landau and Kapitza would be strongly linked by events, with important consequences. Indeed, Landau would receive the Nobel Prize for explaining Kapitza's work before Kapitza himself got the Prize. But all of this is far in the future in 1930. Landau worked on his English, and enjoyed evenings with Kapitza and his wife Anna in Cambridge. Landau attended Kapitza's weekly seminar 'club,' and gave at least one talk himself, on his theory of the Landau levels of electrons in a metal in a magnetic field.[49]

---

[45] Maiya Bessarab, *ibid.*, p 29.

[46] L.D. Landau, Zeitschrift fur Physik **64**, 629, (1930). Landau levels are discussed in introductory quantum mechanics courses.

[47] George Gamow, "My World Line, an Informal Autobiography," Viking Press, NY, 1970, p 83.

[48] D. Schoenberg, *ibid.*, p 344.

[49] See reference 46.

Figure 2.2 George Gamow, Abram Ioffe, and Rudolf Peierls in Odessa in 1930. Credit AIP Emilio Segre Visual Archives, Frenkel Collection.

After some months in Cambridge Landau went to Zurich in the fall of 1930 to meet Wolfgang Pauli. Pauli was thinking about a puzzle in nuclear β decay: the electron energy spectrum was inconsistent with the conservation of energy in the nuclear transition. Something else had to be emitted together with the electron to conserve energy; something undetected. This is in contrast with α or γ emission, where energy was conserved without difficulty. Remember our calculation of the kinetic energy of the α particle on page 34. On December 4, 1930, Pauli wrote a famous letter to attendees at a conference in Tubingen, which began "Dear radioactive ladies and gentlemen," and contained the neutrino hypothesis – that a new elementary particle with very small mass and no visible interaction with matter was emitted with the electron in nuclear beta decay.[50] At that time there were only two elementary particles, the proton and the electron. There was much resistance to this idea, but Landau immediately understood Pauli's hypothesis.[51] The direct observation of the neutrino had to wait until 1956.[52]

In 1931 Landau attended a seminar in Berlin organized by Erwin Schrödinger. Rudolf Peierls gave a talk on work done together with Landau to extend the uncertainty principle to relativistic quantum electrodynamics.[53] Landau then went back to Copenhagen before returning voluntarily to Leningrad in March, 1931. He was to visit Copenhagen again in 1933, but otherwise remained in the Soviet Union before WWII. His stated purpose on returning was to establish the best science in the world in his homeland. He said that he was indifferent to the higher standard of living in the West. Stalinist repression was perhaps not as obvious in 1931 as it would become a few years later.

2.13. Landau Returns Home

Leningrad turned out not to be a good fit for Landau after working in Western Europe. A.F. Ioffe, shown in Figure 2.2, the senior Soviet physicist most responsible for creating a vibrant research institute in Leningrad, one of the best in the Soviet Union, had an idea about the electrical properties of crystals, which, unfortunately, young Landau proved wrong. Landau was never one to tell someone gently of a perceived mistake, and Ioffe was greatly offended. So Landau had to go elsewhere. Fortunately, I.B. Obreimov invited him to Kharkov, to the newly organized Ukranian Physico-Technical Institute (UFTI).[54]

---

[50] The actual letter is available at www.library.ethz.ch/exhibit/pauli/neutrino_e.html (in German).

[51] Maiya Bessarab, *ibid.*, p 32.

[52] Frederick Reines, Nobel Address, 1995.

[53] L. Landau and R. Peierls, Zeitschrift fur Physik **69**, 56 (1931).

[54] Maiya Bessarab, *ibid.*, p 34.

Landau was sympathetic to the ideals of Communism, but was not politically active in the early 1930's. The widespread famine in Ukraine caused by the drive towards collective farming did not seem to concern him. Khalatnikov was a teenager in Dnepropetrovsk, an industrial city in Ukraine, south of Kharkov. He describes some of the living conditions at that time.[55] "There was hunger in Ukraine in the thirties. I started school in 1926, and in 1932, when there was famine, I got my first lesson in hypocrisy. Imagine the scene – swollen bodies having died of starvation lay on the streets, and we, 13-14 year olds not only saw them every day with our own eyes, but also we did not have as much to eat ourselves. And at school they told us of the advantage of the socialist system and soviet life. After each phrase our teacher, turning to the hungry children, asked us 'Children, tell the truth, are you not satisfied?'"

In August, 1932, Landau was named head of the theory division of UFTI. He promptly set about in cooperation with Kharkov University to train talented graduate students in theoretical physics. He established a 'theoretical minimum' for all aspiring students to achieve. This so-called 'Landau minimum' consisted of nine examinations: two in mathematics and seven in theoretical physics. Landau considered knowledge of the material covered by the minimum to be absolutely essential for independent research in theoretical physics.[56] He held weekly seminars on new developments in physics research. Between 1934 and 1961, Landau's career in training students, 43 people passed the 'theoretical minimum' test. Seven of these were subsequently elected to the Soviet Academy of Sciences.[57] Thus did the Landau school in Kharkov become a hot spot for theoretical physics in the Soviet Union. Landau was largely responsible for inspiration and training of several physicists who were to play an important role in the Soviet Atomic Project. Landau's own reputation and personality attracted bright young minds to physics who might otherwise have gone into other fields.

---

[55] I.M. Khalatnikov, "Dau, Kentavr, i Drugie," *ibid.*, p 5.
[56] Maiya Bessarab, *ibid.*, p 49.
[57] E.M. Lifshitz, *ibid.*

Back in Cambridge, Kapitza was doing well in low temperature research. A new facility, the Mond Laboratory, was constructed for his equipment, and opened in February, 1933. Stanley Baldwin, three-time British Prime Minister, and a graduate of Trinity College, Cambridge, attended the dedication ceremony.[58] Rutherford took good care of Kapitza, but Kapitza never attempted to become a British citizen. He retained his citizenship of the USSR, even though he was elected to the Royal Society, and was highly regarded in England.

2.14. The USSR Retains Kapitza

As mentioned above, Kapitza routinely travelled back to the Soviet Union on summer vacations. In October, 1934, after attending a conference in Leningrad and lecturing in Kharkov, Kapitza and his wife were about to leave Leningrad for England when he was told that his exit was denied, and he must remain in the Soviet Union.[59] His wife Anna was allowed to return to England for a few days to collect the children and close out the house. She told Rutherford what had happened on October 10. It was several months before the family was reunited in Moscow.

Kapitza had recently been offered a position at the Physico-Technical Institute in Kharkov, where a new low temperature group was getting started, and where Landau was forming the theory group, but he declined, saying that it was more efficient in the short term for him to continue exploiting his equipment in Cambridge. George Gamow, the talented Soviet theoretical physicist noted for his theory of α decay, had recently failed to return to the USSR after a visit to the West, which may have left the authorities nervous.[60]

---

[58] I.M. Khalatnikov, "Dau, Kentavr, I Drugie," (Dau, the Centaur, and others) Fizmatlit, Moskva, 2012, p28. Centaur was a nickname for Kapitza, and Dau for Landau. Isaak Markovich Khalatnikov (1919 -) is a theoretical physicist of the Landau school.

[59] D. Schoenberg, *ibid.*, p 349.

[60] D. Schoenberg, *ibid.*, p 350.

2.15. George Gamow Emigrates to the USA

Georgy Antonovich Gamov, to give his Russian name, was a bright star in contemporary physics.[61] He was born in Odessa in 1904 to a prosperous family. He then attended Leningrad University, and joined Ioffe's Institute, LFTI, where Landau was also a student. Vladimir Fok was the group leader. The apartment of Eugenia Kannegeisser, a young student, was a gathering place for physics graduate students. The young German Rudolf Peierls was also there, and Rudy and Genia were soon married. Peierls, shown in Figure 2.2, was to emigrate to England, and to lead the British mission to Los Alamos in 1943. Among the members of this mission was Klaus Fuchs, who will play an important role in our story. But back to Gamow. As a student, he had severe financial difficulties, and when he graduated it was decided to send him outside the USSR for further education in 1928. He then formulated his theory of α decay of heavy nuclei – the first application of quantum mechanics to the atomic nucleus. See Appendix A for a brief description of this idea. This work made a big impression both within and outside the USSR. Until 1931 Gamow returned annually to Moscow for holiday. Guglielmo Marconi organized an international conference in Rome on nuclear physics for October, 1931, and invited Gamow to be a plenary speaker. Gamow received the invitation in the summer in Copenhagen, while he was at the Bohr Institute. It would be convenient for him to stay in Europe for the summer, attend the conference in Rome, and return to the USSR afterwards. He accordingly went to the Soviet Embassy in Copenhagen with this proposal. After some negotiation Gamow was told to return to Moscow and wait the appropriate passport there.

Gamow took a plane from Copenhagen to Moscow in June, 1931. Upon his return he found the atmosphere in the USSR regarding science and scientists to have changed dramatically in the two years that he had been away. Russian science was now one of the weapons for fighting the capitalist world. It became a crime for Russian scientists to 'fraternize'

---

[61] This account of George Gamow comes from I.M. Khalatnikov, "Dau, Kentavr, i Drugie," *ibid.*, p 59ff. The spelling of his last name, used in USA, comes from German.

with scientists of capitalist countries.[62] Russian scientists, writers, composers, and artists were expected to demonstrate Marxist ideology in their works. Marxist philosophy in the form of dialectic materialism was enforced on scientific institutions by the appointment of 'philosophers' who presided over the usual laboratory directors to insure that the scientific work had the requisite purity. (This is reminiscent of the political commissars who were placed in the Red Army during WWII).

Einstein's theory of relativity was scorned as inconsistent with dialectic materialism. Relativity was created in part to refute the existence of an 'ether,' a medium for the propagation of light. But dialectic materialism required an ether.[63] Einstein was out, but so was a version of quantum theory called matrix mechanics, while the wave mechanics of Schrödinger was considered OK.[64]

After visiting his family in Odessa, Gamow stayed in Moscow in September, 1931, in order to get his passport to attend the Rome conference. While he was waiting, he met a young woman graduate student at the University of Moscow named Lyubov Vokhmintseva, whom he soon married. But in the meantime, he was denied permission to attend the conference by the Soviet Foreign Ministry. Gamow's address was delivered at the conference in his absence by Max Delbrück.[65]

The tightening grip by the Soviet government on science, as indicated by travel restrictions, publication oversight, and 'political philosophers,' convinced Gamow that it was necessary to leave the USSR, and he convinced his new wife to accompany him. Over the next two years they made at least two attempts to escape the Soviet Union.[66] The idea was to

---

[62] George Gamow, "My World Line an Informal Autobiography," Viking Press, New York, 1970, p 91.

[63] I am quoting Gamow on this. I know absolutely nothing about dialectic materialism.

[64] Another mystery, since the two versions of quantum mechanics are identical, as was proven by Schrödinger himself.

[65] Max Delbrück was a physicist from Göttingen who won a Nobel Prize for his work in genetics in 1969.

[66] George Gamow, "My World Line an Informal Autobiography," *ibid.*, Chap 5.

escape by sea, in a kayak obtained from a sports club in Leningrad. The first attempt was made in Crimea – 170 miles on the Black Sea to Turkey. This sounds foolhardy, and indicates how desperate Gamow was to get out of there. The Leningrad sports club was a part of the State Commission for Helping Scientists, which ran Scientists Houses[67] in major cities, and also had resort accommodations in the South, like in Crimea. So early in the summer of 1932 Gamow and his new bride arrived in Crimea with their collapsible kayak and food supplies that they had managed to hoard from rations in Leningrad. They figured on a 5-6 day trip in the open boat.

They actually paddled for two days, and were blown off course by a storm, landing ashore in Soviet territory new Balaklava, some 70 miles from where they started. They claimed that their trip, an overnight excursion, had been disrupted by bad weather, and this explanation was accepted, without suspicion. A second attempt was made in the North, from Leningrad to Finland, which also was unsuccessful. Landau was aware of this second attempt, although he did not participate. Apparently, the Soviet authorities remained unsuspicious.

In September, 1933 Gamow was invited to Brussels to the Solvay Conference. He was already planning to remain abroad, and wanted to take his wife with him to Brussels. He received permission to go, but without his wife. He appealed to Molotov, who asked 'Can't you live without your wife for two weeks?' Gamow replied that his wife was his secretary, and he would be unable to work without her. Molotov thought it over. Then the Foreign Ministry issued only one passport. He refused to go without his wife, and in the end both did indeed receive passports. Gamow did not return to the USSR. From Western Europe Gamow attended a physics summer school in Ann Arbor, Michigan. While he was there he got an offer of a permanent faculty position at George Washington University in Washington, D.C., with a joint research appointment at the Carnegie Institution. He needed a different visa in order to accept the appointment, so he returned to Cambridge later in the summer of 1934, in order to re-

---

[67] Colleagues at Leningrad took the author to the Leningrad scientists' house, which was one of the mansions of the Russian aristocracy along the Neva River.

enter the United States. After his return to the US in the fall of 1934 he registered with the Soviet Embassy. His Soviet passport soon expired, and was not renewed. He worked at GWU for 22 years. He will be a major player in theoretical physics in the USA, although he did not participate in the Manhattan Project during WWII.

Master spy Pavel Sudaplatov says that Ioffe mentioned to the NKVD that George Gamow had relatives in the USSR, and that insuring their safety might be an offer to persuade Gamow to furnish intelligence information to the Soviet operatives in Washington, D.C.[68] The idea was that he was in Washington, and was well connected with top physicists in the country, so he would be well informed on what was going on. One of the Washington agents, Lisa Zarubin, approached Gamow through his wife. The NKVD assumed that Gamow's acquaintances in physics, who were working on the Manhattan Project, would keep him informed of progress, and that they, the NKVD, could profit by this knowledge. There is no evidence that this link produced any intelligence information. What is striking regarding this episode is that Ioffe, the distinguished academician, would inform against Gamow to the NKVD.

Many sources say that Gamow's departure influenced the Soviet Foreign Office in its decision to force Kapitza to stay in the USSR after his summer sojourn in 1934. Gamow denies this, saying that the Foreign Office had been watching Kapitza closely for years, and finally decided to crack down. It is not clear whether the Soviet Foreign Office had any knowledge of the attempts to leave by Gamow prior to the fall of 1934. In any event, Gamow states that the retention of Kapitza certainly added weight to his own decision to remain in the United States.[69]

Whatever the cause of Kapitza's retention, a torrent of protest fell on the Soviet Union in Kapitza's defense, to no avail. Kapitza was initially quite depressed, but soon recovered his stamina, and managed to purchase

---

[68] Pavel Sudaplatov, "Razvedka I Kreml," *ibid.*, p 227. Sudaplatov's book is very interesting to read, but many have questioned its accuracy.
[69] George Gamow, "My World Line," *ibid.*, p 131.

and transfer much of his Cambridge equipment, for £30,000 to the new laboratory in Moscow, built on Stalin's orders. Rutherford in Cambridge assisted Kapitza in moving the equipment, saying that Kapitza was more important than apparatus, and that it would be more useful under his control in Moscow than in Cambridge without him.

The new laboratory was located at the beginning of Vorobevsky street, on the Moscow River, south-west of the Kremlin, and not too far from the present location of Moscow State University (MGU). This was the edge of town in 1935. Further south-west was rural, with small farms and gardens.[70] This is remarkable, considering how enormous modern Moscow is, but 75 years ago it was a lot smaller. Laboratory space, workshops, and housing were constructed on the grounds of the new institute. Kapitza proved to be an effective organizer and manager, and the Institute for Physical Problems attracted skilled technical personnel and began research in solid state and low temperature physics in 1935. Kapitza had spent a long time out of the country, and he used his 'visitor' status to his advantage. He was very skilled at appealing directly to the government, bypassing the Academy of Sciences, when he felt it necessary to do so in order to get things done. He knew how to be just outrageous enough to get the commissar's attention, without getting in trouble.[71] He was also fortunate to have Olga Alekseevna Stetskaya as administrative assistant.[72] She joined the Communist Party in 1917, and was a close associate of Lenin's wife, Nadezhda Krupskaya. Her former husband was propaganda minister for the Bolshevik Party. As such, she had good connections to government circles, and was able to help Kapitza solve many problems that arose daily in getting the Institute going. In the first months of operation of the new Institute Kapitza showed his talent for innovative engineering. He invented a revolutionary method of obtaining industrial quantities of oxygen by liquefying air.[73] At that time the industrialization

---

70 Maiya Bessarab, *ibid.*, p 62.
71 Alexei Kojevnikov, "Stalin's Great Science," Imperial College Press, London, 2006, p 114.
72 I.M. Khalatnikov, *ibid.*, p 29.
73 At atmospheric pressure oxygen boils at -183 °C, and nitrogen at -195.8 °C, so oxygen liquefies out first as the temperature is lowered.

of the USSR was a high national priority, and oxygen was needed in particular for metallurgy.[74] Kapitza was successful in selling his idea to the Soviet government, and during WWII in 1943 he was appointed State Commissar for industrial oxygen, giving him full exposure to the Soviet bureaucracy! All the while he was busy doing fundamental physics experiments at very low temperatures.

Landau returned to Leningrad, and Kapitza to Moscow, both to their native land. There was only one political party, the Communist Bolsheviks, led with an iron grip by Joseph Stalin. How did these two outstanding physicists relate to the Soviet Government? It was not possible to be apolitical. Both were patriotic. They did not want to forsake the motherland. On the other hand, both were very individualistic, in conflict with Communist ideas of collectivism. Landau went home partly to teach the new physics to the next generation of Soviet scientists, a goal which he admirably accomplished. Landau, together with E.M. Lifshitz, wrote a sequence of sophisticated graduate texts in theoretical physics, all of which have been translated into English and many other languages. Many of the Soviet Atomic Project scientists were graduates of the Landau school. Kapitza was to contribute substantially to the level of physics research in the USSR. Yet it is hard to say to what extent they supported Stalin's repressive government. A few years later both would become victims of the party's wrath; Landau in 1938 as a result of political unrest at UFTI, and Kapitza in 1945, partly because he made many enemies as State Commissar for industrial oxygen.

2.16. American Leaders, Oppenheimer and Lawrence

J. Robert Oppenheimer was born in the USA, educated at Harvard, and did graduate work in Europe, both at Cambridge where Rutherford was at that time, and in Göttingen working with Max Born. He learned quantum mechanics, and brought it home to Berkeley. Splitting his time between UC Berkeley and Cal Tech, Oppenheimer trained the next generation of

---

[74] I.M. Khalatnikov, *ibid.*, p 30.

American theoretical physicists. Thus Oppenheimer and Landau played similar roles in graduate education in their respective countries. Oppenheimer did not take the time to write a series of graduate texts, but he had a strong influence through his lecture notes on the content of some texts that were written, and he certainly had a 'school' of devoted followers. The graduate level qualifying examinations, the equivalent of the 'Landau minimum,' were prepared and administered by the graduate faculty, rather than by one individual, but Oppenheimer undoubtedly had considerable influence over their content.

Ernest Lawrence was also born in the USA, and educated at Yale. He did not go to Europe to study. Instead he went out West to Berkeley, where he foresaw opportunities for success in physics research in a growing university environment. Lawrence and Kapitza had similar skills. Both had boundless energy and enthusiasm, a talent for raising money for research, and engineering ability. Both liked big equipment. Lawrence was thinking of the next larger cyclotron before the first cyclotron was working. He was a builder rather than a user. The Joliot-Curies' first discovered artificial radioactivity, even though the Berkeley cyclotron was producing radioactive isotopes earlier. The cyclotron and the radiation counters were turned on and off simultaneously, so that residual radioactivity, which manifests itself after the cyclotron is off, could not be observed.[75]

By 1939 Lawrence was promoting a 100 MeV proton cyclotron.[76] However, there was a problem with building bigger cyclotrons with higher energies, other than cost and scale of the magnets. The cyclotron frequency formula $\omega = eB/m$, is energy independent only in the non-relativistic limit. As the proton kinetic energy approaches 50 MeV or so, the rotation frequency of the protons begins to noticeably decrease, so that the oscillator is no longer synchronized with the particles, and acceleration

---

[75] Nuel Pharr Davis, *ibid.*, p 57.

[76] The 184" cyclotron was actually built, and used for isotope separation in the Manhattan Project. It became a proton accelerator only after the war.

ceases. For this reason, a cyclotron is not useful for the acceleration of electrons. A way around this difficulty was discovered after WWII.[77]

A few words are in order about the politics of our two prominent US physicists, to match what was said about Kapitza and Landau. It is perfectly acceptable for citizens of the United States to love their country and hate their government. We do it all the time.[78] In Stalin's Soviet Union such sentiments were not tolerated. So our protagonists did not have similar boundary conditions. But we knew that. That said, Oppenheimer and Lawrence were both loyal US citizens, and for most of their lives close friends, with about as divergent political views as can be imagined. Lawrence grew up in South Dakota, where his father was president of a teachers' college. Oppenheimer grew up in New York City, where his father was a prosperous merchant. Lawrence's family was protestant, Oppenheimer's was Jewish. Lawrence was what we would call conservative in his politics. Oppenheimer was left wing.

The next development in nuclear physics was the discovery of the fission of Uranium.

---

[77] Nuel Pharr Davis, *ibid.*, p 251.
[78] I am indebted to Prof. E. Friedman for pointing this out to me.

Chapter 3

The Discovery of Fission
of Uranium

3.1. Fusion and Fission

Light nuclear fusion, climbing up the left side of the curve in Figure 2.1, Chapter 2, which is the source of energy in the sun, and in the hydrogen bomb, is a garden variety nuclear reaction. For example, using the atomic mass number table in Tipler and Llewellyn[1] for the reaction $^2H + {}^3H \rightarrow {}^4He + n$ the energy release is:

M(1,2) = 2.014012 atomic mass of deuterium
M(1,3) = 3.016049 atomic mass of tritium
M(2,4) = 4.002602 atomic mass of helium 4
M(0,1) = 1.008665 atomic mass of neutron (no electron!)

$$M(1,2)+M(1,3) - M(2,4) -M(0,1) = 0.01879 \text{ amu} = 17.5 \text{ MeV}.$$

Where we have used 1 amu (atomic mass unit) = 931.5 MeV. The recoil α particle takes about 3.5 MeV of energy, leaving 14 MeV for the neutron.[2] Note as usual that the charges and the mass numbers add up, and the electrons (two of them) cancel out. This is a very exothermic nuclear reaction. It generates a lot of energy. This reaction does not take place in a mixture of deuterium and tritium gas at room temperature.[3] To make it go, you have to overcome the coulomb repulsion of the two initial state

---

[1] Tipler and Llewellyn, *ibid.*, Appendix A.
[2] The recoil energy is $E_2 = E/(1+A)$, where A= N + Z of the recoil nucleus.
[3] Cold fusion, which is this kind of thing, does not exist.

61

nuclei, which requires kinetic energy of 1 MeV or so. Inside the sun the temperature is high enough to give kinetic energies of this order to the plasma (ions and electrons) to allow the reactions to proceed. On earth, an electrostatic accelerator or cyclotron is used to impart sufficient energy to a deuteron beam. In the hydrogen bomb an atomic bomb is used as a trigger, to heat the material to be fused to a high enough temperature. Nuclear fission of heavy nuclei, which we are about to discuss, has its own peculiar terminology: thermal neutrons, critical mass, and chain reaction. These terms do not apply to fusion.

Fission is different. It occurs only in the heaviest nuclei.[4] Some heavy nuclei spontaneously fission without neutron absorption, while others require an extra neutron to make a new isotope. Fission results in two mid–range nuclei, which are very neutron rich, because the stable nuclei in the middle of the periodic table have a smaller neutron excess than does uranium. This results in a lot of residual radioactivity, as the neutron rich nuclei β decay to convert neutrons into protons. Some of the decays have half-lives of thousands of years, which leads to the radioactive waste problem for nuclear power reactors. The existence of fission was not anticipated theoretically, and its experimental discovery came only after careful chemical analysis of the elements remaining in a sample of uranium after neutron bombardment.

3.2. Absorption of Slow Neutrons

Fermi's group in Rome studied the absorption of slow neutrons by a number of elements, including uranium.[5] Physicists realized that neutrons could cause nuclear transmutations in even the heaviest elements, for which charged beams from electrostatic accelerators and cyclotrons were not energetic enough to overcome the coulomb barrier. Emilio Segre, a close colleague of Fermi, has described the experimental work in Rome.[6]

---

[4] There are some exceptions. For example, $^8\mathrm{Be} \to \alpha + \alpha$.

[5] E. Fermi, Nobel Address, 1938. A complete set of reports can be found in "Collected Papers of Enrico Fermi" Vol I University of Chicago Press, 1962.

[6] Emilio G. Segre, "The Discovery of Nuclear Fission," Physics Today **42**, 38 (1989).

Work with the lighter elements was straightforward. The element was exposed to the radon-beryllium neutron source,[7] and then the sample was removed to a Geiger counter for measurement of artificially induced radioactivity. The group proceeded through the periodic table, finally reaching uranium. Here the prospect of making the transuranic element Z=93, hitherto unknown, was exciting. Uranium was already radioactive, without any neutron exposure, and this fact complicated the study of the results, not only for Fermi's group, but for all of the others who were working in the field at that time. Parallel work was in progress by Otto Hahn and Lise Meitner in Germany, and Irene and Frederic Joliot-Curie in Paris. Steps were taken to reduce the natural radioactivity that in turn also precluded the observation of fission products. Chemical separation of the radioactive residue was very difficult, because of the very small concentration of daughter elements. The goal was to observe $^{238}U + n \rightarrow$ $^{239}U \rightarrow {}^{239}Np + \beta + v$. This reaction does indeed occur, but there is so much else going on when uranium is bombarded by neutrons. Neptunium is the modern name for element Z=93. All of the research groups used a carrier technique to precipitate out element 93 together with known elements lower in the periodic table thought to have similar chemical properties. Some of these precipitate elements matched the fission products, but nobody realized it at the time. They thought they were chemically isolating Z=93, when in fact they were doing no such thing, but rather isolating Ba, La, Ce – elements in the middle of the periodic table.[8]

Fermi thought, and the Nobel Committee agreed, that transuranic elements heavier than uranium had been made. While that may have been true, the radioactivity observed was from fission products. Hahn and Strassmann published their results after the Nobel ceremony.[9] Fermi modified the written version of his Nobel address to reflect their discovery. Hahn and Strassman's careful chemical analysis identified barium, $Z = 56$, in the residue after bombarding uranium with neutrons. Meitner and Frisch

---

[7] Source strength was about 10^7 neutrons/sec.

[8] There are many examples in science of difficulties caused by too much going on. Once it is all sorted out, one sees how rich in information the experiment was.

[9] O. Hahn and F. Strassmann, Naturwiss. **27**, 11 (1939).

then described the splitting process in terms of a liquid drop model of the nucleus, and adopted the word 'fission' from biology.[10]

3.3. Fission Crosses the Atlantic

Niels Bohr visited the United States, and brought the news of the fission of the Uranium nucleus to Princeton in January, 1939. Fermi was already at Columbia University in New York City, having emigrated from Mussolini's Italy after the Nobel Ceremony in Stockholm. Bohr and John Wheeler wrote a seminal paper on the mechanism of nuclear fission, based on the liquid drop model of the nucleus.[11]

It cannot be overemphasized that fission is an isolated phenomenon, and not really of fundamental significance. Its main contribution to fundamental research in nuclear physics is through the invention of the nuclear reactor, which furnishes copious fluxes of neutrons and neutrinos, and manufactures radioisotopes for research and medical purposes. Texts on nuclear physics do not devote a lot of space to the fission process. That it has such terrible and profound implications is really a bit of a fluke. Bohr and Wheeler described the process in terms of the deformation of a liquid, with surface tension tending to hold it together, and coulomb repulsion trying to tear it apart. The parent nucleus is assumed spherical, and when excited by neutron absorption or otherwise, it separates into two spheres with a distance between their centers which increases with time. There are three naturally occurring isotopes of Uranium: ^{234}U (0.0055%), ^{235}U (0.72%), and ^{238}U (99.27%). The numbers in parentheses are the natural abundances. All three isotopes are radioactive (all nuclei above $Z = 82$ – lead – are radioactive), but the half-lives are sufficiently long that they have not decayed away in the age of the earth. ^{234}U is so rare that it can be

---

[10] L. Meitner and O.R. Frisch, Nature **143**, 239 (1939). Fission comes from biology, and chain reaction comes from chemistry.

[11] N. Bohr and J.A. Wheeler, Phys. Rev. **56**, 426 (1939). This is probably the best paper on the phenomenon of fission. The liquid drop model is the basis for the semi-empirical mass formula, which has five adjustable constants to fit the binding energy B (Z,A) to the data for stable nuclei. It works surprisingly well. See Appendix A.

neglected. The hypothesis that the fission of uranium by slow neutrons was due to the isotope ^{235}U was due to Bohr, based on subtle differences in the energetics of the compound nuclei ^{236}U and ^{239}U formed after neutron capture. This hypothesis was soon shown to be correct. The isotope ^{235}U could fission after capture of fast or slow neutrons, while only fast neutrons (bringing in extra excitation energy) could stimulate fission in ^{238}U. The probability of fission by slow neutrons in bulk natural uranium is low because only the rare isotope ^{235}U participates.[12]

Richard Rhodes describes Leo Szilard's idea of a neutron induced chain reaction.[13] According to Rhodes, Szilard, a Hungarian Jewish physicist in exile in London in 1933, was incensed to read of Lord Rutherford' speech to a meeting on the transformation of the elements. Rutherford said that "anyone who looked for a source of power in the transformation of the atoms was talking moonshine." Szilard was not quite sure what 'moonshine' meant, but he was annoyed that an authority like Rutherford would publicly state that something was impossible.

Not everyone knows that the term 'chain reaction' came from chemistry, but Szilard was familiar with the idea.[14] As Szilard waited to cross the street at Russell Square it occurred to him that if an element could be found which is split by a neutron, and would subsequently emit two neutrons, then a nuclear chain reaction could be sustained in a sufficiently large mass. He proceeded to patent this idea in 1934.[15] This all occurred well before the discovery of the fission of uranium.

Since the fission process yields two very neutron rich isotopes in the middle of the periodic table, it is likely that extra prompt neutrons are

---

[12] Numerical calculations of the energies based on the semi-empirical mass formula are given in Appendix A.

[13] Richard Rhodes," Making of the Atomic Bomb," *ibid.*, p 28.

[14] In a neutron chain reaction suppose n + A $\rightarrow$ A' + 2n (a net gain of one neutron, since one is lost), and both final state neutrons do the same thing. Then starting with one reaction, you have two, then four, then eight, 16, 32 etc. - a diverging geometric series in a large enough sample of material.

[15] Recollections of Eugene Wigner, *ibid.*, p 188.

emitted. The term 'prompt' refers to neutrons emitted at the same time as the fission, while delayed neutrons come from the fission products. Fission products are radioactive. They stop in the material of the reactor, and β decay into daughters with half-lives of typically several seconds. The daughters then emit delayed neutrons. Measurements soon confirmed that the average number of fast prompt neutrons was approximately 2.3.[16] This number encouraged work towards a controlled chain reaction.

In the meantime, Szilard emigrated to the United States, joining several Hungarian colleagues. This collection included Edward Teller, Eugene Wigner, and John von Neumann. They were so remarkable that Fritz Houtermans, whom we met in Chapter 1, claimed that they were actually from Mars, masquerading as Hungarians to avoid suspicion. Everyone knew that Hungarians could not speak good English, so their Martian origin was camouflaged. A post war Hungarian known to the author was Valentine Telegdi, who ably carried on the tradition of mystical powers.

Szilard agitated for secrecy in uranium research. He tried to stop publication by Frederic Joliot-Curie, who was working in Paris, but Joliot-Curie preferred to publish.[17] His first success was the doctoral dissertation of Herbert Anderson.[18]

Szilard wanted to warn Belgium regarding its uranium mines in the Belgian Congo, to keep the ore out of German hands. He also wanted to alert the US government to the possibility of nuclear weapons. Wigner in his memoirs described what happened next.[19] Szilard and Wigner were

---

[16] This number is not easy to measure. Delayed neutrons from the fission products are one complication. See a series of Manhattan Project reports in The Collected Papers of Enrico Fermi, Volume 2, University of Chicago Press, 1965.

[17] Recollections of Eugene Wigner, *ibid.*, p196. An example was a letter to Nature: H. von Halban, F. Joliot, and L. Kowarski," Number of Neutrons Liberated in the Nuclear Fission of Uranium" Nature 143, 680 (1939). They reported $\nu=3.5\pm0.7$, a bit high.

[18] "The Collected Papers of Enrico Fermi," Vol 2, *ibid.*, p 31. In those days, a thesis at Columbia University had to be published in order to satisfy the degree requirements, so delaying publication put Anderson in a bind. Dean Pegram came to his rescue, and Anderson got his degree. The thesis was published in the Physical Review after the war.

[19] Recollections of Eugene Wigner, *ibid.*, p 197ff.

both at Princeton in the summer of 1939. The physics faculty there had no connections with the government or the armed forces. Dean Pegram at Columbia University did know some people in the federal government, and he agreed to organize a meeting between Fermi and naval officers. That meeting took place, but had little impact. Fermi got $2000 from the navy for isotope research.

3.4. Einstein Writes to Roosevelt

Wigner and Szilard decided to get help from Albert Einstein, the country's most famous physicist. Einstein somehow knew the Belgian royal family, and could help alert Belgium to the danger of German acquisition of uranium ore. Wigner and Szilard decided that the State Department should contact the Belgian government, rather than private citizens – indeed private individuals who were not citizens. In the end, a letter to Belgium was never sent. However, Einstein could serve as a direct contact to President Roosevelt.

Einstein was normally in Princeton at the Institute for Advanced Study, but that summer he was vacationing on Long Island. They set out for Long Island to find Einstein, and proceeded to get thoroughly lost. A pedestrian walking along the road told them how to get to Einstein's house. Einstein's English was never very good, and Wigner spoke with him in German. Einstein grasped the situation, agreed to write a letter to President Roosevelt, and dictated it to Wigner that afternoon. Wigner and Szilard returned to Princeton, where Wigner translated Einstein's letter into English. Wigner wrote: "I am largely responsible for the English wording of the famous Einstein letter." They returned to Long Island, Einstein signed the letter, and it was given to a government economist, Alexander Sachs, to hand deliver to President Roosevelt. The date on the letter was August 2, 1939. Hitler invaded Poland on September 1.

3.5. A Uranium Committee is Created

In the aftermath of the letter a Uranium Committee was set up to monitor developments in atomic energy. Lyman Briggs, director of the National Bureau of Standards in Washington, was named chairman. The first meeting of this committee took place in Washington on October 21, 1939, and is described by Rhodes and by Wigner.[20] In attendance were Briggs, Sachs, Szilard, Wigner, Teller, Richard Roberts (an experimental nuclear physicist, and alternate for physicist Merle Tuve), Lt. Col Keith Adamson from the Army, and Commander Gilbert Hoover from the Navy. Both military men were ordinance experts. The sum of $6000 was appropriated from Army funds to work on the idea of a controlled chain reaction using uranium. This grant, and the Navy money mentioned above, were among the first US government funds towards the development of atomic energy. In Wigner's view, the Manhattan Project started with this meeting, although the National Bureau of Standards did not in the end play an important role in the project. The work of the Uranium Committee increased when it was learned, in April, 1940, that the Kaiser Wilhelm Institute in Berlin had begun an ambitious program in uranium research.[21] Regarding the interesting question of whether Roosevelt ever drafted and signed a specific memorandum to the effect that development of nuclear weapons should proceed with highest priority, Rhodes stated in his book that no such document has ever been found.[22]

Experimental work proceeded to determine design parameters for a chain reacting pile – a primordial nuclear reactor at Columbia University under Fermi. Wigner's theory group at Princeton worked on calculations. In the meantime, Lawrence's 60" cyclotron at Berkeley was used to accelerate deuterons and bombard uranium targets to make transuranic elements. This exercise was successful. McMillan and Abelson chemically isolated element 93, which they named neptunium in the spring of 1940.[23]

---

[20] Rhodes, *ibid.*, p 315ff. Wigner, *ibid.*, p 202ff.

[21] Leslie M. Groves, "Now it Can be Told," *ibid.*, p 7.

[22] Richard Rhodes, "Making of the Atomic Bomb," *ibid.*, p 387.

[23] E. M. McMillan, Nobel Address, 1951.

Soon Glenn Seaborg and coworkers at Berkeley determined that element 94 was created as a daughter from the β decay of element 93.[24] Interest grew in element 94, named plutonium, as an alternate candidate to ^{235}U for fission.[25]

It is fortunate for suppression of nuclear proliferation that the use of either isotope to create a nuclear weapon is extremely difficult. The early workers must have wondered if there was not an easier way, given that they were determined to make a practical atomic bomb. The answer turns out to be no, there are no other more accessible fissionable isotopes. The energy considerations are discussed in Appendix A. So one is forced to use either ^{235}U, which requires isotope separation from natural uranium, or ^{239}Pu, which has to be manufactured in some way, because it does not occur in nature at all.[26] When Seaborg started studying plutonium, no one had built a nuclear reactor. Nevertheless, Seaborg 's group worked with microgram samples of plutonium to study the chemistry of the new element, and to devise a chemical scenario that could be used on an industrial scale to extract plutonium from the fission products of a yet to be designed and built nuclear reactor.

The National Defense Research Committee (NDRC) was created in June, 1940, with Vannevar Bush, an electrical engineer and president of the Carnegie Institution, as chairman. The Uranium Committee was one of its subcommittees. Uranium research was expanded. By November, 1941, sixteen projects were underway, with a budget of $300,000.[27]

---

[24] Glenn T. Seaborg, Nobel Address, 1951. See also Seaborg, et al., Phys Rev **69**, 366 (1946). This letter was received on Jan 28, 1941, but not published until after the war.

[25] The Wall Street Journal, October 30, 2013, has a story about the radioactivity that lasted for decades in UC Berkeley's Gilman Hall, where plutonium was discovered.

[26] The half-life of ^{239}Pu is 24,000 years, long enough to work with the metal, but too short to remain in the earth's crust. Some short-lived isotopes, like those of Ra, occur naturally because they are made by decay of ^{238}U. ^{239}Pu decays by α emission.

[27] Leslie M. Groves, "Now it Can be Told," *ibid.*, p 7.

3.6. Activity in Great Britain

Efforts had been underway in Britain since the discovery of fission. Rudolf Peierls came to Britain in 1933, and elected to stay after Hitler came to power. He became professor of mathematical physics at the University of Birmingham in 1937.[28] He was joined at Birmingham by Otto Frisch, Lise Meitner's nephew, in 1939. Frisch was in Sweden visiting Meitner during the Christmas holidays of 1938, when Hahn and Strassman announced the fission of uranium, and the two wrote a paper explaining the phenomenon. Frisch then returned to Copenhagen, to the Niels Bohr Institute. In the summer of 1939 he went to Birmingham, and the outbreak of WWII prevented his return. Both Frisch and Peierls were 'enemy aliens.' Peierls from Germany and Frisch from Austria. Peierls became a British citizen in late 1940, which expedited his significant participation in the Manhattan Project as a member of the British mission to Los Alamos in 1943. In 1940 Frisch and Peierls became interested in the possibility of the application of fission to nuclear weapons. They began studying fissionable materials, cross sections, critical masses, isotope separation, and all of the aspects of what was to become the Manhattan Project. They wrote a report in 1940, called the Frisch-Peierls memorandum, outlining their calculations, and concluding that a uranium bomb was feasible. It is mentioned in Appendix C in the context of isotope separation by thermal diffusion, a technique recommended for uranium enrichment. As a result of their studies, Sir Henry Tizard, a chemist and the civilian chairman of the Committee on the Scientific Survey of Air Defense, decided to form a committee.[29]

Thus the MAUD committee was created. MAUD stands for something - the 'U' is for uranium - but it is better just to use 'MAUD.'[30] The MAUD committee consisted of G.P. Thomson (chair), Marcus Oliphant, Patrick Blackett, James Chadwick, Philip Moon, and John Cockcroft. GP was the

---

[28] en.wikipedia.org/wiki/Rudolf_Peierls
[29] Richard Rhodes, "Making of the Atomic Bomb," *ibid.*, p 325.
[30] Rudolf Peierls, who should know, said that MAUD does not stand for anything. See Rudolf Peierls, "Bird of Passage" Princeton University Press, Princeton NJ 1985 p 156.

son of JJ Thomson, the discoverer of the electron. GP shared the Nobel Prize in physics with C.J. Davisson (of Bell Labs) for electron diffraction in 1937. He was professor at Imperial College London.[31] Thomson pere showed that the electron was a particle, and Thomson fils showed that the electron was a wave. Marcus Oliphant was an Australian, who came to Cambridge, the Cavendish, and Rutherford (who was from New Zealand), to do experimental nuclear physics. Oliphant also did important war work in developing microwave radar. Chadwick from the Cavendish was the discoverer of the neutron, and won the Nobel Prize in 1935. Cockcroft developed the electrostatic accelerator named after him, also at the Cavendish, and was to win the Nobel Prize in 1951. Blackett contributed to the British war effort in many ways, anti-submarine warfare in particular, and was to win the Nobel Prize in 1948 for 'discoveries in nuclear physics and cosmic radiation.'[32] Moon was an experimental nuclear physicist, trained at the Cavendish, and a colleague of Oliphant at Birmingham. The MAUD committee was an impressive list of prominent British experimental physicists. The committee first met on April 10, 1940, at the Royal Society, London.[33]

Frisch and Peierls were not on the committee because of their prior enemy alien status. But the Frisch-Peierls memorandum became the basis for what is known as the MAUD report. The top secret report, dated July, 1941, was written by James Chadwick. It outlined several steps necessary to achieve a nuclear weapon. It is a very prescient document, only a few pages long. The last meeting of the MAUD committee was July 2, 1941, at the Royal Society.[34]

The MAUD report dismisses the practicality of slow neutron fission, that is nuclear reactors.[35] It emphasizes rather fast neutron fission of ^{235}U. It estimates the critical mass. It recognizes the necessity of enriching natural uranium, which contains 0.7% ^{235}U, to essentially 100% ^{235}U.

---

[31] en.wikipedia.org/wiki/George_Paget_Thomson
[32] www.nobelprize.org/nobel_prizes/physics/laureates
[33] Richard Rhodes, "Making of the Atomic Bomb," *ibid.*, p 330.
[34] Graham Farmelo, "Churchill's Bomb," Basic Books, New York, 2013, p 183.
[35] Chadwick was to prepare a shorter report on nuclear power. I have not seen it.

Gaseous diffusion, thermal diffusion and centrifuges are all considered for isotope separation. Frisch was interested in thermal diffusion, and had done some experiments with it. However, gaseous diffusion was adopted as the most favorable. Some detailed estimates of the scope of a gaseous diffusion plant were included in an Appendix, prepared by Franz Simon. Simon was a German refugee chemist who was working at the Clarendon Laboratory in Oxford.[36] Compared to what was actually accomplished at Oak Ridge the analysis seems very optimistic, forecasting 1 kg/day of ^{235}U from natural product. Nevertheless, the report chooses one of the techniques actually used to enrich uranium. The weapon must weigh not more than 1 ton, as to be deliverable by existing bomber aircraft. The critical mass was to be assembled by firing two sub critical hemispheres at each other in a gun type assembly. Some estimate was made of the physical destruction of the blast, based on data from a 2400 metric ton TNT explosion of a munitions ship in the harbor of Halifax, Nova Scotia in 1917. That blast was the strongest man made explosion until nuclear weapons. Estimates of the biological hazard presented by the residual radioactivity were also presented. Although the Manhattan Project deviated from the MAUD report in some important aspects – no mention is made of ^{239}Pu production or its use as a fissile material – on the whole the outline is eerily correct. It was sent to Lyman Briggs, the chairman of the US Uranium Committee. Briggs locked it in his safe without circulating it for evaluation by the committee.

Oliphant flew to the United States in August, 1941, on military aircraft, to discuss advances in radar, and to champion the MAUD report.[37] He was totally flummoxed by Briggs in Washington, DC, who admitted that he had not let anyone see the MAUD report. Oliphant then went to Berkeley, to appeal to Ernest Lawrence, where he received support for his efforts to stimulate the United States into more vigorous activity regarding the uranium problem. Partly as a result of Oliphant's insistence, the Manhattan Project started to roll.

---

[36] Richard Rhodes, "Making of the Atomic Bomb," *ibid.*, p 339.
[37] Richard Rhodes, "Making of the Atomic Bomb," *ibid.*, p 372.

Back in England there was a disagreement regarding the feasibility of an atomic bomb project on British soil. Frederick Lindemann, a long time confident of Winston Churchill, and a professor at Oxford, was leading scientific advisor to the British government during Churchill's prime ministry, and attended meetings of the War Cabinet. He naturally had access to the MAUD report, and felt that the United States could not be relied upon to accomplish the goal of a nuclear weapon in a timely fashion. Therefore, he supported the implementation of a nuclear weapons program in Britain. At first, he was supported in this by James Chadwick.[38] Churchill approved, and was the first national leader to officially endorse the atomic project. Churchill wrote to Lord Ismay:

> "Although personally I am quite content with the existing explosives, I feel we must not stand in the path of improvement, and I therefore think that action should be taken in the sense proposed by Lindemann, and that the Cabinet Minister responsible should be Sir John Anderson."[39]

Anderson was trained as a chemist, but had spent his career as a civil servant. He decided after reading the MAUD report and discussing the situation with Blackett and others that Lindemann was incorrect, and that the project should be pursued on US soil. On 12 October, 1941, while Oliphant was in the US, Roosevelt sent Churchill a message, proposing cooperation and exchange of information and scientific personnel regarding the project proposed in the MAUD report. Later that month the British government decided to turn the MAUD project over to the Imperial Chemical Industries, ICI, to be administered by the ICI director of research, Wallace Akers. Oliphant resigned in protest when he learned of the transfer. ICI named the project 'Tube Alloys,' an irrelevant expression intended, like 'Manhattan Project' to avoid suspicion. Akers, from British industry, was not a good choice to head Tube Alloys. The Americans, Groves in particular, were uncomfortable with full cooperation, because they feared that the British were in it to exploit US resources for an industrial advantage in nuclear power after WWII. The British delegation

---

38 Graham Farmelo, "Churchill's Bomb," *ibid.*, p 186.
39 Graham Farmelo, "Churchill's Bomb," *ibid.*, p 190.

of scientists, including Frisch, Peierls, and Chadwick arrived in Los Alamos in 1944.

3.7. Manhattan Project Begins in USA

The Manhattan Project pursued two parallel developments toward obtaining enough fissile material to make a nuclear weapon. One was extracting kilogram amounts of ^{235}U from uranium ore by various methods of isotope separation, predominantly at plants in Oak Ridge, Tennessee. The other was manufacturing kilogram amounts of ^{239}Pu produced by neutron bombardment of uranium in high power reactors built at Hanford, Washington. Then the weapons were assembled and tested at Los Alamos, New Mexico.

As Wigner told the story, on December 8, 1941, one day after the attack on Pearl Harbor by the Japanese, Arthur Compton called Wigner at Princeton to say that the Uranium Project was being reorganized, and that he (Compton) was heading a study at the Metallurgical Laboratory at the University of Chicago to design and build a graphite-uranium lattice reactor.[40] Fermi moved from New York City to Chicago, to continue his studies on reactor design.

Power generating nuclear reactors are now spread all over the world, and have a host of problems, from radioactive waste handling to nuclear accidents caused by the reactor itself (Chernobyl), or by natural catastrophe (Fukushima). But in 1942 there were no such objects, and making one work was considered an essential step towards building nuclear weapons. A reactor is in a sense a test bed for a weapon. It generates power by fission, and it works by chain reaction. It can go 'critical.' In the days before digital computers, it was necessary to test experimentally as many of the ideas as possible, since the modern tool of computer simulation did not exist. Fermi called the first nuclear reactor a 'pile.' It was constructed under the West Stands of the University of

---

[40] E.P. Wigner, *ibid.*, p 208.

Chicago football stadium, across Ellis Avenue from the present Research Institutes building (the West Stands have been replaced by the Regenstein Library. A sculpture by Henry Moore titled 'Nuclear Energy' now sits where the West Stands used to be, in recognition of the success of Fermi's project.)

The US Army Corps of Engineers was given overall responsibility for construction of the industrial facilities needed to make an atomic bomb in June, 1942. In September of that year General Leslie M. Groves, the builder of the Pentagon and an officer in the Corps of Engineers, assumed overall responsibility for the direction of the Manhattan Project.[41] General Groves is shown in his office in Figure 3.1. He reported to the Army Chief of Staff, General George Marshall. He obtained AAA rating for any and all requests for resources from the War Production Board. The highest ranking member of the Roosevelt cabinet who retained constant contact with the Manhattan Project throughout the war was Secretary of War Henry Stimson, the civilian in charge of the US Army, and therefore General Marshall's superior. Stimson was 73 years old in 1940 when Roosevelt offered him the Secretary of War in his cabinet. He was trained as a lawyer, and had served previously in the cabinets of William Howard Taft as Secretary of War, and Herbert Hoover as Secretary of State.[42]

---

[41] Leslie M. Groves, "Now It Can be Told," *ibid.*, p 22.

[42] Obituary of Henry L. Stimson in the New York Times, October 21, 1950. Stimson was born in 1867, two years after the Civil War ended.

Figure 3.1 General Leslie Groves, who managed the Manhattan Project for the US Army. Courtesy US Department of Energy.

Figure 3.2 CP1 group in front of Eckhardt Hall at the University of Chicago for the fourth reunion on December 2, 1946. Front row left to right: Enrico Fermi, Walter Zinn, Albert Wattenberg, Leona Marshall and Leo Szilard (both one step back), and Herbert Anderson. Back row second from left: Sam Allison. Photo courtesy US Department of Energy.

3.8. Chain Reacting Pile in Chicago

A description of the first nuclear reactor written by Fermi appeared in the American Journal of Physics in 1952, and is reprinted in Fermi's collected works.[43] A photograph of the group that achieved the first chain reaction is shown in Figure 3.2. Secrecy had already begun. The Metallurgical Laboratory was a cover name for atomic bomb research. As Laura Fermi said, there were no metallurgists at the Metallurgical Laboratory.[44]

Natural uranium will not generate a fission chain reaction by itself, no matter how much of it you have.[45] The nuts and bolts of the fission process in a reactor are described in Appendix B. The bottom line is that only the rare isotope ^{235}U can be fissioned by slow neutrons. It emits fast neutrons, which have to be slowed down for efficient capture by the next ^{235}U, creating a chain reaction. The material used to slow the neutrons down is called the moderator. Light nuclei like hydrogen make good moderators.

Natural uranium is 99.3% ^{238}U which is a very heavy nucleus and hence a poor moderator. To make matters worse, ^{238}U captures the fast neutrons, so they disappear. Therefore, a chain reaction requires some sort of a mixture of fissile material embedded in a moderator, so that the fast neutrons from one fission can be slowed down before encountering the next ^{235}U. A homogeneous mixture of uranium and moderator does not go critical, so the reactor has to have uranium concentrated in some locations, separated by moderator. [46]Most modern reactors have enriched ^{235}U to 5% or so, which substantially improves performance. The uranium is contained in fuel slugs that can be inserted and removed from the reactor core without tearing it apart, separated by moderator, which can be water, heavy water, or graphite.

---

[43] E. Fermi, Am J of Phys **20**, 536(1952). Reprinted in Collected Papers of Enrico Fermi, Vol 2, *ibid.*, p 272.

[44] Laura Fermi, "Atoms in the Family," University of Chicago Press, 1954, p 176.

[45] There is geologic evidence in Oklo, Africa, of a naturally occurring chain reaction in uranium ore. This 'reactor' went critical some 1.7 billion years ago, using ground water as a moderator. ^{235}U was about 3% at that time. It was an unusual coincidence of fissile material and moderator which made it work. See en.wikipedia.org/wiki/ natural_nuclear_fission_reactor.

[46] S. Esposito and O. Pisanti, eds "Neutron Physics for Nuclear Reactors," World Scientific, Singapore, 2010, p 112.

In the first nuclear reactor, which was called the 'pile,' natural uranium was used. Leona Marshall Libby told us where the uranium came from.[47] The answer was the Belgian Congo. At the start of the war the head of the Union Miniere, Katanga, had 1000 metric tons of high-grade uranium ore loaded into drums, shipped to the US, and stored in a New York warehouse.[48] The Manhattan Project learned of this supply, and the ore was converted into fuel elements for the Chicago pile.

Graphite (carbon) was chosen as the moderator, because it was inexpensive and available. Water is now used as a moderator in commercial reactors, but because neutrons are captured by hydrogen, forming deuterium and removing a neutron from the flux, such reactors require enriched uranium, typically 5% ^{235}U. Enriched uranium was not available for the very first try, and time pressure precluded waiting for it to be produced. Heavy water, that is D_2O, works with natural uranium, but heavy water was, and is, very costly. The uranium was formed into clumps of about 2 kilograms, and placed at the corners of cubes, 21 cm apart inside the graphite bricks. Most of the uranium could not be removed without dismantling the structure. There was no provision for cooling, and no radiation shielding.

The pile went critical on December 2, 1942, soon after the top layer of graphite blocks was added. It was operated sporadically afterwards, generating about 200 watts. Since there was no cooling by either circulating air or water, overheating was an obvious concern. So was radiation leaking out on to Ellis Avenue, where no one walking by had any knowledge of what was going on inside the adjacent West Stands. Consequently, after a few months it was disassembled and shipped to a new site outside of Chicago, which would become Argonne National Laboratory. Libby described thermal neutron cross section measurements made by her and Fermi, using the original Chicago reactor reassembled at

---

[47] Leona Marshall Libby, "The Uranium People," Charles Scribner's Sons, 1979, p 83. This is a terrific book about life in the Manhattan Project. You cannot put it down.

[48] It was on Staten Island. Residual radioactivity still remains. WSJ Oct 30, 2013.

Argonne.[49] They would drive out from Chicago every day to use the thermal neutron column on top of the pile. The pile was finally dismantled and buried in the 1950's in wooded park land of southwestern Cook County, Ill.[50]

Design was ongoing in parallel for a power reactor which would make ^{239}Pu from the reactions $^{238}U + n \rightarrow {}^{239}U \rightarrow {}^{239}Np + \beta + v$; $^{239}Np \rightarrow {}^{239}Pu + \beta + v$. Seaborg's group was already working on the separation of plutonium from the fission products. In order to do this on an industrial scale, the reactor had to operate at megawatt intensity, had to have cooling, radiation shielding, and the means to extract the spent fuel elements for chemical analysis. This was a potentially dangerous operation, which had to be thoroughly planned in advance.

3.9. Oak Ridge is Built in Tennessee

Also in parallel were developments in isotope enrichment, needed for the fuel for high power reactors and for weapons grade uranium, which had to be 80% pure ^{235}U. Isotope separation and chemical extraction of plutonium are discussed in Appendix C.

The mass difference between the two uranium isotopes is only 1.2%. Originally the Manhattan Project considered four possibilities, for isotope separation, none of which was particularly effective:

1.   Gaseous diffusion – one of the techniques actually used.
2.   Thermal diffusion – tested but not initially adopted.
3.   Electromagnetic separation – also used.
4.   High speed centrifuges – tested but not adopted.

Gaseous diffusion uses the fact that the lighter isotope of the gas, which was UF_6 (uranium hexafluoride), has a slightly higher average speed in thermal equilibrium at temperature T. $^{235}UF_6$ moves through small holes

---

49   Libby, *ibid.*, Chap 5, p 140ff.
50   Wall Street Journal Oct 30, 2013.

slightly faster than its heavier partner ^{238}UF$_6$, so on the other side of a penetrable barrier there are more molecules of the lighter isotope. A single stage of the separator consists of two concentric cylinders. The wall of the inner one is the penetrable barrier, and contains the mixed gas at high pressure.[51] Between the two cylinders is the 'enriched' gas at a lower pressure. But the enhancement is very small, and a large factory full of such cylinders is required to produce an enrichment of ^{235}U from 0.7% to the 5.0% needed for most power reactors. The pumps needed to move the gas were the main power requirements of a gaseous diffusion plant. Uranium hexafluoride is a highly reactive chemical compound, and as a result is very corrosive to container walls, valves, pump seals, etc. Much of the plumbing was coated with nickel, and exposed to fluorine gas to form a fluoride surface that would resist corrosion.

Thermal diffusion uses two concentric cylinders also, with the inner one heated and the outer one cooled.[52] Superheated steam under pressure flowed in the inner pipe. UF$_6$ in liquid form (the substance has a very interesting phase diagram) flowed between the pipes, and the outer one was water cooled. The lighter isotope tended to move towards the hot pipe, and the heavier one to the cool pipe, resulting in enrichment of the liquid on the inside. This technique worked, but consumed a lot of energy to make the superheated steam. It was used in the Manhattan Project as a first step of initial enrichment, before gaseous diffusion. It was implemented at Oak Ridge only in late 1944.[53]

The principle of the mass spectrograph is the basis for electromagnetic separation. Here uranium atoms have to be ionized and accelerated by a voltage across the ion source placed inside a large magnet. The charged ions then move in circular arcs in the magnetic field. The lighter isotope will bend more, giving a spatial separation between ^{235}U and ^{238}U ions. In

---

[51] It is hard to believe, but it seems that the exact nature of this barrier is still classified, after 75 years. Perhaps it is a newer version that is secret. Libby, *ibid.*, p 94 says that the barrier was developed at Columbia University, and was not compromised by Soviet espionage because of the compartmentalization of information initiated by Groves.

[52] R. Rhodes, "Making of the Atomic Bomb," *ibid.*, p 550.

[53] Leslie Groves, "Now It Can be Told," *ibid.*, p 120ff.

a large magnet, where the radius of the orbit can be 1 m, the separation of the two isotopes is about 1 cm for 180^0 bend. This sounds great.

All you need to do is put two open collection cans in the magnet, separated by 1 cm. One can has the light isotope, and the other the heavy one. However, we are considering one ion at a time, and there are lots of problems in trying to separate macroscopic amounts (kilograms) of material. This was Lawrence's favorite technique, so it was used in the uranium production for the Manhattan Project. The spectrographs were called 'calutrons' after UC Berkeley. Copper for the magnet coils was not available during the war, so Groves borrowed silver from the US Mint to make the coils![54]

Robert Wilson, then an assistant professor at Princeton, invented an isotope separator which he called an 'isotron,' which was a radio frequency device. He had been a graduate student in Lawrence's laboratory at Berkeley, working on the early cyclotrons before the war, and getting in trouble with Lawrence for being overzealous in his experimental techniques.[55] Wilson at Princeton was competing with Lawrence's calutron at Berkeley, which was also being implemented on an industrial scale at Oak Ridge. Under pressure to succeed or join Lawrence, Wilson went to Berkeley to confront Lawrence head on, and agreed to leave Princeton for Los Alamos, which was just beginning to get started.

The centrifuge method of isotope separation is like a cream separator. In a hand operated cream separator a bowl full of milk is rotated at several hundred rpm. The heavier skim milk goes to the outside, and the lighter cream moves to the middle. The two isotopes of uranium have almost the same mass, so to get much separation you need very high speed spinning cylinders, together with the elaborate plumbing required to separate the two samples of gas. Accordingly, there were stringent requirements on the strength of materials used to fabricate the rotors, and on the bearings to

---

[54] Leslie Groves, "Now It Can be Told," *ibid.*, p 107.
[55] Oral History Transcript, Robert R. Wilson. www.aip.org/history/ohilist/

suspend the axes. An industrial array of these devices obviously contains a large amount of rapidly moving machinery. These difficulties have since been overcome after WWII, as we will describe later, but centrifuges were not used for the Manhattan Project.

This is a good time to discuss decision making and authority in the Manhattan Project. Silver from the mint for magnet coils is only one story relating to the seemingly limitless resources accorded the secret project. Who decided what to do, and how was permission to commit vast amounts of money and manpower towards the necessary initiatives obtained from the government? This question will arise again in the Soviet Union. The project was top secret. The US Congress did not know about it.[56] In most cases the responsible administrators had no assurance that anything was going to work. And yet the project forged ahead at warp speed. Branching out from Chicago to Argonne for reactor development, to Oak Ridge for isotope separation, to Hanford, Washington for plutonium production, and finally to Los Alamos for bomb assembly and testing, the project seemed to occupy most of the country and its resources. Even with a blank check, there had to be compromises and limits on what was done. There were committees: the Uranium Committee and the National Defense Research Committee chaired by Vannevar Bush. Bush had direct access to the President, who retained a surprising amount of control. Negotiations with the British, who collaborated in the Manhattan Project in all respects, were carried out at the highest level.

Groves had direct access to General Marshall. Even under the continual pressure of the war and its demands, the Manhattan Project proceeded along several expensive avenues unencumbered. As mentioned earlier in Chapter 1, military procurement for 1942 was $160 billion, and the total cost through 1945 of the Manhattan Project was $2 billion. While

---

[56] The President, the vice president (Henry Wallace), the secretary of war Henry Stimson, and General George Marshall, chief of staff, all knew of the project. Harry Truman, who was chair of a special senate Committee to Investigate the National Defense Program from 1941-44, was aware of the project, but not the details until he became president! (Smyth, *ibid.*, p 87).

substantial, it was less than 1% of the annual budget.[57] The Project made extensive use of private industry. Groves had dealt before with a firm called Stone and Webster, which designed and built many industrial plants for war production. The firm still exists, based in Massachusetts.[58] It was principally responsible for the construction of the isotope separation and electromagnetic separation plants at Oak Ridge, Tennessee. Westinghouse built the magnets, and Eastman Kodak operated them in the electromagnetic separation plant. There was opposition from the scientists for the choice of Stone and Webster to construct the plutonium production reactors at Hanford, Washington. As a result, Groves assigned this task to the DuPont Company, which had more scientific and technical expertise.[59] A young chemical engineer named Crawford Greenewalt headed the DuPont effort at Hanford with distinction.

Groves instituted a compartmentalized security system, in which the various laboratories and factories did not know what was going on elsewhere. Only a few high level people knew the whole picture. The workers at Oak Ridge had no idea what they were trying to do. They were trained to operate the machinery without any appreciation of the goals of the production line. It is a tribute to the site engineering that these complex devices were sufficiently user friendly that the operators were successful despite a lack of technical knowledge. As Libby pointed out, the compartmentalization meant that local spies could only learn local secrets, which had a limiting effect on Soviet espionage.[60] The US Army in Groves furnished an able leader, who knew how to get things done in Washington, and never lost sight of the final goal. The scientific establishment on the whole adapted to army life.

There is at least one documented case of a scientist who declined to work on the Manhattan Project because of the imposed secrecy and

---

[57] Ed Cray, "General of the Army George C. Marshall," WW Norton & Co., NY, 1990, p
       315. This biography describes some interaction between Groves and Marshall.
[58] en.wikipedia.org/wiki/Stone_%26_Webster.
[59] Libby, *ibid.*, p 90.
[60] Libby, *ibid.*, p 94.

compartmentalization. Edward U. Condon, a distinguished American theoretical physicist, at that time associate director for research at Westinghouse,[61] resigned from Los Alamos.

In the letter which he wrote to J.R. Oppenheimer, reprinted in Groves' book,[62] Condon expresses dismay at Groves' scolding of Oppenheimer for discussing technical matters with Arthur Compton. Compton was at Chicago, and Oppenheimer at Los Alamos, and the two were not supposed to exchange information. But this seems to be an isolated incident, and not typical of the relations between the scientific staff and the Army.

Hanford for plutonium production and Oak Ridge for isotope separation were both started in 1942, before the first pile went critical, and before any defining evidence for an isotope separation method had been achieved. The scope of the project, given the thin evidence in its favor, has to have been daunting.

Oak Ridge, Tennessee, existed as a small town before the US Army Corps of Engineers started building the laboratories – factories really – to create sufficiently pure ^{235}U to cause a nuclear explosion, and incidentally to supply nuclear reactors with enriched uranium for plutonium production. As Groves describes it, there was concern that an accident in one factory could affect the other, and thus the gaseous diffusion plant, called K-25, was located 17 miles from the electromagnetic separation plant, called Y-12.[63] The Tennessee Valley Authority was created by the US Congress in 1933 for the economic development of the region. It controlled a complex of power plants, which were used in WWII for aluminum refining and for Oak Ridge.

The population of Oak Ridge grew from 3000 in 1942 to 75000 in 1945. At the height of production there were nearly 100,000 workers.[64] A

---

[61] National Academy of Sciences Memoirs: condon-edward-u-1902-1974.pdf

[62] Leslie Groves, "Now it Can Be Told," *ibid.*, p 429, Appendix VII.

[63] Leslie Groves, "Now it Can Be Told," *ibid.*, p 94ff.

[64] theatlantic.com/thesecretcity. This web site has a good collection of contemporary photographs of Oak Ridge.

large number of them were women, recruited from the surrounding area.[65] There were of course construction workers, electricians, plumbers, office workers, and everybody else needed to construct and operate the factories.[66] The place was called the Clinton Engineering Works, had a fence all of the way around it, and gates manned by military police. A small city had to be created, with streets, sewers, houses and barracks, lights, parks, schools, hospitals, and all the rest. The city was given independent governance in 1959.[67]

A brief technical description of the isotope separation processes is given in Appendix C. The gaseous diffusion plant, K-25, was a continuous building 44 acres in area. The thermal diffusion plant, which came on line later, was attached to K-25. The choreography of the uranium enrichment process is impossible to reconstruct. What material went where when we will never know, but in the end all of the ^{235}U that was used in the Hiroshima bomb was processed at Oak Ridge, and went through both K-25 and Y-12 plants. A photograph of an alpha calutron is shown in Figure 3.3.

The first pilot reactor for plutonium production, indeed the second reactor in the world after CP-1 in the Chicago West Stands, was built at Oak Ridge. The original plan was to have the plutonium production facilities at Oak Ridge, together with the isotope enrichment. However, this seemed too risky for several reasons. A large amount of real estate would be needed for the plutonium reactors, as well as a large amount of electric power and cooling water. There was concern about the safety of the operation, which had never been attempted, and the dangers to the surrounding population near Oak Ridge.[68] So a new remote site near water and electric power had to be found for the plutonium production reactors. Groves knew that, if a reactor explosion should occur, endangering the

---

[65] Denise Kiernan, "The Girls of Atomic City," Touchstone Books, New York, 2013. This recent book covers the daily lives of the inhabitants of Oak Ridge during WWII.

[66] Leslie Groves, "Now it Can Be Told," *ibid.*, p 100, recounts Groves' satisfaction with organized labor.

[67] theatlantic.com, *ibid.*

[68] Leslie Groves, "Now It Can Be Told," *ibid.*, p 69.

population of Knoxville, the security of the Manhattan Project would be lost, and progress would cease for an extended period. Endless Congressional hearings would begin.[69]

Figure 3.3. An alpha calutron 'race track' at the Oak Ridge Y-12 plant. The magnetic field runs around the ring. 'C' shaped vacuum chambers are visible on the left. Courtesy of US Department of Energy.

3.10. Plutonium Reactors go to Hanford, Washington

So they decided to look for another site, which met all of the requirements for security and safety. A large isolated area had to be found, near water and electricity. An area bordering a bend in the Columbia River was chosen. The original site was 1740 km^2, mostly desert sand, but with ample hydroelectric power from surrounding dams. The indigenous population of small farmers was bought out, and construction of the

---

[69] Leslie Groves, *ibid.*, p 69. Chernobyl comes to mind. Groves knew what he was doing.

plutonium production complex was begun in early 1943, soon after CP-1 went critical.[70]

Plutonium has become the element of choice in the nuclear weapons industry. Operation of high power nuclear reactors for plutonium production, and extraction of the metal from the uranium fuel rods removed from the reactors have proven, in the course of the cold war, to be very dangerous operations, resulting in some of the worst radioactivity contaminations in the world. Hanford, Washington, and its Soviet twin Chelyabinsk-40, have both suffered major environmental damage as a result of plutonium production. Part of the radioactivity was caused by accidents, some were simply due to unanticipated consequences of the urgency to achieve the goals of the weapons program, and some due to insufficient knowledge of the biological hazards of radiation exposure.

In 1943 these problems were far in the future. Groves gives credit to the DuPont Company for careful attention to safety.[71] The operation had two distinct components: high power, water cooled, graphite moderated uranium fuel reactors; and chemical plants where the fuel rods were taken after removal from the reactors for extraction of the plutonium made by neutron bombardment of ^{238}U, via the β decay chain given on p 80. The first reactor, called 105-B, was designed to run at 250 megawatts. When Fermi and Wigner were figuring the specifications for 105-B, the only reactor in the world ran at 200 watts, a factor 1 million lower. So they had to scale things up a lot. The interim test graphite reactor at Oak Ridge was air cooled, and operated at 1 megawatt, still much less than the Hanford reactors, although it did supply valuable information.[72]

Libby described a problem that arose as soon as 105-B was brought up to full power for the first time.[73] Almost as soon as the reactor reached full power, it started to fade, and the operators extracted more control rods to

---

[70] en.wikipedia.org/wiki/Hanford_Site
[71] Leslie Groves, "Now It Can Be Told," *ibid.*, p 82ff.
[72] Leslie Groves, "Now It Can Be Told," *ibid.*, p 78.
[73] Leona Marshall Libby, "The Uranium People," *ibid.*, p 179ff.

keep it going. Soon that did not work, and the power level dropped until the reactor was shut down. Libby wrote calmly about it all, but those present, including her, must have been scared witless to see the monster fizzle out before their eyes. It took a couple of days to figure out what was wrong. It is now a well known phenomenon in thermal neutron reactors – xenon poisoning.[74] The isotope ^{135}Xe is produced directly by fission of ^{235}U, and is also a daughter from the β decay chain of other fission products:

^{135}Te $\rightarrow$ ^{135}I $+ \beta + \nu$; ^{135}I $\rightarrow$ ^{135}Xe $+ \beta + \nu$. The half-life of ^{135}Te is 19 sec, and of ^{135}I is 6.7 hours. ^{135}Xe is itself radioactive, decaying to ^{135}Cs with a half-life of 9.2 hours. ^{135}Xe has an enormous capture cross section for thermal neutrons – larger than that of cadmium, which is used as a control rod. So only a small amount of xenon gobbles up all of the neutrons, and the reactor dies. The low power reactors studied prior to the operation of 105-B did not produce enough xenon for it to matter. The high power reactor recovers after a day or so – the xenon decays away. Of course, a restart simply leads to a repeat collapse. The only simple fix is to add more uranium fuel, which could be done for 105-B, since it was designed with extra slots. This was done, and 105-B started operating smoothly.

What Libby called the Wigner disease appeared soon after the beginning of full operation of 105-B.[75] Wigner anticipated a type of radiation damage to the graphite, which causes it to expand. This effect is independent of ordinary thermal expansion, which happens too in a hot reactor. Carbon atoms in graphite form well defined planes. The distance between the planes can be altered by dislocation of the atoms as a result of neutron bombardment. The graphite swells, generating pressure inside the moderator assembly. The swelling does not go away when the reactor is shut down. However, subsequent heating of the graphite causes it to relax.

---

[74] John R. Lamarsh, "Nuclear Reactor Theory," Addison Wesley, Reading MA, 1966. p 467ff.

[75] Leona Marshall Libby, "The Uranium People," *ibid.*, p 187.

Thus the cure was to allow the temperature to rise in the graphite periodically. In this way, the expansion was controlled.

A Table in Appendix C shows what happens to the uranium fuel in a reactor after three years of operation at 16 Megawatts. The input reactor fuel is enriched to 3.3% ^{235}U, all of which is burned, resulting in 0.89% plutonium isotopes, and 0.35% fission products. The fuel started out 96.7% ^{238}U, and dropped to 94.3% at extraction. While uranium is itself mildly radioactive, it is the fission products that create huge radiation hazards for people. Handling spent fuel rods is a very dangerous operation, which has to be carried out periodically at every nuclear reactor. Plutonium extraction requires even more handling, and was the cause for much concern in the design of the Hanford plant.

The chemical plants were called 'Queen Marys' after the steam ship. Hanford eventually had three reactors going for the Manhattan Project, and each had its own Queen Mary. The reactors were in the 100 area, along a bend in the Columbia River, and the chemical plants were in the 200 area, some 15 km away from the reactors. The two were connected by a special railroad. The uranium was cased in aluminum cans, about the size of a roll of US quarters (2 cm × 10 cm). Upon extraction from the reactor, the cans were allowed to 'cool' by radioactive decay for a few days. Libby described the blue glow from Cherenkov radiation[76] in the water surrounding the fuel slugs.[77] Gamma rays from fission products scatter off electrons in the water, and the electrons radiate visible light. No doubt such proximity to the fuel element storage would be strictly forbidden today.

After cooling, the slugs, still highly radioactive, and always immersed in water, were loaded on rail cars and transported to the Queen Mary, which was a continuous concrete structure 240 meters long, divided into individual cells for the stages of the chemical process. The cells were surrounded by 2 meters of concrete shielding on all sides to protect

---

[76] Pavel A. Cherenkov, Nobel Prize Address, 1958.
[77] Libby, *ibid.*, p 185.

personnel from the intense radioactivity.[78] It was known that direct handling for repairs or any other purpose of any part of the system exposed to radiation was not permissible. Accordingly, all of the works had to be designed so that manipulations could be made remotely. There was a heavily shielded overhead crane in each plant. A brief description of the chemical process is given in Appendix C.

Handling and storage problems did not end with the extraction of plutonium. Everything which came in contact with the spent fuel rods – the water, the pipes, the handling tools – was dangerously radioactive. Some of it is still around Hanford today.[79]

People who worked at Hanford as it was being built and as it came up to operation have vivid memories of the living conditions. Oak Ridge was New York City – or San Francisco if that is your favorite place – in comparison to Hanford. Hanford was desert, and the wind blew so strong that sand got into sealed tin cans. It was a huge place, about 2300 km^2, with vast open territory, partly for safety reasons. There were several thousand construction workers, who had been attracted to the site from all over the USA. They had nothing to do but work, and after work there were lots of fights. Libby said that liquor was rationed, so there was not much drinking, but even so occasionally a body was found stuffed in a garbage can the next morning.[80] The town of Hanford was built for the construction workers, and became a ghost town after they left. Groves describes some of the attempts to supply amenities and creature comforts to the women who were employed as office workers for DuPont. They were recruited to go to the state of Washington, famed for its green forests and beautiful scenery, only to be stuck in a godforsaken desert, living in a wartime dormitory.[81]

---

[78] Leslie M. Groves, "Now It Can Be Told," *ibid.*, p 85.
[79] Madame Curie's cookbooks, in the Curie Museum in Paris, are radioactive.
[80] Leona Marshall Libby, "The Uranium People," *ibid.*, p 167.
[81] Leslie M. Groves, *ibid.*, p 90ff.

We will leave the description of Los Alamos, the bomb lab, until after bringing the Soviet operation up to speed after 1938.

Chapter 4

The Soviet Union and Stalin's Terror 1937-1939

4.1. Widespread Oppression Begins

Stalin's terror impacted virtually everyone in the Soviet Union. The lives of two prominent Soviet physicists, Lev Landau and Peter Kapitza, are of special interest to a history of the Soviet Atomic Project. Landau and Kapitza were both in the USSR, in Kharkov and Moscow respectively, after their sojourns to western Europe in the early 1930's. A dispute with the director of the Leningrad Institute, Academician A.F. Ioffe, obliged Landau to leave Leningrad, but he obtained an appointment in the theory group at the Ukranian Physico-Technical Institute in Kharkov, and moved to Ukraine. In 1934 Kapitza was denied permission to return to England where he had been working in Ernest Rutherford's laboratory, but was awarded his own institute in Moscow, and was able to retrieve his research equipment from Cambridge in order to continue his low temperature research. The revocation of an exit visa was a painful sign of Soviet authority, but Kapitza settled in to work the Soviet system to his advantage. Earlier, in 1933, Niels Bohr had invited Landau to attend a conference on theoretical physics in Copenhagen, and Landau was able to visit the West once again. Edward Teller was also there. Landau and Teller argued about the structure of molecules.[1] Everything seemed perfectly normal.

---

[1] Maiya Bessarab, "Lev Landau," *ibid.*, p 48.

93

However, repression in the Soviet Union under Joseph Stalin was increasing. George Gamow had already sensed a change in government procedures and a tightening of restrictions between 1929 and 1931, several years before the widespread terror began. Robert Tucker dates the beginning of the dramatic rise in domestic terror with the murder of Sergei Kirov on December 1, 1934.[2] Kirov was a top Communist Party official in Leningrad. He was shot by a man named Nikolaev, but on whose orders is still a matter of speculation. Most historians blame Stalin himself. Stalin, on the other hand, immediately seized the murder as evidence of a broad plot against the Soviet Union, confirmation of the danger of those he called "Trotskyites."

Leon Trotsky has been described as the genius of the Russian revolution. He was an associate of Lenin and the first People's Commissar of the Red Army during the 1918 Russian civil war. Trotsky opposed Stalin's policies after Lenin died in 1924. Among other things, they differed regarding world revolution. Trotsky wanted to extend communism to other countries, while Stalin was determined to build communism in Russia, surrounded by capitalist economies. Trotsky was exiled from the Soviet Union in 1929, and eventually was murdered in Mexico City in 1940. In 1934 anyone under suspicion for any reason could be labeled a "Trotskyite," a most heinous traitor to the cause.

In 1934 Lenin had been dead ten years, and revolutionary fervor was fading. Party bureaucrats were getting comfortable, and feeling good about themselves. They might even have occasional thoughts of replacing Stalin with someone else, "someone more tolerant, more loyal, more polite, less capricious, etc."[3] These were all worrisome signs for Stalin, but if nothing obviously bad had happened,[4] how do you know whom to eliminate? The obvious answer is you eliminate everyone, but then you

---

[2] Robert C. Tucker, "Stalin in Power," W.W. Norton & Co, New York, 1990, p 291.

[3] Vladimir Lenin, Last Testament. www.duel.ru/200246/?46_5_1(in Russian). In his secret last testament, Lenin recommended that Stalin be replaced with someone having these characteristics.

[4] Many Russian Tsars were assassinated. Most dictators, and even some US presidents, likewise. Stalin was the exception. There are no known plots on his life after the 1920's.

have no one left to run the country. Nevertheless, this is the approach that Stalin adopted.

4.2. No One Was Safe

Widespread arrests began with a vengeance in 1936, affecting virtually all responsible people in the Soviet hierarchy. Ministers and those who reported to them, on down to minor administrators and factory managers, were all threatened with arrest. This happened all over the Soviet Union. Many were arrested, tried, and executed. Others spent 10 or 15 years in prison camps. Tucker quotes the historian Volkogonov, who placed the number arrested in 1937 and 1938 at between four and one half to five million people.[5] About 900,000 prisoners were executed. Those who were not directly affected had relatives or close friends who were. Since the state controlled everything, families of arrested workers lost their place to live and means of survival. Entire apartment buildings in Moscow were emptied.[6] Inna Gaister, the daughter of a minister of agricultural economics, was 12 years old when her father was arrested. Marshal Tukhachevsky, a famous Red Army commander, fell victim to the purge in May, 1937, a few months before Inna's father. Inna and Tukhachevsky's daughter were friends. Inna asked her mother how it was possible that such a national hero as Tukhachevsky could be an enemy of the people? Her mother had no satisfactory answer.[7]

Tukhachevsky's arrest started a purge of high ranking Red Army officers. Georgy Zhukov escaped arrest, but Konstantin Rokossovsky was not so lucky.[8] Perhaps because he was born in Poland, Rokossovsky was accused of being a Polish spy and was arrested in August, 1937. He lost

---

5 Robert Tucker, "Stalin in Power," *ibid.*, p 474.
6 Inna Shikheeva-Gaister, "Deti Vragov Naroda," (Children of enemies of the people) Moskva, Vozvraschenia, 2012. Her father was executed in 1937.
7 Inna Shikheeva-Gaister, "Deti Vragov Naroda," *ibid.*, p 38.
8 Konstantin Rokosssovsky, "Soldatsky Dolg," Olma Press, Moskva, 2002, p 8. Like many people who were imprisoned, and later restored to prominence, Rokossovsky himself does not discuss his prison experience. It is mentioned in the introduction, written by a historian, Aleksei Basov.

his fingernails under 'questioning,'[9] then was sent to Vorkuta, a camp above the Arctic Circle, where he was a servant to a camp manager.[10] He was released and fully reinstated as an officer in the Red Army in March, 1940, as if nothing had happened. He was to become a marshal of the Soviet Union in WWII.

4.3. The Terror Visits Kharkov

The Ukranian Physico-Technical Institute at Kharkov (UFTI) was one of the first laboratories to experience the terror. UFTI came under NKVD suspicion because of the large number of foreign visitors to the laboratory. Visiting scientists, communists or communist sympathizers from Germany, Austria and Britain came to the USSR in general, and UFTI in particular, during the depression in the West.[11]

Fritz Houtermans, whose interesting fate was discussed in Chapter 1, was among the German nationals at Kharkov. UFTI was a busy place before the purges. Landau was there working on solid state theory and training the future generation of Soviet theoretical physics, Shubnikov was experimenting with low temperature phenomena, and Leipunsky was doing nuclear physics with a new electrostatic accelerator. Figure 4.1 shows the group in the mid 1930's, looking perfectly normal, with Shubnikov, Leipunsky, Landau, and Kapitza, a visitor there at that time, on the front row. The group photograph could have been motivated by Kapitza's presence. Bessarab quotes a telegram sent to Stalin from UFTI in the fall of 1932: "The Ukranian Physico-Technical Institute in Kharkov as a result of persistent work, wishes to report in honor of the 15th anniversary of the October Revolution the first success on the disintegration of atomic nuclei. On October 10, the high voltage group

---

[9]   Simon Sebag Montefiore, "Stalin, The Court of the Red Tsar," Alfred A. Knopf, NY, 2004, p 331.

[10] Anne Applebaum, "Gulag," Anchor Books, New York, 2004, p 266.

[11] The UFTI affair, cripo.com.ua/print.php?sect_id=9&aid=64379. The account of UFTI's time of troubles, in quotes, comes from this reference. Translation from Russian by the author.

disintegrated the lithium nucleus. The work continues."[12] Progress in experimental research was a sign of normalcy.

Figure 4.1. Physics group at the Ukranian Physico-Technical Institute in the mid 1930's. Front row: Lev Shubnikov, Aleksandr Leipunsky, Lev Landau, Peter Kapitza. Second row: Nisson Finkelstein, Olga Trapesnikova, Kirill Sinelnikov (Igor Kurchatov's brother in law), and Ryabinin. Credit AIP Emilio Segre Visual Archives, Frenkel Collection.

The fate of UFTI during the terror is of considerable interest because it involves many prominent Soviet physicists, including Lev Landau and members of his group. A translation of the account on the cripo.com.ua web site follows:

"Many physicists from Europe were visitors, including Niels Bohr from Copenhagen, Paul Dirac from Cambridge and Paul Ehrenfest from Holland. UFTI hosted several international conferences and seminars in the 1930's. Lev Landau was a prime mover in these activities. An international physics journal,

---

[12] Maiya Bessarab, "Lev Landau," *ibid.*, p 38.

'Physikalische Zeitschrift der Sowjetunion'[13] was established at Kharkov so that Soviet scientists, who had previously published in foreign journals, could use a physics journal edited in the USSR. Since UFTI was under the Ministry of Heavy Industry, approval by the Academy of Sciences was not required for changes in administrative structure, and in 1933 A.I. Leipunsky, a talented physicist and a member of the Communist Party replaced Academician Obreimov as director. However, as a sign of things to come, Leipunsky was replaced the next year by S.A. Davidovich, who had no scientific reputation. In March, 1935, UFTI received several technical assignments pertaining to national defense. Strict security measures accompanied this shift in mission. Foreign visitors were no longer welcome. International conferences were not allowed and travel by local scientists was severely limited. Not surprisingly, these administrative restrictions were not cheerfully accepted by the scientific staff, an atmosphere which, when coupled with Stalin's terror, was bound to lead to disaster.

"A resistance movement was started by Landau and the head of low temperature physics Alexander Weisberg. The first protests made fun of the newly issued passes that were required to access the laboratory grounds. The passes were worn on dog leashes, or pasted on the seats of their pants. Then letters of complaint were sent to "Izvestia" and to the UFTI local news. To increase the level of protest, complaining letters were sent to Communist party bosses Nicolai Bukharin and Georgy Pyatakov [both of whom were to perish in the terror]. Some members of the scientific staff had mixed feelings, torn between personal freedom to exchange scientific knowledge, and the importance of national security. One such was Lazar M. Pyatigorsky, an associate of Landau, who had lost a hand fighting in the civil war in 1919, and was an active participant in Communist party operations. He co-authored a textbook on mechanics with Landau, which was to be published in 1938.

"In 1935 Pyatigorsky wrote a letter informing the NKVD. He identified Moisei Abramovich Korets as the root troublemaker. Korets was a graduate student who came from the Urals region to work with Landau. According to Pyatigorsky, Korets sensed the revolt upon his arrival, and helped create an unworkable atmosphere, making it impossible to fulfill assigned defense work.

---

[13] It had a short life, 1932-1938. The UFTI time of troubles canceled the journal in 1938, but it published many papers in theoretical and experimental physics. The working languages were German and English. There were no papers in Russian, the native language of most of the authors. The journal circulated internationally.

"Pyatigorsky's letter produced results. In the winter of 1935 Korets was arrested, accused of disrupting activities at UFTI necessary for national defense, and in February, 1936, he was sentenced to one and one half years in prison. In the meantime, Landau and Weisberg had written to the Communist Party Central Committee requesting that Davidovich be replaced by Leipunsky.

"Landau's letter also produced results. In the midst of the crisis, Leipunsky was restored as director of the Institute by order of the Commissar of Heavy Industry. Encouraged by his success, Landau wrote yet another letter, this time to the Ukranian NKVD, claiming Korets' innocence of all wrongdoing, and requesting his reinstatement. This request too was granted! Korets was absolved of all guilt at the end of July 1936, and his case was closed. It seemed that the conflict was settled. However, after 1936 came the year of terror, 1937.

"What is known as the UFTI affair in 1937-38 was really several criminal cases. In August, 1937, the experimental physicist Lev V. Shubnikov, the director of the low temperature laboratory at Kharkov, L.V. Rozenkevich, the director of the nuclear physics laboratory, and V.S. Gorsky, the director of the X-ray section were arrested, accused of participation in the activities of anti-Soviet counter-revolutionary groups. All three were subsequently executed in October, 1937."

Shubnikov's widow, Olga Trapeznikova, was also an experimental physicist at UFTI, and was not informed of his fate until much later.[14] His date of death was falsely recorded in the Great Soviet Encyclopedia as 1945.[15]

Shubnikov was a very distinguished experimentalist in what we now call material science. He discovered more than one physical effect that bears his name. Lev Landau wrote to the military prosecutor in support of his posthumous rehabilitation in 1956: "Lev Vasilevich Shubnikov was undoubtedly one of the strongest physicists working in the low temperature field, not only in the USSR, but also in the whole world."[16] His needless execution was a great loss to Soviet science. Davidovich, the

---

14 This was a common phenomenon. See Inna Gaister, "Deti Vragov Naroda," *ibid.*, p 60, "Her husband was sentenced to 10 years with no correspondence privileges. At that time we did not know it, but that meant that he was already shot."

15 en.wikipedia.org/wiki/Lev_Shubnikov

16 russcience.euro.ru/repress/lfti/shubnikov.htm.

former director who oriented the Laboratory towards classified work, was also arrested in 1937, and executed. Pyatigorsky was not arrested, but was subsequently black-listed by Landau. His actions against unrest affected his reputation and scientific career.

German nationals Weisberg and Houtermans were both detained, and repatriated to Nazi Germany – Weisberg in 1937, and Houtermans in 1939.

Landau's research group was at UFTI, but he also held a faculty position at Kharkov University, where he lectured to students. In 1937, when things were getting uncomfortable at UFTI, the rector of the University called Landau into his office to criticize his teaching techniques. Landau was fond of asking students in physics courses about a wide range of subjects, like classical literature, or the seven wonders of the ancient world. The rector found this approach to teaching unsatisfactory, and dismissed Landau forthwith.[17]

Aleksandr Akhiezer passed the Landau minimum examination in 1934 and was one of Landau's successful graduate students.[18] He was to write a number of physics books, both technical and popular, and to spend his entire career as a theorist at UFTI. In discussing the time of troubles at UFTI, Akhiezer gives a slightly different account. He does not mention the start of secret work for the military as a cause for dissent within the ranks of scientists. He says that everyone was dismissed for anti-Soviet activities: Landau, Lifshits, Akhiezer, Shubnikov, Gorsky. They were accused of protesting as a group, and going on strike at the Institute. Someone unnamed (Pyatigorsky?) had written an incriminating statement. They were all ordered to report to the Commissar of Education in Kiev, which had become the Ukranian capital, replacing Kharkov, in 1934.[19] The Commissar was a chemist named Zatonsky who was an old Bolshevik, an acquaintance of Lenin himself. Zatonsky stuttered, so he spoke slowly and

---

[17] Maiya Bessarab, "Lev Landau," *ibid.*, p 60.

[18] The Landau minimum examination was a series of physics problems composed and administered by Landau to individual candidates.

[19] www.aip.org/history/ohilist/images/foreign/4400/4400_0013.jpg Oral history of Aleksandr Akhiezer taken in Kharkov in 1995.

was easy to understand. He asked what caused the UFTI scientists to go on strike? Idealism? Akhiezer replied that idealism had nothing to do with it. He said that Landau had made many enemies by ridiculing the die hards who resisted modern physical theories, and that the criticism of the group was no more than that. Zatonsky understood, and told them to go home and continue working. Zatonsky was later shot.

In the meantime, Kapitza was looking for a physicist to head a new theory group at the Institute of Physical Problems in Moscow. His first choice was Max Born, the Göttingen physicist who had worked with Robert Oppenheimer on a quantum theory of diatomic molecules. Born was seeking an appointment outside of Nazi Germany.[20] When Born accepted a professorship at the University of Edinburgh Kapitza offered the position to Landau, who accepted for at least two reasons: UFTI was becoming dangerous, and he was fired from Kharkov U! Not many institutions have the distinction of firing someone as famous as Landau, although it does happen now and then. Akhiezer replaced Landau as head of the theory group at UFTI.

The second wave of arrests at UFTI included the former directors Obreimov and Leipunsky, both of whom were eventually released and rehabilitated. By 1939 Leipunsky was back doing nuclear physics experiments with the UFTI electrostatic accelerator.[21] Isaak Pomeranchuk, one of Landau's star students in theoretical physics, was excluded from the local Komsomol 'because of his connection with Landau,' an example of guilt by association.[22] Landau had left Kharkov when Obreimov and Leipunsky were arrested, but the NKVD followed him to Moscow.

---

[20] I.M. Khalatnikov, *ibid*, p 58.
[21] Some of the Kharkov Institute work was published in the *Physical Review*. For example, *Phys Rev* **56**, 891 (1939) "Scattering of photoneutrons from deuterium by light nuclei" by Goloborodko and Leipunsky – Kharkov.
[22] I.M. Khalatnikov, *ibid.*, p 64.

4.4. Landau is Arrested

Landau was arrested at his apartment on the grounds of Kapitza's Institute of Physical Problems in the early morning hours of April 28, 1938.[23] Kapitza was notified as soon as Landau failed to show up for work. Kapitza wrote a letter to Stalin that same morning:

"Comrade Stalin,

"L.D. Landau, a scientific worker at the institute, was arrested this morning. Despite his young age of 29 years, he and Fok[24] are the best theoretical physicists in the USSR. His works on magnetism and the quantum theory are frequently cited in foreign scientific journals. Only in the last year he published a fine piece of work, pointing to a new source of energy in stars. He proposed a solution to the question: 'What is the source of the energy in the sun and stars? Simple thermodynamic considerations would require the sun to burn out in a short time.' Bohr and other leading physicists agree that Landau's proposal is sound. There is no doubt that the loss of Landau as a scholar for our institute, as for Soviet physics, and indeed for world science, will not go unnoticed and will in fact be strongly felt. Of course his scientific talent, no matter how great, does not give him the right to violate his country's laws, and if Landau is guilty, then he must answer for it. But I ask you, given his remarkable talents, to give suitable instructions so that his case is carefully managed. Also, it seems to me necessary to note that Landau's character is, simply speaking, bad. He is a bully and a trouble-maker. He loves to find mistakes in the work of others, and when found, especially in more senior members of the Academy of Sciences, to tease them mercilessly. He has made many enemies.

"His membership in our Institute has not been easy, although he has agreed to our rules, and things are improving. I have excused his faults in view of his remarkable talents. However, despite his faults, I cannot believe that he could do something dishonorable.

"Landau is young. He still has great promise of accomplishment in science. No one except another scientist could write to you about his skills.

Peter Kapitza"[25]

---

[23] Maiya Bessarab, "Lev Landau," *ibid.*, p 72.

[24] Academician Vladimir Alexandrovich Fok (1898-1974) was a theoretical physicist at LFTI and Leningrad State University. Known for the Hartree-Fok method of treating the electronic structure of complex atoms.

[25] Quoted by Maiya Bessarab, "Lev Landau," *ibid.*, p 73. From the collected papers of Peter Kapitza. Kapitza wrote many letters to Stalin. Stalin is said to have read them all, although most did not elicit a written response from the Kremlin.

This letter produced no immediate results. When Niels Bohr learned of Landau's arrest he wrote a letter to Stalin from Copenhagen on September 23, 1938. Bohr's letter praised Landau's many contributions to physics. Bohr was deeply troubled by Landau's arrest, and he hoped that it was all a terrible mistake.[26] Bohr's letter also had no immediate effect. The NKVD had implicated Landau with the UFTI affair. Korets, for whom Landau had obtained a release in Kharkov, reappeared in Moscow. His name was linked to Landau regarding a political leaflet, the authorship of which is still questioned. Some say that Korets wrote it, and Landau edited it. Some say that the NKVD made the whole thing up, and that no such leaflet ever existed outside the confines of Lubyanka Prison. The leaflet in question, authentic or not, is as follows:

"Proletariat of all nations, unite!

Comrades!

"The great October Revolution has gone astray. The country is full of blood and dirt. Millions of innocent people have been imprisoned, and no one knows when their turn might come. Industrial infrastructure has broken down. Famine is drawing near.

"Do you not see, comrades, that the Stalin clique has achieved a fascist coup? Socialism remains only in the pages of lying newspapers. Stalin is the equal of Hitler and Mussolini in his hatred of genuine socialism. Stalin has destroyed socialism in order to preserve his power over the country, making socialism easy prey for brutal German fascism.

"The only exit for the working class, indeed all workers of our country, is a decisive struggle against the fascism of Stalin and Hitler, a struggle for socialism.

"Comrades, unite! Do not fear the executioners of the NKVD. They are only able to beat defenseless prisoners, to capture innocent people, to loot state property, and to fabricate judicial processes against non-existent conspiracies. Comrades, join the Antifascist Workers Party. Make contact with the Moscow committee. Organize groups of AWP at your workplace. Exercise your underground skills. Use

---

[26] Maiya Bessarab, *ibid.*, p 74.

propaganda and agitation to prepare for a massive movement for socialism.

"Stalin's fascism lives only because of our disorganization.

"Proletariat of our country, having overthrown the Tsar and the capitalists, can also overthrow the fascist dictator and his clique.

"Celebrate the first of May, the day of struggle for socialism!

Moscow Committee

Antifascist Workers Party"

The authorship of this leaflet remains unknown. Landau admitted in his signed deposition after questioning in Lubyanka prison to knowledge of the leaflet and to participation in its preparation, but he denied any role in its distribution.[27] He named Korets as the principal author of the leaflet, and he stated that the Antifascist Workers Party was fictitious. Landau's admissions under pressure do not necessarily reflect the truth, and the correspondence between his confessions and the leaflet do not preclude the preparation of both documents by the NKVD. One may wonder why would the NKVD bother? People were arrested every day without any case of evidence against them, fabricated or genuine. The leaflet is provocative, but would it be illegal in the USA of that time if some substitutions were made: Roosevelt for Stalin, FBI for NKVD, and democracy for socialism? Or would it be protected by the first amendment to the Constitution? The answer would be decided by the courts, but the document would be questioned.[28]

---

[27] Maiya Bessarab, "Lev Landau," *ibid.*, p 77ff.

[28] The author is indebted to Earl Munson of the Boardman Law Firm, Madison, Wisconsin, for an opinion in this regard. The most incriminating sentence is: "Proletariat of our country, having overthrown the Tsar and the capitalists, can also overthrow the fascist dictator and his clique." Treason is defined by the United States Constitution as "Treason against the United States shall consist only in levying War against them or in adhering to their Enemies, giving them aid and comfort."

In the course of writing her biography of Landau, Maiya Bessarab interviewed a physicist named Abram K. Kikoin, the brother of Isaak Kikoin – leader of isotope separation by gaseous diffusion in the Soviet Atomic Project. He doubted that Pyatigorsky wrote any slanderous letter, in contradiction to the description of the UFTI affair quoted above. He wrote to Bessarab in 1968 as follows: "The NKVD of that time had a far from crystal clear reputation, and I suspect their agents of foul play. It is possible that the sudden release of Landau after one year of confinement stimulated the NKVD to give him some clarification of the cause of his arrest, and to produce this forged document. The 'signature' of the NKVD is obvious in the character of the charges: Landau, a Jew, has spied for the Germans! In the 30's! For agents of the NKVD at that time this was a standard, completely ordinary, trivial charge."[29]

Landau was confined in Moscow, in Lubyanka prison, the main headquarters of the NKVD. He was not sent to one of the remote camps, and was never tortured, but was interrogated for extended periods of time. There is no record of his attempting to work on theoretical physics. His parents had moved from Baku to Leningrad, and his sister Sonya came from Leningrad to Moscow after she learned of his arrest. She was unable to learn anything about her brother from the authorities. She went to the publishing house where the mechanics text by Landau and Pyatigorsky was in preparation, to learn that Landau's name had been removed, leaving Pyatigorsky the sole author. Such was the fate of enemies of the people.[30] Another side effect of being arrested during the terror, discussed by Inna Gaister,[31] and experienced by Sonya, was ostracism by friends and acquaintances. No one wanted to associate with a relative of an enemy of the people, for fear that the guilt would rub off. The only exception was Peter Kapitza, who was a real hero in the Landau case. Sonya met with Kapitza and his wife several times, and she was assured that he would not rest until Landau was released.

---

[29] Maiya Bessarab, "Lev Landau," *ibid.*, p 84.
[30] Maiya Bessarab, "Lev Landau," *ibid.*, p 75.
[31] Inna Gaister, "Deti Vragov Noroda," *ibid.*, p 46.

Aleksandr Akhiezer told another story about the effect of Landau's arrest on publications bearing his name as an author.[32] Together with Landau and Lifshits he wrote a book entitled 'Mechanics and Molecular Physics,' which was ready for the publisher in 1938. Soon after Landau's arrest he and Lifshits were ordered to report to Moscow from UFTI by the Central Committee of the Communist Party.

Understandably somewhat intimidated, the two went upstairs to an office in a government building on Staraya Square in Moscow. Staraya Square is near the Kremlin, but outside it, in a district called Kitai Gorod (China town). To quote Akhiezer:

"There sat a very intelligent looking man with a beard and pince-nez glasses, who said: 'Here is the situation. Your book should be published. The Soviet Union needs such texts in order to educate our young students. However, one of the authors is Landau, who is an enemy of the people. We therefore request your permission to remove his name from the list of authors, leaving Akhiezer and Lifshits.'

"Lifshits and I looked at each other, and said that this was not possible. In the first place, Landau has not been convicted, and could be freed tomorrow. How then could we look him in the eye? Are we betrayers? There is no way for us to do this.

"The intelligent man with a beard then said: 'Well, OK, let it be your way. Give me your permits.' He signed the permits and we left the building for the Staraya Square."[33]

Persisting in his efforts to get Landau released, Kapitza visited Molotov in the Kremlin. Kapitza said that Landau was the only theorist who could explain Kapitza's discovery of superfluidity in liquid helium. Molotov said "OK, we will free Landau, but first you must talk to the NKVD." Two very scary midnight visits to Lubyanka by Kapitza followed his conversation with Molotov. The NKVD came to his apartment and picked him up, and took him off to the prison. Kapitza's wife Anna was

---

[32] Oral History of Aleksandr Akhiezer, *ibid.*, p 13.

[33] Permits were required to enter government buildings. Indeed, permits were required for many parts of Moscow. The Kremlin was not open to the public. The book was not actually published until Landau was released, leaving the question moot. Akhiezer does not recall the official's name, but does say that he was later executed in the purge.

terrified that he might not return. But in the early morning hours, after conversations with the authorities, he was delivered home again. Nighttime interrogations were standard procedure, because Stalin was fond of working at night.

Early in 1939 Lavrenty Beria replaced Nikolai Yezhov as head of the NKVD. Yezhov had managed the most severe period of Stalin's purge. Beria had a more balanced approach, and the purge began to subside. Kapitza wrote another letter in behalf of Landau. This one was addressed to Beria on April 26, 1939:

"I request that physics professor L.D. Landau be released from confinement and placed under my personal trust. I certify that Landau will not carry out any counterrevolutionary activities against the Soviet state in my Institute, and I will take all necessary measures to assure that he will not carry out such activities outside of the institute as well. In case I notice any actions on the part of Landau intended to harm the Soviet state, I will promptly notify the NKVD.

P. Kapitza"[34]

Note that Kapitza made no statement regarding Landau's guilt or innocence. Landau was an important member of Kapitza's Institute, and he was willing to take responsibility for Landau's behavior. On these grounds he requested Landau's release. This tack was clever psychology on Kapitza's part, because it worked. One can only imagine, under the circumstances, the courage this request must have taken. April 29 was the anniversary of Landau's arrest, and on this date in 1939 he was released into Kapitza's custody by order of L.P. Beria. Without Kapitza's determination on Landau's behalf, he would not have been freed. Despite his unwavering support for Landau, Kapitza was never again his close friend after the prison episode. When asked later about Kapitza's rudeness, Landau responded, "Kapitza changed my condition from negative to positive, and I am not able to object to him."[35]

---

[34] Quoted by Bessarab, *ibid.*, p 85.
[35] Maiya Bessarab, Lev Landau, *ibid.*, p 87.

Landau's mother sent him small sums of money to buy food in prison, but he never received any of it. After his release they questioned the whereabouts of the money, and an NKVD employee who had kept it returned the money to her. Had Landau not been released, the employee would not have been caught.

4.5. Terror Spreads to Leningrad

A.F. Ioffe's Leningrad Institute fared better than Kharkov, but did not escape entirely. The distinguished theorist and member of LFTI Vladimir Aleksandrovich Fok was arrested twice, once in 1935, and then again in 1937.[36] He was released in 1935 on the day of his arrest. In 1937 Kapitza appealed to Stalin again, this time in defense of Fok, and he was released four days later. The reason for his second arrest was suspicion regarding a connection to the 'Pulkovo Affair.'

Pulkovo Astronomical Observatory was founded in 1839 on the outskirts of St. Petersburg, and is still an active organization of the Russian Academy of Sciences today. The St. Petersburg airport has the same name. The Pulkovo Affair (1936-37) refers to criminal charges brought by the NKVD against a group of Soviet scientists accused of 'participation in a fascist Trotskyite-Zinoyev terrorist organization founded in 1932 through the efforts of German espionage groups, and having as its goal the overthrow of Soviet power and the establishment of a fascist dictatorship in the territory of the USSR.'[37] Pulkovo astronomers were arrested first, but the net was widely cast, affecting many different disciplines – geology, geophysics, mathematics, etc. Moscow and other cities were affected as well as Leningrad. Astronomers aroused the suspicion of the NKVD in 1934, when plans were being made to invite foreign observers to the USSR to watch a solar eclipse predicted for 19 June 1936. In the summer of 1936 a series of articles was published in the Leningrad press, claiming 'unhealthy conditions' at the Pulkovo Observatory, and accusing the astronomers of 'foreign contacts' because they published papers in

---

[36] ru.wikipedia.org/wiki/Fok_Vladimir_Alexandrovich
[37] ru.wikipedia.org/wiki/Pulkovskoe_delo.

western scientific journals. This would all be comical if it were not so deadly serious.

Between November, 1936 and September, 1937, 13 Pulkovo astronomers were arrested, including the director of the observatory, Boris P. Gerasimovich. Seven wives were also arrested. Fok was caught in a later sweep, along with about 100 others. Boris Gerasimovich was shot in November, 1937. The affair affected the entire intellectual community of Leningrad.

There is a poignant story in the popular science magazine 'Priroda' about three young theoretical physicists who were executed in 1938.[38] M. P. Bronshtein was a graduate student at LFTI, and S.P. Shubin and A.A. Vitt were graduate students at MGU. They were born between 1902 and 1908, and were contemporaries of Landau (born in 1908). They were trained by Igor Tamm, who taught theoretical physics at MGU, and in Leningrad. They came from 'middle class' families – a doctor, a journalist, and a businessman. While Vitt concentrated on physics, Shubin was also interested in social work, and Bronshtein was a writer of fiction for young readers. Nikolai Bukharin, an early supporter of Stalin and editor of 'Pravda,' founded a periodical called 'Socialistic Reconstruction and Science' in 1932. Called 'Sorena' for short,[39] the journal lasted until Bukharin was liquidated in 1938, and served as one of the media – in addition to the daily papers 'Pravda' and 'Izvestia,' local party meetings, etc - for the ongoing debate between modern physics and Marxist philosophy. The electron energy distribution in nuclear beta decay did not seem consistent with energy conservation. An unseen extra particle was needed to conserve energy. Pauli came to the rescue with his neutrino hypothesis in 1933, but this solution was not universally accepted. Curiously enough, the Marxist philosophers liked the idea of non-conservation of energy. Despite the laws of thermodynamics (which hardly count as modern physics), they thought that perpetual motion machines would solve the energy needs of a growing communist society!

---

[38] G.E. Gorelik, 'Those who did not become academicians,' Priroda, 1990 #1 p 123.
[39] Nauka is science in Russian, which explains the last syllable in 'Sorena.'

Entering this debate was a dangerous business for young theoretical physicists. However, Bronshtein submitted an article to 'Sorena' in which he argued that the law of conservation of energy was a 'bourgeois' concept that had no place in Marxist society. Shubin, on the other hand, wrote that physics and Marxist philosophy were in harmony, and the conservation of energy was OK. The editors juxtaposed the two articles in one issue of 'Sorena.'

Whether this debate called attention to Bronshtein and Shubin during the terror is unknown. There are many examples of terror victims with no known guilt of any kind towards the Soviet state. On the other hand, Bronshtein was unfortunate that his last name was the same as Trotsky's, although there was no relation.[40] Shubin had been banished from the Komsomol for being a Trotsky supporter while in graduate school in the late 1920's, but he later swore allegiance to the Bolsheviks, and was reinstated. Vitt had a solid reputation in quantum theory, and did not participate in philosophical debates, but was accused of being a German spy, a net that caught many workers at Kharkov. All three were arrested in 1937, and executed in 1938.

NKVD interest was often stimulated by some individual writing a note to a newsletter or some other publication accusing someone else of disloyalty to the USSR. Then the typical modus operandi of the NKVD was to arrest the accused, force the victim to implicate others in some fictitious plot, and then arrest all of those mentioned under duress. Thus starting from nothing more than an intemperate disposition on someone's part, an entire operation could be brought to a halt. Sometimes the process was stopped before it could get going by unanimous resistance on the part of everyone. LFTI had one such experience in May-June, 1938. A certain A. Moiseev wrote a slanderous article for Leningradskaya Pravda about the institute under the title 'Pure physics and real life.'[41] Many prominent physicists were criticized, A.I. Ioffe in particular. There was some

---

[40] Trotsky was born Lev Bronshtein, and took the name Trotsky during his imprisonment
     in 1900.
[41] www.famhist.ru/famhist/ap/00023ecf.htm

agreement within the Institute regarding the criticism of Ioffe's management, but Ioffe himself successfully refuted the arguments point by point, and a lab wide discussion led by Alexandrov gave him full support of the staff.[42] No harm came to LFTI because of the unfavorable article.

4.6. Modern Physics is not Compatible with Marxist Theory

Another headache for Ioffe, not caused by the NKVD, but nevertheless requiring considerable effort in response to criticism, came from, of all places, The Soviet Academy of Sciences. 'Modern physics,' a term referring to the revolution in scientific thought that started with the discovery of the electron, the photoelectric effect, and naturally occurring radioactivity in the late 19th century, and led to the theory of relativity and to quantum mechanics, was not universally accepted by physics professors who had been trained in the 'old school' of classical physics - mechanics, electromagnetism, thermodynamics, theory of sound, hydrodynamics, etc. For the most part the senior people were quietly retired, or shunted off to a corner somewhere, and replaced by younger scientists who understood and supported modern physics. This changeover took place throughout the Western world. University students deserved an education in the latest developing fields. But there were some individuals who claimed that relativity and quantum mechanics were baloney. Experimental confirmation of the new theories assured that they did not stand a chance, but they could be annoying. In contrast with, say, the United States, where only some very senior physics professors had to be side tracked so that quantum mechanics could be taught, the Soviet Union had to accommodate philosophers as well.

Thus it was that two members of the Soviet Academy of Sciences, V.F. Mitkevich and A.A. Maksimov started a campaign in the Academy against Ioffe and the work at LFTI.[43] Mitkevich was an electrical engineer, and

---

[42] Academician Anatoly Petrovich Aleksandrov (1903-1994), then a staff scientist at LFTI, became a major contributor to the development of Soviet nuclear reactors.

[43] www.famhist.ru/famhist/ap/001b2463.htm#00023ecf.htm

Maksimov was a philosopher. As mentioned in Chapter 2, George Gamow refers to philosophers in his autobiography.[44] According to Gamow, while western philosophers are harmless, Soviet philosophers can be a menace, because modern science can run counter to dialectic materialism. Maksimov seems to have played the role of dissenting philosopher. Together with Professor A.K. Timiryazev of Moscow State University (MGU) they founded a 'Center for reactionary physics.' The group tried to influence the journal 'Uspekhi Fizicheskikh Nauk' (Results of the Physical Sciences) published by the Academy of Sciences. Their activities eventually led to the March Session of the Soviet Academy of Sciences in 1937, where an attempt was made to discredit the work of LFTI and its director, Ioffe, and to stop its support by the Academy. Their main argument seems to have been that if the theories of modern physics made no sense, it was a waste of time to pursue them. Specific objections pertained to the concept of a force field, action at a distance, and the electromagnetic ether – all 19th century topics that had been long buried.[45] Fortunately, Ioffe and his Institute emerged alive and well. But problems with MGU would resurface during the Soviet Atomic Project in the 1940's.[46]

This phenomenon even occurred in Nazi Germany. Johannes Stark won a Nobel Prize in 1919 for his discovery of the Stark effect, the splitting of atomic spectral lines by an electric field. This effect can only be described by quantum mechanics. And yet, after Hitler came to power in 1933 Stark became a strong supporter of the Nazis – against 'Jewish physics.' This included Einstein's relativity and Heisenberg's quantum mechanics. Heisenberg was not Jewish, but Stark accused him of being a 'white Jew.' Stark probably had no real impact on the German uranium project, beyond being a nuisance. He was imprisoned in 1947 for his support of the Nazis.[47]

---

[44] George Gamow, My World Line, *ibid.*, p 94.

[45] G.E. Gorelik, "Natural philosophy physics problems in 1937," Priroda, 1990, #2, p 93.

[46] Sergei Khrushchev, "Nikita Khrushchev Reformator," Vremya, Moskva, 2011, p 111. The 'Center for reactionary physics' at MGU persisted into the postwar period, and continued to be troublesome for Igor Kurchatov. See Chapter 9 for further discussion.

[47] en.wikipedia.org/wiki/johannes_stark

A more common reaction to the proposal of an atomic weapon built from uranium was the dismissal of the idea as impractical and unachievable. Ernest Rutherford expressed this position when he said that anyone talking about practical uses of atomic energy was talking 'moonshine' – a statement that galvanized Leo Szilard. Rutherford clearly understood modern physics, but he did not believe that nuclear energy could be a practical source of power. Gubarev recounts another such incident, after an address entitled "Problems in atomic physics" given by Igor Tamm at the March meeting of the USSR Academy of Sciences in 1936.[48] After his presentation L.V. Mysovsky, a colleague of Kurchatov from the Radium Institute in Leningrad, and a capable experimental physicist, questioned the practicality of nuclear weapons. Such doubts must have been fairly common, given the enormous amount of development work that was necessary to make practical nuclear energy a reality.

4.7. Lasting Effects of the Terror

Stalin's terror of 1937-1939 had a profound effect on Soviet society, particularly those who held responsible positions in government, industry, and science.

The widespread loss of responsible people had an impact on government operations. Those who disappeared had to be replaced with new blood. In the outlying Soviet Republics this often resulted in the replacement of local commissars, who were native to the region, with Russians, who came in from outside.[49]

Andrei Vishinsky was chief prosecutor for the famous show trials, which involved high ranking members of the Communist Party and the Red Army. By 1939 the terror began to ease, partly because of the

---

48 Vladimir Gubarev, "Atomnaya Bomba," Algoritm, Moskva, 2009. p 12. Tamm's report was published: I. Tamm, Uspekhi Fizicheskhikh Nauk, **16**, 922 (1936).

49 Robert Tucker, "Stalin in Power," *ibid.*, p 486.

appointment of L.P. Beria as Minister of Internal Affairs, but extensive damage to the Soviet Union had been done.

How did those who escaped the purge think about what was going on? Were all the victims really guilty of something? Most people felt that they had not done anything to threaten the state. But most of those arrested had not done anything either. So there was no feeling of security. There could be a knock on the door at any time. How do normal people function in such an environment? Part of successful behavior is to say that it simply won't happen to you. We are all mortal, but generally speaking we do not live today as if we were to perish tomorrow. We don't worry about it. Life goes on. And so it was, but the terror had a real effect on the characters in our story, and on the entire population of the country. In contrast with the substantial exodus of talent from Nazi Germany in the 1930's, very few citizens of the USSR managed to emigrate abroad. George Gamow was an exception. The reason was partly that the Soviet government made it virtually impossible to travel abroad, but there was also a strong feeling of loyalty to the Soviet Union and to its pursuit of communist idealism.

There is a brief report on the activities of the 18th Communist Party Congress of the USSR, held from March 10-21, 1939 in Moscow, in the Soviet physics journal Zhurnal Eksperimentalnoi I Teoreticheskoi Fiziki.[50] The presence of such a report in a physics journal, indeed the principal journal of Soviet physics, is an example of the all-encompassing reach of the Soviet government. There is the usual praise of the central committee, and its wisdom in directing the building of a modern communist state. There are a couple of sentences worth quoting, both attributed to Stalin at the Congress. The first is interesting after more than two years of terror: "Through the destruction of enemies of the people and the cleaning of the party organization of degenerates, the party has become more united in its political and organizational work, and more united with the central committee." There are of course no specific references to the terror itself. The second is directed at the audience of physicists reading the journal: "All loyal communist intellectuals must simultaneously pursue two goals:

---

[50] Zhurnal Eksperimentalnoi I Teoreticheskoi Fiziki 8, p 369, (1939).

significant contributions in their respective fields of study, physics, biology, economics, etc.; and steadfast support of the Bolshevik party in its efforts to achieve communism in the USSR."

This environment forged the Soviet scientists, engineers, and technicians who were capable of creating a nuclear weapon. While they enjoyed favored status during the Project after WWII, they did not receive special treatment during the Terror. However, physics as a science and a subject for training students was not neglected. Research in material science was important to the industrialization of the Soviet Union, and was supported by various agencies of the government, the Academy of Sciences in particular. Ioffe's institute at Leningrad, Obreimov's institute at Kharkov, and Kapitza's new institute in Moscow all received government funding for applied research. Nuclear physics was not considered of practical importance, and hence was not initially one of the supported fields, but Ioffe and others with vision figured out how to maintain a competitive research program in nuclear physics below the radar, as it were. Many nuclear researchers also did more practical work, partly as a cover.

While not explicitly part of the terror, the Soviet government placed stringent limits on the activities of artists and writers. Stalin had a firm hand on the controls, and his judgment was sometimes surprising. Stalin helped the writer Mikhail Bulgakov, whose play 'Days of the Turbins' he saw performed several times. Stalin loved the play, even though it is about Russian aristocracy who supported the White Guard against the Bolsheviks in Kiev in 1919, during the Civil War.[51] Stalin hated the opera 'Lady Macbeth from Mtsensk' by the famous composer Dmitry Shostokovich.[52] Shostokovich soon redeemed himself with his magnificent and profoundly ambiguous 5th symphony. Neither Bulgakov nor Shostokovich were ever imprisoned, although many writers were. In

---

[51] Robert Tucker, "Stalin in Power," *ibid.*, p 552.
[52] Pravda, January 28, 1936, p 3. The unsigned article is titled 'Muddle instead of Music.' The author of this book (no music critic) has seen the opera, and agrees with Stalin.

the early days of the revolution several composers and writers left the country, but the flow dried up in the 1930's.

Stalin was not always right in his intellectual judgments. He supported the pseudo-scientific plant genetics of Trofim Lysenko, to the detriment of the Soviet Academy of Sciences and the field of genetics in the Soviet Union.[53] Nonetheless the range of Stalin's interests, the perspicacity of his judgment, and his vast memory are remarkable attributes. He trusted no one, but he was an excellent judge of character. He was a ruthless tyrant, but as we shall see, he chose the correct path in nuclear research without any knowledge of either quantum mechanics or special relativity.

4.8. The Soviet Prison System after 1939

The prison camp system, the Gulag, played an important role in the Soviet Atomic Project by supplying prison labor for mining, construction, and other necessary tasks. The Project physicists were not prisoners themselves, but prisoners were often present on laboratory grounds. There was a rule in the camps against using prisoners in their professional roles of engineer, or geologist, or whatever, for fear of giving them too much privilege. On the other hand there was considerable technical skill in the prison population, and Lavrenty Beria in 1939 developed the idea of a technical prison, called a 'sharashka,' in order to exploit this resource. Inmates who were skilled scientists, engineers, and mathematicians could be obliged to work in sharashkas on R&D projects of interest to the state.[54]

Engineers and technicians who were prisoners in sharashkas had better living conditions than those in the camps, but they were still prisoners. They worked and slept under guard. The famous airplane designer Andrei N. Tupolev, and the father of the Soviet space rocket program Sergei P.

---

[53] Robert Tucker, "Stalin in Power," *ibid.*, p 560.

[54] Valentin Simonenkov, "Sharashki," Eksmo, Moskva, 2011, p 12. Solzhenitsyn's novel "In the First Circle" is about Marfinskaya Sharashka assigned to work on problems in acoustics. It was on the outskirts of Moscow (near what is now Gagarin Square on Leninsky Prospekt) in 1949.

Korolev, both were prisoners in the same sharashka during WWII, working on weapons for the Red Army.[55] The Soviet aerospace program and the Soviet Atomic Project each acquired equipment and personnel from East Germany after the war. Some German prisoners of war with technical training were freed from POW camps in the Soviet Union and moved to one of the research centers managed by German 'guests,' where they worked as free men alongside other technicians. They exchanged better living and working conditions for a longer stay in the USSR.

German rocket scientists were eagerly sought by both the US/British forces invading Germany from the West, and Soviet forces coming in from the East. Peenemunde, the German rocket complex on the Baltic coast where the early V2 rockets were built and launched towards Britain during the last part of WWII, was in the Soviet occupation zone, and became part of East Germany after the war. Allied bombing in 1943 had forced Albert Speer, the Nazi minister of armaments, to move the V2 factory to an extensive underground complex built by slave labor at Nordhausen in central Germany, south west of Berlin. Nordhausen was occupied by US forces in 1945, and several V2 rockets, complete engineering drawings, and many scientists and technicians were captured. General Walter Dornberger and Werner von Braun evacuated from Nordhausen to the south, near Switzerland, where they surrendered to the US Army.[56] The rocket development program was generously supported by the German Air Force during WWII. It is possible that the German atomic bomb program made little progress in part because of resources diverted to the rockets. Walter Dornberger was aware of the 'Uranverein' (German uranium club) and its mission, and he approached Werner Heisenberg about the use of atomic energy for rocket propulsion, without receiving any useful information.[57]

---

[55] Anne Applebaum, "Gulag," *ibid.*, p 111.

[56] Walter Dornberger, "V-2," Viking Press, New York, 1954, p 271. The entire US collection of German rocket scientists were a part of Operation Paperclip. See Annie Jacobsen, "Operation Paperclip," Little, Brown and Co New York, 2014.

[57] Dornberger, "V-2," *ibid.*, p 252.

General Groves observed that the cost of the Manhattan Project would have been prohibitive in peacetime. Similarly, Dornberger wrote: "Never would any private or public body have devoted hundreds of millions of marks to the development of long-range rockets for purely scientific purposes."[58]

Thus the weapons of the cold war, nuclear warheads and rocket delivery systems, were separately and independently developed by the adversaries in WWII. While Germany did not have detailed information about the Manhattan Project, the existence of the V2 program, and other aerospace R&D, like jet aircraft, were known to the Allies, because the weapons were used against Britain. In retaliation rocket factories and launching facilities were subject to Allied bombing attacks. After the war the United States army was very interested in learning more about German rockets. In his book about his grandfather, David Eisenhower wrote: "Using German operating personnel, the OSS (predecessor of the CIA) conducted test firings of V-2 rockets at Nordhausen."[59] Nordhausen was initially in the Western occupation zone of Germany, but was ceded to the Soviets in June, 1945, so that all of the equipment, drawings, and technical personnel that were to be evacuated to the US had to be hastily removed.[60]

While the Soviets did not manage to capture Dornberger and von Braun, they did succeed in acquiring many German rocket and aircraft specialists from the environs of Berlin, and other places in East German territory. They also obtained working hardware, and perhaps more important, detailed drawings. Aerospace hardware is easier to reproduce from the drawings than from the objects themselves, because of all of the technical specifications, tolerances, special alloys, etc., which are not so easy to discern from the assembled, or disassembled, machine sitting in front of you.[61]

---

[58] Dornberger, "V-2," *ibid.*, p 271.
[59] David Eisenhower, "Eisenhower at War 1943-1945," Random House, NY, 1986, p 810.
[60] Annie Jacobsen, "Operation Paperclip," *ibid.*, p100ff.
[61] For these reasons, Israeli Intelligence covertly obtained the drawings for the Dassault Mirage III fighter after the 1967 Arab-Israeli war, even though Israel possessed the aircraft. www.historyofwar.org/articles/concepts_israeli_covert.html.

Because of agreements signed by the Allied Powers, Germany was not permitted to continue weapons work, so much of the German capability which was accessible to the Soviet Union was moved East, rather than exploited in situ. This move was considered part of the war reparations owed the Soviet Union by Germany. The ban on research work also facilitated the departure of many German scientists for the Soviet Union, where they could continue work in their chosen fields. Simonenkov described the living conditions of the German engineers and technicians who were moved with their families and equipment from the Junkers aircraft factory in Dessau to Ivankovo, a town about 100 km north of Moscow. Efforts were made by Soviet authorities to make the environment as comfortable as possible for the Germans – better than for the Soviet workers who were at their side. German language lessons were set up for the Russians, and Russian lessons for the Germans. Documents and drawings had both languages – German on the left, and Russian on the right. Only qualified translators, supplied by the NKVD, could settle technical questions. The location was picturesque, but the housing was miserable.

About 500 German and technical staff and their families had to be housed. During the summer and fall of 1946 there was intensive work to get ready to receive the guests. Shops were enlarged, housing repaired, and 'Finnish' small houses were built for one or two families.[62] But the Germans were not happy with the accommodations, especially the outhouses in cold weather. Local Russian workers were moved out of brick homes and into newly constructed barracks to accommodate the incoming Germans. This naturally caused much anger and tears – "these damned Fascists." But the Germans had left large, comfortable, and well built houses and apartments. Loads of furniture from Germany had to be discarded because there was no place to put it.

Daily life gradually improved. German language schools were set up for the children. The wives of German technicians ran the kindergarten. Germans were issued permits to shop in special stores, where there were

---

[62] Finnish houses were also used in the Soviet Atomic Project. See Chapter 9.

better quality products. (Food ration cards were still in force for the general population.) Cafeterias in the factories had special rooms for Germans. The Russians working beside them were paid 2/3 as much as the Germans. The authorities were acutely aware that, despite widespread wartime destruction on both sides, living conditions in Germany were better than in Russia at that time. German POW's were used as labor to construct the housing facilities for the rocket program 'visitors.'[63] German POW's with technical skills could transfer from labor camps either to sharashkas, or to laboratories of the Soviet Atomic Project.[64] The POW's accepted a delay in repatriation to Germany for better living and working conditions in the USSR. German engineers had annual vacations with pay, but were not allowed to leave the village, so their free time was spent at home. Travel by foreigners was strictly controlled in the USSR.

Metallurgy and alloys are important components of successful manufacture of aerospace hardware, and at least at first were difficult for Soviet industry to reproduce. Items such as turbine blades for jet engines continued to be made in factories in East Germany. German technical capabilities, even after the destruction caused by the war, remained superior in many respects to those of the Soviet Union.

The science prisons were a functional part of the aerospace industry until after Stalin's death in 1953, when the vast state prison system began to be dismantled.

After atomic bombs were dropped on Hiroshima and Nagasaki in 1945, the Soviet Atomic Project achieved highest priority. The physicists and their families, both Soviet and German, who worked on the Project were protected from harassment by the NKVD and were never prisoners either because of the importance accorded to the work by the Soviet government, or because there was a distinction made between scientists and engineers, the work of the former being regarded at a higher level, meriting more

---

[63] This description of the living conditions of German workers is taken from Simonenkov, Sharashki, *ibid.*, p 58ff.
[64] Simonenkov, Sharashki, *ibid.*, p 50.

freedom. In any event, although the locations of the laboratories and factories were top secret, as they were in the US Manhattan Project, the scientists were free to move about, consistent with security.[65] German and Soviet physicists were treated equally. The Soviet government desperately needed the atomic bomb to gain parity with the West, and the physicists were the only ones who could do the job in a reasonably short time span.

---

[65] Andrei Sakharov, writing his memoirs in the 1980's, refers to Arzamas-16, the place near Moscow that was roughly the equivalent of Los Alamos, as the 'object,' a code name commonly used for atomic installations.

Chapter 5

The Soviet Union and Nuclear Research 1934-1942

5.1. Abraham Ioffe and His School at Leningrad

In Chapter 4 we learned that Ukranian Physico-Technical Institute (UFTI) in Kharkov successfully split the lithium nucleus in 1932. UFTI remained an important laboratory throughout the war and the move to the east – to Alma Ata (now Almaty) in Kazakhstan - and the return to the west in time for the push for atomic weapons which began in 1943. Another key player before WWII was Ioffe's Institute in Leningrad, LFTI. Leningrad Physico-Technical Institute (LFTI is the Russian acronym) is located in the Vyborg section of present St. Petersburg, on Polytechnique Street. A.F. Ioffe was a man of great vision. By emphasizing to the government the importance of science to the industrial development of the USSR, he became a spokesman for the development of modern physics in the Soviet Union in the 1920's and early 1930's.

A number of key people in the Soviet Atomic Project were trained in Leningrad. Foremost among them was Igor Vasilevich Kurchatov, followed closely by Yuly Borisovich Khariton, Georgy Nikolaevich Flerov, Isaak Konstatinovich Kikoin, Yakov Borisovich Zeldovich and Abram Isaakovich Alikhanov. Kurchatov, Kikoin, and Flerov were experimenters. Zeldovich, Khariton, and Alikhanov were theorists.

122

5.2. Early Years of Igor Kurchatov

Igor Kurchatov is the one most important individual in the success of the Soviet Atomic Project. He was a man of his time. Well trained at LFTI in experimental nuclear physics, he became the principal leader of the Soviet atomic bomb. He was a competent experimental physicist, and he was also a man possessed. He was determined to overcome all obstacles to assure that the atomic bomb project was successful. He was willing to work the system, and he did so with great skill. He well knew the arbitrary powers of Lavrenty Beria, but he also knew Beria's enormous resources and command. He did not flinch from his goals when things got tough, and he stayed the course from 1943 until his death in 1960.

Igor Kurchatov was born in 1903 in the Urals, in what is today the Chelyabinsk region of the Russian Federation. His father was a surveyor who worked in forestry. His mother was a school teacher.[1] In 1912 the family moved to Simferopol on the Crimean peninsula near the Black Sea coast. The climate was thought to be better for the health of Igor's sister Antonina, who suffered from consumption, but she did not improve in warmer weather, and soon died. Igor and his younger brother Boris grew up and attended school in Simferopol. Igor played the mandolin in the local string orchestra, and acquired a lifelong love of music. He was a healthy, active boy, with innate mechanical ability. He became interested in steam engines, and began to read technical literature, wishing to become an engineer. During WWI he had to work to help support the family. He became a skilled metal worker, but continued his studies. In 1920, he completed high school (Gimnazium) with a gold medal.

In 1918, during WWI, a group of faculty from St Petersburg and Kiev were cut off by German occupation while vacationing in the Crimea. Unable to return home, they decided to establish a local university. By 1920 several distinguished physicists had come to lecture, including Ya. I, Frenkel, I.E. Tamm, and A.F. Ioffe from Leningrad. From 1918 to 1920

---

[1] I.N. Golovin, "Igor Kurachatov," Atomizdat, Moskva, 1967. Details of Kurchatov's life are taken from this biography.

the Crimea was under considerable political stress. The German army, the Red army and the White army all invaded and retreated back and forth, until the Red army finally prevailed in 1920. Living and working conditions in the area were marginal. Jobs were scarce, and Igor's father was out of work. Igor worked as a night watchman in a movie theater until the theater closed after a robbery.

In September, 1920, Igor Kurchatov enrolled as a student in the physics-mathematics faculty of the local university. There were only a small number of students, and the facilities were limited. Many of the physics texts were in German, which did not make studies any easier. Igor had access to the instruments of the university's modest physics laboratory, where he could use his considerable laboratory skills. In the summer of 1921 his expertise earned him an extra 150 g of bread[2] to be added to the very modest student ration. Students were served soup for lunch, free of charge, which was probably the best meal of the day. Student quarters were not heated, and although the climate was milder than in Moscow, winters could be cold. The political situation in the early 1920's could not have been very stable either, but Igor and his contemporaries survived these very trying times. Many professors had returned to Leningrad, but enough activity remained in Simferopol to afford a college education. The students in their laboratory were able to reproduce the Zeeman effect in helium, and to measure the polarization states of the spectral lines.[3] These results were very satisfying to the young students. The interferometer used to observe the splitting of the lines required considerable experimental skill to adjust, and Kurchatov was up to the task. He was acquiring the reputation of a talented experimental physicist.

---

[2] About 1/3 of a pound loaf of black Russian bread, which can be excellent, but at that time was probably not the highest quality. We do not know the daily ration without this addition. Golovin refers to 'shrapnel soup,' which was probably made with barley, served with small fish for lunch. Nobody had lots to eat.

[3] The Zeeman effect is the splitting of atomic spectral lines in a magnetic field. Quantum mechanics gives a complete description of the splitting, including the polarization of the emitted light. It is a nice experiment in advanced physics lab.

The students were not satisfied with the university resources in Simferopol. They decided to finish the local requirements as quickly as possible, and to then transfer to Leningrad, to the Polytechnic Institute. Kurchatov led the group through the examinations. He chose a theoretical topic for his senior thesis, and completed the four year course requirements in three years in the summer of 1923. That autumn he departed for Leningrad to study engineering to become a ship designer. The Polytechnic Institute admitted him without support. In order to earn some money, Kurchatov took a job at the meteorological observatory in Pavlovsk, about 50 km south of Leningrad. He had come from a mild climate into a harsh winter. The commute from Leningrad to Pavlovsk took two hours by train, streetcar, and on foot, and his Russian great coat was not warm enough to make the trip comfortable, particularly after dark. So he spent many nights in the observatory, sleeping on a table. In this way he shifted away from shipbuilding and into physics, because he became interested in measuring the α radioactivity in snow, and all of the implications thereof. He paid less attention to his studies, and the Polytechnic Institute expelled him in the second semester. He knew what he wanted to do, but he did not have the means to do it – experimental research in physics.

Kurchatov returned to the Crimea in 1924 to work in a weather station in Feodosya. His friends invited him to Baku, where they were studying properties of dielectrics for electrical engineering. A.F. Ioffe was interested in this work, and that led to an invitation to return to Leningrad, which Kurchatov accepted in the spring of 1925. Ioffe, who was always looking for young talent, invited him to join LFTI. The hard times of the revolution and civil war began to ease. There was a sense of optimism about the future. In 1927 Kurchatov married Marina Dmitrievna Sinelnikova, the sister of one of his close friends from Simferopol.

Kurchatov demonstrated his technical prowess in preparing experiments on dielectrics and semiconductors. Much of the work was done together with his brother in law, Kirill Sinelnikov. Igor's brother Boris also turned up in Leningrad, doing research in chemistry. One substance of interest was Seignette's salt, or Rochelle salt – potassium sodium tartrate. This material exhibits piezoelectricity, in which an electric field causes a deformation of the crystal, and conversely pressure or tension on the crystal produces an electric field (opposite signs for pressure and tension). Charged ions in the crystal are brought to the surface by deformation, and the surface charge creates an electric field. Such substances have many practical applications.[4] Igor Kurchatov was becoming an accomplished experimenter. The working groups were small, and he led by example, spending long hours in the lab. He devoted a long monograph to Rochelle salt in 1933. He was always interested in industrial uses of scientific developments. After all, the middle name of LFTI was 'technical,' and applications were encouraged, as potential contributions to the growing Soviet state.

LFTI was a busy place. There were weekly seminars on current physics problems, which challenged Kurchatov. Ioffe was a very good organizer, and he had at that time the best physics institute in the Soviet Union. We have already noted Ioffe's interest in keeping current with research abroad, and his arrangement of foreign travel for members of his institute, including Landau. But Kurchatov never took advantage of these opportunities to visit laboratories in Western Europe. He always had something interesting going on in the laboratory at home, and declined the invitation to travel. Whether the absence of western exposure had anything to do with his loyalty to the Soviet Union is a matter of speculation.

5.3. Kurchatov Switches to Nuclear Physics

In 1934 Igor Kurchatov was 31 years old. He had a good reputation as an experimental physicist, would soon obtain the degree of doctor of science, and could become a professor, writing a paper or two a year, and preserve his health. He had recovered from tuberculosis, and could use a rest. But he again became interested in nuclear physics, recalling his early days studying the radioactivity of snow. LFTI did not have a program in

---

[4] En.wikipedia.org/wiki/Potassium_sodium_tartrate. Piezoelectric crystals have been used as phonograph pickup cartridges, and in microphones.

nuclear physics, partly because of Ioffe's interest in material science, and partly because of the apparent lack of applications of nuclear physics to daily life. The Leningrad Radium Institute, established in 1922 by academician Vladimir I. Vernadsky, was the local laboratory for nuclear studies. Vernadsky was interested in obtaining uranium ores from Siberia, and he and chemist Vitaly Khlopin had organized a small factory in the Ural mountains for the extraction of uranium and radium in 1921.[5] Lawrence had begun accelerating protons in an early cyclotron at Berkeley in 1932, and was producing neutrons in 1933 soon after their discovery by Chadwick.[6] There were no cyclotrons in Europe until 1937, but in the mid '30's one was under construction at the Radium Institute.

Kurchatov realized that the neutron was a powerful tool for studying reactions in the heavier nuclei, because it had no electric charge, and would not be repelled by the nucleus as were protons or α particles. Accordingly, he began neutron bombardment experiments using Ra-Be neutron sources, following the work of Fermi's group in Rome. Studying the radioactivity of bromine bombarded by Ra-Be neutrons, Kurchatov's group noticed a new state of the nucleus, which is now called an isomeric state.[7] Excited nuclear states normally decay promptly down to the ground state by γ emission. Half-lives of nuclear excited states are 10^{-16} sec or less. But occasionally there is an excited nuclear state that has a much longer half-life – seconds, or hours, or in some cases even years. Decay by γ emission is inhibited by what are called selection rules. These are isomeric, or metastable states.[8]

---

5 Alexei B. Kojevnikov, Stalin's Great Science, *ibid.*, p 129.

6 Nuel Pharr Davis, Lawrence and Oppenheimer, *ibid.*, p 40.

7 I.V. Kurchatov, B.V. Kurchatov, L.V. Myssovsky, and L.I. Rossinov, Comptes Rendus Acad des Sciences, v **200**, p 1201 (1935).

8 Technetium-99m is a metastable isomer that emits a 140 keV γ ray with a half-life of 6 hours. It is the most commonly used medical radioisotope, with millions of diagnostic procedures annually. It was discovered in 1938 by Seaborg and Segre at Berkeley. (en.wikipedia.org/wiki/technetium-99m). This is just one unforeseen example of a practical application of nuclear physics.

Neutron sources were prepared by mixing beryllium with a heavy element α emitter, either radium or polonium, with both substances in powder form. The α particle emitted by the heavy element was captured by a beryllium nucleus: $^4\text{He} + {}^9\text{Be} \rightarrow {}^{13}\text{C}$. The excited ^{13}C nucleus decayed into ^{12}C plus a neutron. Polonium had a lower background of γ rays than radium. The α particles had a very short range – they traveled only about 50 μm in the material before stopping – hence the use of a powder mixture. Radioactive sources of this type produced modest fluxes of neutrons with a continuous energy distribution averaged around 4 MeV. A moderator like paraffin could be used to reduce the neutrons to thermal energy (0.025 eV on average), and an energy range could be selected by judicious choice of foils which had known absorption cross sections. Low energy neutrons from sources could be selected by passing a beam through a mechanical velocity selector, and Kurchatov designed such a device. Some of these techniques are described in Appendix B. Utilization of neutrons from radioactive sources for systematic studies of neutron - nucleus interactions was limited by the low neutron intensity, and the difficulty in neutron energy selection.

Despite these limitations, Kurchatov's group was quite successful in a number of experiments using neutrons from radioactive sources. They measured the n + p capture cross section for deuterium formation with thermal neutrons: $\text{n} + \text{p} \rightarrow \text{d} + \gamma$, with an energy release of 2.2 MeV. And they observed sharp resonances in the absorption cross sections of neutrons on complex nuclei, where the cross section increases dramatically for a specific incident neutron energy. The velocity selector was useful for observing sharp resonances. Such resonance phenomena became standard tools in monitoring neutron fluxes, as discussed in Appendix B.

There was a natural reluctance among experimenters working with sources to change techniques and adapt to the accelerator environment.[9] Fermi's group in Rome, Rutherford's group in Cambridge, and Kurchatov's group at LFTI were very productive using radioactive decays

---

[9] Oral history transcript with Maurice Goldhaber; www.aip.org/history/ohilist/4632.html.

as sources for the neutrons. These 'table top' experiments could be carried out in a short time. Laura Fermi describes how Enrico would run from the room where the neutron source irradiated the sample under study into another room where the Geiger counter was located to count the residual radioactivity. The Geiger counter had to be isolated from the radioactive source.[10] A stopwatch was used to measure the time between removal from the neutron source, and the decay activity. Individual counts were recorded by an electric register, like an automobile odometer, which made loud clicks for each detected decay. One could draw conclusions from today's measurements, and plan tomorrow's accordingly. It was fun. Kurchatov had groups at LFTI and the Radium Institute all doing similar experiments. Some of the induced radioactivity was observed in a Wilson cloud chamber.[11] Kurchatov's group had common interests with researchers at the Radium Institute, and they joined forces to collaborate.

5.4. Early Particle Accelerators in the USSR

Eventually, however, table top experiments with radioactive sources of neutrons became obsolete. The accelerators were simply too powerful to overcome by cleverness. But the lifestyle for accelerator experiments was different. After the very early days, when the experimenters also built the accelerators, machine operation and experimental activity divided into two camps. The machine operators were always there. When the machine broke, they fixed it. Beams of protons, deuterons, α particles, and neutrons were supplied to experimenters who came with their experimental apparatus, used the beams for measurements, and went home. War research ushered in big physics, and the field was never the same.

Kurchatov knew about the advantages of particle accelerators over radioactive sources. Lawrence's cyclotron was producing more intense neutron fluxes, with better collimation and energy definition than possible with Ra-Be or Po-Be sources. He was not nostalgic about the simple life

---

[10] Laura Fermi, "Atoms in the Family," *ibid.*, p 89.
[11] There are several papers by Kurchatov and coauthors in Physikalische Zeitschrift der Sowjetunion in 1935 and 1936.

of table top experiments. He wanted to remain competitive, and accordingly he developed an interest in the cyclotron under construction at the Leningrad Radium Institute. L.V. Mysovsky initiated the construction of the cyclotron in 1932.[12] This machine had problems from the start. It was decided to tip a conventional cyclotron on its side, so that the median plane of the magnetic field, where the beam is accelerated, was vertical. Perhaps there were plans to use the magnet for cosmic ray studies.[13] In any event, the geometry was problematic, since everything that was placed in the magnet gap had to be suspended by wires somehow, rather than just sitting on the pole tip.[14] Not surprisingly, it took a while to get it to work. A brief description of cyclotrons, based on the Berkeley design, is presented in Appendix D.

A report in very poor English about first operation of the Radium Institute cyclotron was published in the Physical Review in 1937.[15] In March, 1937, a 1.2 MeV proton beam was achieved, reaching 3 MeV by the summer. A beam of helium ions was also accelerated. Kurchatov started nuclear physics experiments as a joint LFTI-Radium Institute venture as soon as the cyclotron started to work. Kurchatov and Alikhanov, with Ioffe's support, proposed to build another, larger cyclotron at LFTI. Kurchatov went to Moscow to secure resources for the new machine. He succeeded in procuring 10 metric tons of copper for the magnet coil windings, which was no small feat, considering the industrial and armaments demands for copper. However, WWII intervened as the project was getting started, and the cyclotron was completed after the war. In the end, the various parts became two cyclotrons, one in Leningrad, and one in Moscow. In the meantime, Ya. Khurgin at LFTI worked out the

---

[12] HISAP96, Vol **2**, *ibid.*, p 31.

[13] A picture of the device is in A.V. Kojevnikov, Stalin's Great Science, *ibid.*, p 128.

[14] They were not alone. Oak Ridge National Laboratory built several vertical cyclotrons after WWII. They converted their calutrons to cyclotrons, and the calutron magnets sat on end, with a horizontal magnetic field. See Oak Ridge National Laboratory Review, Vol **25**, nos 3&4 p 58, available on ornl.gov.

[15] V.N. Rukavichnikov, Phys Rev **52**, 1077 (1937). The editors must not have paid any attention to this letter. The magnetic field has to decrease with increasing radius in order to focus the beam, so that it does not get lost inside the chamber. It is not clear from this letter that this design feature was implemented.

relativistic effects on the cyclotron orbit, leading to the post war synchrocyclotrons, which could accelerate protons to several hundred MeV.[16]

In the mid 1930's two large electrostatic generators were built: a 2 MV generator at the Leningrad Electro-physical Institute, and a 6 MV machine at UFTI. The 6 MV machine had two spheres, one at +3 MV, and the other at -3 MV, connected by a 12 meter long horizontal beam tube.[17] This arrangement was not the most convenient, since all of the experimental apparatus had to sit at high voltage inside one of the spheres. For protons, the source sits at +3 MV, and the target and detectors sit at -3 MV. This made access to the target area difficult. The horizontal beam tube, with high voltage on both ends, could be supported in the middle, which was at ground. Further discussion of electrostatic generators is presented in Appendix D.

As mentioned in Chapter 2, the two accelerators, cyclotrons and electrostatic machines, had complimentary features for nuclear physics studies. Cyclotrons avoided direct high voltages by cumulative acceleration of the particles, but the resulting beam energy was not sharply defined. Electrostatic accelerators had the full beam energy stored on a sphere – if you put a voltmeter on it, and multiplied by the elementary charge, you got the energy. Absolute calibration and sharp energy definition are very useful parameters for some nuclear physics experiments. While it is possible to artificially introduce timing structure into the beam of an electrostatic machine, the beam current is normally continuous. But a cyclotron beam is automatically pulsed at the RF frequency – making it a natural source for measuring neutron velocity by time of flight.[18]

---

[16] The author worked on such a machine at the University of Chicago in the 1950's.

[17] The beam tube, made of glass or ceramic sections spanned 6 MV, or 5 kV per cm, which is a modest voltage. Similar tubes were used to support the two spheres. Modern beam tubes are designed for ~20 kV per cm.

[18] As an example, accelerate deuterons in a 1 Tesla magnetic field cyclotron, to make neutrons by 'stripping' $d \rightarrow n+p$. The cyclotron frequency is 8 Mhz, so you get a pulse of neutrons every 0.12 μ sec. A I MeV neutron travels about 2 m in that time.

It is worth mentioning that these accelerators, electrostatic or cyclotron, were hand made, and no two were really alike. As Appendix D attempts to make clear, each machine was very sophisticated, and several crucial parts were left to the designer to figure out how to fabricate. The beam tube in electrostatic accelerators, and the RF system in cyclotrons were examples. There were no blue prints. While the performances of the machines were published in journals, this was far from enough information to tell the reader how to make one in his/her basement. So it took some time and determination to make these things work.

Maurice Goldhaber, an outstanding experimentalist in nuclear physics, was at the Cavendish Laboratory from 1933 to 1938, when he emigrated to the US. He did table top experiments at the Cavendish, and did not use the accelerator built by Cockcroft and Walton. In his oral history Goldhaber listed the most important centers for nuclear research before WWII in his opinion: the Rome school of Fermi, the Joliot-Curie group in Paris, the Cavendish Laboratory of Rutherford, and the LFTI group of Kurchatov.[19] Goldhaber placed Kurchatov in the top tier, but did not list any German laboratory. This was just one person's opinion, but it was borne out by history. For although fear of Germany's capabilities in the development of atomic energy was a strong motivation for the Allies, in the end there was no danger, and Goldhaber's ranking was one indication of a shortfall. Goldhaber thought that the Berkeley group under Lawrence showed promise with operating cyclotrons, but was just short of first rank. He said that the published work was often flawed by hasty preparation. When he visited Berkeley himself, he realized that the group was very able, but his criticism had some merit. Lawrence built cyclotrons; study of the nucleus was of secondary importance to him.

5.5. Soviet Physics Journals of the Pre-War Period

Articles in theoretical and experimental physics in the USSR in the 1930's were published in the Kharkov journal Physikalische Zeitschrift

---

[19] Oral History transcript of Maurice Goldhaber, *ibid.*

der Sowjetunion (PZS for short) in Zhurnal Eksperimentalnoi I Teoreticheskoi Fiziki published in Leningrad, and Uspekhi Fizicheskikh Nauk published in Moscow. The Kharkov journal lasted for only six years, from 1932 to 1938. The articles were written in German or English, obviously aimed at non-Russian readers. Landau published many articles in German in PZS, while Leipunsky's experimental group tended to publish in English. Both were at UFTI at the time. Kurchatov and his Leningrad group also published experimental results in nuclear physics, mostly about slow neutron activation, in English in PZS. The English, to the authors' credit, was reasonably good. PZS was a valuable source for Soviet physics. Publication in German journals of work done in the USSR diminished during this time. Zhurnal Eksperimentalnoi I Teoreticheskoi Fiziki was founded in 1931, and edited by A.F. Ioffe and L.I. Mandelstam, with Yu. B. Khariton as deputy. All were at LFTI at the time. This journal survived the war, and became the principal physics archive for the USSR, like the Physical Review in the USA. The title in English is Journal of Experimental and Theoretical Physics, abbreviated as JETP. Virtually everything in JETP is in Russian, although the table of contents and some of the abstracts are also in English. All of the prominent scientists published in JETP. A survey of the contents in 1939 shows that about 2/3 of the articles pertained to material science and low temperature physics, both experimental and theoretical, and only about 1/8 were devoted to nuclear theory and experiment. The rest were atomic and molecular physics, plasma physics, and early calculations in quantum electrodynamics, using Dirac's theory of the positron. PZS had a similar mix. Material science, with obvious practical applications, led the way. Akhiezer's solution to the problem of light – light scattering was an early calculation to use the new field of quantum electrodynamics. It was published in PZS.[20] Uspekhi Fizicheskikh Nauk, founded in 1918, was, and is more of a review journal, with single author articles reviewing progress in a particular sub field, similar to the US journal Reviews of Modern Physics. Uspekhi is published in Russian. A typical nuclear physics article of this period is "Artificially Induced Fission in Heavy

---

[20] A. Akhiezer, PZS **11**, 263 (1937).

Nuclei" by E.V. Shpolsky of Moscow.[21] JETP periodically printed a list of the journal holdings of the library of LFTI. It is interesting to note that the outputs in nuclear physics from the US and the USSR as reflected in journal publications were comparable. The USSR had about 20% of the world output. So despite all of the difficulties experienced in Soviet society, the physicists were competitive. This fact will play a crucial role in the Soviet Atomic Project. Leningrad and Kharkov were the most productive, with Moscow in third place, and a few papers from other laboratories in central Asia and elsewhere.

Abram Ioffe received a letter towards the end of 1938 from Frederic Joliot-Curie, informing him that Hahn and Strassman had shown that uranium when bombarded by slow neutrons splits into two nearly equal halves – fission. This news caused immediate excitement in the Leningrad Institute. Every week brought new discoveries. Soviet physicists and radiochemists could think of nothing else. Investigation of the fission of uranium by slow neutrons occupied center stage in Kurchatov's laboratory, as well as in the Radium Institute headed by V.G. Khlopin, I.M. Frank in Moscow, and Leipunsky in Kharkov.[22] One important question, addressed by Joliot-Curie, was how many neutrons were released on average when uranium fissions. They realized the possibility of a chain reaction, and the release of very large amounts of energy, which could be used to generate electric power or for weapons purposes. A protégé of Kurchatov named Georgy Flerov was to play a crucial role in future developments.

5.6. Georgy Flerov

Georgy Nikolaevich Flerov was born in 1913 in Rostov-on-the Don. He was ten years younger than Kurchatov. His paternal grandfather was a priest in the Russian Orthodox Church,[23] and his mother was Jewish. In 1929, he finished high school and started work first as a lab assistant, and

---

[21] E.V. Shpolsky, Uspekhi Fizicheskikh Nauk, **21**, p 253 (1939).

[22] I.N. Golovin, "I.V. Kurchatov," *ibid.,* p 39.

[23] Russian Orthodox Priests could marry.

later as a mechanic and electrician. [24] In 1931 Flerov went to Leningrad to work in a factory, and in 1933 he entered Leningrad Polytechnic Institute as a graduate student. He finished his studies at the Polytechnic Institute in 1938, and joined the research laboratory of Igor Kurchatov at LFTI. He started experimental work on the absorption of slow neutrons by various elements.[25] Flerov quickly rose to one of the more competent experimenters in Kurchatov's group. By 1938 Kurchatov was wearing several hats. He directed the nuclear physics laboratories at LFTI and the Radium Institute, and managed the physics department at the Leningrad Pedagogical Institute as well. As a result, he had access to many graduate students and technicians. Among the students were G.N. Flerov and K.A. Petrzhak. A young Georgy Flerov is shown in Figure 5.1.

5.7. Petrzhak and Flerov Discover Spontaneous Fission of Uranium

When Hahn and Strassman reported the fission of uranium, Kurchatov gave Petrzhak and Flerov the task of repeating the experiments and verifying the results. In 1969, Petrzhak told the story[26] of the discovery of spontaneous fission of uranium made by him and Flerov, published in 1940.[27] This is Petrzhak's account:

"In the pre-war years only a small number of people studied nuclear physics. And the number who, like Kurchatov, believed in the practical applications of nuclear physics was even smaller. Virtually all of the equipment needed for research was home-made: particle counters, amplifiers, sources, and targets were built by hand.

---

[24] Ria.ru/spravka/20111201/503658377.html

[25] G. Flerov, "Absorption of slow neutrons in Cadmium and Mercury," JETP, **9**, 143 (1939).

[26] n-t.ru/ri/ps/pb092.htm. This web site, on the element uranium, reprints Petrzhak's description of the discovery of spontaneous fission originally published in "Khimiia I Zhizn"

[27] K.A. Petrzhak and G.N. Flerov, "Spontaneous fission of uranium," JETP 10, p 1013 (1940). (in Russian). This paper is very nice. A brief note, with no experimental detail, was published in English in Phys Rev **58**, 89 (1940). Petrzhak is spelled Petrjak in Phys Rev. The JETP paper is translated and annotated in Appendix E.

At this time Kurchatov did not have a beard.[28] After two years Kurchatov sent another student to me for consultation, a combative and self centered individual by the name of Georgy Flerov. We had similar thesis projects, and in the end worked together for years, but we were from different institutes.

"After the excitement of the Hahn-Strassman discovery, Kurchatov assigned me and Flerov the task to repeat and verify the experiments. We had some uranium in the form of pitchblende, and a radon-beryllium neutron source. We also were proficient in building and operating the necessary charged particle counting equipment. Along with the interest in neutron induced fission there naturally arose the question of spontaneous fission. Could uranium simply break up into two nearly equal parts without the excitation of a captured neutron? Bohr and Wheeler[29] had calculated this effect, and estimated the branching fraction to be about 10^{-13}, too small to be observed. Willard Libby[30] attempted to observe the effect, and placed a limit on the branching fraction to be less that 10^{-6}, a long way from Bohr's estimate.

"Undaunted, we began our search for spontaneous fission of uranium. We started by looking for a threshold energy for neutrons which could stimulate fission. Is there a neutron kinetic energy below which fission in uranium will not occur? All of the equipment was hand built. The object was to determine a null effect – fission would cease if the neutrons dropped below some threshold energy. But there was no null effect observed. Fission continued to occur, even after the neutron source was removed.

"There were background sources to worry about. Street cars outside generated electrical and mechanical noise. There were neutron sources elsewhere in LFTI, and the cosmic radiation generated an ever present neutron background. Then there were instabilities in the ionization chamber electronics. We moved away from the street cars and LFTI's other neutrons, and the fission signal persisted. We talked to Kurchatov, who said that if this is true, it is a new phenomenon, and people only discover such things once in a lifetime if at all, so you better be sure that it is correct.

"One needs to increase the vertical shielding in order to get rid of the cosmic ray background. We thought of setting up the experiment in a submarine, and going to the bottom of the Baltic Sea, but the Baltic is too shallow to attenuate the cosmic ray background sufficiently. Kurchatov pointed out that Moscow now has a subway system, and that some of the stations are buried deep enough under ground that the

---

[28] Later, after the Soviet Atomic Project was in full force, Kurchatov wore flowing chin whiskers, and was called 'the Beard."

[29] Bohr and Wheeler, Phys Rev **56**, 426 (1939). Footnote added by LGP.

[30] Libby, Phys Rev **55**, 1269 (1939). Footnote added by LGP.

cosmic ray flux would be suppressed significantly relative to sea level. Abram Ioffe, director of LFTI, wrote a letter to the people's commissar for Moscow transport, requesting that we be accommodated in one of the subway stations to carry out our experiment.

"The commissar approved, and gave us full support. So we took the train to Moscow with our equipment, and set up in the Dinamo metro station, on Leningradsky Prospekt.[31] We set up in a room off the main station, and worked there six – eight months. We only worked nights, when things were relatively quiet. [32]The earth overburden was about 60 m, sufficient to attenuate the cosmic ray background by 95%. We repeated the measurements made at sea level, and the effect persisted. Having eliminated all known sources of background, we reported the discovery of spontaneous fission. Ioffe sent a telegram to the Physical Review."

There is much interesting information in Petrzhak's account. For one thing, it was business as usual in research at that time. The experimenters could have been anywhere. The Soviet system did not present them with insurmountable problems. In fact, it helped them, both in research support and in accommodations in the Metro. They were not hampered by poor working conditions. Everyone's equipment was hand made. Commercial suppliers of physics research equipment did not exist anywhere until after WWII. Most of the research worldwide was done with radioactive sources rather than the new particle accelerators. Of course for the spontaneous fission work they didn't need an external neutron source, although such sources were used to get started, and to make the transition from stimulated to spontaneous fission. This is certainly one of the first, if not the first, instance of the use of an underground laboratory to shield the cosmic radiation.[33] Then Kurchatov cautioned them to be careful, that a new discovery was worth extra effort. This is very true. Only a few scientists

---

31 This stop is on the #2 Green line "Zamoskvoretskaya," opened in 1938, one of the two oldest lines of the Moscow Metro. The #1 Red line is the other. Mayakovskaya on the Green Line and Chistye Prudy on the Red Line were both used as government air raid shelters in 1941. See en.wikipedia.org/wiki/Moscow_Metro.

32 The modern Moscow Metro runs from 5:30 am to 1:00 am, so if the same schedule obtained in 1939, they only had 41/2 hours a day of quiet time.

33 Underground laboratories are in common use today for experiments requiring very low background levels. Some are in mines, both active and inactive, where the experiment is reached via an elevator. Others are under mountains, where access is available via ground level tunnels.

ever make a significant new discovery, and fewer still ever repeat the achievement by making a second one. Then, as is often true in research, they were lucky. Bohr and Wheeler's estimate of the branching fraction quoted by Petrzhak was wrong. The true branching fractions for ^{235}U and ^{238}U are given in Table II in Appendix A as 2.0×10^{-9} and 5.4×10^{-7} respectively, about six orders of magnitude larger than Bohr and Wheeler's number. In their 1940 paper in JETP Petrzhak and Flerov quote a branching ratio range for ^{238}U between 5×10^{-7} and 5×10^{-8}, in good agreement with modern results. If Bohr and Wheeler had been correct, they would never have seen anything. Libby's limit was only a factor of 10 away from making the discovery, but he did not pursue the matter further. Libby's note in the Physical Review was published before Bohr and Wheeler, so he was probably not aware of their calculation. Petrzhak and Flerov directly counted fission fragments in an ionization chamber, while Libby looked for the β decay of ^{131}I, one of the daughters of fission products, which was not as sensitive. Libby also looked unsuccessfully for secondary neutrons accompanying spontaneous fission. The Bohr Wheeler paper is on the whole quite good, but the estimate of spontaneous fission, based on an extension of Gamow's barrier penetration model for α decay, is so far off as to be totally useless. But Bohr was a very famous man, so Petrzhak and Flerov, who were graduate students at the time, had to be super sure that they were correct.[34] Petrzhak and Flerov's complete paper in JETP **10**, 1013 (1940) is translated and annotated in Appendix E. The conclusion is that the work was solid, and that the Leningrad group of nuclear physicists was as competent and capable as anyone before WWII.

---

[34] P&F's colleagues at LFTI, Zeldovich and Khariton, published a paper in 1941, after the results of the spontaneous fission experiment were known, in which the Bohr-Wheeler numbers were adjusted to agree better with the experimental result. 20-20 hindsight. Ya. B Zeldovich and Yu. B. Khariton, Uspekhi Fizicheskhikh Nauk, **25**, 381 (1941). In the Gamow formula the transition rate is proportional to an exponential, and relatively small changes in the exponent lead to large changes in the rate. Z&Kh decreased the barrier thickness to bring the prediction into agreement with experiment.

Figure 5.1. Georgy Flerov in 1940. Credit JINR Dubna.

In the summer of 1940 the Soviet Academy of Sciences established a uranium committee. Izvestia reported on a meeting of the geophysical section of the Academy of Sciences in which Academicians Vernadsky and Khlopin gave presentations on the possibility of extracting large amounts of energy from the fission of uranium, and the necessity of procuring significant quantities of uranium ore from known sources in the Kirghiz mountains near Tashkent.[35] During the war years, however, little progress was made on uranium mining.

5.8. The German Invasion

In 1939-40, while Petrzhak and Flerov were at work in the Dinamo metro station in Moscow, war had come to Europe. Germany invaded Poland on September 1, 1939. As part of the Molotov-Ribbentrop pact the Soviet Union participated in the partitioning of eastern Poland, and occupied the Baltic states of Latvia, Lithuania, and Estonia. In the spring of 1940 Germany invaded France, and in the summer-fall of 1940 the Battle of Britain was waged in the air over England by the Royal Air Force against the Luftwaffe. Stalin believed that the neutrality of the USSR was protected by the Molotov-Ribbentrop pact. The cross channel invasion of

---

[35] Izvestia, 26 June 1940, page 4 "Use of Atomic Energy."

Britain by German infantry did not take place. Hitler called it off on the grounds that air superiority had not been gained by the Luftwaffe. Most historians believe that the air Battle of Britain was at best a draw. Hitler may have had other reasons for not invading the British Isles. In any case, nothing much happened between the fall of 1940 and the summer of 1941, a period referred to as the 'phony war.'

Then on June 22, 1941, Hitler invaded the Soviet Union. Many books have been written about the disaster suffered by the Red Army and the Soviet people. Whether the imprisonment and execution of many competent Red Army officers during Stalin's terror in 1937-38 contributed to the unpreparedness of the Soviet defenses or not is a matter of debate. It seems clear, however, that Stalin himself did not anticipate the invasion, despite warnings from his spy network, and that he did his best not to provoke the Germans. Three main German army groups, labeled North, Center, and South crossed the eastern border into the Soviet Union, under the command of field marshals von Leeb, von Bock, and von Rundstedt respectively. The German invasion front was over 600 km long. In the far south the Rumanians under Antonescu invaded Ukraine. By mid-July General Guderian and his panzer divisions were attacking Smolensk, 190 km west of Moscow. In September, 1941, Leningrad was encircled, blockading the city for 900 days. By October the panzer divisions were on the outskirts of Moscow. In the Moscow region the German army was stopped, and the city was not occupied. Stalin remained in the capital throughout the war. Chapter 1 describes the disruption caused by the invasion, the mass movement of factories and people to the east, the loss of agriculture, and other aspects of the catastrophe which impacted the country.

5.9. Kurchatov Joins Naval Research

We now turn to the impact of the war on the physics laboratories and researchers in the occupied territory, the western Soviet Union. As the invading army approached Leningrad, the laboratories began to empty. Draftees left the laboratory for the army. Many who were not drafted volunteered for the home guard. All scientific equipment, books, and papers were evacuated to the east, out of range of German aircraft.[36] The war plan of Leningrad Physico-Technical Institute was managed by Abram Ioffe, and included work in radar, protection against mines at sea, and the possibility of a uranium bomb. Radar and disarming the threat of mines against shipping were the two highest priorities. A.P. Aleksandrov was already working with the Baltic fleet. He successfully demagnetized some ships as a protection against magnetic mines, and then guided them through a mine field. He then returned to LFTI, but continued his work in mine safety. The anti-mine defense work by Ioffe's Institute made a good impression on the Soviet government. Kurchatov decided to join the effort, and to postpone the work in nuclear physics. The LFTI cyclotron was not yet built, and could not be finished in war time, so it was better to do more direct war research. While the rest of LFTI evacuated to Kazan, Kurchatov and Aleksandrov stayed in Leningrad working on anti-mine protection.

Then in early August, 1941, Kurchatov and Aleksandrov were called to Sevastopol to help combat the destruction of Black Sea shipping by mines. They left Leningrad by air to fly to Sevastopol, and on the way they were intercepted by German aircraft and Soviet anti-aircraft fire, and were forced to land in the Moscow region. After a brief encounter with Red Army troops who made them lie on the ground at bayonet point, they were taken to a local hospital, which was the only place where guests could be accommodated, and fed and lodged for the night. The next day they were taken to Moscow, and given renewed orders to fly to Sevastopol. The Germans had not yet reached Sevastopol, but had managed to plant several mines from aircraft.

The magnetic mine fuses were built to exploit the fact that the earth's magnetic field is enhanced by the iron hull of the ship.[37] Thus as the ship sailed around, it carried with it a stronger field than would be present due to the earth alone, and this field could be used to detonate the mine. This

---

[36] I.N. Golovin, "I.V. Kurchatov," *ibid.*, p 42ff. This description comes from Golovin.

[37] en.wikipedia.org/wiki/Degaussing. There is a photograph of the Queen Mary showing the degaussing coil wrapped around her. Wooden ships did not have this problem.

could be fixed by sending current in a coil around the hull of the ship to oppose the earth's field, resulting in a ship invisible to mines. The current could be adjusted for latitude, and even reversed in the southern hemisphere. (The geographic north pole is a magnetic south pole – that is the bar magnet is not only tilted with respect to the axis of rotation, but is also upside down, i.e. opposite to the angular momentum.) The opposing current flowed in a plane parallel to the water surface, which for a large ship was a lot of wire. The earth's field is only about 0.00005 T, but because of the large area to be protected substantial currents were required. Kurchatov designed demagnetization equipment for several ships, and trained the crews in techniques of mine avoidance. They rigged a captured German mine so that it would not explode, but transmit its firing signal to land. Then they ran ships by the mine, some with the demagnetization equipment, and some without. Sure enough, the demagnetized ships gave no trigger signal to the mine's firing mechanism, while those without did.

This experiment was followed by an actual flotilla of 'fixed' and 'unfixed' ships sent out into the mined harbor of Sevastopol on the Black Sea. Two unfixed ships were lost, but no fixed ones. This demonstration convinced the sailors of the value of the physics professors' analysis. British sailors visited Sevastopol, and were impressed with the mine avoidance techniques developed by Kurchatov and Aleksandrov. The method was called the 'LFTI system,' and was awarded a Stalin Prize in 1942.

In September, 1941, Aleksandrov was sent back to the northern fleet. Demagnitization work kept him busy returning to the Black Sea, to the Caspian Sea, and even to the Volga River region near Stalingrad. In August, 1942, with the German Sixth Army less than 60 km west of the city, Aleksandrov was on a cutter in the river looking for mines. A warrant officer on the cutter witnessed Aleksandrov, the professor, retrieve a German magneto-acoustic mine from the river, and disarmed it himself. This made a vivid impression on the sailors.[38]

---

[38] Vladimir Gubarev, "Atomnaya Bomba," Algoritm, Moskva, 2009, p 23.

Kurchatov remained in Sevastopol for two more months, until November, 1941. The Soviet navy then decided to send him and his team to Poti, a town on the Black Sea coast of Georgia. They went by sea from Sevastopol. It was a short but harrowing voyage, thanks to German aircraft attacks. Kurchatov did not have much to do in Poti. Some members of his team were then sent to the Caspian Sea to work on demagnetization of oil tankers. Kurchatov himself was dispatched to Kazan, where LFTI had been already evacuated. This time travel by air was not authorized, and the overland trip to Kazan by rail in wartime Russia was long and hard. Troop trains and supplies had priority, and Kurchatov spent more than one overnight waiting on a train platform. It was very cold, and he caught a cold, which when he arrived in Kazan at the end of December, 1941, became pneumonia. His wife Marina Dmitrievna was there, as was Abram Ioffe, both of whom nursed him back to health.

5.10. Leningrad, Moscow, and Kharkov Institutes Evacuated to Kazan

In addition to LFTI, the Radium Institute, the Institute for Physical Problems, and the Physics Institute of the Academy of Sciences, from Leningrad and Moscow, were all evacuated to Kazan in August, 1941.[39] Peter Kapitza was there. Vladimir Khlopin requested that the supply of radium be evacuated from Leningrad to Kazan. The cyclotron of the Radium Institute was left in Leningrad. Sergei V. Kaftanov was the responsible administrator for science in the Ministry of Defense during the war. He was also the chairman of the all union committee for higher education in the Council of People's Commissars. He played an important role in relocating and refurbishing Soviet science in wartime. Kaftanov was trained as a chemist, but spent most of his career in the government. He had survived Stalin's purges.

---

[39] www.ras.ru/atom/72d6e070-60a2-4e17-ace5-48574827d414.aspx shows written documents of the period, requesting electricity and water for experimental facilities in Kazan.

Kazan is a city on the Volga River about 800 km east of Moscow with a very interesting history. The Volga was the eastern boundary of Russia at the time of the Mongol invasion in 1237.[40] Whether Kazan existed as a city at that time is a matter of debate. The local capital was Bulgar, south of modern Kazan, and was totally destroyed by Mongol hordes. The Bulgars emigrated to the Black Sea region where Bulgaria now is, but many local tribes, like the Chuvash, still inhabit the Kazan area, with their own languages and customs. By 1438, two hundred years later, the Tatars dominated the city, and it was a Muslim capital. Then in 1552 Kazan was captured by Russian armies of Ivan IV (Ivan the Terrible), and its Muslim population was killed or exiled. The Tatar presence never really vanished, however, and the city today retains an eastern flavor.[41]

Living conditions in the Kazan region in 1942 were not comfortable. The influx of refugees from the western Soviet Union, combined with wartime shortages, led to severe problems with the food supply. Nadezhda Orlova-Korotkevich, a widow with two small children from Leningrad, was among the displaced persons, and was having difficulty feeding her family. Her husband Igor Korotkevich had been an instrument maker for Igor Kurchatov at LFTI, had remained in Leningrad after the blockade, and perished there during the starvation winter of 1942. Kurchatov in Kazan learned of Korotkevich's death, and notified his widow, who was then living and working as a school teacher in a small village in the Kazan region. Demonstrating the compassion for the wellbeing of his coworkers and employees that was to contribute to his future success as leader of the Soviet Atomic Project, Kurchatov obtained a 5000 ruble award for Nadezhda from the Soviet Academy of Sciences, and continued to look after her and her children.[42]

During his illness in Kazan Kurchatov grew a beard, which he retained for the rest of his life. Beards were unusual in the Soviet Union, as they were in the mid-twentieth century west. Perhaps a beard was considered

---

[40] James Chambers, "The Devil's Horsemen," Athenium, New York, 1979. Chapter 6.
[41] en.wikipedia.org/wiki/History of Kazan.
[42] Raisa Kuzntsova, "Kurchatov v Zhizni", *ibid.*, p336.

bourgeois. In any event, Kurchatov was to become a very busy man, and having a beard saved time in the morning. At first the beard was only temporary, until victory over Nazi Germany was achieved; but shaving it was postponed until after the first atomic bomb test, and then indefinitely.

Kurchatov continued for a time to work on the problem of protecting ships from mines. Then after the death of the local director of the laboratory for tank armor, Kurchatov was named his replacement, and he began work in the physics of explosions. But his attention would soon be drawn back to nuclear physics, and the fission of uranium.

By March, 1942, the projections regarding the war situation began to change. It was no longer obvious that the Soviet Union would be defeated, and indeed the discussion was centered on whether one year or several years would be necessary for victory. The Red Army learned to fight better, and supplies, which had been severely curtailed during the 1941 invasion, began to recover because of new factory capabilities in the east. As the German army advanced on Stalingrad it began to outrun its resources. A weakened rear guard left the German troops exposed to encirclement by Soviet Operation Uranus. Uranus led to the surrender of the German 6th army at Stalingrad in January, 1943, and stopped German advance to the east. After Stalingrad the Red Army steadily pushed the Germans westward, reaching Berlin two and a half years later.

5.11. Military Service Absorbs Younger Physicists

Younger members of the LFTI research group were drawn into the Red Army. K.A. Petrzhak joined the anti-aircraft artillery. Mesheryakov and Panasyuk became infantrymen, and in July, 1941, Georgy Flerov volunteered for the Leningrad home guard and saw some action in the defense of the city. Flerov's technical skills were rewarded by a transfer out of the home guard and into training for combat aircraft maintenance.[43] He became a student in the Air War Academy in Leningrad, which was

---

[43] Raisa Kuznetsova, "Kurchatov v Zhizni," Izdatel'stvo Glavarkhiva, Moskva, 2007, article by Georgy Flerov, p 472.

soon evacuated to Ioshkar-Ola, which is not far from Kazan.[44] His mother remained in the blockaded city of Leningrad. Flerov was not just an ordinary air cadet, and his classmates and instructor noticed his unusual ability. He talked to his instructor in the school about his pre-war work in LFTI on the uranium problem, the chance of obtaining a new source of energy, and the concern that the German scientists were working on the same subject, with an obvious threat to the USSR. Flerov was already aware of the apparent censorship of scientific papers on the uranium problem in the West.

The instructor was sufficiently impressed that he suggested that Flerov write a letter directly to Stalin calling his attention to the situation. Flerov said that he was already thinking about such a letter. In the meantime, he requested permission to visit Kazan, where members of LFTI and other physics institutes had been evacuated, creating a 'small' Academy of Sciences. In December, 1941, his instructor obtained permission from the army for Flerov to visit Kazan and meet with Kurchatov. Although far removed from the front lines, there was still some concern about a young lieutenant travelling by himself from one place to another. There was the possibility that he be arrested as a spy. But he arrived in Kazan without incident, only to learn that Kurchatov was still enroute from Poti in the south. Flerov talked to Ioffe, Kapitza, and other physicists about the necessity of obtaining critical equipment from Leningrad for continuation of research in nuclear chain reactions. He argued that the light isotope of uranium was the favored one for a nuclear explosion, and he outlined a research program which, when pursued, would achieve the desired goal. Kapitza received his outline with enthusiasm, although some others doubted that the goal was really practical in a finite number of years with finite resources.

---

[44] www.warheroes.ru/hero/hero.asp?/Hero_id=9275.

5.12. Lieutenant Flerov Wrote Letters to Moscow

Flerov was a prolific letter writer. His first letter to Stalin about the uranium problem has been lost. It was referred to in his first letter to Kaftanov on the same subject. The Kaftanov letter was dated November, 1941.[45] Flerov began by saying that in his view the war had interrupted physics research, but he proposed a series of measures which would lead to the solution of the uranium problem. He estimated the success of the project between 5-10% (In later letters this was raised to 10-20%). He admitted that, because of the time required for development, a uranium weapon would have no impact in the present war. On the other hand, if there were such a weapon, one bomb would wipe out the entire city of Berlin. He noted that Germany and Britain were both working on the uranium problem, and that no foreign journal carried research reports on the subject. This absence was clear evidence of government secrecy. He suggested that the Soviet foreign office request from allies (Britain and USA) summaries of research successes on the uranium problem. Realizing that this might not work, he offered to write the overseas laboratories himself, asking for comments on the spontaneous fission of uranium, in that way opening an avenue for discussion. Also, perhaps one or two people could be sent to England. But in parallel with seeking information from the allies, Flerov wanted to proceed with R&D in the USSR. He proposed enlisting Kapitza, Kurchatov, Alikhanov, and others from LFTI and the Radium Institute. He requested that he and Petrzhak be released from the army for this purpose. He closed with the observation that the state which first solved the uranium problem would be able to dictate its conditions to the whole world.

The letter to Kaftanov urged action, but had no technical details. The next letter, to Kurchatov, was quite different, for it discussed several technical problems in uranium weapons development. Of course Flerov and Kurchatov knew each other very well. Some excerpts from the six page printed version of the letter to Kurchatov follow:[46]

45　The letter is quoted in Russian by Yu. N. Smirnov, HISAP96, Vol **2**, *ibid.*, p 139ff.
46　This letter is published in full in HISAP 96, Vol **2**, p 160. (in Russian).

"Dear Igor Vasilivich!

"I write you from Kazan, in anticipation of your arrival here. I have finished the air academy. I am now a second lieutenant, and have received orders from Kuibishev to train aviation mechanics in physics and electrical technology.

"I write to tell you my reason for coming here. I must do physics research. I do not mean research directly related to defense applications, but rather the continuation of our work on the uranium problem. It seems to me that it is shortsighted to neglect this question.

"I gave a seminar here yesterday, outlining my thoughts on the uranium problem. I don't think I made much impression…"

Flerov then presented some preliminary theoretical results on the design of a nuclear reactor, based on natural uranium or uranium enriched to 10% ^{235}U, and deuterium or helium as a moderator. The numbers were not encouraging. The first Soviet reactor was built 5 years later, using graphite as the moderator. His discounting of the practicality of nuclear reactors parallels that of the British MAUD report discussed in Chapter 3. Of course Flerov, a lieutenant in the Red Army, had no knowledge of the top secret MAUD report, although Beria did. Flerov continued:

"The predicted amounts of material are so large that even enthusiasts like A. F. Ioffe are skeptical.

"What follows is a brief outline of my thoughts on this problem. I am convinced that sooner or later we will be again occupied with uranium. Success will depend on a number of factors, including improvements in the war and the economic situation of the country, as well as enlistment of a large number of reluctant physicists.

"In any case, the ground is not yet adequately prepared, and that which I write further must be for the future.

"1. The energy release from fission of 2 kg of natural uranium is roughly equal to the energy from burning 10 tons of coal.[47] This assumes that the energy source is the small component ^{235}U. The yield per kg is much larger if the ^{235}U is enriched. But the enrichment process itself will require considerable energy, more than 10 tons of coal...Therefore further calculations on uranium-deuterium or uranium-helium combinations do not make sense, and it is necessary to find other possible paths of the utilization of nuclear energy.[48]

"2. Nuclear fuel differs from chemical fuels in that it has the ability to burn slowly – slow neutron fission – or quickly – fast neutron fission. The first generates heat like coal, and the second leads to explosions like dynamite. Coal requires oxygen to burn, while dynamite has its own oxygen built in. For an explosion, the energy release must take place in one to ten microseconds.[49] ...Force is more important than energy for most weapons applications, placing a premium on a short reaction time. From an economic perspective, it seems more auspicious to pursue the explosive option.

"3. The heat content per kg of dynamite is less than that of coal, so that 2.4 kg of uranium equals 10^5 tons of dynamite.[50] ... Can we actually obtain complete explosive burning of ^{235}U, and convert all of the energy of fission (200 MeV) into a shock wave?

"4. The basic question is: Can we establish a fast neutron chain reaction either with ^{235}U or $^{239}_{94}$[51]? We do not know the exact number of neutrons emitted between one and three MeV kinetic energy per fission. From the work of Petrzhak and myself it seems that the energy spectrum of secondary neutrons starts at one MeV, and that the fission cross section is larger than 2×10^{-24} cm^2, and less than 3×10^{-24} cm^2.[52] We take the number between 2 and 3...It is necessary to measure this number, but until then let us assume two possibilities for the cross section: $\sigma = 2\times10^{-24}$, and 3×10^{-24} cm^2. These two cases give corresponding numbers for the critical mass of a sphere,

---

[47] The quoted mass ratio for equal energy content is coal/(natural uranium) = 5000. The actual number is more like 40,000, for complete burning of the ^{235}U at 200 MeV per nucleus. The energy content of 1 kg of coal is 24 Megajoules.

[48] Flerov discounts the practicality of nuclear reactors.

[49] Flerov clearly understood the two applications of nuclear power – reactors and weapons.

[50] The energy content of 1 kg of ^{235}U is about 10^8 MJ, and 1 kg of dynamite is about 4 MJ, so energy content ratio dynamite/^{235}U = 4×10^{-8}, about 60% larger than Flerov's value.

[51] Plutonium of course. Flerov did not know the name, but he knew of the transuranic element, and that is a substitute for ^{235}U.

[52] www.kayelaby.npl.co.uk/atomic_and_nuclear_physics/4_7/4_7_2.html quotes the fission cross section for ^{235}U = 1.2×10^{-24} cm^2, averaged over the fission neutron energy spectrum. So Flerov is a bit high, but close. Impressive.

assuming that all of the neutrons are reflected back into the fissioning material: $\sigma = 2$, $M = 3$ kg, and $\sigma = 3$, $M = 0.9$ kg. Neither value is too large. If one can obtain 0.9 kg of fissionable material, one can also obtain 3 kg...[53]

"5. How does one establish the conditions for an explosion - an explosion in which a noticeable part of the uranium nuclei successfully fission? The quantity K depends on several factors: the number of secondary fast neutrons emitted per fission; the number of uranium nuclei in a neutron mean free path from the place of creation to the boundary of the uranium sphere; and the coefficient of reflection of neutrons in uranium. For the process of preparation of a uranium bomb K can be anywhere between zero and one. For an explosion K-1 must be greater than unity. One of the possible schemes for a bomb explosion is shown in Fig 1."[54]

Figure 5.2 depicts Flerov's "gun type" arrangement, similar to the one suggested in the MAUD report, and actually used in the Hiroshima bomb. The assembly is a gun barrel, presumably of iron. The region between the two hemispheres is evacuated. Conventional explosive A drives hemisphere B_1 into hemisphere B_2, creating a super critical mass, and an explosion. He imagines a gun barrel five or ten meters long.

---

[53] Critical masses for ^{235}U and ^{239}Pu are 52 kg and 10 kg respectively. See Wikipedia/critical_mass. Flerov is a bit optimistic.

[54] This drawing was copied from Fig 15 in David Holloway, "Stalin and the Bomb," *ibid.*, where page 12 of Flerov's hand written letter to Kurchatov is reproduced.

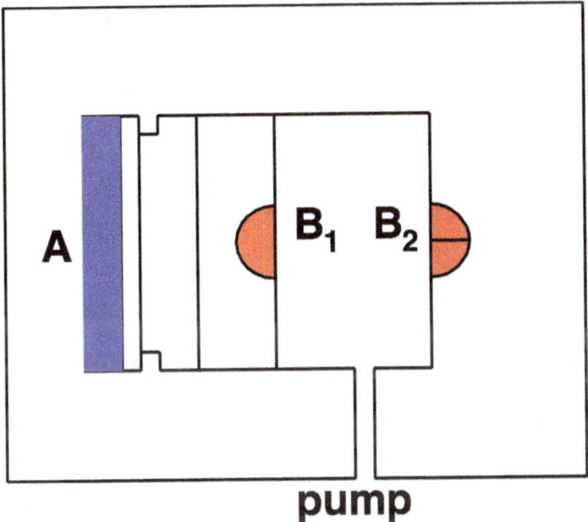

Figure 5.2. Author's rendition of Flerov's drawing. A is chemical explosive, and B$_1$ and B$_2$ are halves of a sphere to be shot at each other for the creation of a critical mass.

Flerov then considers triggering the chain reaction with stray neutrons, including neutrons from spontaneous fission, and the neutron multiplication factor K which determines the timing: small K-1 means no explosion. He estimates the total energy and the maximum temperature, and asks Kurchatov for theoretical help from Pomeranchuk and Migdal[55] regarding how the energy of the explosion is shared among pressure waves, X-rays, residual radioactivity, etc. Flerov concludes:

"If you think the work should be pursued, it would be useful to ask the English and Americans about the results that they have obtained. In particular, how many neutrons come out of the fission reactions? (!)

"I have a personal request for you, Igor Vasilevich – could you try to extract K.A. Petrzhak from army service?

---

[55] Theorists in Landau's group in Moscow.

"Please excuse the incoherence and smudges in this letter. I have no time to re-write it, since I must return to Ioshkar-Ola in two or three hours. I am not in a good mood. Mother is still in Leningrad, and things are going badly for her. And I have been studying in the Air Academy for five months without learning anything substantive.

"Well, that is all. Write to me at the address I left with the Institute. I do not want to lose contact with the Institute.

Regards, G. Flerov, December 21, 1941.

Kurchatov did not read the letter until after his recovery from pneumonia in Kazan. He never replied to the letter, but kept it on his desk throughout the Soviet Atomic Project. After Kurchatov's death the letter, written on school notebook paper, was found in his desk drawer.[56] The letter is remarkable for its detail and accuracy, in many ways comparable to the MAUD report, which was written in England under much different conditions.

Life in Kazan was difficult for the scientific refugees. Electricity was unreliable. The experimental equipment from Leningrad sat around in unpacked boxes.[57] The experimental program that Flerov called for required a large mobilization of forces, and before that it was necessary to have some assurance that the goal was achievable. Kurchatov knew that Flerov had no proof of principle, and he wondered whether it was not more useful to work towards victorious tank battles than to seek the creation of nuclear weapons. The navy solved Kurchatov's dilemma, by ordering him to Murmansk for anti-mine operations. However, two years later Kurchatov was to write: "Uranium must be separated into two parts. At the moment of detonation the two parts must be fused at high relative velocity. This procedure is not new. The same idea was proposed by G.N. Flerov, who calculated the necessary relative velocity of the hemispheres, results agreeing well with present data."[58]

---

[56] HISAP96, Vol **2**, *ibid.*, p 152.
[57] I.N. Golovin, "Igor Kurchatov," *ibid.*, p 47.
[58] Vladimir Gubarev, "Atomnaya Bomba," *ibid.*, p 31.

Flerov was a man possessed. Returning to the front, Flerov wrote to Ioffe: "We must not forget the hope of making uranium weapons, but it is necessary to separate the lighter isotope of uranium to achieve this goal."[59] Flerov's reconnaissance squadron was dispatched to various places, and he found himself in Voronezh. The university had been evacuated, but the library was open, and despite the war, it had a current collection of American physics journals. Flerov was especially interested in learning what had transpired regarding the uranium problem since he had been out of touch in the army. To his surprise there was nothing on the subject published since 1941. He asked himself: "Why should the flow of publications on nuclear studies cease?" His answer was that the work had become classified, and that meant that the USA was seriously pursuing nuclear power. The Germans could be doing the same things, and they certainly had many experts in the field, and plenty of raw materials and industrial capacity.[60] The prospects were very scary. This was not the first time he had made this observation; he mentioned it in the letter to Kaftanov in November, 1941. But his resolve was reinforced. He correctly surmised that this absence was due to secrecy that had been applied to all such publications, beginning with the dissertation of Herbert Anderson, as recounted earlier. At first in the US and England the censorship was voluntary, but soon became mandatory after the US Army took over management of the Manhattan Project. A similar phenomenon took place in the USSR around 1940. The first Soviet paper affected seems to have been a discussion of the uranium chain reaction by Zeldovich and Khariton, a follow-up of previously published papers by the same authors.[61]

The next letter of interest from Flerov, motivated in part by the absence of western publications on uranium, was addressed to Stalin. This is a famous letter, on par with the Einstein letter to Roosevelt discussed earlier. There is a long tradition in Russia of writing letters to the Tsar to

[59]　I.N. Golovin, "Igor Kurchatov," *ibid.*, p 48.
[60]　Raisa Kuznetsova, "Kurchatov v Zhizni," *ibid.*, p 472.
[61]　Vladimir Gubarev, 'Atomnaya Bomba,' *ibid.*, p 10. Earlier papers are in JETP, Vol **10**, p 29 and p 477 (1940).

address various grievances. The Tsar was the father who would clearly see injustice and assuage any wrongdoing. Stalin was, for this purpose, the Tsar, and must have accordingly received many letters every day from throughout the Soviet Union, in time of war or peace. The Soviet Atomic Project would have begun, and succeeded, without Flerov's letter. The same is true of the Einstein letter to Roosevelt. The projects are too vast, too important for their respective countries, and have too many supporters, to be the product of one letter, no matter how timely and persuasive. Before writing Stalin, Flerov learned of his mother's death of starvation in Leningrad around March 12, 1942. His mother did not work, so she had a very small bread ration, 125 grams per day. She fainted in the street, and was carried home by two neighboring women. She died at home.[62] Flerov wrote a cover letter to Stalin's secretary (Aleksandr Poskrebyshev?), urging that the uranium question be brought to the attention of Comrade Stalin. He stated that the probability of success is 10-20%, an increase over the estimate given to Kaftanov. He then said that there was evidence that foreign powers were at work on the uranium problem, and he correctly stated the explosive power of a uranium bomb as equivalent to 20,000-30,000 tons of TNT. He pointed out that one bomb would destroy all of Berlin or all of Moscow, depending on who used it. He mentioned that Kaftanov had not answered his letter, and only Stalin in person could get uranium research going again in the Soviet Union.

Here are excerpts of Flerov's letter to Stalin:[63]

"Dear Joseph Vissarionovich!

"Here we are, 10 months after the beginning of the war, and it seems to me that we are beating our heads against a stone wall. Am I mistaken?

"Do we value the significance of the 'uranium problem'? I don't think so. Some say that the project appears fantastic, and the time required appears too long for a practical solution. But they are only people who fear the unknown, who have had

---

[62] HISAP96, Vol **2**, *ibid.*, p 152.
[63] HISAP96, Vol **2**, *ibid.*, p 168.

unfortunate experiences with exaggerated projects – a burned child avoids fire.

"I beg to differ. Maybe I am incorrect – there is always risk in science, and in the case of uranium the risk is even bigger. In my letter to Comrade Kaftanov I quoted a 10-20% chance of success.[64] Let me tell you what 'uranium' means. It does not require a revolution in technology – this has been demonstrated by work done in the last months before the war. But a uranium bomb would be a revolutionary change in the technology of warfare. If it proceeds without us, we will pay dearly. Do you know the principal argument against uranium? "It would be too expensive even if it were achieved. Nature rarely indulges mankind." In answer I would point out that an important difference between man and animals is the ability to overcome difficulties, to snatch from nature that which is necessary.

"It could be that, being on the front lines, I have lost my perspective regarding what science should be doing at the present time, and that the uranium problem should be postponed until after the war. This is what academician A.F. Ioffe thinks, but he is making a big mistake. It seems to me that unless we can raise the level of activity to equal that of the foreign scientists, and even to surpass them in some areas, we will make a big mistake, and voluntarily give up a winning position. In one or two years we will be so far behind that Ioffe will throw up his hands, but eventually it will be time to turn from war development towards the uranium task. The best of intentions can lead to the largest stupidity.

"We all want to do everything possible to annihilate the fascists. But we must not dampen enthusiasm for uranium by emphasizing only those questions which pertain to 'pure' wartime applications."

Flerov then writes that he and Petrzhak, nuclear physicists from LFTI, and co-discovers of spontaneous fission of uranium, are both lieutenants in the Red Army, and not able to pursue the research tasks at hand. He notes that Stalin no doubt receives two types of letters: those with suggestions to help end the war; and those with personal requests for aid. And he is aware that Stalin could place the present letter in the second category; Flerov wants out of the army. Therefore, Flerov recommends that Stalin organize a presentation.

---

[64] There was a second letter to Kaftanov, which has not been located.

"Thus it seems to me necessary that a meeting be arranged with Academics Ioffe, Fersman, Vavilov, Khlopin, Kapitza, and Leipunsky, professors Landau, Alikhanov, Artsimovich, Frenkel, Kurchatov, Khariton, Zeldovich, and doctors Migdal, and Gurevich, and also K.A. Petrzhak. I request a 1 and ½ hour report. Your presence would be very valuable, but if not practical, then one of your representatives. This is not the time to have a scientific circus, but I see this as a singular opportunity to demonstrate the truth – the necessity to concentrate on the uranium problem…

"I have sent Comrade Kaftanov a letter and five telegrams, but have received no reply.[65] The Academy of Sciences has not produced a plan for uranium. But I hope that you can break this wall of silence, and that this is the last letter necessary to solve the problem…

"I insist on my first request, to acquire from all of the candidates of the meeting mentioned above a written evaluation of the uranium problem…"

G.N. Flerov, April, 1942.

Flerov's letter to Stalin was anticlimactic. It had no technical information – although the cover letter did say something about the uranium bomb's destructive power. This was reasonable, given the intended recipient. It also rambled on more than seems necessary in a letter to the leader of the country. The Einstein letter was more to the point, and shorter. The Einstein letter called for the appointment of a uranium committee, which Flerov had mentioned in earlier letters, but did not bring up in the letter to Stalin.

5.13. A Notebook is Found on a Dead German Officer

Sergei Kaftanov was also aware that the Germans were interested in atomic weapons. This came about through a bizarre incident. NKVD Colonel Ilya Grigorevich Starinov was an expert in land mines, and specialized in mining operations behind the German lines. He began this

---

[65] Kaftanov confirmed the receipt of Flerov's correspondence. S.V. Kaftanov, "Po Trevoge," "Khimiia i Zhizn," #3, 1985, p 6ff.

work during the Spanish civil war. The German Eleventh Army, under the command of General Erich von Manstein, had been in the Crimea since December, 1941. The military situation was fluid, and Starinov was busy with partisan mining operations. On one of the patrols in late February, 1942, on the southern bank of the Gulf of Taganrog, at the northeast corner of the Sea of Azov, a notebook was retrieved from a German officer who had been killed. A staff officer of the 56th Soviet Army (under the command of Major General V.V. Tsyganov) gave the notebook to Colonel Starinov. He ruffled through the notebook, which was in German, and noticed what looked to him like diagrams and chemical formulas, accompanied by explanations. One of his officers who could understand German said that it mentioned atomic energy. Starinov returned to Moscow, and called Stepan Balezin, the scientific representative of the State Defense Committee (GKO). Starinov had interacted earlier with Balezin regarding calculations for shaping charges, and recipes for homemade plastique explosives. They had been introduced by Kaftanov. Balezin promised to share the notebook with his boss, Kaftanov, and to send the notebook off for expert examination. The formulas looked to Kaftanov like diagrams of nuclear transformations of uranium.[66]

Starinov hardly paid any attention to Balezin's response at the time regarding the notebook contents, but later came to appreciate that what he had thought were chemical formulas in the notebook were in fact diagrams of uranium processing [which could also be chemical formulas]. Balezin and Kaftanov theorized that the German officer killed at Krivoy Spit had been sent to search for uranium.[67] On the other hand, he could have been a physics graduate student or post doc, like Georgy Flerov, who was drafted into the army, but who retained an interest in nuclear physics and nuclear energy, and thought about these things in his spare time. In any event, the notebook was given to the nuclear physicist from Kharkov, Aleksandr Leipunsky, and a military expert in explosives, General Georgy Pokrovsky. Leipunsky and Pokrovsky both recommended that the USSR

---

[66] Sergei V. Kaftanov, "Po Trevoge," "Khimiia i Zhizn," #3, 1985, p 6ff.

[67] Ilya G. Starinov, "Over the Abyss," Mass Market Paperbacks, New York, 1995, p 265ff. Russian version "Proidi Nezrimym," Moskva, Molodaya Gvardia, 1988, p 110ff.

not pursue nuclear weapons during the war, as they believed that the development time required would be too long to affect the war, and that limited resources would be better applied to more pressing defense needs. This recommendation countered Flerov's. Many other scientists also had their doubts, but in the end Flerov prevailed.

In the meantime, evidence from Soviet espionage in England, and pressure from the Soviet Academy of Sciences, together perhaps with Flerov's incessant clamoring, produced an order from the State Defense Committee. On September 28, 1942, while the Germans were still advancing on Stalingrad, order No. 2352ss appeared on the subject of organization of uranium work.[68] The Academy of Sciences, through Academician Ioffe, was ordered to begin work to study the utilization of atomic energy as a result of the fission of uranium, and to report to the State Defense Committee by April 1, 1943, on the possibility of making a uranium bomb or uranium fuel. It outlines a number of tasks and goals, only some of which actually happened. It was signed by Stalin. It got things rolling. Georgy Flerov was discharged from the army on orders of Kaftanov. Ioffe appointed Igor Kurchatov to lead the effort.

5.14. Early Soviet Espionage

The Soviet spy network in the West was well established and humming along in the 1930's, before WWII. It concentrated on diplomatic and political information, but it also had an interest in scientific and industrial data. It operated principally out of the Soviet embassies and consulates, although there were some operatives who had 'legitimate' jobs in the private sector. As we saw earlier, the Soviet Union at that time, partly because of the depression in the capitalist countries, and partly because of the noble experiment of communism, enjoyed an aura that attracted some people to its cause. There have been a number of history books and

---

[68] Order of the State Defense Committee #2352ss available in Russian on rusarchives.ru.

memoirs written about Soviet espionage in the pre-war period, and into the early part of WWII.[69]

According to Yury Modin, John Cairncross informed the NKVD that the British and Americans had been working on the joint manufacture of an atomic bomb since 1940.[70] The first entry in Beria's diary on the subject was dated December 24, 1941, where he notes material from London on atomic energy. He decides not to report this information to Stalin at that time, but to wait for more confirming evidence.[71] The second entry was on March 28, 1942. Beria wrote:

"I reported data to Koba (Stalin) regarding the atomic energy of uranium. Malenkov and Molotov were present. There is evidence that it is possible from a 10 kg uranium bomb to obtain the force of 1600 tons of TNT. But it will take lots of factories and cost lots of money.

"He asked: 'Is the information accurate?' I answered that the sources have been reliable up to now, and the same information has come from different sources. We have watched this for a year already. We will try harder.[72] We have prepared a report for you, Comrade Stalin.

"He said: 'Don't send the report. The affair may be important, but there is no time for it now…. We do not have enough tanks and aircraft, and how the summer campaign goes is still unknown. But I will not ignore the subject. So wait a bit with your atom. But don't forget about it, Lavrenty. When we are able to breathe easier, we will look into it. Until then, gather information, and keep me up to date on the subject. Before the war we hoped to start work on the use of atomic energy to generate electricity.'

---

[69] Allen Weinstein and Alexander Vassiliev, "The Haunted Wood," Random House, New York, 1999. Yury Modin, "My Five Cambridge Friends," Farrar, Straus Giroux, New York, 1994.

[70] Modin, *ibid.* p 109.

[71] L.P. Beria, Secret Diaries 1937-1953, Yauza Press, Moskva, 2012, p 265.

[72] On January 27, 1941, the NKVD sent instructions to G.B. Ovakimyan, deputy chief of the New York 'station,' requesting scientific-technical information regarding the uranium problem. Information from Beria's Diary, *ibid.*, p 291-292.

"We did not talk long after that. I must notify Fitin, stir him up."[73]

Beria's diary tends to be terse and abbreviated, and his record of what was said in any one of these meetings in Stalin's office is probably incomplete.[74] Stalin's response is nevertheless interesting. He does not dismiss the uranium problem out of hand, nor does he plead total ignorance of a very technical subject. He wants to stay informed, even in very trying times, with lots of other things on his mind.

On October 3, 1942, just after the issue of order No. 2352ss referred to above, Beria again discussed the subject with Stalin. [75]

"Molotov has begun to prepare the task outlined in the order. I said: 'Comrade Stalin, we already have lots of information regarding the uranium project. I guarantee its reliability.' He said that he did not have time to read it now, but give him the summary sheet.

"The main thing is that we are winning. The Germans will not come any farther than they already have. Stalin knows that we will soon start rolling towards the finish line. We have not perfected the art of war, but we will continue from Stalingrad to Berlin.

"Nothing of substance has been done on uranium. The Academy management is not effective. We need a strong organization, with full participation of industry. But maybe it is too early."

Thus, although the highest levels of the Soviet government were now involved, and scientists were beginning to move back to the West together with their laboratories, a fully organized push to solve the uranium problem was still a few years.

---

[73] Pavel Fitin was trained as an engineer, and served as an NKVD operative from Moscow. He became a lieutenant general in 1945.

[74] There is a log of people coming and going from Stalin's office. 'Vypiska iz zhurnalov zapisi litz prinyatykh I V Stalinym.' The early days of WWII are available online.

[75] Beria's Diary, *ibid.*, p 331-332.

Chapter 6

The Manhattan Project Creates Los Alamos

6.1. Need for a Weapons Lab

Chapter 3 covered the Manhattan Project through the spring of 1943, when two very large factory facilities were coming into operation: the Clinton Engineering Works at Oak Ridge, Tennessee, and the Hanford Engineering Works at Hanford, Washington. Oak Ridge was for isotope separation to produce weapons grade ^{235}U by gaseous diffusion and magnetic mass spectrographs, and Hanford had three high power nuclear reactors to breed ^{239}Pu by neutron absorption on ^{238}U followed by β decays. Each engineering works had the job of producing enough weapons grade material to make a nuclear bomb. They were separated geographically for both safety and security reasons. The Metallurgical Laboratory at the University of Chicago still existed under Arthur Compton, some facilities were starting up at what was to become Argonne National Laboratory, and there was a hotbed of activity at UC Berkeley, where Lawrence had his cyclotrons, and Seaborg did his chemistry.

If all worked out according to plan, the Manhattan Project would get its weapons grade product in a timely fashion. The next step would be to build the bombs themselves. Many books have been written about Los Alamos. Of all of the laboratories of the Manhattan Project, it is the most luminous. The Los Alamos laboratory director was J. Robert Oppenheimer, the subject of several biographies, and one of the most interesting people of the 20th century.[1] Although much of the work of the laboratory was technical, or engineering, there were many physics and chemistry problems to be solved. The scientific staff was a galaxy of luminaries. In addition to Oppenheimer himself, there were Hans Bethe, Vicki Weisskopf, Enrico Fermi, Richard Feynman, Robert Wilson, Edward Teller, Robert Serber, Rudolph Peierls from Britain, and many others. Rabi and Lawrence were frequent visitors, but were not members of the lab staff. The senior scientists were famous before the Manhattan Project, and the junior ones were famous afterwards. The Manhattan Project, and the subsequent post-war surge in government support of science, created a cadre of high profile physics departments, and spawned many careers. In the aftermath of WWII physics had an aura which it lost during the extended cold war.

The number of books on Los Alamos and Los Alamos lore is truly daunting. In addition to the usual suspects – histories of the Manhattan Project – there have been a number of accounts of lives of individuals and groups at the laboratory during the critical period 1943-1945. Laura Fermi has described living conditions.[2] Jennet Conant has written a book about Oppenheimer's 'gate keeper' in Santa Fe, Dorothy McKibbin.[3] McKibbin is shown with Oppenheimer and Weisskopf in Figure 6.1. There have been several books about Edith Warner, a writer and artist who ran a tea room at an old bridge over the Rio Grande River, where Oppenheimer loved to go for dinner off the lab site.[4] The list goes on and on, because Los Alamos is a very interesting subject.

---

[1] Kai Bird and Martin J. Sherwin, "American Prometheus," Alfred A. Knopf, New York, 2005. This is the biography I have used.

[2] Laura Fermi, "Atoms in the Family," University of Chicago Press, Chicago, 1954.

[3] Jennet Conant, "109 East Palace," Simon and Schuster, NY 2005. 109 East Palace was the address in Santa Fe of the Los Alamos office, where all visitors had to report before going to the laboratory.

[4] Edith Warner, "In the Shadow of Los Alamos," University of New Mexico Press, Albuquerque, 2008.

Robert Wilson, who became the first director of Fermi National Accelerator Laboratory in the mid 1960's, where the author of this book worked for 40 years, was one of the group leaders at Los Alamos. In an oral interview, he throws some cold water on the whole idea, saying that if all you wanted to do was build a gun type uranium bomb, this could have been done without creating a new laboratory in the middle of nowhere.[5] He has a point. The gun type bomb was proposed by Frisch and Peierls, mentioned in the MAUD report, and sketched in the letter Flerov sent to Kurchatov. So it was not news. All it really needed was enough ^{235}U to exceed a critical mass, and the separation plants at Oak Ridge were already functional. While some design parameters needed to be worked out, these could have been measured at the Metallurgical Laboratory at

Figure 6.1. Dorothy McKibbin, Robert Oppenheimer, and Victor Weisskopf at an evening gathering at Los Alamos. Credit Los Alamos National Laboratory.

---

5   Oral History Transcript – Dr. Robert R. Wilson, www.aip.org/history/ohilist/

Chicago.[6] However, it is a moot point. Los Alamos was created, and, while the Hiroshima bomb was a ^{235}U gun type weapon, the Trinity test at Alamogordo, and the Nagasaki bomb were both implosion type ^{239}Pu weapons, which required substantial R&D at Los Alamos, and proved to be the fission design of choice in future weapons of the cold war.

Arthur Compton and the Metallurgical Laboratory at the University of Chicago had the over-all responsibility for bomb development. In June, 1942, he appointed J. Robert Oppenheimer as the head of this particular phase of the project. Oppenheimer accordingly set up a small group at Berkeley, which included Robert Serber.[7] Groves, in discussions with Compton, Conant, and Bush, decided that it was necessary to initiate a new laboratory devoted to bomb R&D. The laboratory had to be located in a remote place, for security and safety reasons. Near Oak Ridge or Hanford would not do, and certainly the highly populated environs of Chicago were not appropriate either.

6.2. Selection of the Director of the Laboratory

The choice of laboratory director was, and is a much discussed subject. Oppenheimer was a theorist, and much of the work of the laboratory was to be technology, not theoretical physics. Oppenheimer was a university professor, with no managerial experience whatsoever, but was an effective organizer within his theory group. While charismatic to those in close association with him, he had no record of public relations or political skills. The Soviet Union had many sympathizers at the University of California in the 1930's, and Oppenheimer, while never a member of the Communist party, had ties to people who were. Groves knew all of this, and took personal responsibility for Oppenheimer's loyalty. Other possibilities, Lawrence, Compton, and Urey were considered and

---

[6] The gun design required ordnance testing, which would have to be done outside of the city of Chicago.

[7] Leslie M. Groves, "Now It Can Be Told," *ibid.*, p 60.

discarded for various reasons.[8] So Oppenheimer was selected to direct the laboratory, but where should it be located?

6.3. Selection of the Site

Oppenheimer had always loved New Mexico. Oppenheimer was born in New York City in 1904. After high school graduation in New York he set out in 1922 on a trip out west with one of his school teachers.[9] They went to Albuquerque, New Mexico, and from there to a dude ranch near Santa Fe called Los Pinos. On a horseback trip out from the ranch, they visited a Spartan boys' school called the Los Alamos Ranch School. He subsequently enrolled in Harvard, and after graduation studied under Max Born in Göttingen, as was described in Chapter 2. While still working in Europe, Oppenheimer and his younger brother Frank in 1928 persuaded their father to sign a four-year lease on a ranch in the mountains above Los Pinos, which they called Perro Caliente. They continued to lease the property until 1947, when Robert purchased it for $10,000.[10] In the 1930's Oppenheimer had professorial appointments at Berkeley in the fall and Cal Tech in the spring. Oppie, as he was called, would move back and forth to Pasadena every year, with his entourage of students and post docs. Then everyone in the summer was invited to the New Mexico ranch. One of the post docs after 1934 was Robert Serber, who that year graduated with a PhD from the University of Wisconsin. His thesis advisor was John H. Van Vleck. Serber and Oppenheimer became lifelong close friends. Another New York boy, Serber became a proficient horseman, as was Oppenheimer.[11] As discussed in Chapter 2, Oppenheimer was the American version of Landau, educating the next generation of physicists in the modern theories of quantum mechanics and relativity. Serber remembers exciting times in quantum electrodynamics before WWII, where Oppie's group at Berkeley and the various theory groups

---

[8] Groves, *ibid.*, p 62ff.

[9] Bird and Sherwin, *ibid.*, p 25.

[10] Bird and Sherwin, *ibid.*, p 73.

[11] R. Serber, "Los Alamos Primer," University of California Press, Berkeley, 1992, p xxvii.

abroad were in constant competition for the successful application of the new theory to basic problems in electrodynamics.[12]

So it should not be surprising that, after choosing the laboratory director, and looking for a site somewhere in New Mexico for its isolation, Groves was introduced by Oppenheimer to the mesa where the Los Alamos school was located.

Libby describes the scenery.[13] The valley of the Rio Grande River is 1500 m above sea level. The distance between Santa Fe, to the east, and Los Alamos is about 40 km as the crow flies, but in the 1940's there was no heavy loading bridge over the Rio Grande until Espanola, which was 40 km north of Santa Fe on the east (wrong) side of the river. At Espanola, the road crosses the Rio Grande. One then drives south on the west side of the river. About half way down, one picks up what is now called state highway 502 'The Los Alamos Highway.'[14] This road crosses the Rio Grande at the Otowi suspension bridge, the location of Edith Warner's tea room. The bridge is now in the National Register of Historic Places. It was not considered robust enough to handle the loads anticipated for Los Alamos. A modern highway bridge has since been constructed next to it, but the suspension bridge is still there. One continues south on 502 on the west side of the river, and climbs in a winding fashion through changing vegetation. It was a typical mountain gravel road, with hairpin turns, switchbacks, and precipices off to nowhere. Finally, one emerges on the mesa, a fairly flat tongue of land at 2100 m, 600 m above the Rio. The mesa area is 180 km^2.[15] The mesa is surrounded by a canyon on all sides except the road, making it ideal for security. It also has a number of crannies that could be used for testing explosives – something that the laboratory was to do often. A series of arroyos, or dry creek beds, run from the Jemez mountains immediately to the west of Los Alamos eastward to

---

[12] R. Serber, "Los Alamos Primer," *ibid.*, p xxv.

[13] Leona Marshall Libby, "The Uranium People," *ibid.*, p 194ff.

[14] There are good topo maps available on line. See www.pickatrail.com/topo-map/l/los-alamos-new-mexico.html

[15] William L. Laurence, "Dawn over Zero," Alfred A. Knopf, New York, 1953, p 181. The modern city of Los Alamos has no doubt expanded to exceed this area.

the Rio Grande River. The Sangre de Cristo mountain range lies immediately to the east of Santa Fe, not far away. The country is very rugged. In winter, there is plenty of snow. Ashley pond, in the middle of the ranch school property, was used for ice skating.[16] Sawyer's Hill, off the mesa to the south, was used by the Los Alamos Ranch School as a ski area. The new laboratory adopted the practice.[17] When breaks from the work schedule allowed, Los Alamos scientific staff enjoyed hiking in the surrounding countryside, and winter sports.

Leslie R. Groves was promoted to brigadier general just before assuming overall control of the Manhattan Project. During the years of the project he received a second star, becoming a major general. The US army in WWII had at least a thousand major generals, so Groves was not outstanding in that regard. His responsibilities were similar to those of Lavrenty Beria in the Soviet Atomic Project, but his position in the government was in no way comparable. Beria was number two to Stalin, with wide powers in the Soviet Union. Stalin and Beria met virtually every day. The power in the US government is more diffuse below the president, so there really is not a number two man. Groves certainly did not meet with Roosevelt on a daily basis. In fact, they met only once or twice if at all. Nevertheless, Groves managed the task with remarkable agility and overall success. He had the necessary authority to mobilize the resources.

6.4. Lab Staff and Facilities

Oppenheimer travelled around the country to the various physics labs, looking for recruits for Los Alamos. One of the negative factors in his sales pitch was that Groves wanted all of the scientists to be in the army. Oppenheimer was amenable to this idea, but he met strong opposition. To quote Leona Libby: "A meeting on this subject was held at the Waldorf Astoria in New York City, attended by I.I. Rabi, Robert Bacher, Robert Oppenheimer, R.R. Wilson, and others…Wilson was enormously relieved

---

[16] Ashley Pond was a man's name. He was the founder of the Ranch School.
[17] There is a photograph of Niels Bohr on cross country skis climbing Sawyer's Hill.

that Rabi and Bacher succeeded in talking Oppenheimer out of the idea of having everyone conscripted into the army. Of course, Oppenheimer still had to face Groves, but he obviously did."[18] The scientific staff at Los Alamos was not inducted into the US Army. The Soviet Atomic Project scientists retained civilian status also, partly because the war was over when the Project got underway, and partly because the Red Army was not a key player.

Once the site was selected, bulldozers from the U.S. Army Corps of Engineers began scraping the mesa, setting up unnamed streets and barracks type structures for housing. Everything was painted olive drab, except for the original log houses that were a part of the Los Alamos School. There was a large lodge, used by the lab as a reception area, and six individual log homes that were assigned to the highest ranking members of the Los Alamos staff. The log homes had no kitchens, since meals were supposed to be served to everyone at the lodge. But they did have indoor plumbing, and bathtubs. New construction contained only stall showers. So the log homes were called 'bathtub row.' At least some of the historic buildings still exist on bathtub row in modern Los Alamos. The absence of kitchens was addressed.

Lab space had to be constructed from scratch, since there were no laboratories at the Los Alamos Ranch School. Initially, there were few scientists on site to tell the engineers what to build, and security considerations prevented them from being too specific about what was going to take place. Groves assigned Robert Wilson the task of buying/borrowing/stealing the Harvard cyclotron.[19] So Wilson went to Harvard, and met with Percy Bridgeman about taking away the cyclotron for an undisclosed purpose for the war effort. Wilson quotes Bridgeman: "And you have the nerve to bring some physicist from Princeton here to steal our cyclotron." After some hemming and hawing the deal was struck, and the Harvard cyclotron began its long journey up the canyon road to Los Alamos, going by train as far as Santa Fe, and then by truck in pieces.

---

[18] Leona Marshall Libby, "The Uranium People," *ibid.*, p 197.
[19] Oral History of Robert R. Wilson, www.aip.org/history/ohilist/

John Manley was another experimentalist, a faculty member at the University of Illinois, Urbana, who was at the Metallurgical Laboratory at the University of Chicago in 1942 when Oppenheimer came looking for recruits. Manley and Wilson were the first experimenters on site at Los Alamos in the spring of 1943. Manley moved the Illinois Cockcroft-Walton electrostatic accelerator from Urbana to Los Alamos via the same tortuous route – train to Santa Fe, followed by trucks in pieces up the canyon road. The two accelerators were housed in newly constructed buildings painted olive drab like everything else. Two electrostatic accelerators of the Van de Graaff type were also moved from the University of Wisconsin, Madison, to Los Alamos. The inventory of lab accelerators was completed by a second machine from Illinois, an electron ring machine called a betatron, which was used to generate x-ray pulses for high speed photography.[20]

Oak Ridge and Hanford had post offices, but Los Alamos did not. All mail was addressed to PO Box 1663, Santa Fe, New Mexico. Officially, Los Alamos did not exist, and all of its inhabitants were stuck in that PO Box. Outgoing mail was deposited in unsealed envelopes for the censors. It could not contain any information regarding the secret laboratory. Any letter deemed unsatisfactory was returned to the sender, with the offending parts highlighted. Incoming mail was supposed not to be screened, but various tests by residents showed that it was, and the army admitted to it in 1944. There were no telephones, except for the office of the laboratory director. The site was fenced all around, with two gates: the east gate was the main gate link to Santa Fe, and the west gate led into the mountains out back. Both gates were guarded 24/7, and everyone had to have a pass to go either out or in. This was a real pain for everyone concerned.

Dorothy McKibbin was hired by Oppenheimer in March, 1943 to be the gatekeeper for the laboratory. She occupied an office in downtown Santa Fe at 109 East Palace. Everyone with business at the lab had to see her first, and obtain credentials to get past the gate. All laboratory

---

[20] Leona Marshall Libby, "The Uranium People," *ibid.*, p 194.

employees and their spouses and families came through Dorothy's operation.

Oppenheimer remained in Berkeley in the spring of 1943, as Los Alamos was getting started. Wilson and Manley pestered him constantly about getting organized. On one of their visits Oppenheimer produced an organization chart for the laboratory. It called for four divisions: experimental physics, theoretical physics, chemistry and metallurgy, and ordnance. Group leaders within each division would report to the division chiefs, who would in turn report to Oppenheimer.[21] Oppenheimer initially placed himself as head of the theory division, but quickly learned that he had too much on his plate, and Hans Bethe was appointed theory division leader.

6.5.   Calculation Before Computers

There was no computing division – there were no computers! This seems unbelievable today, but digital computers assembled with vacuum tubes only began to exist after WWII, and only became really practical after the invention of the transistor in 1948, and the development of integrated circuits, that dramatically increased the computing power per unit volume. Electric powered mechanical calculators existed, and everyone had a slide rule, and that was it.[22] Mechanical calculators were the size of a large typewriter. The common brands were Marchant and Friden, with the former being dominant at Los Alamos.[23] Some calculating machines could take the square root of a number. In doing so, the machine jumped up and down on the desk until it was finished. The absence of digital computers and the ability to perform computer modeling placed a big burden on the experimental physics division, which had to perform dangerous modeling experiments with real active materials.

---

[21] Kai Bird and Martin Sherwin, "American Prometheus," *ibid.*, p 208.

[22] The author grew up in the slide rule era. They disappeared the day pocket calculators were introduced.

[23] R.P. Feynman, "Los Alamos from Below," Engineering and Science 39, #2 (Jan-Feb 1976), p 12. A transcription of a talk given by Feynman.

Herbert Anderson in an address to a conference in 1986 reminisced about the pre-computer calculators.[24] Fermi in Rome used a Brunsviga, a German hand crank mechanical calculator, and brought it with him to Columbia University in 1939. In New York he soon learned that the Marchant, powered by electricity, could outpace his hand crank, and he switched over to the upgraded machine. By 1944 Los Alamos had a very large number of Marchant calculators humming day and night, and repairs became a work stoppage matter. It was not practical to ship the machines back to the manufacturer. Richard Feynman and Nick Metropolis set up a repair shop in their spare time. They learned how the machines worked, and became very adept at locating mechanical malfunctions and fixing them.

As the program progressed towards an actual bomb, the numerical solutions of non-linear differential equations describing shock waves became of interest. Feynman describes the techniques that were developed at Los Alamos to solve the necessary equations by numerical methods, using first Marchant calculators, and then IBM punched card machines.[25] There were the four basic operations, add, subtract, multiply, and divide. Initially an operator with a Marchant would do a sequence of arithmetic manipulations. To quote Feynman: "A rather clever fellow by the name of Stanley Frankel realized that it could possibly be done on IBM machines." Frankel had been one of Oppenheimer's graduate students at Berkeley. He was recruited by Oppie to work in the Los Alamos T-Division (Theoretical, headed by Hans Bethe). The hand calculating group was called T-5. The IBM machine group was T-6. Army draftees in the Special Engineering Detachment handled the IBM machines 24 hours a day. These machines were built for banks and other businesses, used punched cards, and could perform arithmetic operations – one per machine. Handling the input and output on punched cards removed the need to write down

---

[24] H.L. Anderson, "Scientific Uses of the MANIAC," *Journal of Statistical Physics*, **43**, 731 (1986).
[25] R.P. Feynman, "Los Alamos from Below" *ibid.*, p 25.

numbers, and speeded things up.[26] To test the idea, the Marchant calculations were changed, so that one operator performed only one arithmetic task, and sent the result on to the next operator. This proved to be much more efficient than doing several operations at one station. Three IBM 601 multipliers, a 402 tabulator, a reproducer, a verifier, a sorter, and a collator were ordered, and duly shipped to Los Alamos, in huge wooden boxes.[27] Uncrated, the 601 was about the size of a large mahogany desk. For a single arithmetic operation, the 601 was not much faster than a Marchant, but it could work around the clock without getting bored. Frankel, Feynman, and others developed sequences of operations from one machine to the next to solve successfully the shock wave problems encountered in the bomb detonation.[28] The Computing Division arose from such humble beginnings.

The first programmable electronic digital computer at Los Alamos was called MANIAC, based on a similar machine built at the Institute for Advanced Study in Princeton, under the guidance of John von Neumann.[29] By 1948 Metropolis had left Los Alamos for the University of Chicago, but was invited back by Carson Mark, who at that time was head of the Theory Division, to start development of a digital computer facility. Several electronic digital computers, based on vacuum tubes, were being built around the world at that time. MANIAC programs were written in machine language, in hexadecimal arithmetic. Original logic flow diagrams and some of the code written by Fermi for a phase shift analysis of pion-proton scattering are shown in Anderson's article.

---

[26] The 'IBM' punched card has disappeared as a machine readable data storage device, but it lasted almost 100 years – impressive longevity.

[27] George Dyson, "Turing's Cathedral," Pantheon Books, New York, 2012, p 60.

[28] The life and work of Stan Frankel, who made many contributions to the early days of computers is described in the Hewlett-Packard history web site. www.hp9825.com/html/stan_frankel.html.

[29] H.L. Anderson, "Scientific Uses of the Maniac," *ibid*.

6.6. Daily Life

At first Oppenheimer thought that the work of the laboratory could be accomplished by five or six scientists and their support staff. By the summer of 1945 the population consisted of 4,000 civilian and 2,000 military personnel, and construction work on site continued unabated throughout the war.[30] The town was guarded, and inside the town was the Tech Area, which was also guarded. The Tech Area had two layers of fences. Laboratory passes were of two kinds: blue badges were for access to the town; only white badges admitted the wearer inside the Tech Area, where the work of the laboratory was carried out. Outside were the apartments and dormitories, the PX, the laundry, gas station, medical center, canteen, and the water tower. Water supply was an issue, which at times became critical. Drinking water was sometimes blamed for upset stomachs. Since the streets had no names, the water tower served as a landmark to aid in navigation about town – rather like the radio tower in Tokyo. Tokyo streets have names, but this does not help the visitor who cannot read Japanese, and the city looks the same in every direction when one emerges from the subway. In physics language Tokyo is homogeneous and isotropic. Los Alamos had no subway – only unpaved roads, and olive drab barracks like buildings.

The Oppenheimers arrived on the mesa at the end of March, 1943, and occupied one of the houses on bathtub row. Kitty Oppenheimer had a kitchen installed at once. The houses were small, but comfortable. There were only six of them, enough perhaps for Oppie's original estimate of the size of the scientific staff, but nowhere near enough to accommodate the crowd that would soon arrive. Robert and Charlotte Serber moved into a sort of duplex newly built by the Army Engineers at about the same time. Robert and Jane Wilson occupied the other half of the duplex. Jane Wilson used to say, "The Serbers were our nerbers." Stoves for heating and cooking in the new construction were problems. Initially the cook stoves were iron wood burners, like one might find in a mountain hut. Some were oil fired. There was no natural gas on site. Furnaces for the buildings were

---

[30] William L. Laurence, "Dawn Over Zero," *ibid.*, p 181.

overzealous, and the living quarters were often too hot. People worried about the whole place burning down.

6.7. The Assigned Task

Oppenheimer insisted that all Tech Area workers, with white badges, should be informed regarding the progress lab wide. This ran counter to Groves' security partitioning, but Oppie managed to convince the general of the benefits of the exchange of scientific information. A weekly colloquium started as a result.

The first series of talks began in mid-April, 1943, and were delivered by Robert Serber. These lectures were a technical description of the mission of the laboratory, using all of the pertinent information available at that time. The lectures have since been declassified and published.[31] They begin with the sentence:

> "The object of the project is to produce a practical military weapon in the form of a bomb in which the energy is released by a fast neutron chain reaction in one or more of the materials known to show nuclear fission."

Hugh Richards has written an account of the nuclear physics measurements carried out using the two Van de Graaff electrostatic accelerators moved to Los Alamos from the University of Wisconsin, Madison.[32] Richards was a recent PhD graduate from Rice Institute in Houston, with experimental expertise in measuring the energies of neutrons from the range of recoil protons in photographic emulsions. After graduation in 1942 Richards headed north to the University of Minnesota, and on the way he dropped by the Metallurgical Laboratory at the University of Chicago, where he met John Manley. Early in 1943 the entire staff of experimental nuclear physics at Minnesota, including Richards,

---

[31] Robert Serber, "Los Alamos Primer," *ibid.*, p 3ff. Parts of these lectures are explicated in Appendix F.

[32] H.T. Richards, "Through Los Alamos, 1945," The Arlington Place Press, Madison, WI, 1993.

was moved to Los Alamos. He had a car, which was fairly unusual. When he got to Santa Fe he learned that there was no permanent housing available at Los Alamos, but he was fortunate to be quartered temporarily on site in the Ranch School main building, the lodge, which he refers to as 'the big house.' Later in April, 1943, he moved into a newly completed dormitory.

Richard Feynman arrived from Princeton at about the same time with a brand new PhD in theoretical physics. After the war Feynman made huge contributions to quantum electrodynamics, and won a Nobel Prize in Physics in 1965, but in 1943 he was just another post-doc at Los Alamos, although he stood out by being what can only be called a 'smart-ass.' His experience with housing was similar to that of Richards.[33] Feynman was married, but his wife was ill in a hospital in Albuquerque, so he was a dormitory bachelor when the dorms became available. Before that he stayed in buildings of the old ranch school. He was so obviously smart, and of value to the project, that he was cut some slack in his antics by Laboratory management, although patience must have worn very thin at times. His ability to open combination safes holding top secret documents of the Manhattan Project in anyone's office has become part of Los Alamos lore. He was so capable that when someone wanted in their safe, but couldn't remember the combination, Feynman would obligingly open it for them. Then there was the hole in the security fence, which according to Feynman was cut by construction workers who were tired of going in and out through the guards at the main gate. Feynman tested the robustness of Los Alamos security by going out through the main gate, but then coming back in through the hole in the fence. After he repeatedly exited the lab, but on no occasion returned to it, the guards got suspicious.[34]

After the accelerators were up and running, Richards worked on the measurement of neutron energy spectra from fission, using the emulsion technique. Richards writes: "For a couple of weeks in July (1943) our group had the world's supply of plutonium, a barely visible speck (a few

---

[33] R.P. Feynman, "Los Alamos from Below," *ibid.*, p 15.
[34] R.P. Feynman, "Los Alamos from Below," *ibid.*, p 19.

micrograms) prepared by cyclotron bombardment at Washington University in St. Louis."[35] The accelerators were operated 24 hours a day, and the experimenters worked in shifts, as is typical in peace time research. When an accelerator broke down, the experimenters fixed it. To quote Richards again: "With proper choice of nuclear reactions, these accelerators could then provide monoenergetic fast neutrons whose energies could be varied from a few keV to several MeV. Our priority activities related to the number of neutrons emitted per fission event for both ^{235}U and ^{239}Pu, the energy spectrum of fission neutrons, and the dependence of the fission cross section on the energy of the fast neutrons...These were important measurements: they showed that the number of neutrons emitted per plutonium fission was in fact larger than that for ^{235}U."[36] Nuclear physics research at Los Alamos was focused on a few specific problems, and had continuous time pressure to produce results, but was otherwise quite like laboratory research at universities. Richards kept doing what he had done as a graduate student, and there were many like him. He met and married Mildred Paddock, secretary to the head of the Metallurgical Division, Cyril Smith. The wedding was in Santa Fe. The newlyweds were moved into a one bedroom apartment. The government issued furnishings did not include a bed. Mildred's mother supplied a double bed, mattress, and springs, which were shipped to PO Box 1663, Santa Fe, NM.[37]

Later in 1943 the reactor at Oak Ridge, which was a pilot plant for the Hanford reactors, began to produce plutonium in larger amounts than the cyclotrons. By December, 1943, two milligrams of plutonium had been delivered to Los Alamos.[38] Techniques for irradiating uranium, extracting it from the reactor, and chemically separating the plutonium were first developed at Oak Ridge. As part of its responsibility for bomb construction, Los Alamos was assigned the task of developing the special methods necessary to purify the plutonium after chemical separation.[39]

---

[35] H.T. Richards, "Through Los Alamos, 1945," *ibid.*, p 60.
[36] H.T. Richards, "Through Los Alamos, 1945," *ibid.*, p 60.
[37] H.T. Richards, "Through Los Alamos, 1945," *ibid.*, p 65.
[38] Leslie Groves, "Now It Can Be Told," *ibid.*, p 41.
[39] Leslie Groves, "Now It Can Be Told," *ibid.*, p 163.

The Hanford Engineering Works was described in Chapter 3. The first 250 MegaWatt plutonium production reactor, 105 B, was built from June, 1943 to September, 1944, and began operation after recovering from the Xe poisoning problem in November of that year. After one month of operation the uranium fuel rods were removed for chemical extraction of the artificially produced plutonium, and the first delivery of plutonium was made to Los Alamos from the 221-T chemical refinery on February 5, 1945.[40] Earlier samples of plutonium from the Oak Ridge reactor had already shown in the summer of 1944 an excess number of neutrons. This came as a surprise to Los Alamos, where it was assumed that uranium and plutonium bombs could each be of the 'gun-type.' The laboratory was reorganized in the fall of 1944 in order to develop another, faster method of assembling a super-critical mass of plutonium.

The isotope ^{240}Pu, formed by neutron capture on ^{239}Pu, was the source of the extra neutrons. ^{240}Pu has a very large spontaneous fission rate.[41] Both isotopes of plutonium are produced from neutron capture: ^{238}U + n $\rightarrow$ ^{239}U $\rightarrow$ ^{239}Np + β + v; ^{239}Np $\rightarrow$ ^{239}Pu + β + v; ^{239}Pu + n $\rightarrow$ ^{240}Pu. A reactor supplies a bath of slow neutrons, so the longer the uranium fuel rod stays in, the more ^{240}Pu is produced. The two isotopes of plutonium are too close in mass to allow any practical isotope separation, so that the ^{240}Pu content is fixed after chemical extraction of plutonium from the fuel rod, and nothing can be done about it. Cyclotron production of plutonium did not subject the target to a sustained neutron flux, so the amount of ^{240}Pu was much smaller. The spontaneous fission table in Appendix A shows that 920 neutrons per gram-sec from ^{240}Pu is 50,000 times larger than its nearest competitor, ^{239}Pu. This built-in neutron source causes the bomb to pre-detonate. This phenomenon is discussed in Appendix F. So called 'weapons grade Pu' has ^{240}Pu < 7%, while 'reactor grade Pu', the plutonium content in power reactors not operated to produce nuclear weapons, has ^{240}Pu > 20%.[42] To manufacture weapons grade plutonium,

---

[40] En.wikipedia.org/wiki/Hanford_Site.

[41] Spontaneous fission is discussed in Appendix E.

[42] www.nuclearweaponarchive.org/Library/Plutonium

the reactor has to be operated in a periodic fashion.[43] The reactor operation table reproduced in Appendix C is for a power reactor, where in three years of running at 16 Megawatts, about 5 kg of a mix of plutonium isotopes was produced from 600 kg of uranium.

Plutonium production does not scale linearly with time, or with reactor power, because ^{239}Pu, just like ^{235}U, undergoes fission when bombarded by thermal neutrons. The fission cross sections are 700 and 505 barns respectively. Hence plutonium is burned as fuel. A crude estimate of the plutonium yield from a 250 MegaWatt reactor is about 200 gms per day, which would give a critical mass in a month.[44] From February until August, 1945, six or seven months, Hanford delivered enough plutonium for two bombs, the Trinity test and the Nagasaki weapon.

6.8.   The British Group

We left the British effort in Chapter 3. Britain did not want to be left out of the development of the atomic bomb. Opinion was divided in Britain regarding cooperation with the US or going alone. In August, 1943, Churchill and Roosevelt met at Quebec, hosted by Canadian Prime Minister Mackenzie King.[45] One result was an agreement that an Allied landing in France would take place in 1944 (Overlord). Another result involved US-British cooperation on the Manhattan Project. In response to this accord, a team of British scientists arrived at Los Alamos. The first was James Chadwick, who moved with his wife to North America in January, 1944. Their two daughters had been living in Nova Scotia for three years. They were evacuated from England during the Blitz. The parents had a chance to reunite in Halifax enroute to New Mexico.[46]

---

[43] There are proposals to use neutrino detectors to monitor reactor operations from a distance, thus determining whether or not the reactor is making weapons grade plutonium.

[44] See Appendix B.6 for a derivation of this number.

[45] Graham Farmelo, "Churchill's Bomb," *ibid.*, p 239.

[46] Graham Farmelo, "Churchill's Bomb," *ibid.*, p 275.

The Chadwicks were quartered in one of the houses on bathtub row at Los Alamos, considered luxurious compared to the alternatives. But Mrs Chadwick was never happy in the middle of the New Mexico desert. The Chadwicks were followed by the Peierls, James Tuck, Klaus Fuchs, and Otto Frisch. Later G.I. Taylor came from Cambridge, an expert in hydrodynamics. The Rayleigh-Taylor instability is named after him, and is one of the causes of the mushroom cloud. Otto Frisch was a skilled experimenter, and also a talented pianist. While at Los Alamos he played classical piano music over the local radio station.[47]

6.9. Niels Bohr and Los Alamos

Niels Bohr's visit to Los Alamos was a consequence of the British mission. Bohr had returned to Europe in May, 1939, after his visit to Princeton earlier that year, and the excitement over the discovery of the fission of uranium. There he remained until October, 1943. The German Army occupied Denmark as a prequel to the invasion of France in April, 1940.[48] Bohr and Heisenberg met in Copenhagen in September, 1941. This meeting has received a considerable amount of attention in post war discussion and documentation.[49] Michael Frayn's award winning play has stimulated public interest.[50] Carl Friedrich von Weizsäcker, a physicist, friend of Heisenberg, and son of a high official in the German Foreign Office, was doing theoretical work on the practicality of various designs for a nuclear reactor.[51] The use of graphite as a moderator had been rejected because of neutron absorption measurements made on graphite that was not sufficiently pure. The boron content was too large, but this was not realized at the time. So the German effort did not pursue graphite-natural uranium, which was successfully used by Fermi and Kurchatov. The natural uranium and heavy water approach seemed possible, and the

---

[47] H.T. Richards, "Through Los Alamos," 1945, *ibid.*, p 70.

[48] Denmark.dk/en/society/history/occupation.

[49] Two online sources are: Werner-heisenberg.unh.edu for Heisenberg's correspondence; and www.nba.nbi.dk/papers (the Niels Bohr archives)

[50] Michael Frayn, "Copenhagen," London: Methuen, New York, Anchor Books, 1998.

[51] Von Weizsäcker had earlier proposed the semi-empirical nuclear mass formula, discussed in Appendix A.

question arose in the Uranverein whether they should proceed with such a program, with its obvious large investment in resources, long time scale, and possibly a very dangerous weapon at the end. They were aware of the production of element 94 in a reactor, which could become an explosive substance, like ^{235}U. Uranium isotope separation did not seem realistic in wartime Germany. Heisenberg and von Weizsäcker came up with the idea of talking to Niels Bohr about the question of the practical achievability of energy from the atomic nucleus. Accordingly, they decided to visit Copenhagen, to attend an astrophysics conference at the German Scientific Institute. The conference was a cultural outreach on the part of the German occupation, to try to engage the local populace – in this case scientists. Heisenberg and Weizsäcker were invited to speak at the conference. Niels Bohr and his colleagues boycotted all meetings at the German Institute as a matter of principle – they did not wish to appear to be collaborating with the German occupiers.[52] An informal meeting with Bohr had to be separately arranged, but that presented no problem, since Bohr and Heisenberg were old friends. Weizsäcker wrote to Bohr, telling him of their proposed visit, with the possibility of an ancillary meeting between him and Heisenberg. The Germans could not hop on a train and go from Leipzig, where Heisenberg was working, to Denmark during WWII without special permission, but the astrophysics conference supplied the reason for the trip, and official sanctions from the Foreign Office were forthcoming. On Monday, September 15, 1941, with the German army advancing on Moscow on the Russian front, Weizsäcker and Heisenberg got off a train in Copenhagen.

Heisenberg gave his talk at the conference early in the week. He had lunch several times during the week at Bohr's institute – apparently not in Bohr's company. The lunchtime conversation often turned to the war. Heisenberg is quoted as saying that it was very important that Germany win the war.[53] This remark angered the Danes; Heisenberg seemed to be insensitive to Danish frustrations regarding the German occupation of

---

[52] Thomas Powers, "Heisenberg's War," Alfred A. Knopf, New York, 1993, p 120ff.

[53] Letter of Stefan Rozental, Bohr's assistant, quoted by Abraham Pais, "Niels Bohr's Times," Oxford University Press, 1991, p 483.

their country. Heisenberg's remarks got back to Bohr, and as a result, Bohr was reluctant to meet Heisenberg at all, but changed his mind, and he and his wife Margrethe invited Heisenberg to dinner. The conversation between the two, which is the subject of much speculation even today, took place after dinner. There are no contemporary written documents regarding the subject matter of the talks. Heisenberg wrote to his wife from Copenhagen, but did not say anything of substance regarding his talks with Bohr. He did say that he would carry the letters back to Germany and mail them from there, in order to avoid censorship.[54] Bohr talked to his family, but nothing was recorded until later. Recollections of the conversation only appeared after the war, and were inevitably affected by personal memory. The letters which Bohr wrote to Heisenberg after the war, but never sent, are available in English on the Bohr archive web site. They do not address the questions regarding the Bohr-Heisenberg conversations. The Bohr archive web site quotes a statement from Bohr's son, Aage, taken from an article in a book about Niels Bohr edited by Stefan Rozental:[55]

After the outbreak of war and especially after the occupation of Denmark we in Copenhagen were completely cut off from following the allied nations' efforts in the field of atomic energy. Various rumors reached us, however, of the German efforts, and the impression that in Germany great military importance was given to these possibilities was strengthened by the visit to Copenhagen in the autumn of 1941 of Werner Heisenberg and C.F. von Weizsäcker. There were in Copenhagen on other business, but in a private conversation with my father Heisenberg brought up the question of the military applications of atomic energy. My father was very reticent and expressed his skepticism because of the great technical difficulties that had to be overcome, but he had the impression that Heisenberg thought that the new possibilities could decide the outcome of the war if the war dragged on.*

*In the book "Brighter than a Thousand Suns" by Robert Jungk it is asserted that the German physicists submitted a secret plan to my father, aimed at preventing the development of atomic weapons through a mutual agreement with colleagues in the allied countries. This account has no basis in the actual events, since there was no

---

54　"My Dear Li" Heisenberg to Heisenberg, ed by Anna Maria Hirsch-Heisenberg, Yale University Press, 2016, Copenhagen letters are p 168-170.

55　Stefan Rozental (ed) "Niels Bohr, His Life and Work as Seen by his Friends and Colleagues," North Holland Publishing Co, Amsterdam 1967 p 193.

mention of any such plan either during Heisenberg's visit, or during a later visit to Copenhagen – also mentioned by Jungk – of the German physicist Hans J.D. Jensen. On the contrary, the very scanty contact with the German physicists during the occupation contributed – as already mentioned – to strengthen the impression that the German authorities attributed great military importance to atomic energy.

This all seems clear enough. However, there is Heisenberg's own version, written in a letter to Robert Jungk regarding the contents of his book. Heisenberg writes:[56]

January 18, 1957

Many thanks for your letter, asking me to write in a little more detail about my Copenhagen conversations with Bohr during WWII. In my memory which may, of course, deceive me after such a long time, the conversation roughly unfolded in the following way…As a result of our experiments with uranium and heavy water, we in our "Uranium Club" had come to the following conclusion: It will definitely be possible to build a reactor from uranium and heavy water which produces energy. In this reactor (based on the theoretical work by V. Weizsäcker) a decay product of 239-uranium will be produced which just like 235-uranium is a suitable explosive in atomic bombs. We did not know a process for obtaining of 235-uranium with the resources available under wartime conditions in Germany, in quantities worth mentioning. Even the production of nuclear explosives from reactors obviously could only be achieved by running huge reactors for years on end. Thus we were quite clear on the fact that the production of atomic bombs would be possible only with enormous technical resources…This situation seemed to us to be an especially favorable precondition as it enabled the physicists to influence further developments. For, had the production of atomic bombs been impossible, the problem would not have arisen at all; but had it been easy, then the physicists definitely could not have prevented their production. The actual givens of the situation, however, gave the physicists at that moment in time a decisive amount of influence over the subsequent events, since they had good arguments for their administrations – atomic bombs probably would not come into play in the course of the war, or else that using every conceivable effort it might yet be possible to bring them into play…In this situation we believed that a talk with Bohr might be of value…Because I knew that Bohr was under surveillance by German political operatives, and that statements Bohr made about me would most likely be reported back to Germany, I tried to keep the conversation at a level of allusions that would not immediately endanger my life. The conversation probably started by me asking somewhat casually whether it were justifiable that physicists were devoting themselves to the Uranium problem right now during times of war, when one had to

---

[56] Werner-heisenberg.unh.edu/Jungk.htm

at least consider the possibility that progress in this field might lead to very grave consequences for war technology. Bohr immediately grasped the meaning of this question as I gathered from his somewhat startled reaction. He answered, as far as I can remember, with a counter-question "Do you really believe one can utilize Uranium fission for the construction of weapons?" I may have replied "I know that this is possible in principle, but a terrific effort might be necessary, which one can hope, will not be realized anymore in this war." Bohr was apparently so shocked by this answer that he assumed I was trying to tell him Germany had made great progress towards manufacturing atomic weapons. In my subsequent attempt to correct this false impression I must not have wholly succeeded in winning Bohr's trust...

I then asked Bohr once more whether, in view of the obvious moral concerns, it might be possible to get all physicists to agree not to attempt work on atomic bombs, since they could only be produced with a huge technical effort anyhow. But Bohr thought that it would be hopeless to exert influence on the actions in the individual countries...

Aage Bohr and Werner Heisenberg obviously do not agree on what was discussed, and Niels Bohr doesn't say. Heisenberg was under pressure after WWII to explain his activities in the Uranverein during the war – did he intend to put nuclear weapons in the hands of the Nazis? This pressure may have influenced his response to Jungk.[57] Heisenberg always claimed to be a German national, but not a Nazi. His support of Germany's war aims soured the personal friendship with Niels Bohr. Despite the ambiguities surrounding the meeting, perhaps one can definitely say that Bohr knew something about the existence of the German Uranverein, if not any details about what specific progress had been made, as a result of Heisenberg's visit. One may perhaps also state that Heisenberg neither sought nor obtained any information from Bohr regarding the British-American efforts at that time.

There is one more curiosity regarding the 1941 Copenhagen meeting. Hans Bethe has said that Bohr took with him to Los Alamos in 1943 a sketch that Heisenberg had given him in 1941. It was supposed to show the reactor development work in progress in Germany at that time. It was

---

[57] A thoughtful analysis of this meeting is given by Mark Walker, "German National Socialism and the quest for Nuclear Power," Cambridge University Press, 1989, p 222ff.

a box with some rods sticking out of it.[58] According to Bethe, it was scrutinized diligently at Los Alamos. Aage Bohr has claimed that no such sketch ever existed.

In February, 1942, Albert Speer, Minister of Armaments and War Production for Nazi Germany, attended a meeting in which Heisenberg among others discussed the prospects for nuclear weapons in WWII. The result of this meeting was that the Uranverein was given enough money to stay afloat, but not to grow, so the German atomic project at that point was dead in the water.[59]

According to Pais, Bohr's first contact with the Allied atomic weapons program began early in 1943, when he received a visit from Captain Volmer Gyth of Danish intelligence.[60] What follows is about as cloak and dagger as it gets. Bohr was asked if he would receive a message from England that was on ultramicrofilm, hidden inside a key on a ring of other keys. Bohr agreed, but requested that Gyth copy the content of the microfilm on paper. The microdot was a letter from James Chadwick, inviting Bohr to visit England. Bohr declined, saying that it was better if he stayed in Denmark for now. Bohr's reply to Chadwick was duly reduced to the size of a pinhead, wrapped in foil, and inserted by a dentist in the hollow tooth of an agent! Later that same year Bohr changed his mind.

The German occupation of Denmark was relatively benign at first, but by 1943 Danish resistance began to increase, as the Germans threatened to deport local Jews. Bohr's mother was Jewish, so he and his family could be affected.[61] One evening Niels and Margrethe Bohr left their

---

[58] Thomas Powers "Heisenberg's War," *ibid.*, p 127 and note 31 to Chapter 12. Powers quotes Jeremy Bernstein, "Hans Bethe, Prophet of Energy," Basic Books, New York, 1980, p 77. Bernstein learned of the sketch from Bethe.

[59] Thomas Powers, "Heisenberg's War," *ibid.*, p 132.

[60] Abraham Pais, "Niels Bohr's Times," Oxford University Press, New York, 1991, p 485ff.

[61] Graham Farmelo, "Churchill's Bomb," *ibid.*, p 245. The account of Bohr's escape is adopted from this chapter.

Copenhagen home, went to the coast, and travelled that night by boat about 15 km to Limhamn, just south of Malmo in neutral Sweden. Copenhagen and Sweden are very close together, separated by only a small stretch of water. Bohr was no ordinary refugee, and was soon talking to the Swedish King and Foreign Minister about the threats to Danish Jews. The British learned of his presence in Sweden, and arranged to fly him to England in an unarmed military aircraft. Military aircraft were not outfitted with creature comforts in WWII. Bohr's head was too large for the oxygen mask to fit properly, so he passed out at high altitude as the plane flew to Scotland, where it landed on October 6, 1943. He was then flown to London, where he was met by James Chadwick and his wife.

Bohr remained in London for two months. He was introduced to the Tube Alloys project (the code name for British atomic weapons research), and given an office in Tube Alloys headquarters and housing in a nearby apartment. His son Aage, who had also escaped Denmark via Sweden, joined Bohr in London. Bohr became interested in the post war implications of atomic weapons, and became convinced that world peace could be secured, and an arms race avoided, if world powers, including the Soviet Union, shared the technical details of weapons construction. The Quebec agreement between Roosevelt and Churchill had motivated closer ties between the Manhattan Project and Tube Alloys, and Groves soon learned of Bohr's presence in England. Groves proposed that Bohr visit Los Alamos, and on November 29, 1943, Bohr sailed for New York. Niels and Aage Bohr were given code names Nicholas and James Baker, which probably fooled no one.[62] It was proposed that Bohr join the Institute for Advanced Study in Princeton, where Einstein was, but that did not work out. Soon Bohr was on a train heading West. In Chicago Groves and Richard Tolman joined him for the rest of the ride to New Mexico. Bohr arrived at Dorothy McKibbin's Santa Fe office to check in on December 31, 1943, and was then driven up the canyon road to Los Alamos. Oppenheimer brought Bohr up to speed on the Manhattan Project. Bohr was overwhelmed by the vastness of the undertaking. Although he did not

---

[62] This charade was applied to all prominent scientists in the Manhattan Project. Enrico Fermi was Eugene Farmer.

personally tackle any of the problems associated with the project, his presence at Los Alamos had a very positive effect on the morale of the laboratory. He discussed his ideas about a post war arms race and world politics in the nuclear age with Oppenheimer.[63] Bohr gave at least one of Oppenheimer's site wide colloquia. He was famous for many attributes, one of which was a mumbling speech impediment that made him impossible to understand.[64] The audience was on the edge of its collective seat in an attempt to hear what the great man had to say. Later on, in the summer of 1944, Bohr had the opportunity to present his ideas on politics in the nuclear age to Roosevelt, and then back in England to Churchill. Roosevelt was friendly and noncommittal, but Churchill was vehemently opposed to the sharing of any information with the Soviet Union. Churchill and Roosevelt met in Quebec in September, 1944, and afterwards decided that the Manhattan Project should remain top secret, with no information given to the Soviet Union, despite Bohr's entreaties. Churchill considered Bohr potentially dangerous.[65] Bohr traveled in and out of Los Alamos for a year, contributing to the work in small ways. At the end of the war in Europe in May, 1945, Bohr was in England. He returned to Copenhagen in August, 1945.

6.10. The Fermis at Los Alamos

Enrico Fermi was constantly in motion, and did not land at Los Alamos, and direct his own Division, until 1944. Laura Fermi moved from Chicago to Los Alamos in the summer of 1944, to be joined there soon by Enrico.[66] The Fermi's were assigned apartment D in building T-186. They were upstairs, with the Peierls' downstairs. Genia (Evgenia) Peierls, Rudolph Peierls' wife, was, as we learned in Chapter 2, a Russian mathematician. They met at a conference in Odessa in 1930. Peierls became a British citizen in 1940. In Chapter 3 we discussed the Frisch-Peierls memorandum, which became the MAUD Report, of great

---

[63] Graham Farmelo, "Churchill's Bomb," *ibid.*, p 259.
[64] H.T. Richards, "Through Los Alamos 1945," *ibid.*, p 61.
[65] Graham Farmelo, "Churchill's Bomb," *ibid.*, p 272.
[66] Laura Fermi, "Atoms in the Family," *ibid.*, p 200.

historical significance in the early development of nuclear energy. The Peierls and the Fermis each had two children. Like the Chadwicks, the Peierls children had been living in Canada, and were now reunited with their parents. Laura and Genia had met in Rome years ago. On one of their first afternoons in Los Alamos, Genia Peierls proposed that they all go to a picnic in Frijoles Canyon:

"You must take car. We'll be large group. You'll always be in large groups here."[67]

Laura Fermi was writing in English, while her native language was Italian. She said that all of the foreign born Los Alamos crew enjoyed noting the grammar and pronunciation mistakes in the English of others, but not of themselves. Italian has articles, but Russian does not. Laura's ear was sufficiently sharp that she realized that Genia did not use articles. Fluent English can be spoken without any articles at all. You really don't miss them, unless you listen closely. On the other hand, written English without articles is atrocious.

Los Alamos wives could have household help only if they, the wives, were employed on site. Laura Fermi had never been without household help. Her father was an admiral in the Italian navy. So she took a part time job in the local hospital in order to qualify. Enrico had a bodyguard, John Baudino, assigned by General Groves. Baudino appeared while Fermi was working in Chicago, followed him to Hanford, and then to Los Alamos. Baudino was trained as a lawyer. He was with Fermi everywhere he went for the duration of the Manhattan Project. Leona Libby and Laura Fermi both have several Baudino stories.[68]

---

[67] Laura Fermi, "Atoms in the Family," *ibid.*, p 209.

[68] Leona Marshall Libby, "The Uranium People," *ibid.*, p 183 (Baudino at Hanford). Laura Fermi, "Atoms in the Family," *ibid.*, p 212 (Baudino at Los Alamos).

6.11. Ordnance Division

It is interesting that the Smyth Report for security reasons does not discuss the work of the Los Alamos Ordnance Division at all.[69] Ordnance worked on two problems, both pertaining to the assembly of a super critical mass. One was the design of the gun, projectile, and target for the assembly of the Hiroshima ^{235}U bomb (Little Boy). The other was the design of a spherically converging shock wave, created by simultaneous firing of an array of explosives, in order to compress a sub critical sphere of ^{239}Pu to trigger a nuclear chain reaction. The gun design had several problems of muzzle velocity, weight, and overall length, in order to fit in the bomb bay of a B-29 bomber, the largest carrier in the US Air Force. Since the gun was closed at the far end, the end plug had to be sufficiently robust to hold together long enough for the nuclear reaction to proceed. The spherically converging blast had to solve the problem that from a spherical distribution of explosives the blast wave naturally spreads out from the source, rather than being focused into a coherent converging spherical wave. Converging lenses had to be developed for the blast wave. Several people at Los Alamos contributed to solving this problem. Seth Neddermeyer, John von Neumann, and Robert Christy all worked on the problem, and James Tuck from the British group developed the explosive 'lenses.'[70] The implosion scheme had to be tested with a real bomb. A test was feasible because of the relatively low critical mass for plutonium (about 10 kg, but less before compression[71]), and the efficiency of plutonium production at the Hanford reactors. There was never enough ^{235}U to allow an analogous test of the gun assembly scheme. People were confident, however, that the ^{235}U bomb would work, and they were right.

We have already noted the absence of a computing division, and the necessity of doing dangerous experiments instead of using computer simulations for much of the weapons design work. These experiments were carried out at Los Alamos in a remote location called Omega

---

[69] H.D. Smyth, "Atomic Energy for Military Purposes," *ibid.*, p 222.
[70] Robert Serber, "The Los Alamos Primer," *ibid.*, pp 59-60.
[71] The critical mass depends inversely on the square of the density.

Canyon.[72] A small, high flux reactor called the 'Water Boiler' was built as an intense neutron source. It was Fermi's third nuclear reactor, after the Chicago pile and the graphite reactor at Oak Ridge. The first water boiler, called LOPO for low power, had no cooling and no shielding, like the first Chicago pile, but it operated with an aqueous solution of uranyl sulfate enriched to 14% ^{235}U, supplied by Oak Ridge. It was very compact – the active volume was a sphere only 30 cm in diameter - in contrast with the behemoth reactors using graphite moderators. The water of the aqueous solution served as the moderator. It was the intense neutron source needed for Los Alamos experiments. It was assembled by Don Kerst, who was a colleague of the author's for many years after the war at the University of Wisconsin.

Otto Frisch proposed a bomb test which became known as the 'dragon experiment.' A plug of enriched uranium hydride was at the top of a guillotine arrangement, and a ring of the same material was at the base. The plug was dropped through the ring, forming a supercritical assembly for a few milliseconds. The neutron production was measured. To quote the Los Alamos History from Rhodes:

These experiments gave direct evidence of an explosive chain reaction. They gave an energy production [sic] of up to 20 megawatts, with a temperature rise in the hydride of 2 K per millisec. The strongest burst produced 10^{15} neutrons.[73]

Another experiment assembled metal spheres of ^{235}U from small, accurately machined metal cubes. To quote Libby:

"These cubes were put together by hand. Fred deHoffmann and Harry Daghlian agreed with each other never to work alone at night, and always to add the last uranium piece from the bottom, never from the top. Despite this agreement, Harry had an accident working alone at night. Adding the last uranium piece from the top, he formed a low grade critical assembly that irradiated him lethally before its internal heat caused it to disassemble and stop the chain reaction."[74]

---

[72] Leona Marshall Libby, "The Uranium People," *ibid.*, p 202.
[73] Richard Rhodes, "Making of the Atomic Bomb," *ibid.*, p 612.
[74] Leona Marshall Libby, "The Uranium People," *ibid.*, p 202.

This was the evening of August 21, 1945, after the Hiroshima and Nagasaki bombs. Laura Fermi did office work in the local hospital, run by a Dr. Hempelmann. Dr. Hempelmann treated Daghlian for radiation exposure, and Laura Fermi observed.[75] Distinct from the victims of the atomic bombs, Daghlian's damage was purely from radiation, rather than a combination of heat, radiation, and shock waves. By measuring the radioactivity in his blood, it was possible to calculate that his right hand had received over two hundred thousand times the allowed average daily dose. He died on the twenty-fourth day after the accident.

A similar accident occurred in May, 1946, when Louis Slotin was making bomb assemblies by pushing parts together slowly with only a screwdriver in between to prevent super criticality.[76] The screwdriver slipped, and five people, including Slotin, were exposed. Slotin was the closest to the assembly, and died a few days later. The others survived. Slotin's experiment, with the spherical assembly and the hemispherical tamper, is shown in Figure 6.2.

6.12. The Trinity Test

The Trinity Test of the implosion type plutonium bomb occurred before these accidents. Ken Bainbridge assumed responsibility for the test at the request of Oppenheimer in March, 1944. First they had to pick a site for the test that was sufficiently remote that it could safely accommodate a nuclear explosion. They wanted to be conservative, since the force of the explosion was not predictable with confidence. Fortunately, there was an air force gunnery range, 48×96 km in area, already owned by the US government, large enough to contain the blast, and evacuated. Groves staked a claim to a 29×38 km parcel, near the northern boundary of the range.[77]

---

[75] Laura Fermi, "Atoms in the Family," *ibid.*, p 233.
[76] Leona Marshall Libby, "The Uranium People," *ibid.*, p 204.
[77] Richard Rhodes, "Making of the Atomic Bomb," *ibid.*, p 652.

The range was near Alamogordo, New Mexico, in the desert about 320 km south of Los Alamos – a five hour drive by car in 1945.[78] The area is called the 'White Sands Missile Range' today, and is still closed to the public. The Trinity Test site, where the weapon was detonated, is 80 km north of Alamogordo, 16 km below the northern site boundary. The Trinity Test site is open to the public one day a year – the first Saturday in April. The McDonald ranch house, where parts of the bomb were assembled, was 3 km away from ground zero. All of the working ranches on the site were abandoned in January, 1942, at the request of the War Department.[79] Figure 6.3 shows the terrain.[80] US highway 380 is the northern boundary of the white sands test area. In the summer of 1945 the New Mexico desert was hot and dry, and by definition an air force gunnery range is the middle

Figure 6.2. Plutonium critical assembly with tamper. Courtesy Los Alamos National Laboratory.

---

[78] Leona Marshall Libby, "The Uranium People," *ibid.*, p 221.

[79] www.abomb1.org/trinity/trinity1.html

[80] Osti.gov/manhattan-project-history/images/TrinitySiteMapCircleLarge.jpg

of nowhere, so it is easy to believe that setting up the first test of an atomic bomb was a difficult task. There was no hardware store. Everything had to be brought in by truck from Los Alamos.

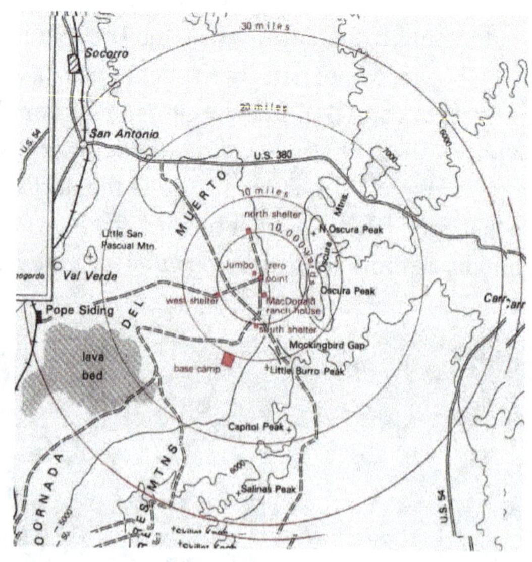

Figure 6.3. Trinity test site in New Mexico. Alamogordo is off the map lower right. Credit US Department of Energy.

Another former student of Oppenheimer at Berkeley, Philip Morrison, was working at Los Alamos. He got the job of accompanying the two nickel plated plutonium hemispheres from Los Alamos to the Trinity test site. He rode in the back seat of an army sedan, next to a box containing the plutonium, for the five hour journey. There was an armed escort convoy for security.[81] General Thomas Farrell, Groves' deputy, received the delivery at the McDonald ranch house. When asked to sign for the parcel, Farrell put on rubber gloves, and picked up one of the plutonium

---

[81] Stephen Walker, "Shock Wave," Harper Collins, New York, 2005. p 25.

hemispheres, which he noted was heavy for its size (2 or 3 kg), and warm to the touch because of its radioactivity.[82]

A base camp of tents and barracks, located 16 km south of ground zero, was constructed by the army engineers, and was temporary housing for the bomb assembly crew of scientists and soldiers. [83] Three shelters were built 9 km away from zero, labeled north, west, and south. The shelters were bunkers covered by concrete roofs and heavy wooden beams, with slits oriented towards zero. North and West contained various measuring equipment, while South was the command post. All of the control and monitoring cables from ground zero and the intermediate stations were connected to the command post.

Richards describes his role in the test. He was in charge of group TR-3B, assigned the task of measuring the neutron flux from the explosion. They wanted to learn the neutron flux as a function of distance from ground zero, and also as a function of time at particular points away from ground zero. They set up data recorders in three locations, one on a balloon 300 meters above ground zero, a second on the ground at 300 meters, and a third at 600 meters on the ground. Only the 600 meter instrument survived the blast.[84] The instruments were called 'cellophane catcher cameras.' An electric motor pulled a cellophane tape between two plates coated with ^{235}U. The neutrons from the blast caused the uranium to fission, and some of the fission products would stick to the tape. The motor was started just before the blast. Since the fission products are radioactive, measuring the activity as a function of position on the tape gave a relative time dependent flux. Absolute calibration was achieved with known neutron sources in the lab at Los Alamos. The camera at 600 meters gave useful data regarding the neutron flux.

---

[82] Stephen Walker, "Shock Wave," *ibid.*, p 26. Leona Libby says that plutonium feels warm, like a live rabbit. ("The Uranium People," *ibid.*, p 171).

[83] Richard Rhodes, "Making of the Atomic Bomb," *ibid.*, p 653.

[84] H.T. Richards, "Through Los Alamos 1945," *ibid.*, p 74.

Another scheme for measuring the efficiency of the bomb was to collect radioactive fission fragments from the area near ground zero shortly after the blast. Two army tanks were obtained, lined with lead, and outfitted to serve as scavengers to scrape up any radioactivity – like a Mars lander, only the tanks were to be manned by Fermi and Herb Anderson.[85]

Ken Bainbridge wanted a conventional explosion before the Trinity test in order to learn more about blast effects. Accordingly, a 100 ton TNT calibration shot was detonated from a 6 meter tower on May 7, 1945.[86] The Trinity Test itself was scheduled for July 16, 1945.

Iron workers erected a 30 meter high tower – an eight story building - to support the bomb.[87] The test was to be a ground blast, in contrast with the air blasts which were soon to follow over Hiroshima and Nagasaki. The tower had an electric winch installed at the top to hoist components up to a platform where the bomb was to be readied for detonation. There was a roof and three sides of corrugated iron surrounding the platform. The open side faced west, towards the west shelter (see Fig. 6.3).[88] There was no elevator for people, only an iron ladder of welded cross bars.[89]

Cars and trucks went back and forth every day from Los Alamos, bringing people and supplies. Uncensored mail without a 3 cent stamp was hand carried from workers at the Trinity site back to families who stayed at Los Alamos.[90] Communication by telephone was possible, but only just

---

[85] Richard Rhodes, "Making of the Atomic Bomb," *ibid.*, p 655. A photo of one of the tanks is shown in a very nice online article about the Trinity Test on the Los Alamos Historical Document Retrieval and Assessment web site: www.lahdra.org.

[86] The Halifax explosion in 1918 was 200 tons of TNT.

[87] Taller than a typical oil drilling rig, but shorter than the Eiffel Tower (300 m). Said to be a surplus fire ranger tower from the Forest Service.

[88] Richard Rhodes, "Making of the Atomic Bomb," *ibid.*, p 654.

[89] Leona Marshall Libby, "The Uranium People," *ibid.*, p 223.

[90] H.T. Richards, "Through Los Alamos 1945," *ibid.*, p 78. Richards publishes letters to his wife.

barely. For security the connection from Trinity to Los Alamos went through a special operator in Denver.[91]

Parts were trucked from Los Alamos. The Po-Be initiator rode in the same car as the hemispheres of plutonium, as described earlier. The uranium tamper shell, and the sphere of conventional explosives for compression were shipped separately. In order to minimize traffic, the conventional explosive rig was trucked at night.[92] The assembly, about 1.5 meters in diameter and weighing over three metric tons, sat in a tent at the base of the tower. One of the explosive lenses was missing, leaving an access hole for insertion of the active material. There was some panic when the insertion did not fit, which turned out to be due to thermal expansion, overcome by equilibrium temperature. Then the sphere was raised onto the platform floor at the top of the tower, through a trap door. The door was closed, and the bomb placed in the center of the platform. There the electrical connections were made to the detonators to fire the implosion lenses.

It was the middle of July in the New Mexico desert, but it started to rain. In fact, the rains started a couple of weeks before the test, causing floods and muddy roads.[93] There was concern that the bad weather would delay the test, as it did, but only for an hour and a half.

Leona Libby describes one of the many measurements made on the bomb on top of the tower.

"Next the little group of scientists had to monitor the neutron background coming from the Po-Be interface, and this they did with a manganese wire that they periodically inserted into a long tube in the bomb assembly and pulled out and measured its radioactivity induced by neutron absorption. Because the lifetime of radioactive manganese is about 2.7 hours, they measured the radioactivity every 4 hours until the shot was fired."[94]

---

91 Leona Marshall Libby, "The Uranium People," *ibid.*, p 221.
92 Richard Rhodes, "Making of the Atomic Bomb," *ibid.*, p 659.
93 H.T. Richards, "Through Los Alamos, 1945," *ibid.*, p 78.
94 Leona Marshall Libby, "The Uranium People," *ibid.*, p 223.

After hoisting the rig to the top, these people had to climb the ladder every 4 hours to perform the check. The neutron source remained constant, as it should, unless an accident caused the Po and Be to mix prematurely.

A second sphere of implosion lenses was assembled at Los Alamos for a test. The results of the test were not encouraging, and Oppenheimer was informed the day before Trinity that the bomb would likely fail. The weather continued to be uncooperative. Keeping steady nerves was challenging.

The importance of the weather illustrates a feature of nuclear explosions. The destructive effects of the bomb depend to a great extent on the atmosphere. A nuclear explosion in outer space, where there is no atmosphere, would be quite different, and less effective. The atmosphere ignites to form the fireball, dissipating the heat generated by the chain reaction. The pressure from the explosion creates shock waves, which destroy buildings. The fireball is less dense than the surrounding air. As the lighter fireball rises, like a hot air balloon, material from the blast is drawn into a column, forming a chimney, and the fireball becomes a toroidal shape, resulting in a mushroom cloud. The cloud ceases to rise when the density balances with the surrounding atmosphere. All of these phenomena depend on the atmosphere, and would presumably be about the same for a 'conventional' explosion of similar energy, if it were possible to achieve comparable energy density. A unique property of a nuclear device is the radioactive debris in the cloud, which may drift over long distances, and fall to earth, creating 'fallout.' Rain may cause the fallout to affect a limited area, creating a radiation hazard to living things.

These effects depend on the energy release of the bomb, which for the Trinity test was not really known, but many of the consequences were anticipated. Evacuation of populated areas which could be threatened by fallout was planned.[95] Winds aloft and rain were critical parameters in determining the risk. For these reasons there was close attention paid to the weather. The forecasters at the Trinity test site were the best in the

---

[95] Leslie M. Groves, "Now It Can Be Told," *ibid.*, p 298.

business that the US Army had to offer. Chief among them was Jack Hubbard, who was on location at the test site on the evening before the planned detonation. One of the reasons for requiring clear weather was the flight of observation planes to view the blast from the air. Louis Alvarez was to be on board one of the planes. The weather remained disagreeable, but Alvarez' plane was airborn when the detonation occurred. His written description of the event from the air is available on line.[96]

6.13. Truman Goes to the Potsdam Conference

President Truman arrived in Antwerp, Belgium on the heavy cruiser USS Augusta on July 15, 1945, having departed from Hampton Roads, VA on July 7. He then flew to Germany for the Potsdam Conference with Winston Churchill and Joseph Stalin, which was to start on July 17.[97] He was accompanied by his Secretary of State, James Byrnes, and Secretary of War, Henry Stimson. Stimson was to receive a coded message from Groves regarding the outcome of the Trinity test. Stalin, Truman, and Churchill are shown at Potsdam in Figure 6.4.

---

[96] www.slate.com has Alvarez's notes at :history_of_the trinity_test_luis_alvarez.

[97] General elections in Britain ousted Churchill's government during the conference, and he was replaced as prime minister by Clement Attlee. Early Potsdam photos are Stalin, Truman, and Churchill (Figure 6.4). Later ones are Truman, Attlee, and Stalin.

Figure 6.4. Joseph Stalin, Harry Truman, and Winston Churchill at Potsdam, July, 1945. Credit US Army Signal Corps, Harry S. Truman Library and Museum.

On the morning of Monday, July 16, another heavy cruiser, the USS Indianapolis, loaded parts of the atomic bomb to be used on Hiroshima at the Hunters Point Naval Shipyard in San Francisco Bay.[98] This was the 'gun type' ^{235}U bomb called 'Little Boy', which would be deployed untested. The Indianapolis carried the gun with a closed muzzle end, and half of the weapons grade uranium. The other half would go by air. The Indianapolis would depart that day for Tinian, in the Mariana Islands of the Pacific Ocean. Many of the Los Alamos scientists at work on the Trinity test would soon fly to Tinian also.

6.14. Trinity Test Waits on the Weather

Oppenheimer was in the base camp, stewing over what to do. His nerves were ragged. Groves, Bush, and Conant arrived by car from

---

[98] Stephen Walker, "Shock Wave" Harper Collins, New York, 2005, p 55.

Albuquerque on July 15.[99] The three of them had been travelling by air to Hanford, Washington, then to San Francisco to visit Lawrence at Berkeley, then to Pasadena, and on to Albuquerque. Groves says that air travel in the US has improved a lot since 1945, and their landing in Pasadena was scary, but as they say, "Any landing you can walk away from is a good landing."

Groves tried to calm Oppenheimer. The two agreed to a procedure to set the time for the test. The original time set was 4 am Monday, July 16, 1945. They would meet at 1 AM to evaluate the weather situation and make a decision.

Earlier in the day Oppenheimer was worried about the safety of the bomb sitting 30 meters in the air in an electrical storm, with wind and rain. So he delegated Don Hornig,[100] a young chemist who had worked on the design of the implosion explosives, to baby sit the bomb until he was called back to base camp 30 minutes before the test (there was a telephone atop the steel tower, connected to the command post 9 km to the south. Jeeps were waiting at the base of the tower for evacuation.) There was a plan in case the jeep broke down – run. He sat there all by himself. He had a book to read, but the light was not very good, and besides, there was all that thunder and lightning, and the world's first nuclear bomb at his elbow. It is not clear exactly what he would have done if the lightening had caused a pre-detonation, or some other disaster had befallen the test. But as it turned out, he was spared the need for emergency response.

Groves, Oppenheimer, and the weatherman Jack Hubbard met at 1 am, as planned. Hubbard said that the weather was improving, and should be OK by sunrise. Groves and Oppenheimer delayed the test from 4 am to 5 :30 am July 16, 1945. Once the decision was made, the sequence of events was determined.

---

[99] Leslie M. Groves, "Now It Can Be Told," *ibid.*, p 290.

[100] Obituary of Donald Hornig, New York Times, January 27, 2013. After surviving his assignment on the Trinity tower, he served as Science Advisor to President Lyndon Johnson, and was president of Brown University.

Groves was reluctant to delay the test any longer. He was aware of the Potsdam Conference, and the value of sending news of a positive experiment through Henry Stimson to President Truman. Groves was also mindful of the costs of prolonging the war. Technically everything was ready to go, and a delay would inevitably cause problems in bringing all of the components up to speed again. The crew was also exhausted, and would require some recovery time before starting the readiness process all over again.

6.15. Trinity Test Proceeds Successfully

With 30 minutes to go, Hornig was called down from the top of the tower, and driven to the command post. Everyone at base camp and the command post were told to lie down facing away from ground zero. Oppenheimer was at the command post in the dugout. Fermi, Groves, Bush and Conant were all at base camp, lying on the floor with one minute to go.[101] Feynman was with another Los Alamos group in a bus about 32 km northwest of ground zero, at a place called Compania Hill. Several physicists without duties at the test site were here, including Teller, Serber, Bethe, McMillan, and Chadwick.[102] William Laurence, a New York Times reporter on loan to the Manhattan Project, was also in the bus. They had a radio for contact with the command post, by which they were supposed to be informed of the timing, but the radio did not work. Feynman was trying to fix it by flashlight.[103] Feynman got it working just a few minutes before the blast, and they were told that there was 20 seconds to go. Feynman got into the bus, and sat behind the windshield, using the glass as an ultraviolet filter. Everyone had been issued welder's goggles, but Feynman decided that he could look directly through the windshield. They were, after all, 32 km away.[104]

---

[101] Leslie Groves, "Now It Can Be Told," *ibid.*, p 296.

[102] Richard Rhodes, "Making of the Atomic Bomb," *ibid.*, p 668.

[103] Stephen Walker, "Shock Wave," Harper Collins, New York, 2005, p 54. See Also R.P. Feynman, "Los Alamos from Below," *ibid.*, p 29.

[104] R.P. Feynman, "Los Alamos from Below," p 29.

Figure 6.5 shows the trinity test bomb assembly with its lifting rig and Norris Bradbury.

Rhodes gives a description of the sequence of events after the trigger signal was sent to the bomb: "The X-unit discharged; the detonators at 32 detonation points simultaneously fired; they ignited the outer lens shells of Composition B; the detonation waves separately bulged, encountered inclusions of Baratol, slowed, curved, turned inside out, merged to a common inward-driving sphere; the spherical detonation wave crossed into the second shell of solid fast Composition B and accelerated; hit the wall of dense uranium tamper and became a shock wave and squeezed, liquefying, moving through; hit the nickel plating of the plutonium core and squeezed, the small sphere shrinking, collapsing into itself, becoming an eyeball; the shock wave reaching the tiny initiator at the center and swirling through its designed irregularities to mix its Beryllium and polonium...to start the chain reaction."[105]

---

[105] Richard Rhodes, "Making of the Atomic Bomb," *ibid.*, p 670.

Figure 6.5. Trinity test implosion bomb with Norris Bradbury, who succeeded Robert Oppenheimer as director of Los Alamos, for scale. Credit US Department of Energy.

Figure 6.6 shows the fireball photographed by remote camera after 0.006 seconds.

Figure 6.6. Trinity test explosion fireball after 0.006 sec. Courtesy US Department of Energy.

By 16 millisec the fireball had expanded into a sphere about 250 m in diameter, for an average expansion rate of 8 km/sec. This is much less than the speed of light (300,000 km/sec), but faster than the speed of sound in air at STP (0.35 km/sec). The shock wave reached the base camp 16 km away about 50 sec later than the light signal, consistent with the speed of sound.

The initial flash of light was felt by everyone, and has been described in personal memoirs many times. That an object about the size of a grapefruit could release so much energy in such a short time was difficult to grasp. The magnitude of the achievement was overwhelming. There was much debate before the test regarding the energy released by the blast. Herb Anderson's scheme of collecting samples of debris and analyzing them for fission fragments was implemented.

Perhaps the most bizarre technique for estimating the energy of the blast was carried out by Fermi. The story has been told by so many people, including Laura Fermi, that it probably really happened. Leona Libby is the best source to quote, as she often is:

> "Fermi made the first measurement of the strength of the blast. From the equations for conservation of energy and momentum of a shock wave, he computed how far the arriving shock front would blow scraps of paper at his observation point 9 km away for various bomb yields. Then after the fireball formed and before the shock front arrived, he stood up and began dribbling paper scraps from his hand. When the front arrived, it blew them a short distance, which he measured using his shoes, which he knew were 22.5 cm long, looked in his previously prepared table, and announced that the yield was equivalent to 20 kilotons of TNT."[106]

After the surrender of Germany in May, 1945, the goal of the Manhattan Project became the end of the war with Japan. The Soviet Union was a US ally against Germany, and at Potsdam would agree to commence hostilities against Japan. Soviet espionage kept the USSR informed on the Manhattan Project, and Stalin at Potsdam was fully aware

---

[106] Leona Marshall Libby, "The Uranium People," *ibid.*, p 229. Fermi was famous for his 'back of the envelope' estimates of complex phenomena. He was always right.

of the success of the Trinity Test. Some historians date the beginning of the cold war on July 16, 1945.

Chapter 7

The Soviet Union Creates
Laboratory #2

7.1. The Kurchatov Institute Today

The Kurchatov Institute is located south of Kurchatov Square, about 10 km northwest of the Kremlin in modern Moscow. The Institute is a National Research Center, occupying an area of more than 1 km^2 in city real estate. It is a big, busy place, with fences and guarded gates and internal streets and many buildings. It is a multipurpose laboratory, composed of research reactors, tokamaks for plasma physics, a 'hot lab' for handling radioactive materials, a synchrotron light source for material science, a biotechnology complex, nuclear medicine, etc. The tokamak was invented here, and is still one of the leading contenders for electromagnetic containment of plasmas to generate fusion power. The first nuclear reactor in Europe, which went critical in 1946, is still on site in its original building, and is still occasionally operated. A two-story house, built on the laboratory grounds by a grateful Soviet government for the founder of the Soviet Atomic Project, Igor Kurchatov, still stands, and is used as a museum. There is also a historical display museum in one of the laboratory buildings, and there is an archive of historical documents pertaining to the Soviet Atomic Project in the laboratory directorate building.

The Kurchatov Institute was established in 1943 as Laboratory #2 of the Soviet Academy of Sciences, and celebrated its 70th anniversary in 2013. Before it was named after its founder, the Laboratory acquired in 1949 the title 'Laboratory for Instrumentation of the Academy of

Sciences,' with the Russian acronym LIPAN. Laboratory #2, LIPAN, and the Kurchatov Institute are all the same place.

7.2. Beginnings of the Soviet Atomic Project

Even before the Order #2352ss of the State Defense Committee on September 28, 1942, discussed in Chapter 5, some wheels had started to turn. Georgy Flerov was called to Moscow from the southwest front in July, 1942, just before the German attack on Stalingrad.[1] He was met by S.A. Balezin, a senior aid to S.V. Kaftanov, the administrator for science in the Ministry of Defense. Balezin brought up Flerov's letters to Stalin regarding the uranium problem – the one dated April, 1942 being the most recent. Balezin had heard one of Flerov's lectures before the war on uranium fission, given at Moscow State University. Balezin asked Flerov to draw up a plan for renewal of the uranium project. Flerov responded that the first thing to do was get K.A. Petrzhak released from the Red Army.

By the fall of 1942 scientists in the evacuated laboratories in Kazan, Ufa, and Alma-Ata, as well as those remaining in Moscow and Leningrad, renewed interest in chain reactions, production of heavy water, and separation of uranium isotopes.[2] Order #2352 specified Abram Ioffe as the member of the Academy of Sciences responsible to the government for progress on the uranium problem. Ioffe in turn chose Igor Kurchatov as the person in charge. By the end of 1942 the State Defense Committee found the strength to mount the counter attack at Stalingrad, and at the same time to begin work on the atomic bomb. Kurchatov was authorized to get things going in Moscow. Physicists were excited to learn the news that Kurchatov was establishing a uranium institute.[3]

---

[1] HISAP96, Vol 2, *ibid.*, p 155.
[2] I.N. Golovin, "I.V. Kurchatov," *ibid.*, p 50.
[3] I.N. Golovin, "I.V. Kurchatov," *ibid.*, p 51.

7.3. State Defense Committee Endorses Atomic Energy

On February 11, 1943, the State Defense Committee of the USSR published Special Resolution # 2872, signed by Stalin, to organize the scientific-technical work for the utilization of atomic energy with the goal of building an atomic bomb. I.V. Kurchatov was named scientific director of the uranium project. The people's commissar for chemical industries, M.G. Pervukhin, was named the deputy for this project to V.M. Molotov, member of the Council of People's Commissars. Pervukhin is shown in Figure 7.1. He was trained as electrical engineer. Molotov was assigned overall responsibility.[4] The Soviet Academy of Sciences in response to the resolution of the State Defense Committee in March, 1943, created Laboratory #2, and named Kurchatov the director.

Figure 7.1 Mikhail Pervukhin, a scholarly looking man, was Minister of Chemical Industries. Credit en.wikipedia.org.

---

[4] HISAP96, Vol **2**, p 99, article by N.V. Knyazkaya.

Pervukhin was the responsible manager, handling all of the daily requests, reports, and assessments of the project as it began to gain momentum. He controlled the movement of scientific and technical personnel from one place to another, and was responsible for obtaining discharge from the Red Army of many people for work on the uranium problem. He also served as intermediary between Kurchatov and the other scientists, and members of Stalin's Central Committee, Beria, Molotov, Mikoyan, et al. From 1943 until the project went on the national fast lane in August, 1945, Pervukhin was the Soviet Atomic Project's General Groves.

Kurchatov left his defense work in Kazan and returned to Moscow in February, 1943. He began at once to assemble the scientific work force necessary to pursue the uranium problem with vigor. Among those in the first group were G.N. Flerov, I.K. Kikoin, Ya. B. Zeldovich, and A.I Alikhanov, all of whom shared Kurchatov's view that the work should proceed without delay. But they had to agree on scientific and industrial courses of action. They decided to build a thermal neutron reactor, and at the same time to study methods of isotope separation. Isotope separation by thermal diffusion, and by gaseous diffusion through a porous barrier were under consideration. Both of these methods are discussed in Appendix C. They were known in the open literature before the war, and were used in the Manhattan Project. Gaseous diffusion was favored as a more practical approach. Kurchatov forged ahead, urging calculations regarding the possibility of a fast neutron chain reaction in ^{235}U, even though he had no sample of the separated light isotope. Interest in the artificial element 94, its production in a power reactor, and its use as fissionable material in a bomb, was growing in the theory group, but had not yet impacted the experimental program.

Kurchatov was given broad powers to collect the necessary personnel for the project. He could obtain the release of individuals by the State Defense Committee who were in the Red Army, or were working in defense factories. I. Ya. Pomeranchuk, Yu. B. Khariton, and Igor's brother, the chemist Boris V. Kurchatov joined the group. He soon had about 20 scientists. They came to Moscow with only the clothes on their backs and maybe a small suitcase. They had no books, scientific papers, or instruments, all of which had been lost either during the evacuation to the east in 1941 or subsequent destruction by bombardment.[5] Kurchatov's first duty was to feed and house the newcomers, which he did. His concern for their wellbeing had a very positive effect on people experiencing the heavy burdens of wartime existence.

They took advantage of the fact that many institutions had left Moscow for the east in order to house the new Laboratory#2. The presidium of the Academy of Sciences gave it temporary space in a building previously occupied by the Institute of Seismology, on Pyzhevsky Pereulok, just across the Moscow River from the Kremlin, not far from the Tretyakov Art Gallery. Kurchatov also obtained space in the Institute for Inorganic Chemistry, located on Bol'shaya Kaluzhskaya ulitsa (now Leninsky Prospekt). Here Flerov began experiments on the resonance capture of neutrons during the moderation process. It was at Flerov's lab that armed guards first appeared at the doors as a sign of national security and secrecy.

Kurchatov was a quiet, but thorough manager. He paid close attention to the researchers and their work, verifying each step carefully, as he had done at LFTI before the war. As the workers made steady progress, the Communist Party Central Committee gave Kurchatov broad powers to begin an attack on the uranium problem on a wide front. This led to a reassessment of what had to be done, and to plans for expansion away from table top experiments and into industrial scale operations. Like the Manhattan Project, the Soviet Project should pursue redundancy, because it was not obvious which techniques would work out. Isotope separation was an obvious candidate for a multi-task approach, but questions of chain reactions with fast or slow neutrons, how to slow down the MeV neutrons from fission, how to design a working slow neutron reactor to generate power, and how to use the fast neutrons to create an explosion were all on the table.

---

[5]   I.N. Golovin, "I.V. Kurchatov," *ibid.*, p 52.

Kurchatov's space needs were not to be satisfied by empty buildings in the capital. He wanted room to expand, to do large scale experiments, and to be safely outside the central city. He needed new space, constructed according to his specifications. He obtained support from the State Defense Committee in the summer of 1943, and started looking around for a new construction site in the Moscow region.

7.4.   Laboratory #2 is Sited

Kurchatov selected a plot of land 10 km northwest of the Kremlin (sound familiar?), out in the country. It was near a railroad that at that time circled Moscow, and not far from the Moscow River, which bends around the city to the west. There was a three-story brick house, and within a few hundred meters two more dilapidated one story houses with two storage cribs without roofs. There was an unused potato field, and a rifle range for target practice. A small factory for X-ray equipment was about 500 meters away.[6] In 1943 the laboratory site was out in the countryside. Modern Moscow is a lot bigger place.

Kurchatov thought it had possibilities. He planned to use the three-story house for small lab and office space, and expand outside as needed. The site was designated Laboratory #2 of the USSR Academy of Sciences. Around this time elections to membership in the Academy of Sciences for 1943 were in progress. There was one vacancy, and two candidates, Alikhanov and Kurchatov.[7] The Academy selected Alikhanov. Kaftanov appealed to Molotov to create a new slot, and assign it to Kurchatov. Molotov complied, and Kurchatov was elected to full membership in the Academy – bypassing corresponding membership - by direct intervention of the Soviet leadership, and despite some grumbling by the Academy that he had not done enough significant research. His membership was justified by his management skills in organizing such a huge undertaking as the development of nuclear energy. Obviously, being a member of the

---

6   I.N. Golovin, "I.V. Kurchatov," *ibid.*, p 54.
7   S.V. Kaftanov, "Po Trevoge," "Khimiia i Zhizn," #3, 1985, p 6ff.

Academy was a big political advantage, given the moving and shaking that he had to do.

As described in Chapter 1, Kharkov changed hands four times during the German invasion and retreat. It was liberated for the last time in August, 1943, after the battle of Kursk. Sinelnikov and coworkers from UFTI returned to the city. UFTI became Laboratory #1. On February 9, 1944, Pervukhin wrote a letter to the secretary of the Kharkov Communist Party requesting help for UFTI in restoring the laboratory's research capabilities, in particular the high voltage electrostatic accelerator.[8] There were now two centers in the USSR for nuclear research, aimed at weapons development. As events unfolded, Laboratory #2 became the more important facility.

Laboratory #2 needed a cyclotron for neutron bombardment. This was 1943, the middle of WWII, and everything was in short supply, especially copper which is used for machine gun bullets and artillery shells. As mentioned in Appendix C, the Manhattan Project faced similar problems, and Groves borrowed silver bars from the mint to make the coils for the calutrons because he could not get the copper. Kurchatov's cyclotron did not require as much copper as the Oak Ridge calutrons, but it was still hard to get. He remembered that before LFTI was evacuated from Leningrad in 1941, copper and brass parts for the accelerating chamber of the Leningrad cyclotron (the new one being built at LFTI, not the one at the Radium Institute) were buried on the LFTI grounds, in the Vyborg section of the city.

7.5. Nemenov Goes to Leningrad in 1943

In the fall of 1941 the German army surrounded the city of Leningrad. The city faces the Gulf of Finland to the west, and the Karelian Isthmus and Finland itself to the north, so the city could be cut off from the

---

[8] work.atomlandonmars.com/APSSSR/Atomnyy_proekt_SSSR_T.1.Ch.2.(2002).pdf report #212. Referenced as /APSSSR/ hereafter.

mainland Soviet Union by a semi-circle which stretched from the coast of the Gulf of Finland in the west to Schlisselburg at the southern tip of Lake Ladoga in the east. During the starvation winter of 1941-42 supplies were trucked into the city over ice roads across the frozen Lake Ladoga. Rail lines coming from the south, from Moscow, were completely useless, being in German hands. Rail lines from the east terminated near the ice roads. No railroad reached the city. In January, 1943, Operation Iskra (spark) breached this hermetic seal at Schlisselburg, and a small window by rail into the city opened. The city had been completely surrounded for 526 days.[9] It was to be another year before the city was accessible on all sides, when the siege was lifted on January 27, 1944. Until then Leningrad remained within German artillery range. Kurchatov sent Leonid Nemenov into this cauldron in March, 1943, to obtain parts of the Leningrad cyclotron.[10]

Nemenov tells a really good story.

"Moscow at the beginning of March, 1943: the workers at Kurchatov's laboratory could be counted on your fingers. The lab director conducted one of the very first scientific meetings in his modest office. Alikhanov, Kikion, and myself were present. Igor Vasilevich gave us the assignment to obtain the transuranic element 94, the sooner, the better. Obtain it even in trace amounts. The theorists have predicted that this element, not found in nature, will have important properties; namely that it will fission under neutron bombardment, just like ^{235}U, and it may be a second substance which could undergo an explosive fast chain reaction. Microscopic quantities of plutonium could be used for chemical purification experiments.

---

[9]   Harrison Salisbury, "The 900 Days," *ibid.*, p 535ff.

[10] Leonid Nemenov, "Tekhnika-Molodezhi," June, 1975, p 18. Nemenov's fascinating article in a popular science magazine is entitled "Poslezavtra Nachnem Obluchenia" (The day after tomorrow we will begin irradiation). Its publication date implies early declassification.

"Element 94 may be obtained by bombarding a lithium target with 4-5 MeV deuterons from a cyclotron. The reaction $^7Li(d,n)^8Be$ will give fast neutrons[11], which can irradiate a paraffin coated uranyl nitrate target to cause the capture reaction $^{238}U + n \rightarrow {}^{239}U \rightarrow {}^{239}Np + \beta + v$; $^{239}Np \rightarrow {}^{239}Pu + \beta + v$. The transitions U→Np→Pu are short half-life β decays.

"We have 16 months to design, build, and commission a cyclotron. That seems unbelievably short, but that is what the government wants. A current of deuterons with energy above 4 MeV can be obtained with a cyclotron of diameter 0.73 m, and a magnetic field strength of 1.4 T.[12] The wave length of the high frequency oscillator would be 28.3 m.[13] Before the war such a cyclotron was partly constructed at LFTI in Leningrad. Kurchatov wanted its high frequency oscillator, still in Leningrad, brought to Moscow. There was no other way; it would not be possible to obtain such a complicated device in Moscow at this time.

"They sent an engineer, P. Glazunov, and myself to Leningrad. Our authority was confirmed by two letters addressed to the secretary of the local communist party, A. Zhdanov, and the president of the executive committee of the city council, P. Popkov. These letters were signed by the deputy president of the Council of Peoples Commissars of the USSR, M. Pervukhin. The letters requested the full cooperation of the Leningrad city management in our endeavors. As he sent us on our way, Kurchatov took personal responsibility for placing orders in Moscow factories for the manufacture of the cyclotron electromagnet.[14]

"The trip to Leningrad was not easy. Although the blockade had been lifted in January, 1943 in one place (Schlisselburg – see above), the Fascists still surrounded the city. We flew as far as Khvoinoe,[15] and there we waited for darkness. After dark

---

[11] Specific nuclear reaction is my hypothesis. Nemenov does not give it. Eq. (D.14) in Appendix D gives the neutron KE at 90^0, ie radially out from the cyclotron.

[12] Eq. (C.12) of Appendix C comes in handy to check this. p (MeV/c) = 300 B (in Tesla)×R (in meters), gives p = 153 MeV/c, and $E = p^2/(2m) = 6.2$ MeV kinetic energy

[13] Eq. (D.4) of Appendix D gives this number, after some manipulation. $f = eB/(2\pi m) = 10.7$ Mhz, and the wave length $\lambda = c/f = 28$ meters, where $c = 3\times10^8$ m/sec is the speed of light. Coaxial line drivers would be $\lambda/4 = 7$ meters, twice as long as a machine designed for protons. See Figure D.9 in Appendix D.

[14] A window frame magnet design is described in Appendix D: 16 metric tons of iron and 1.0 tons of copper for the coils, with a cylindrical gap 80 cm dia and 20 cm high, 1.4 Tesla. Coil has 100 turns, and carries 2200 amperes. 100 kwatts-water cooled.

[15] I have been unable to locate this place, but it must be east of Leningrad. Khvoiny refers to a pine forest, which in Russia could be anywhere.

we flew in a light plane low over Lake Ladoga and into Okhtinsky airport.[16]

"We took with us more that 100 packages for relatives of our co-workers who lived in blockaded Leningrad. Each one was less than 1 kg, but they were priceless treasures for the recipients. There were two bags of packages. A problem arose at the first airport, where the dispatcher refused to load the bags. The pilot asked what the problem was, and looked in one of the bags. He saw neatly wrapped packages, and told the dispatcher that the plane would not fly without them.

"I do not remember the pilot's name, but I learned that he was killed a few days later, flying from this same airport.

"We landed safely, and hitched a ride to the physico-technical institute (LFTI). We met again our coworkers in Leningrad with unforgettable joy.

"The first day passed in conversation, delivering the packages and letters. The next day we went to the Smolny Institute and delivered our authorization letters from Pervukhin. The Leningrad city council gave us permits to go anywhere in the city 24 hours a day, even during air raid warnings and artillery fire.[17] These permits helped us a lot. Our comrades at LFTI were weakened from hunger, so we had to perform the necessary physical work ourselves.

"We regularly reported to I. Kurchatov in Moscow over the government telephone line in the office of P. Popkov. Often we talked to the responsible member of the Council of People's Commissars A. Vasin. It is not possible to overstate the amount of help he gave us. He connected us with many organizations, made sure that the required tasks were performed, found solutions to the most intricate problems, and always remained calm and supportive.

"We stayed in Leningrad more than two months. We prepared the high frequency oscillator for shipment, acquired the necessary insulators, vacuum grease, and putty. We dug up from the Institute grounds copper tubes and brass sheets, which were buried before the evacuation of LFTI in 1941. The party secretary, Ya Kapustin, recommended that we hide our supply of non-ferrous metal, saying:'After the war the Institute will need this metal, but for war purposes it is a drop in the bucket.'

---

[16] This airport no longer exists. St. Petersburg airport is at Pulkovo, which was under German occupation.

[17] There was a police enforced curfew. People were not allowed on the streets after dark.

"This buried metal was very useful! The accelerating chamber of the cyclotron, except for the lid, was made entirely of brass and copper, which was practically unobtainable in wartime.

"I went to the 'Electrosila' factory to verify the condition of the electromagnet for the LFTI cyclotron, which was built in 1941. Getting there from LFTI was not easy. Most of the trip was on foot. The magnet was there, but it was in parts scattered about the plant, mostly without labels. We had to make new labels, and to assemble the parts in one place. The coils were covered with metal caps. The day we were there, 35 shells fell in the neighborhood of the factory – the front lines were only 3 km away.

"We decided to load the equipment bound for Moscow on two trucks, and haul it to the railroad, for transportation out of Leningrad to Tikhvin.[18] A group of partisans helped us with the loading. A section of the railroad was under fire by the Germans, so we decided to let the wagons loaded with our equipment travel unaccompanied, and we flew out by air.

"A long list of unfinished work awaited us at the capital. We had to expedite the forging of the electromagnet of Armco iron at the 'Hammer and sickle' factory in the Moscow region. We ran into problems with the top forging, which had gotten too hot, and had not cooled properly. A special container was erected for forced air cooling.

"We started constructing the accelerating chamber immediately after our return from Leningrad. Design drawings were made by myself and L. Kondrashev, The chamber was made in the shops of the Institute for flammable minerals of the USSR Academy of Sciences, under the direction of M Egorov, who not only completed the job, but also took an interest in the assembly and commissioning of the device itself. The body and lids were made in the 'Projector' factory on a vertical lathe – a technique new to the workers. But the work proceeded smoothly, and the accelerating chamber was finished in two weeks. We had a modest banquet in the work room of the vertical lathe when all was said and done. [19]

---

[18] Tikhvin is due east of Leningrad, past Lake Ladoga, on one of the main Leningrad rail lines.

[19] A vertical lathe turns the work in a horizontal plane – that is, the axis is vertical.

"The day came for moving the 25 ton magnet out of the factory.[20] A hole was cut in the lab wall, and the ground was reinforced outside with an array of thin boards and metal clamps. A T-34 tank from the local armored detachment obliged in rolling over the corduroy and smoothing it out for us.

"On the appointed day early in the morning the best mill-wrights in Moscow took the load from the factory and towed it through the city streets designated by the police. By noon the magnet was at Laboratory #2, and by 4 pm that afternoon it was sitting on its base. We lost no time in mounting the rest of the equipment.

"The accelerating chamber had to be vacuum tight.[21] There were about 100 rubber seals, and we did not have a leak detector, so we clamped the lids on with iron cross braces, and lowered the rig into a water tank. Then we filled the chamber with pressurized air, and looked for bubbles. We carefully patched the bubble spots.

"The cyclotron facility occupied three rooms: One for the machine itself – magnet, accelerating chamber, vacuum pumps, the high frequency oscillator, etc.; One for the control room, and in between the two was a room full of shielding blocks and tanks of water. [22] The power supplies were in the basement. After checking the components separately, it came time to get everything working together. This required a round the clock effort, with only a 4-5 hour break for rest. Occasionally someone fell asleep at the desk, and we went into another room so that he could rest.

"Kurchatov was a busy man, but he kept an eye on our progress with the cyclotron. He would call on the telephone from time to time, asking what we had achieved. We were aware of the deadline, and of the importance of the cyclotron for progress on the way to a solution of the big picture. Finally all of the hurdles were jumped, and it was time to accelerate deuterons. Igor Vasilevich went at 8 that evening to a meeting with Boris Vannikov.[23] He asked us to call him if there was any news.

---

[20] My design is only 18 tons – skimpier on the iron. Russians like hefty magnets. Before constructing a large magnet one usually makes a ¼ scale model to test.

[21] The vacuum chamber contained the D's – hollow cans with a gap. Across the gap is the RF voltage, synchronized to the cyclotron frequency. See Figs 9 & 10 Appendix D.

[22] Nemenov doesn't say, but the cyclotron was located in the original 3 story building.

[23] Boris Vannikov was introduced in Chapter 1 as People's Commissar for Armaments. He was in prison at the beginning of WWII. During the war, he was a deputy to the People's Commissar for Armaments. After 1945 he worked on the Soviet Atomic Project.

"We had a slow start up, because we had to 'train' the vacuum system to withstand the high voltage, which took a couple of hours.[24] After the D's would hold voltage, we turned up the deuteron source.[25] We checked the oscillator frequency and the strength of the magnetic field. There were a few minor problems which we fixed. We adjusted the magnetic field, and accelerated deuterons struck an internal target placed between the D's. Some distance away from the cyclotron a Geiger counter fired, indicating the presence of neutrons!

"Everyone became excited – was the effect real? We increased the voltage across the D's, and the deuteron current increased, indicating that accelerated beam was reaching the periphery of the magnet. 'Training' of the vacuum system continued. We measured the deuteron current on a probe – 50 μ amp! This was victory. Everything seemed to be working fine. I suggested that we extract the beam outside the vacuum chamber. We replaced the vacuum chamber window with a phosphorescent screen, and turned everything back on. The screen glowed brightly.

"Then we removed the screen, and installed a thin aluminum window, which would allow the beam to escape into the room. This all took about one hour. Then we turned everything back on. Everyone stood still, frozen in suspense. Hurrah! The beam of deuterons came right out into the room. One could see the beam quite well. We turned off the lights, and in the darkness a blue-violet tongue of light came out of the vacuum chamber.

"We stopped the cyclotron, and I called Vannikov. He answered the phone himself. I asked to speak to Kurchatov. Vannikov asked: 'Have you got the cyclotron working?' I answered: 'No, but Igor Vasilevich wanted us to call him.' Kurchatov came to the phone and asked: 'Does it work? What is the current?' I answered: ' The internal current is more than 50 μ amps. We steered the beam outside, and observed the light created in the air.' Kurchatov answered: 'Turn everything off and let it rest. I will be there in about one hour. Congratulations to all.' I then asked Kurchatov to apologize for me to Vannikov for my concealment of the truth. [Kurchatov had to be the first to know.] (added by LGP)

---

[24] This is a common problem. Metal surfaces, no matter how carefully cleaned, have to be 'cooked' to burn off the remaining dirt, which causes electrical breakdown, and spoils the vacuum. The D voltage was quite high – 30 kV or so.

[25] Deuterium gas was leaked in through a tube to the center of the machine, where a tungsten filament emitted electrons to ionize it. Gas and ions flowed into the vacuum chamber. The pumps had to handle this leak.

"It was 2 am. We wrote in the lab notebook: '25 September 1944 a beam of deuterons was extracted from a cyclotron for the first time in the USSR.'

"In one hour, as he promised, Kurchatov arrived happy and smiling, full of congratulations. His first words were: 'Is everything OK?' We started the cyclotron up, and the long sought beam reappeared. The vacuum had improved, and the current probe now read almost 100 μ amps. Then Igor Vasilevich asked us to irradiate with neutrons a silver foil covered with paraffin.[26] We carried the irradiated foil to a Geiger counter. The counting rate jammed the scaler at a distance of 2 meters.

"It was 4 am. Kurchatov congratulated everyone again, and invited us over to his house for a champagne toast. His wife, Marina Dmitrievna, was surprised to see us to say the least, but she recovered when she learned of our success. Finally, Kurchatov said: 'Tomorrow we will make additional measurements with paraffin blocks. The day after tomorrow we will begin irradiation of uranyl nitrate."

Nemenov then says that the irradiation lasted until December, 1945, and that Boris Kurchatov, Igor's brother, did the chemical analysis to separate out the plutonium. Boris and his coworkers used the sulphate coprecipitation method in a water solution.[27] In this way the first 'cyclotron plutonium' was produced in Europe in 1946.[28] Thus the task assigned by Igor Kurchatov in March, 1943, had been fulfilled.

Nemenov's article is interesting for a number of reasons. First of all, it is a tale of experimental physics. It is not weapons work per se. As mentioned in Chapter 6, much of the work at Los Alamos was a continuation of experimental work done at university labs. Much of the work of Laboratory #2 was the same. The second half of the story reads like a laboratory notebook. 'We glued this to that, and then we fixed the leaks.' Cyclotrons were not new – they had been built before WWII, even in the

---

[26] Natural silver is a 50-50 mix of mass number 107 and 109. Each isotope captures neutrons to form 108 and 110 respectively, which beta decay with short half-lives into isotopes of cadmium.

[27] A technique invented by the founder of the Auer Co, and pioneered by Marie Curie. See Marie Curie, Nobel Lecture, 1911.

[28] A very small amount – about 6×10^{11} atoms – was made earlier by exposure of natural uranium to neutrons from a Ra-Be source. See /APSSSR /*ibid.*, report #261.

USSR, in Leningrad. But they were complicated pieces of experimental apparatus, involving a large magnet, high voltage, radio frequency oscillator, vacuum system, ion source, and various detectors. Industrial shops were used for some of the heavy manufacturing, but the machines were individually designed and hand built. No two were alike. The experimenters, probably about 6 people, had to work on the device around the clock to get it working, and once it worked, around the clock to keep it working. Stuff breaks, vacuums leak, high voltages arc, magnet currents drift, etc. There were never enough diagnostics. The lesson of Nemenov's article is that experimental physics is much the same everywhere. Everywhere, that is, except in Leningrad in 1943, with the German army a few kilometers away, and shells falling all around you. The impressive part of this half of the story I believe is the clout which Nemenov had to operate in war-torn Leningrad, thanks to letters supplied by Pervukhin, who was a People's Commissar. This illustrates the importance that the Soviet Government ascribed to the Atomic Project even in its infancy in the middle of the war, with German forces still occupying much of the western part of the country. They travelled in and out of Leningrad by air, and they shipped parts of the cyclotron out by rail, under circumstances when transport that was neither military nor humanitarian aid had to have highest priority in order to go anywhere. They were not generals – they were just two guys from the lab.

The late night meeting with Kurchatov is not surprising. Physics experiments usually run better at night, when the power grid is quiet, and some of the experts have gone home. That Kurchatov and Vannikov were meeting in the Kremlin at 2 am is not surprising either, because Stalin's nocturnal habits caused all of the ministers to be up all night. A formal portrait of Boris Vannikov is shown in Figure 7.2.

Figure 7.2. Boris Vannikov, Minister of Armaments. Credit Istoria Rosatoma.

7.6. Leningrad is Liberated

In January, 1944, the Leningrad blockade was completely lifted by the Red Army. Both the Leningrad Physico-technical Institute and the Radium Institute were split into two parts, one remaining in Leningrad for the three years of the blockade, and the other temporarily evacuated to Kazan, where many of the institutes of the Moscow region (Kapitza's Institute for Physical Problems, for example) were also located. Laboratory equipment and personnel were divided between the two places. Heavier apparatus, like cyclotron magnets, tended to stay put in Leningrad, and some researchers remained under starvation conditions, as Nemenov observed. Among the many problems facing Ioffe and Khlopin, the directors of LFTI and the Radium Institute respectively, was a ban on relocation of personnel from remote places like Kazan back into Leningrad. The ban could have been implemented due to bad conditions inside the city, or other reasons. If there was a similar ban for the Moscow region, Kurchatov, Kapitza, and others had enough influence to override it. In any event, as late as September, 1944, Ioffe wrote to the Leningrad Communist Party head, A.A. Zhdanov, stating that all of the LFTI equipment and personnel in

Kazan was loaded on rail cars and ready to come to Leningrad, but was not permitted to travel. The Radium Institute had perhaps a worse predicament, in that some of its equipment and personnel was allowed to return before the ban became effective, leaving the laboratory stuck in two places, disassembled in Kazan, but not allowed to go anywhere else. [29]

The Leningrad story in 1944 involves two cyclotrons, both in need of work. One had operated before the war at the Radium Institute. The other was at LFTI, where Kurchatov began assembling parts before the war. It was left incomplete in 1941. Both machines were raided for parts by Moscow's Laboratory #2. One might ask what happened to the parts that the Moscow raiding party left behind? A.I. Alikhanov, one of the original scientists at Laboratory #2, and the candidate for Academy membership chosen over Kurchatov, requested a transfer from Moscow to LFTI in Leningrad in a letter to M.G. Pervukhin in March, 1944.[30] This request was granted, and Alikhanov began restoration work on the LFTI cyclotron. The building, constructed before the war, was still standing, although it had lost most of its windows. Foundations for the heavy magnet were still in place. The overhead cranes needed to move things around and assemble the components had disappeared, as had storage batteries, motor-generator sets, and other electrical parts. The rusted magnet parts were still in the 'Electrosila' factory, where Nemenov had found them the previous year. Iron workers who had left for the war were needed in the factory to expedite the completion of the magnet. The RF oscillator and vacuum chamber had been shipped to Laboratory #2, but there was a chance that some equipment not needed in Moscow could be returned to Leningrad. The vacuum chamber was one of the most complex parts of the machine. Alikhanov was optimistic that, if technical personnel could be restored to LFTI, the cyclotron could be up and running in 1945.[31] His request for financial support was endorsed by Kurchatov in a letter to Molotov in June, 1944.[32] The LFTI cyclotron was larger than the one commissioned

---

[29] /APSSSR/, *ibid.*, reports #262 and #269.

[30] /APSSSR/, *ibid.*, report #217 Alikhanov to Pervukhin.

[31] /APSSSR/, *ibid.*, report #232.

[32] /APSSSR/, *ibid.*, report #245.

in Moscow, with a 1.2 m diameter pole tip, a 14 cm gap, and a 1.7 Tesla magnetic field. The magnet weighed 75 metric tons. Using the cyclotron formulas in Appendix D, the LFTI cyclotron could accelerate deuterons to 25 MeV. It would be the largest cyclotron in the USSR when brought into operation. Pervukhin and Kurchatov submitted a list of goals for the coming year to the State Defense Committee that specified full operation of the LFTI cyclotron in April, 1945.[33] The cyclotron actually was brought on line at full energy in December, 1946.[34]

The other Leningrad cyclotron, the one used by Kurchatov in the Radium Institute in the 1930's, had suffered during the war from the general destruction of infrastructure of the laboratory, which lost generators, storage batteries, buildings, etc. The director, V.G. Khlopin requested resources from Pervukhin to begin restoration in April, 1944.[35] By October of that year Khlopin said that most of the cyclotron was in working order, but that they had not been able to pump out the vacuum chamber, because Laboratory #2 had 'borrowed' their large vacuum pump in the fall of 1943. It was expected to return, however, and the USSR's very first cyclotron should be up and running again by the end of the year. An order issued by Beria on November 25, 1944, lifted the travel ban from Kazan to Leningrad, ordered the necessary rail transportation for the Radium Institute, and told Khlopin to get the cyclotron going by January 1, 1945.[36]

There was an amazing request from the Ukranian Academy of Sciences to Nikita Khrushchev, then the head man in Ukraine, to ask Mikoyan, the Minister of foreign trade, to purchase a complete cyclotron laboratory from the USA. The energy of the machine was not specified, but the cost was estimated at $500,000 (1944 dollars – probably reasonable). The request states that it would be useful for the USA to supply a few physicists

---

[33] /APSSSR/, *ibid.*, report #249.
[34] /APSSSR/, *ibid.*, report #302.
[35] /APSSSR/, *ibid.*, report #226.
[36] /APSSSR/, *ibid.*, report #283.

along with the accelerator to get things up and running![37] Khrushchev endorsed and forwarded the request to Mikoyan, but nothing happened after that.

7.7. R&D Program for Laboratory #2 in 1945

In May, 1944, Kurchatov outlined to Pervukhin, and Pervukhin subsequently forwarded to Stalin, the R&D program for Laboratory #2 for 1944-45.[38] Kurchatov knew that the isotope ^{235}U could be used for a nuclear weapon, and probably the isotope ^{239}Pu of the artificial element plutonium would work as well. He knew that the separation of ^{235}U – only 0.7% of natural uranium – would be a formidable task. He also knew that a working high power nuclear reactor was needed to manufacture plutonium via neutron capture on the abundant isotope ^{238}U. The USSR had no working nuclear reactor of any kind, no functional isotope separation plant, and, to make matters worse, no appreciable amount of uranium. There was no UF_6 gas for a gaseous diffusion plant, and only very early thoughts on how to build an isotope separator based on the principles of mass spectrometry (calutrons).[39]

On the reactor front, there were three types which had possibilities: a natural uranium reactor with graphite moderator, like Fermi's pile; an enriched uranium reactor with a 'light water' moderator; and a natural uranium reactor with a 'heavy water' moderator. Kurchatov understood the advantages and requirements of the different designs. He knew that, because of the large neutron capture cross section, $n + p \rightarrow d + \gamma$, ordinary water had to be fueled by uranium enriched in ^{235}U, which gives a greater neutron density. Deuterium also captures neutrons: $n + d \rightarrow t + \gamma$, where t is tritium.[40] This reaction however has a smaller cross section, so it would be possible to build a heavy water reactor with natural uranium. Without enrichment of ^{235}U by isotope separation, only graphite and heavy water

---

[37] /APSSSR/, *ibid.*, report #209.
[38] /APSSSR/ *ibid.*, #233, 234, and 235.
[39] /APSSSR/ *ibid.*, #238.
[40] One could write $n + {}^1H \rightarrow {}^2H + \gamma$, and $n + {}^2H \rightarrow {}^3H + \gamma$.

designs were feasible. Heavy water is a more efficient moderator than graphite, so less of it would be needed for a reactor to reach criticality. However, the USSR had no heavy water, so that left only graphite, and Kurchatov knew that a graphite – natural uranium reactor would need about 500 tons of pure graphite, and 50 tons of pure uranium. That amount of pure uranium did not exist.

7.8. Reactors and Espionage

Kurchatov and Pervukhin were well informed. Their knowledge came from three sources: 1.) The Manhattan Project through Soviet espionage; 2.) The Soviet theory group, particularly Khariton, Zeldovich, and Pomeranchuck; and 3.) Experimental studies underway in a tent constructed beside the main building at Laboratory #2. In the middle of 1944 espionage was most important, since the Manhattan Project had solved many of these same problems. The information was accurate, and in many cases very detailed and quantitative. Kurchatov knew all of the specs for Fermi's chain reacting pile – amount of graphite, uranium, matrix layout, overall size, critical layer, access ports and control rods, you name it. [41] It is no accident that the F-1 reactor at Laboratory #2 and Fermi's CP-1 are virtually identical – carbon copies if you will forgive the expression. Libby thinks that the source for this information was the British physicist Alan Nunn May, who worked at the Metallurgical Lab at Chicago in 1943. [42] May was arrested in Britain after WWII, and served a prison term for espionage. However, I have not found a direct link between him and Fermi's CP-1.

Vsevolod Merkulov, people's commissar for state security, outlined in a memorandum to Beria in February, 1945, the knowledge then in hand regarding the status of the Manhattan Project. He reported that there were three principal centers:

---

[41] /APSSSR/, *ibid.*, #251.
[42] Leona Marshall Libby, "The Uranium People," *ibid.*, p 116.

"Camp 1 at Oak Ridge, for production of ^{235}U, with contractors Kellex, DuPont, and Carbide and Carbon Co.; Camp W on the Columbia River in Washington state for the production of plutonium, operated by DuPont; and Camp 2 "Camp Y" at Los Alamos near Santa Fe, New Mexico, under the direction of the US War Department, for the assembly of a workable bomb. Camp 2 is isolated from the outside world…About 2000 people live on the mesa, surrounded by barbed wire and guards. They have comfortable living conditions, good housing, parks and swimming pools for recreation, a club, etc. Mail correspondence with the outside world is controlled. Workers must have special permission from the army in order to leave the camp site.

"There are several firing ranges near Los Alamos for tests. The latest experimental data on the effect of an atomic bomb give new information regarding the scale of destruction. The energy of an atomic bomb weighing about 3 tons would be equivalent to 2000-10,000 tons of chemical explosives, with destructive effects not only from the shock wave, but also the high temperature, and the radioactivity, resulting in annihilation of every living thing within a radius of 1 km. There are two means of initiating an explosion:1. Ballistic; and 2. Implosion. There is not a definite date for the first bomb test – probably in one year minimum, five years maximum. There could be a test of smaller yield sooner, in a few weeks. The Americans have enough active material for one or two bombs not as powerful, but still strong enough to have practical significance as a new form of weapon. The first test of a military explosion is expected in 2-3 months.

"In order to utilize the atomic energy of uranium, a nation must have a stock pile of uranium ore, and access to its source. This is a matter of utmost importance…

"Sources of uranium ore in Czechoslovakia are in the Sudetenland…20 km north of Karlsbad.

"Our intelligence indicates that the British intend to conclude an agreement with the Czechoslovak government in London to exploit these sources of uranium ore."[43]

Merkulov does not give any technical information, but a lot of detail was known, like the number of reactors at Hanford, their power level and plutonium production rate. It is interesting that he talks about living conditions at Los Alamos, which did not seem so great to those who were there. It is not clear what he means by 'a test of smaller yield sooner.'

---

[43] /APSSSR/, *ibid.,* #316.

Perhaps he refers to the critical assemblies at Los Alamos. He is optimistic regarding the Trinity test at Alamogordo – it was about five months off, in July, 1945. He finishes by returning to the problem facing the USSR – not enough uranium production. Other sources have pointed out that the British intelligence service became interested in the Soviet Atomic Project around this time, because of Soviet moves towards Czech uranium ore.

Some higher ranking members of the NKVD, like Merkulov and Makhnev, were critical of the pace of the Soviet Atomic Project in 1944-45. Their concern may have come from the intelligence, which at that time was shared only with Kurchatov.[44] The Manhattan Project had such an enormous scope, that it was clear that one three story house and a tent were not going to get the job done. Later on, after Hiroshima and Nagasaki, the Smyth Report, accessible to most of the ranking physicists on the Soviet project, removed any doubt regarding the scale of the undertaking. Before September, 1945, however, one important consequence of the espionage was that the Soviet government at the highest levels understood the effort required to achieve a nuclear weapon. That knowledge, and the attendant determination for success, resulted in the broad commitment of necessary resources by the USSR.

7.9. Heavy Water

The needs for heavy water and uranium stimulated many different efforts, and much communication between engineers and geologists in Soviet Central Asia and the management in the Kremlin. Kurchatov was peripherally involved in all of it. Some discussion of the uranium ore effort is given below. Although after WWII uranium production by the Soviet Union became substantial, in 1944-45 it was virtually non-existent.

Heavy water has both hydrogen atoms in H_2O replaced by deuterium atoms ($^2H = np = D$), written D_2O. Water with one replaced hydrogen, H-

---

[44] Kurchatov was authorized to share specific information with responsible division heads. For example, Kikoin, in charge of isotope separation, had access to information on Oak Ridge. See /APSSSR/, *ibid.*, #315 discussed below.

D-O, occurs naturally at the level of 1/3200; that is in 6400 hydrogen atoms in water, one is a deuterium atom. Because deuterium is twice as heavy as hydrogen, the water molecular weights increase: $H_2O = 18$; HDO $= 19$; and $D_2O = 20$. Heavy water is 10% heavier than ordinary light water. The boiling points for deuterated water likewise increase with the mass: $H_2O = 100.0^0C$; HDO$= 100.7^0C$; and $D_2O = 101.4^0C$.[45] While ice floats on water, heavy ice sinks. These differences in physical properties allow the isotopes to be separated. Ordinary water can be boiled away, leaving HDO, which has a higher boiling point, behind as a liquid. It was noticed in the commercial production of hydrogen and oxygen by electrolysis that the remaining water had an increased fraction of HDO molecules. Pure HDO becomes ¼ D_2O, ¼ H_2O, and ½ HDO through the continuous exchange of hydrogens when the molecules collide. Subsequent distillation leaves pure heavy water. Since the distillation is usually done by electrolysis, heavy water plants tend to be located near hydroelectric power plants, the famous one in Norway being an example. Hydroelectric plants in the USSR in 1944 were located in Central Asia. Efforts were underway to begin heavy water production at the Chirchiksky electrochemical combine.[46] Chirchik is in Uzbekistan, not far from Tashkent. Pervukhin, the People's Commissar for Chemical Industries, was a natural choice for ultimate responsibility.

Lend-Lease supplied raw materials as well as Studebaker trucks and Willys jeeps to the USSR. Accordingly, on 30 July 1943, Kurchatov wrote to Molotov requesting purchase of 100 tons of uranium from the USA. On July 19, 1943, Pervukhin had requested one kilogram of heavy water. On May 22, 1944, Anastas Mikoyan, Commissar for Foreign Trade, wrote to Molotov and Pervukhin on the status of the uranium request.[47] Mikoyan had learned from the Canadian Lend-Lease administrator that uranium was controlled by a committee composed of the US Secretary of War (Henry Stimson), and his Canadian and British counterparts, and that uranium was being used for guided missiles and jet aircraft (!). Unfortunately, the Soviet

---

[45] En.wikipedia.org/wiki/Heavy_water
[46] /APSSSR/, *ibid.*, #204.
[47] /APSSSR/, *ibid.*, #236.

request was turned down, but could be reconsidered, assuming that the USSR was using uranium for similar purposes. According to Holloway, General Groves was in the loop on the US side, and supported some shipment of uranium to the USSR to give the false impression that it was not considered a particularly special material. In any event, it does not appear that any uranium from the US reached Laboratory #2.[48] The subjects of heavy water and uranium shipments via Lend-Lease will come up again in Chapter 8 on espionage.

7.10. Ballistics Experiments

Several other projects at Laboratory #2 were proceeding in parallel with the cyclotron construction. Pursuant to a gun-type bomb, some physicists began shooting pieces of metal at each other, and recording the results by high speed photography. This exercise started on the second floor of the 'main building' with rifles aimed at each other.[49] Then they got more ambitious, and they proposed using 75 mm cannons instead, but maybe they better do that in a shed outside. Kurchatov had some concerns, so he talked to Boris Vannikov. Vannikov agreed that the experiments were necessary, and he obtained permission to go ahead. Vannikov was, after all, in the Ministry of Armaments. Vannikov and Kurchatov worked well together. In later years, they both recollected that this instance – the 75 mm cannons, represented the transition from primitive laboratory experiments to a broad program of industrial and engineering development.[50] But the work did not get off to a fast start. On February 23, 1945, Kurchatov wrote to Beria about new construction for the planned 'gun barrel' experiments.[51] The buildings were supposed to be ready in December, 1944, but work was suspended due to commitment of construction equipment to other jobs. Kurchatov requests Beria's help in getting the job done. Beria ordered the completion without delay.

---

[48] David Holloway, "Stalin and the Bomb," *ibid.*, p 101.
[49] I.N. Golovin, "I.V. Kuchatov," *ibid.*, p 55.
[50] I.N. Golovin, "I.V. Kurchatov," *ibid.*, p56.
[51] /APSSSR/, *ibid.*, #314.

V.I. Merkin, one of the researchers on the gun barrel project, wrote the following in his memoirs:

"Khariton suggested the first phase of the work, to synchronize the firing of two barrels, aimed at each other one meter apart, so that the two bullets met in the middle within 100 μ sec. The muzzle velocity was about 1500 m/sec, so that the collisions would be spread out over ±15 cm. Khariton helped us out in achieving this goal. We started with army rifles. With the help of Boris Vannikov we obtained several tens of rifles and boxes of ammunition. [One supposes that the rifles did not survive several test shots.] We began with 7.62 mm rifles, and ended with 75 mm cannons."[52]

One may wonder why they did not shoot a projectile at a stationary target, rather than try to shoot two guns at each other to meet in the middle. Perhaps they wanted to increase the collision speed by having the two parts pass through each other. On the other hand, 1500 m/sec is already very fast. The Los Alamos tests, and indeed the Hiroshima bomb, involved a stationary target. The Hiroshima bomb was a gun type nuclear weapon, where a 38 kg projectile was fired at 300 m/sec into a 26 kg stationary target at the closed end of the gun barrel.[53] Both projectile and target were 80% pure ^{235}U, enriched at Oak Ridge. The gun tube was smooth bore, 165 mm in diameter – more than twice the size of Merkin's 75 mm cannon. In fact, almost twice the size of the famous German 88.

7.11. Life at the Lab

Two seminar series on fundamental physics began at the Laboratory in the summer of 1944. Peter Kapitza organized one of the series with topics in all subfields of physics. Kurchatov ran the other series, concentrating on recent results in nuclear physics and cosmic rays. An audience

---

[52] Similar experiments at Los Alamos were carried out at Anchor Ranch off the lab site, where some protection was available. See Richard Rhodes, "Making of the Atomic Bomb," *ibid.*, p 542. Laboratory #2 was in the open countryside. X-ray photography techniques were developed to compliment post-mortem inspections of the collisions. The major design problem was to keep the target from falling apart before nuclear detonation.

[53] John Coster-Mullen, "Atom Bombs," *ibid.*, p 17-27

restricted for security learned about nuclear fission and chain reactions. Kurchatov always attended the seminars, extracting clarity, and encouraging young speakers, with an eye towards recruiting for the Institute. But affairs distracted him, and sometimes he was called by the government and had to leave in the middle of a seminar.

Towards the end of 1944 Laboratory #2 had eight departments: 1.) Atomic Reactors, headed by I.V. Kurchatov 2.) Isotope separation by gaseous diffusion, headed by I.K. Kikoin; 3.) Bomb construction headed by V.I. Merkin; 4.) Heavy water headed by M.O. Kornfeld; 5.) Electromagnetic isotope separation headed by L.A. Artsimovich; 6.) Neptunium and plutonium headed by B.V. Kurchatov; 7.) Cyclotron headed by L.M. Nemenov; and finally, 8.) Theory, headed by I. Ya. Pomeranchuk.[54] Rough correspondences with the Manhattan Project are: (1) Chicago Metallurgical Laboratory; (2) K-25 gaseous diffusion plant at Oak Ridge; (3) Los Alamos; (4) pursued in Canada for reactors;(5) Y-12 calutron plant at Oak Ridge;(6) Hanford, Washington; (7) there were several cyclotrons; and (8) there were several theory groups, at least one at each major installation. As the Soviet Project progressed, many of these departments achieved industrial status, and moved elsewhere in the USSR. For instance, gaseous diffusion was in the engineering model phase, rather than the giant plant eventually needed to do the job. By April, 1944, Kikoin and the gaseous diffusion effort was moved to a new branch of Laboratory #2 in Leningrad, only to return to Moscow the next year.[55]

Nemenov's cyclotron was small, and Kurchatov wanted one for Laboratory #2 that was bigger than the new LFTI machine, perhaps as big as the one that Lawrence had built before WWII.[56] Accordingly, he wrote a memorandum to the Central Committee of the Communist Party requesting approval of a new machine, to be housed in a new building at

---

[54] /APSSSR/, *ibid.*, report #203.

[55] Order #5407ss of the State Defense Committee by V. Molotov, dated 15 March, 1944 /APSSSR/, *ibid.*, report #221.

[56] Lawrence had two working machines, with 37 inch (92 cm), and 60 inch (150 cm) diameter magnet pole tips. The 37 inch, and a magnet with 180 inch (4.5 meter) diameter pole tips were both used as early test rigs for the Oak Ridge calutrons.

Laboratory #2. The magnet pole diameter would be 150 cm, with a 25 cm gap, and a magnetic field strength of 1.6 Tesla.[57] The magnet would weigh 310 metric tons. The building volume of 35,000 m^3 would be spanned by a bridge crane of 40 ton capacity, and would be isolated from other buildings to allow adequate radiation shielding.[58] Kurchatov stated that the cyclotron would be used to accelerate deuterons for neutron bombardment to produce the artificial element plutonium, and could serve as a test rig for an electromagnetic isotope separator (a calutron). This proposal was accepted, and as a result a very interesting work order was signed by Stalin on May 15, 1945, in which all of the steps necessary to implement the complete construction of the new facility were outlined. The Soviet Union was a controlled economy, everyone knows that. But does it then follow that everything that happened had to be signed by Stalin? Perhaps it was a consequence of the importance attached to the Atomic Project that every detail of one of its construction projects was cleared at the highest level. In any event, Resolution #8581 ss/ov of the State Defense Committee specified in six pages the responsible factory or ministry for each component, the individual in charge, the time scale, and the cost. For example, the cyclotron building was assigned to the administration for military airport construction, and broken down into architects for the drawings, steel mills for the framing and iron workers to assemble it, concrete workers, and ordinary builders. The Moscow 'Electrosila' factory was to supply the crane. Mikoyan, the minister of foreign trade, was to import the necessary electric generators. The people's commissar for rubber industry was to supply vacuum hoses and rubber sheets for gaskets. The list goes on and on. Each responsible organization was repeated line after line so that there could be no confusion as to who was supposed to do what.[59] The cyclotron started up in December, 1947, 2 ½ years after Stalin's Order. Whether #8581 was executed as specified is not known, but the end product was made in a timely fashion.

---

[57] Maximum momentum p = 360 MeV/c, or 34 MeV kinetic energy for deuterons.

[58] /APSSSR/, *ibid.*, report #303. A bridge crane moves on overhead tracks along the length of the building, and the hook moves back and forth on a trolley, giving full coverage.

[59] /APSSSR/, *ibid.*, report #350.

Kurchatov got his cyclotron, a valuable research tool for nuclear physics. But a machine of this size was not really necessary for the neutron cross section measurements needed for nuclear weapon design. These measurements were made at Los Alamos with lower energy deuteron beams. For thermal neutron cross sections, reactor neutrons were often used. Deuterons of a few MeV from an electrostatic accelerator could produce a wide range of monoenergetic neutrons by choosing a suitable (d,n) reaction and neutron emission angle. Kurchatov knew this. He was looking to the future, to further work in nuclear physics at Laboratory #2. He was using the weapons project as leverage with the government to obtain research resources for his laboratory. The Laboratory #2 cyclotron was not the last one that he requested. A much larger machine, with a 6 m diameter pole tip and a proton energy of 500-700 MeV was proposed in 1946, and JINR Dubna was founded to house the new cyclotron few years later. Kurchatov was not alone in using the Atomic Project to obtain government support for pure research.

Today the Kurchatov Institute is surrounded by city streets and buses, and is on Moscow Metro Line #7, with the closest stop Shchinskaya. However, in 1944 Laboratory #2 had been going for only about 1 year, and was still a small operation out in the countryside. Transportation for employees must have been a problem. Perhaps the Academy of Sciences operated buses from central Moscow. There were virtually no private cars, and not many roads. The traditional solution for Soviet industry would be to build housing on site, together with a commissary, bath house, house of culture, etc. A few people were housed in the main building, but there was not much support infrastructure. There was no storage space. Several grams of radium (used for neutron sources) were stashed in a hole in the back yard. There is a list of monthly salaries paid to the lab staff in 1944. They are difficult to interpret, since there was very little available in the private sector for purchase at that time. Kurchatov received 3000 rubles, Division heads 2000 rubles, senior scientists 1700 rubles, the librarian 500 rubles, and the chauffeur 800 rubles.[60]

---

[60] /APSSSR/, *ibid.*, report #230.

John Fischer was an American writer who served with an UNRRA mission to Ukraine in 1946, and wrote a book about his experiences.[61] UNRRA was an organization of the United Nations for post-war relief, so he had the opportunity to observe the living conditions in Ukraine at that time. He spent most of his time in Kiev, which was only moderately damaged, but he was also in Kharkov, where the city changed hands four times during the war. Fischer supplied some quantitative data on Ukrainian daily life as he saw it. He claims that, conditions being equal, the standard of living in Ukraine was higher than in the rest of Russia. He quotes an average wage for an industrial worker at 300 to 350 rubles a month. Laboratory #2 workers were better paid, probably because of the importance of the mission, and the location in the suburbs of Moscow.[62]

Any industrial complex in the Soviet Union had its own canteen, where workers could get fed during the day. Food was rationed, and the workers' canteens were part of the rationing system. The ration cards could also be used to make purchases at a shop in town where the card was registered. The bread ration depended on the physical activity required by your job. A heavy worker was entitled to 800 gm of bread a day. All stores in the city of Kiev would have the same stock, with the exception of the privileged stores, which may have a wider selection of goods. The stores were almost empty. Plenty of 1 kg loaves of Russian black bread, priced at 90 kopeks each, some tea, powdered milk (from US army surplus), a barrel of sauerkraut, and a can of peanut butter finished out the inventory. Some meat or fish was also authorized, but not in stock.[63] A kg of good quality Russian black bread is about 2000 calories[64], and although it does not supply all of the necessary nutrients, is enough to keep you alive. Later, in the 1970's, when food was plentiful, and there were many western visitors, the Russian hosts were proud of their white bread, but the visitors all preferred the black, which had unique quality. Unfortunately, as is the

---

[61]   John Fischer, "Why They Behave Like Russians," Harper & Brothers, New York, 1947.

[62]   John Fischer, "Why They Behave Like Russians," *ibid.*, p 45.

[63]   John Fischer, "Why They Behave Like Russians," *ibid.*, p 47.

[64]   These are food calories = 1000 ordinary calories, or 4180 joules.

case with many regional delicacies, currently available black bread is not quite as good as it used to be.

There was not enough food to go around. Workers prospecting for uranium in December, 1944, were given an extra second course food ration and 200 gm of bread without giving up ration tickets. In August, 1947, uranium miners had the following rations in grams/day: rye bread 200 g; meat/fish 300 g for managers, 200 for workers; fats 60 g for managers, 40 g for workers; sugar 60 g for managers, 30 g for workers; potatoes 500 g for managers, 700 for workers. They also received three bars of soap per month.[65] The managers did better than the workers, except for potatoes. Neither fruits nor green vegetables are mentioned. Whether this daily ration actually obtained is not specified. People could and did live on it, but they did not gain weight.

Laboratory #2 was trying to construct more housing, but there were endless delays, and Pervukhin received complaints.[66] One of Beria's deputies, V.A. Makhnev, was particularly critical of the level of activity, considering the magnitude of the task. On November 17, 1944, Makhnev wrote to Beria:

"At the present time the work assigned to Laboratory #2, despite its extreme importance, is poorly organized. The scientists who are occupied with the uranium problem are scattered around the country, in Moscow, Leningrad, Sverdlovsk, and Kazan. Laboratory #2, established at the end of 1942 near Moscow, has only one building, which has to house the experiments, and serve as living quarters for the scientists and guards. There is no housing for technical and service workers. There are no laboratory instruments, no reference materials, and no mechanical infrastructure. Army construction workers were supposed to be building housing and lab space for the past 1.5 years, but they have not done much so far, and have recently stopped altogether. Only 10% of the 800,000 rubles budgeted for construction in 1944 has been spent. In order to move the uranium problem off its primitive state, I suggest: 1.) Concentrate all of the scientists in Moscow, and call for additional effort; and 2.) Establish all of the necessary conditions to get the job done in a timely fashion."[67]

---

[65] HISAP96, Vol 1, *ibid.*, p 291.
[66] /APSSSR/, *ibid.*, #208
[67] /APSSSR/, *ibid.*, #279.

The Kremlin worried about training scientific and technical workers to accomplish the job. On February 21, 1945, Stalin signed an order setting forth goals for output of nuclear physicists![68] Moscow State University was to produce 10 nuclear physicists by December, 1945; 25 in 1946, and no less than 30 every year afterwards. Students were to be exempt from the draft. Student stipends were to be increased to equal that of students training for defense industries. Staff of Moscow State University was permitted to increase to teach the eager students about the atomic nucleus. A subsequent census of existing manpower came up with 4212 total inventory of trained physics personnel.[69] For comparison, the American Physical Society had about 5000 members in 1945.[70]

7.12. The Tent Burns Down

The graphite prism experiments done to check the purity of uranium and graphite samples once they began to arrive were carried out in the army tent, erected next to the main building.[71] Unfortunately, the tent burned down. Kurchatov wrote a letter to Pervukhin on 17 July, 1944, as follows:

"Today at 1700 hours our tent burned down. The fire was started by electrical workers repairing a short circuit caused by lightening in the wiring between the tent and the main building. The loss is approximately 22000 rubles (3000 for the tent, and 18000 for the 1.5 kg of uranium). Damage to measuring equipment was minimal."

Pervukhin notes: "Please try to be more careful."
Kurchatov responds: "Message received."

---

[68] /APSSSR/, *ibid.*, #311.
[69] /APSSSR/, *ibid.*, #312.
[70] www.aps.org
[71] I.F. Zhezherun, HISAP96 Vol 2, p 68.

Much later, after information was declassified, it was learned that this was a cover-up on the part of Kurchatov. A worker in the tent at the time, V.K. Losev, in an article published in 1996, states that the fire was caused by the spontaneous combustion of uranium powder. Metallic uranium powder can catch fire in air, which makes machining of the substance tricky.[72] The powder was on trays in an assembly of graphite blocks, and not easy to access. Losev tried to remove the trays, but they were too hot. He poured water on the fire, but this just supplied more oxygen, and did not extinguish it. Firemen eventually came, but the tent was lost. The fire was doused by dumping sand on the assembly. [73] One may conjecture that Kurchatov, who was notified of the accident at the time, and visited the tent during the fire, decided not to report the true cause to the government because it would indicate carelessness on the part of the experimenters, who were working with uranium powder without knowing of its spontaneous combustion. On the other hand, it probably did not fool anyone, since Pervukhin was not stupid, and had many sources of information regarding Laboratory #2 other than its director. Pervukhin's response, an appeal to safety precautions, would be valid whether or not he knew the true cause.

The industrial work required for the magnet iron, coils, and accelerating chamber for the cyclotron, described by Nemenov, was not especially demanding. Graphite and uranium for a nuclear reactor would be another matter. As mentioned in Chapter 5, Vladimir Vernadsky, the director of the Leningrad Radium Institute in the 1930's, was a pioneer in searching for uranium ore in the USSR. There were some known sources of uranium ore in Central Asia, near Tashkent, and in western Kazakhstan near the Caspian Sea. Vanadium mines were possible sources of uranium ore. Mining and refinement of uranium ore was a major source of correspondence between Kurchatov, who was the prime customer, Pervukhin, who was the project manager, and the geologists working for the people's commissar for non-ferrous metals. However, the eastern

---

[72] Uranium can also burn inside a reactor – the cause of the Windscale fire in England in 1957.

[73] /APSSSR/, *ibid.*, #253 for both Kurchatov's letter and Lesov's story.

Soviet Union was a big place, not easy to get around in, and certainly not easy to prospect for good supplies of uranium ore under intense pressure to find it. So the extraction and purification of the ore was lagging, and the amount on hand in May, 1945, was only seven tons of uranium oxide, while theoretical estimates of natural uranium needed for a controlled chain reaction in a reactor was 25-50 tons of metal.[74] Many memoranda back and forth regarding the absence of refined uranium, accompanied by much finger pointing and excuses, NKVD efficiency and prison mining labor notwithstanding, failed to produce the necessary metal.

7.13. Uranium from Germany-the Soviet 'Alsos'

Chapter 3 discusses the Chicago 'pile,' the reasoning involved in choosing a uranium slug matrix in a graphite moderator, the problems of graphite purity, uranium fabrication and purity, and finally obtaining enough uranium to do the job. Fermi's uranium in 1942 came from the Belgian Congo. The head of a Brussels firm called Union Miniere had shipped to New York, and stored in a warehouse on Staten Island, drums containing 1000 tons of high grade uranium ore. When Groves heard of this, he promptly bought it for the Manhattan Project.[75] Union Miniere indirectly supplied more uranium to the Manhattan Project, and at the same time gave the Soviet Atomic Project a boost in uranium ore towards the end of WWII. This came about because the Germans captured uranium supplies from Belgium. After the Normandy invasion of France in June, 1944, Groves sent the 'Alsos' mission to Europe.[76] Their assignment was to determine the extent of the German nuclear project, and to round up German physicists. Sam Goudsmit, a Dutch-American physicist who was not himself a member of the Manhattan Project, was the scientific leader.[77] They were all issued US Army uniforms,[78] and drove around France, the

---

[74] HISAP96, Vol 2 p103 article by N.V. Knyazkaya. The F-1 reactor used 50 tons.

[75] Leslie Groves, "Now It Can Be Told," *ibid.*, p 33ff.

[76] 'Alsos' means 'grove' in Greek. They first went to Italy, then to France.

[77] Goudsmit's parents had perished in Holland, so he was aggressive in his mission. He was also fluent in German.

[78] Civilians moving with armed forces could be shot as spies. The Red Army had similar rules, and the Soviet 'Alsos' mission was also dressed in uniforms.

low countries, and later on into Germany, just behind the armed forces, in pursuit of their goals. Colonel Boris Pash was the US Army officer in charge. Along the way, they located about 100 tons of uranium ore in two different places in Belgium and France.[79] The ore was shipped back to the US. They also captured Werner Heisenberg, and other prominent German physicists, who were later incarcerated at Farm Hall in England. They determined that the German nuclear program did not amount to much. Goudsmit discovered evidence of German biological warfare while going over records in Strasbourg. Thus began 'Operation Paperclip,' which rounded up German rocket scientists and chemists.[80]

As early as January 25, 1944, Academicians Bardin and Fersman wrote to the Presidium of the USSR Academy of Sciences regarding war reparations from Germany to compensate for the destruction of laboratories, museums, libraries, and other national treasures as a result of the invasion. On July 20, 1944, the Academy of Sciences set up a special committee to participate in reparations activities.[81]

Then on February 21, 1945, an order of the State Defense Committee of the USSR established a permanent commission charged with future war reparations in occupied Germany.[82] Front commanders of the invading Red Army organized 'trophy brigades' to carry out this mission. On March 24, 1945, Beria wrote a letter to Malenkov about sending specialists to Germany to obtain laboratory equipment.[83] Then there is an entry in Beria's diary on April 21, 1945: "I talked to Georgy (Malenkov). His people [the trophy brigades] should cooperate with the physicists regarding the removal of workers and equipment involved with uranium." Other people had similar ideas. Vasily A. Makhnev, one of Beria's

---

[79] Leslie Groves, "Now It Can Be Told," *ibid.*, p 219ff. See also Samuel Goudsmit, "Alsos," AIP Press, New York, 1996.

[80] Annie Jacobsen, "Operation Paperclip," Little, Brown & Co., New York, 2014, p 5. Paperclip became official in a classified memorandum of the Joint Chiefs of Staff on July 6, 1945.

[81] /APSSSR/, *ibid.*, #205.

[82] Lavrenty Beria, "Secret Diary, 1937-1953," *ibid.*, p 415

[83] /APSSSR/, *ibid.*, #324.

deputies, sent him a memo on April 8, 1945, in which he pointed out that the Red Army had occupied the region of Silesia between Czechoslovakia and Poland, where uranium mines were known to exist, and geologists should be sent to the second Ukrainian front to locate sources of uranium ore. Makhnev also suggested attaching physicists to the third Ukrainian front to operate in Vienna, where the Radium Institute has been cooperating with the Germans on the uranium problem. He named G.N. Flerov and I.N. Golovin as likely candidates for the mission.[84] Accordingly, Kurchatov was asked to prepare a list of Soviet scientists who should go into Austria and Germany on a Soviet 'Alsos' mission, and the German scientists who should be rounded up for work in the USSR.

On April 13, 1945, the Red Army entered Vienna. The Radium Institute in Vienna had had contacts with the German Uranverein (uranium club), and some of the scientists were still in the city. Austria furnished the first opportunity for Soviet Alsos. Igor Golovin, Kurchatov's biographer and colleague at Laboratory #2, went to Vienna. Golovin reported his findings to Beria on May 4, 1945. He had learned a lot of valuable information, perhaps the most detailed knowledge of the German program in advance of intelligence work in Germany itself.[85] Here is a translation of the declassified (in 2002) part of Golovin's report:

"The Radium Institute, headed by Professor Gustav Ortner, worked on the uranium problem. He was a member of the Nazi Party since 1938…All of the Institute workers except Ortner departed with the Germans. The uranium problem was the principal assignment of the Second Physics Institute of the University of Vienna, which was organized in 1942 to include the Neutron Institute, financed by the Reichs Treasury in Berlin. Second Physics had all of the apparatus necessary for construction of a uranium reactor, but no reactor was built there. The head of the Neutron Institute was Georg Stetter, a member of the Nazi Party. Other physicists included Willibald Jentschke [after WWII a prominent West German physicist]. Most of the scientists of the Neutron Institute, except for a chemist named Brukl [sic], left Vienna with the Germans.

---

[84] /APSSSR/, *ibid.*, #331.
[85] /APSSSR/, *ibid.*, #339a.

"The project was managed from Berlin, under the general supervision of Reichsmarschall Göring. Before 1943 nuclear physics was under the direction of Professor Abraham Esau. He was replaced in 1943 by the well-known physicist Walter Gerlach. The Neutron Institute performed many experiments with interesting results bearing on the solution to the uranium problem. Most of the technical reports were taken by Stetter when he departed. Remaining documents, containing Institute affairs and preprints from other laboratories working on uranium, were burned by Dr. Gert Vambacher before the Red Army reached Vienna. The following picture comes from the dispositions of Ortner and Vambacher.

"The Kaiser Wilhelm Institute in Berlin is the center for research on the uranium problem, headed by the famous physicists Werner Heisenberg, and Walther Bothe. Other researchers include Carl von Weizsäcker... According to Ortner, the whole group departed Berlin in advance of the Red Army in the direction of Thuringia [actually they went further south, to the small town of Hechingen]. Otto Hahn heads the chemistry group at Kaiser Wilhelm working on uranium. Heisenberg and Bothe have assembled parts for variations of a nuclear reactor, the focus being on uranium-heavy water. Ortner says that they have several tons of uranium, and about 100 kg of heavy water. The results of this exercise are not known. At the present time there is no evidence that they have made an explosive mixture of uranium.

"At least three cyclotrons were built in Germany before the war. A small one was built in Bonn for medical work. The second, for 7-8 MeV protons, was built by Siemens for Heisenberg in Leipzig. It is possible that it has started up in the last six months. The third one belongs to Walther Bothe in Heidelberg. Manfred von Ardenne has built several high voltage accelerators for nuclear research at his institute in Berlin, and may also have a cyclotron there. [When Makhnev's Soviet Alsos group reached Berlin, he determined that von Ardenne did have at least a 60 ton cyclotron magnet. See #345 below.]

"I have not been able to determine the precise amount of uranium. It seems to have been under the control of Professor Gerlach in Berlin. The Auer Company in Berlin and the Shukkard Company[sic] in Gerlitz [Riehl mentions the Degussa Company in Frankfurt, but not Shukkard] are both capable of processing the uranium ore. From the documents we have obtained, it is clear that there was a radium syndicate, devoted to the processing of radium and uranium. One of the members of this syndicate is the joint stock company Trebacher Chemical Works, which owns a factory in Austria, near the town of Klagenfurt. One should keep in mind that there are radioactive sources in the upper reaches of the Danube, near Mount Gaisstein, and there may be radioactive ore in that region. [South of the actual 'Vismut' mines in East Germany, see Section 12.3.]

"The researchers began evacuation of Vienna when the air attacks started. The library and collections of the Radium Institute had been removed two years ago. Total evacuation of scientific equipment to the south of Salzburg began in January, 1945. A building was constructed near the city of Krems for a 1 MeV high voltage generator to make neutrons, but bombing of the railroads made it impossible to transport the machine from Hamburg, where it was assembled by the Muller Company, and ready to go.

"The above report shows that there was intense work in Germany during the war on the uranium problem. It is not possible to confirm the reliability of this information beyond doubt, especially since the sources were members of the Nazi Party, and were only peripherally involved in the uranium research. On the other hand, the two sources (Professor Ortner and Dr. Vambacher) independently gave consistent stories, and it is certain that:

1. The names of the researchers in Berlin are correct.
2. They were working on experiments with a large amount of uranium.
3. Industry to handle the uranium existed.

"To the extent that solution to the uranium problem has fundamental economic and military significance, it is necessary to obtain whatever metallic uranium there is in Germany, to find the industrial capability necessary for the uranium problem, and to obtain complete information on their theoretical and experimental accomplishments and practical achievements."

Signed Colonel of Engineers Ivan Golovin.

Golovin's memorandum is very interesting. It shows how much accurate information on the German Uranverein was obtained shortly before the Soviet Alsos mission began in Germany. It is an almost complete account of the German work. And it highlighted the existence of refined uranium, which the Soviet Project desperately needed. Although Nikolaus Riehl is not mentioned (see below), the Auer Company is.

Beria may have known something about the Uranverein from earlier sources, but Golovin's report gave detailed data regarding who these Germans were, and what they had been doing. Many of the people mentioned by Golovin had already escaped to the West, but not everyone.

Some German physicists preferred to work for the USSR rather than the USA. They may have been in the USSR before the war, and hence knew colleagues there. The political system may still have been attractive. They may also have felt that their services would be valued in the Soviet Union, while the West was not so interested. (This argument did not hold for the rocket people.) Manfred von Ardenne, one of the scientists mentioned by Golovin, was a very productive inventor – the holder of hundreds of patents. He wrote a letter to Stalin on May 10, 1945, volunteering his services and all of his equipment for use in the Soviet Union. This offer was enthusiastically received.[86]

Fredrick Seitz postulated that the NKVD learned of the Uranverein, its members, and their experiments through Soviet spies in Britain.[87] According to his theory, the British learned of the German work through Paul Rosbaud, an Austrian metallurgist who worked for the German publishing house Springer Verlag during the war, and who supplied information to the British. Soviet intelligence had a smoothly functioning system in Britain, but apparently they did not have access to Rosbaud's reports. Soviet intelligence has stated that most of the information obtained regarding German uranium work had come from the Red Army, that is, after the invasion of German territories in 1944-45. The Golovin report confirms this assertion. The source for German uranium was a subject of speculation, and the uranium itself was a natural target for Soviet war reparations.[88]

In any event, in May, 1945, Beria sent a Soviet 'Alsos' mission into occupied Germany.[89] On May 2, 1945, before the German surrender, a plane left one of Moscow's airports with Kikoin, Khariton, Artsimovich, and Nemenov. Avraami P. Zavenyagin, a deputy to Beria in the NKVD with the rank of major general, was in charge. Vasily Makhnev was also on board in a general's uniform. Khariton said that Makhnev was sort of

---

[86] /APSSSR/, *ibid.*, #346.

[87] Nikolaus Riehl and Frederick Seitz, "Stalin's Captive," American Chemical Society, New York, 1996, p 47.

[88] /APSSSR/, *ibid.*, #250, and #276.

[89] Lavrenty Beria, "Secret Diary, 1937-1953," *ibid.*, p 416. Also /APSSSR/, *ibid.* #352

the group secretary. He was being facetious, because Makhnev was a high ranking member of the NKVD, and although he had only modest technical training, he was very astute, and not afraid to make suggestions. He paid close attention to what was going on. In the 1930's, before the war, he worked in the Soviet equivalent of the government accounting office. He will be one of the principals when the Soviet Atomic Project takes off after August, 1945. The physicists, who were not in the military, wore uniforms of colonels in the Red Army, for the same reason as personnel in the Alsos mission. D.L. Simonenko and V.A. Davidenko were in another plane. Simonenko had the rank of major. This was not the only group going into Germany for scientific intelligence. Ya. B Zeldovich was on another mission, to investigate the German rocket industry. Georgy Flerov was there too. He and Artsimovich, both in uniform, enlisted Nikolaus Riehl, a German chemist with the Auer Company in Berlin, to come to the USSR. Riehl remembered that the colonels' uniforms did not fit all that well. Only Khariton's ears prevented his military cap from falling over his face.[90] Sailors on board the cruiser USS Indianapolis headed for Tinian made a similar observation regarding the Manhattan Project staff dressed as army officers to accompany parts of the Hiroshima bomb – their insignia was on upside down.[91]

Kikoin writes in his memoirs that when they started out, the group was troubled by the possibility that the Germans had made more progress in atomic energy than they had.[92] The same sentiment was of course a prime motivator for the Manhattan Project up until the German surrender, although the Alsos mission had put the end to American/British fears a few months earlier.

After the end of hostilities on May 7, 1945, the representative of the Soviet Union in Germany, the person who signed the surrender documents and negotiated with US commander Dwight Eisenhower and British commander Bernard Montgomery was Marshal Georgy Zhukov. Colonel-

---

[90] Nikolaus Riehl and Frederick Seitz, "Stalin's Captive," *ibid.*, p 72.

[91] Stephen Walker, "Shockwave," *ibid.*, p 56.

[92] HISAP 96 Vol 2, *ibid.*, p 181, article by I.S. Drovenikov and S.V. Romanov.

general of NKVD Ivan Serov was Zhukov's deputy, and worked closely with the Marshal. Serov's diaries give some insight into the Soviet Atomic Project in post war Germany and have been posthumously published.[93] Avraami Zavenyagin met with Serov in Berlin in May and informed him of the status of the Atomic Project, the goals of Zavenyagin's mission to gather uranium, heavy water, and scientists from Germany, and the increase in support from Moscow. Serov was to play an important role in the exploitation of German uranium mines.

The success of the Manhattan Project was to become the motivator for the Soviet Atomic Project, dated from the Hiroshima bomb, an event that could not remain secret – only three months after the German surrender.

Laboratory #2 in Moscow was the principal source of Soviet scientists sent to Germany in May, 1945. About 40 people in all were travelling around occupied Germany trying to obtain information, equipment, and researchers. This was a substantial fraction of the total number of physicists at Laboratory #2 at that time. Kurchatov's work must have slowed down to a crawl, awaiting their return.[94]

Kikoin, Khariton, et al landed in Berlin, and went to the Kaiser Wilhelm Institute. The Berlin region is shown in Map #4. They learned that Werner Heisenberg, Otto Hahn, Max von Laue, Carl von Weizsäcker, and others had left for West Germany in February (most of them were found by Alsos near Hechingen in the Black Forest region of southwest Germany).[95] They asked to see the files, which were locked in safes, but were obligingly opened. Due to a mixup in orders, the documents had not been removed or destroyed, and were largely intact regarding German uranium project. They were trying to make a chain reacting pile using heavy water from Norway as the moderator, with metallic uranium and uranium oxide. They found two 5 liter flasks of heavy water, labeled

---

[93] Ivan Serov, "Zapiski iz Chemodana" (Notes from a suitcase), Prosveshenie, Moskva, 2017, p 305-306.

[94] Pavel V. Oleinikov, "German Scientists in the Soviet Atomic Project," The Nonproliferation Review Vol 7 #2, (2000) p 5.

[95] Richard Rhodes, "Making of the Atomic Bomb," *ibid.*, p 609.

'Norsk Hydro.' Some of the equipment, including some crude apparatus for isotope separation, was dismantled and sent to Moscow. According to the documents, Prof. Harteck at Hamburg was working on the centrifuge method of isotope separation, without much success.

Kikoin recalls: "We fulfilled the orders of the Government by inviting Professors Herz, Manfred von Ardenne, and Peter Thiessen to work in the USSR."

Khariton recalls: "We decided with Kikoin to look for German uranium, given that the German army occupied all of western Europe, and that Belgium, through the Belgian Congo, had large supplies of uranium ore. We told Zavenyagin of our plans, and he wholeheartedly supported the idea. He furnished us with a car and driver, so that we could travel freely all over Germany."

Since they were in Berlin already, they decided to look in factories near the city. The first one they went to made gas masks. This may have been a clue, since the Auer Company also made gas masks, and processed uranium, but there is no obvious relation between the two operations. In any event, the first factory was a dead end. The second one, in Grunau, had a basement room with yellow powder, and some heavy metal, plus records which indicated that 100 tons of uranium had been moved from there in February, 1945, to a firm called 'Rohes.' The question was, where did it go? Kikoin and Khariton hopped in the car and went north to the small town of Parchim, on a hint that the uranium had been shipped there. That was a wild goose chase. They found the town, but no uranium. They returned to Berlin. They studied a map to see where a railroad line went between Grunau and Parchim – maybe the stuff was off loaded somewhere along the way. But the rail lines were all destroyed.

Back to looking for 'Rohes.' On the manifest was a Berlin address, but that building had been completely destroyed. They were directed to another building on the outskirts of Berlin that was still standing - seven stories high, and full of German office workers, who were totally uncooperative. But the place had card index files, and Kikoin and Khariton started combing through them. Bingo! They found a card which said 'U 43 0O 48 0' – not too helpful, but it did say 'U.' No other information, like how much or where. Pressing questions on the uncooperative office staff, our 'detectives' were directed to a large warehouse where they found a barrel or two of uranium oxide. They were told that the Red Army had been by, and had obtained some of the yellow powder to paint a building for their headquarters.

In Potsdam, they learned the name of the person in charge of the operation by which the Belgian uranium was shipped to Germany, and they alerted 'SMERSH' to find him.[96] He was delivered in two days. He was a dedicated Nazi, and had been the director of the 'Rohes' firm. Interrogation got them nowhere, so they asked the military to help persuade him. The next day they were informed that the load had been shipped to Neustadt. There were about 20 towns named Neustadt in Germany, 10 of which were in the Soviet occupation zone. Kikoin and Khariton set out to visit them one by one, and the tenth one (naturally the last one they looked at), Neustadt am Glewe, had a tanning factory. The tanning factory was in operation, already shipping hides to the USSR. There was a large storage building. It was full of barrels, labeled 'lead oxide,' which was a chemical used in tanning leather. They did not bother to look at them.

The chief engineer of the tanning factory said that the local authorities had requested that they store a load of barrels from 'Rohes'. He said that those barrels were in another corner, away from the lead oxide. Kikoin and Khariton both tried to look calm, but were in fact pretty excited. Thanking the engineer, they ran to the storage shed and found the long sought missing uranium oxide. There it was![97]

---

[96] 'SMERSH' was an acronym for a military branch of the NKVD, active in the combat zones, and headed by Victor Abakumov, a deputy to Beria.

[97] This story comes from the article in HISAP96, Vol 2, p 179 by I.S. Drovenikov and S.V. Romanov.

The next morning Kikoin reached Zavenyagin by telephone. Zavenyagin thought he was playing tricks, so Kikoin said: "This is COLONEL Kikoin speaking. I request that you dispatch a column of trucks under my command to transport this valuable cargo." The trucks arrived the next morning, and with help from the locals the loading was finished in one day.

Khariton had one more story about the loading. Some barrels were full, but some only half way, and for shipping by rail they should all be full. A group of local women were set to combining uranium oxide from partially filled barrels. Khariton said that they should be sure to wash their hands when they got through work. This caused a panic – are we working with poison? Khariton took off his military jacket, put his hands in one of the barrels, and rubbed both hands with uranium oxide. This seemed to calm the workers, and the filling process continued. More than 100 tons of uranium oxide was shipped to Moscow.

In his old age Academician Kikoin said that the six weeks he spent in Germany in 1945 were the most interesting in his life.

This is a good detective story. Kikoin was 37 years old at the time, and Khariton was 41. They were academics, but dressed in army uniforms with colonel shoulder boards. They were probably not armed – in any event not trained in the use of military weapons - but everyone else was. Eastern Germany was a very dangerous place in the spring of 1945, with widespread destruction and the absence of law and order. However, they ran all over Germany looking for uranium stolen from Belgium, and they found it without getting killed! It would make a good movie. The story as told by Drovenikov and Romanov does not dwell on personal dangers, and support for the mission given by Red Army commanders. Groves explicitly states that the US Alsos mission had at all times excellent support from Allied commanders in the European Theater.[98] It is safe to assume that Kikoin and his colleagues enjoyed similar cooperation. The Soviet Archives contains a memorandum dated May 5, 1945, to the

---

[98] Leslie Groves, "Now It Can be Told," *ibid.*, p 236.

commanders of the first and second Belorussian fronts and the first Ukranian front to furnish all necessary supplies, transport, and protection to the Soviet Alsos mission led by V.A. Makhnev, deputy member of the state defense committee.[99]

Later, early in 1946, Kurchatov wrote that until May, 1945, there was no hope of building a uranium-graphite reactor, because there was only 7 tons of uranium oxide in hand, and no prospect of getting 100 tons before 1948. The Soviet Alsos mission changed all that by sending 300 tons of uranium oxide from Germany to the USSR.[100]

Beria notes in his diary on October 10, 1945, that according to Zavenyagin the Americans in April took 1000 tons of uranium ore from the part of East Germany which is now in the Soviet occupation zone. "Zavenyagin suggests that we request through channels that the US return the ore to us! What an idiot! He doesn't understand that we would be revealing our own interest in the material... They would not give it to us anyway. They are not fools. It would be better to work with our Germans (Riehl and his associates)."[101]

Flerov found himself in Dresden in late May, 1945. Dresden had been thoroughly destroyed by Allied bombing in the war. He was looking for evidence that the German Uranverein had carried out any significant experiments. He had two different approaches to the problem. One was to look for residual radioactivity with Geiger counters and ionization chambers. The other was to interview Soviet citizens who had been prisoners in Germany, and who were being repatriated back to the Soviet Union. The goal of the interviews was to find out if the prison workers had any knowledge of the German uranium project.[102] Given the chaos in Germany at that time, the same filter could catch German scientists who were by chance in the crowd. Flerov said that there were 10,000 to 15,000

---

[99] /APSSSR/, *ibid.*, report #342.
[100] /APSSSR/, *ibid.*, #348.
[101] L.P. Beria, "Secret Diaries 1937-1953," *ibid.*, p 430.
[102] /APSSSR/, *ibid.*, #355. Letter from Flerov to Kurchatov.

people every day flowing through three allocation points in the vicinity of Dresden, and that there were between one and two million former Soviet citizens in Germany. They were returning to the Soviet Union, but most of them were not going home – they would be transferred to camps in the USSR. This cruel policy applied to Soviet soldiers and civilians unfortunate enough to survive years in German prisons, only upon victory to be herded to Soviet prisons after the war. Soviet citizens exposed to western thought and customs – even in a prison camp environment – were considered unreliable. Marshal Zhukov said: "126000 officers who returned from captivity were stripped of their rank and sent to the camps."[103]

Just how Flerov was supposed to interrogate these individuals is not clear. He said that it would be useful to have some coarse screening done by the NKVD during processing, with scientists from Laboratory #2 interviewing a subset that seemed most promising. They were looking for information about German progress, but were also interested in locating technical personnel for employment in the Soviet Atomic Project. Whether or not this exercise was of any use is not documented.

The liberation of Soviet military and civilian prisoners from Germany, only to send them to prison camps in the Soviet Union, was not the only unpleasant mass deportation activity of the NKVD towards the end of WWII.[104] Early in the war nationalism was encouraged as a motivation to fight. Fight for Ukraine, or Georgia, or Armenia, wherever you came from. The German invaders encouraged nationalistic tendencies on local inhabitants. In the Caucasus, for example, the Germans supported independence for Chechens, Ingush, and Kalmucks. But by 1944, after the Germans had departed, these nationalistic feelings became an anathema. There was only one USSR. A new reign of terror, reminiscent of 1937-38, returned to society. Large numbers of people were deported to the east, as

---

[103] Edvard Radzinsky "Stalin," Doubleday, New York, 1996, p 506.
[104] Deported ethnic populations 40 odd years later contributed to the demise of the USSR.

had happened at the beginning of the German invasion.[105] Beria and the NKVD were responsible for expediting this deportation, in which hundreds of thousands of men, women, and children were packed into freight cars and shipped by rail from their homes in Georgia, Crimea, and other regions of the western Soviet Union to Siberia.

7.14. German Scientists Join the Atomic Project

We shall return to the fate of most of the German scientists, and their contribution to the Soviet Atomic Project, in Chapter 13. Nikolaus Riehl, however, is of immediate interest, because he was the chemist who refined the uranium that Kikoin and Khariton found so it could be used in Kurchatov's reactor F-1, built in 1946.

Nikolaus Riehl was born in St. Petersburg, Russia, in 1901. His father was a German engineer with Siemens, working in Russia at the time. He grew up in St. Petersburg, and although he attended German language schools, he became completely fluent in the Russian language as well.[106] His family left Russia in 1918, and he obtained a doctoral degree in chemistry from the Institute of Otto Hahn and Lise Meitner in Dahlem. He then joined the Auer Company, where pitchblende was refined to produce radium. The left over uranium was at that time of no commercial value. The Auer Company refined thorium, and put it in toothpaste![107] The company also made gas masks, an activity that continued after the war when the plant was in East Germany. After the discovery of the fission of

---

[105] Edvard Radzinsky "Stalin Zhizn' I Smert'," AST, Moskva, 2010 p 644 (Russian version)

[106] Nikolaus Riehl and Frederick Seitz, "Stalin's Captive," *ibid.*, p67ff. This is a translation by Seitz of Riehl's original article published in German "Zehn Jahre im goldenen Kafig" (Ten Years in a Golden Cage). Seitz's translation is very bland, considering the subject matter, but this could be due to reticence on Riehl's part.

[107] The late Eugen Merzbacher, author of a popular quantum mechanics text, and former president of the American Physical Society, used radioactive toothpaste as a boy. His father was also a chemist at the Auer Company. Letter to the Editor of Physics Today, 2012, and private communication from Merzbacher to the author. Goudsmit in "Alsos" tells of a hunt for thorium missing from France which was to go to the Auer Company for more toothpaste after the war. ("Alsos" p 64).

uranium, Riehl became interested in uranium chemistry for a potential power industry. Auer was a supplier of uranium to Heisenberg and others in the Uranverein.

Artsimovich and Flerov invited Riehl to a guarded home in Berlin-Friedrichshagen, where he stayed for a week. Then he was taken to the Auer Company factory in Oranienburg, north of Berlin. The factory had been flattened by Allied bombing in March, 1945. Nevertheless, a substantial amount of information and machinery still remained. The Soviets were busy removing all of the equipment that was still usable, and not nailed down. They realized because of the timing of the raid that the Allied bombing was directed against the Russians, not the Germans.

On June 9, 1945, Nikolaus Reihl, his family, and some of his coworkers were flown to Moscow. They were housed in a villa called "Ozyora" that had been the residence of Genrich Yagoda, head of the NKVD from 1934 until he was executed in 1938. It was still called "Yagoda's dacha," and continued to be used by the Soviets to accommodate transient German scientists between 1945 and 1950. During the war, it was the home of Field Marshal Paulus, who surrendered at Stalingrad in January, 1943. Paulus was shipped off to Nuremberg to testify for the prosecution against Nazi leaders at the war crimes trials.[108] Riehl had no time to rest. A suitable location for the uranium purification factory had to be found. As Kurchatov found in 1943, there were still several abandoned factories in the Moscow area – the result of moving industry to the east in 1941-42, and not moving it back. After touring several places, they finally settled on a munitions factory, decommissioned at the end of the war, located at Elektrostal, in the vicinity of the small town of Noginsk, about 60 km east of Moscow. The name Elektrostal was derived from a steel plant, which was built by Germans before the war. In fact, the Auer Company workers were the third German group to occupy the site. Riehl said that it was a dreadful place to live, but they stayed there for five years. One may assume that the Elektrostal Germans enjoyed similar privileges to the aerospace Germans discussed in Chapter 4 – the best housing available, special

---

[108] Eugene Davidson, "The Trial of the Germans," McMillan Co, New York, 1966, p 361.

privileges in stores, and higher salaries. They were able to correspond with relatives and friends in East Germany, but were not allowed to travel, even within the Soviet Union, without special permission. Permission to travel, say to Moscow on official business, would include a human escort. The perimeter of the town was surrounded by barbed wire. According to Riehl, each German family at Elektrostal lived in a prefabricated wooden house made in Finland – 'Finnish' houses mentioned in Chapter 4 for German rocket scientists, and again in Chapter 9 for Igor Kurchatov. Each house in Elektrostal had three rooms, a kitchen, a bath, and an outside garden.[109]

Seitz says that Riehl had no more than a worm's eye view of the Soviet Atomic Project.[110] His responsibility, the refinement of uranium ore, was only one part of the operation. He did, however, attend some high level meetings, and his command of the Russian language gave him an advantage as an outsider interpreting what was going on. One can only speculate on the reasons for his drab account. He was probably sworn to secrecy, and may well have feared retribution by the NKVD, even though he lived in West Germany after 1955.

In June, 1945, there was no uranium purification plant in the USSR capable of supplying Laboratory #2 with the 50 tons of pure uranium needed for their first reactor. Boron and cadmium, and the rare earths had to be removed from the uranium to less than a few parts in 10^6 in order to suppress the capture of slow neutrons which would otherwise be available for fission. High purity can be a challenge. It was Riehl's assignment to produce the required product.[111] He had his German crew, and what equipment could be salvaged from Berlin. Soviet workers joined the effort. Riehl obtained Beria's permission to interview German POW's with technical skills for possible work in the factory. The prisoners were told that they would be doing important work, and that they would have improved living conditions. Their families could join them from Germany.

---

[109] Nikolaus Riehl and Frederick Seitz, "Stalin's Captive," *ibid.*, p 165.

[110] Nikolaus Riehl and Frederick Seitz, "Stalin's Captive," *ibid.*, p 89.

[111] HISAP96 Vol **1**, p 146, F.G. Reshetnikov, "Establishment and Development of Industrial Production of Uranium," confirms that Riehl's plant was the supplier of uranium to the F-1 reactor at Laboratory #2.

But the down side would be that they would not be repatriated to Germany as early as the ordinary POW's – they would enjoy a lighter sentence, but for a longer time. Not everyone agreed to these terms. A Soviet escort always accompanied Riehl on his visits to interview prisoners. The Russian usually did not understand German, so Riehl could discuss the matter freely with the POW's. Two German prisoners joined the Elektrostal operation.

The Elektrostal group consisted of 14 German technicians, plus the two POW's, making a total of 31 people in town.[112] A large vacuum oven was missing from the original equipment from the Auer Company. Riehl appealed to Zavenyagin, who got on the phone, and located the missing oven in Krasnoyarsk, on the Yenisei River in Siberia. It was soon returned to Elektrostal by cargo plane. They started up using a technique called 'fractional crystallization' invented by Auer von Welsbach, the founder of the Auer Company, and also used by the Curies in the discovery of radium. The process was complicated, and could allow undesirable impurities. Two examples with large absorption cross sections for thermal neutrons are boron and cadmium. Riehl's group persevered with fractional crystallization, and managed produced a ton of reactor grade uranium oxide in a few days in the beginning of 1946. But a more efficient method was needed. Then Riehl learned about the Smyth Report. Published in August, 1945, about the time of the Hiroshima bomb, the Smyth Report was a declassified account of the Manhattan Project. It contained no information regarding ordnance, either the design of the gun type uranium bomb or the implosion rig to compress the plutonium bomb. Many specifics were omitted, but the book still contained a lot of useful information on many topics. Uranium processing, isotope separation, and plutonium production were covered in some detail. Appendix 4 described Fermi's first chain reacting pile, CP-1. There is no evidence that the Report contained disinformation intentionally intended to mislead others. Anyone reading the Report was convinced that the Manhattan Project was a vast and expensive industrial undertaking. The Soviet Atomic Project obtained the Report from several sources, and it was translated into Russian and

---

[112] Pavel Oleinikov, "German Scientists in the Soviet Atomic Project," *ibid.*, p 15.

carefully analyzed. On page 93 Section 6.12 begins with the sentences: "Experiments at the National Bureau of Standards by J.I. Hoffman demonstrated that, by the use of an ether extraction method, all the impurities are removed by a single extraction of uranyl nitrate. The use of this method removed the great bulk of the difficulties in securing pure oxide and pure materials for the production of metal."[113] When Riehl read this, he knew what to do to improve the uranium refinement process.[114] By June, 1946, the "ether enterprise" at Elektrostal was up and running.[115]

There were ongoing uranium refinement efforts before Riehl arrived on the scene. One was a metallurgical laboratory at NII-9 NKVD in Moscow. The deputy director of that laboratory, Fedor Grigorevich Reshetnikov was interviewed by the journalist Vladimir Gubarev.[116] Reshetnikov confirmed that the uranium for the F-1 reactor, and also for the 'A' reactor for plutonium production at Combine #817, were obtained using Riehl's refinement technique, which he referred to as the 'German method.' He claimed, however, without giving any technical details, that a superior technique was developed by NII-9, and was used for uranium refinement for later reactors.

7.15. Reactor Grade Graphite

With German uranium refined by German chemists, and graphite produced by the Moscow Electrode Factory, the prism test set ups for quality control of materials to go into the graphite-uranium reactor could get under way at Laboratory #2. The tent that burned had been restored. Moscow Electrode supplied graphite for processing aluminum. Their graphite was initially totally unsatisfactory for a moderator in a nuclear reactor. On February 22, 1945, P.F. Lomako, the people's commissar for

---

[113] Henry D. Smyth, "Atomic Energy for Military Purposes," Princeton, 1945, p 93.

[114] Chemical separation using ether was not new. Fermi used similar techniques in parsing fission fragments at Columbia in 1940, with an occasional ether fire. Collected Papers of Enrico Fermi, Vol 2, p 41.

[115] Nikolaus Riehl and Frederick Seitz, "Stalin's Captive," *ibid.*, p 95.

[116] Vladimir Gubarev, "Atomnaya Bomba," Moskva, Algoritm, 2009, p 396.

graphite.[117] Lomako reported that the factories on hand could not supply the required purity, and that the Moscow Electrode Factory would have to expand its space to develop production techniques. The 600 tons per year needed for the reactor development at Laboratory #2 simply could not be reached without new techniques of manufacturing. He estimated that the new plant could be running by 1946. Impurities were measured at the plant by the ash content, which is the residue after the carbon has been converted to CO_2. Experiments at Laboratory #2 had determined that the ash content must be below 0.04% by weight.[118] This limit was obtained by measuring the neutron absorption cross section for the graphite samples with Ra-Be neutrons, and comparing the cross section to the ash (that is, non-carbon) content. The cross section for pure carbon was $\sigma = 3 \times 10^{-27}$ cm^2. The difference between the pure carbon cross section and the measured (larger) value was linearly related to the amount of ash. Pomeranchuck calculated that a graphite-uranium reactor would work if $\sigma < 5 \times 10^{-27}$cm^2, which allowed for some contamination of the graphite. The Laboratory #2 scientists devised a measurement - the ash content -which could be done in the factory, without the use of neutrons, and which tracked the absorption cross section for small contaminations. Moscow Electrode had to develop a satisfactory industrial process to produce the pure graphite, and could evaluate its own progress towards the goal.

Yefin Pavlovich Slavsky was a metallurgist working in aluminum refining. Moscow Electrode supplied the graphite electrodes for refining aluminum, so Slavsky was well acquainted with the plant, but he knew absolutely nothing about the Soviet Atomic Project. He had read two articles by Igor Tamm on nuclear physics, and was confused by descriptions of the transformation of one atom into another, a process that he was taught in school was quite impossible.[119] Lomako asked him if he knew 'the Beard' (Kurchatov), the man who was demanding pure graphite

---

[117] /APSSSR/, *ibid.*, #314.

[118] /APSSSR/, *ibid.*, #313.

[119] Transcription of an interview by Raisa Kuznetsova of Yefim Slavsky, published in "Kurchatov v Zhizni," p 479ff. He was Minister of Medium Machine Building in 1986 – Chernobyl.

for neutron moderation in a reactor? Slavsky did not know Kurchatov, and he did not know what a neutron was, but he would soon learn all about it – more than he wanted to. Slavsky was first to become engaged in the problem of graphite purification at the factory, and supplying the material to Laboratory #2. Soon after he was recruited by Kurchatov for reactor work, and became a leader in reactor development in the USSR. But during the development of the techniques for graphite purification Slavsky was on both sides – a supplier, and then a consumer.

At first Moscow Electrode shipped graphite blocks out the door, and Laboratory#2 paid for the deliveries. Slavsky kept a record – 10 kg yesterday and 20 kg today, and reported the amounts to Makhnev, the secretary to Beria's committee. Everything was going smoothly until Vladimir Goncharov, one of Kurchatov's deputies at Laboratory #2, told him that Moscow Electrode had, in fact, not delivered a single gram of usable graphite. The samples delivered had been tested to measure the neutron absorption, and none were pure enough for the reactor. This caused a big flap at the factory, since Beria was responsible for their very existence. Kurchatov appeared at the factory, and suggested that Slavsky transfer from non-ferrous metals to the atomic project, as Boris Vannikov's deputy. This personnel change was authorized by Stalin himself, and Slavsky, who was very dubious about going into something at age 46 that he knew absolutely nothing about, was assured that after a year or two he could return to non-ferrous metals. Of course, the new career lasted the rest of his working life.

First the graphite purity problem had to be solved. This was done by heating carbon to a red glow in a chlorine atmosphere. The impurities reacted with the chlorine, formed gaseous compounds, and evaporated. The resulting purified graphite was then sent to Laboratory #2. Slavsky recalled that on his 70th birthday, in 1968, his colleagues gave him two blocks of pure graphite as a memento of past struggles for the Soviet Atomic Project.

Laboratory #2 certified the purity of the graphite by neutron absorption measurements in a graphite prism test. A prism test consisted of a rectangular graphite stack, with square cross section, and slots at various depths to allow the insertion of uranium, BF_3 neutron counters, or foils. Typical setups are shown in Figures B.1 and B.2 in Appendix B, where some measurements of graphite purity and uranium purity for the first Soviet reactor are described. All of the graphite and uranium that went into the reactor, 600 tons of graphite and 50 tons of uranium, passed through these tests. The graphite was machined in blocks 10cm×10cm×60cm, weighing about 10 kg each, so the reactor contained about 60,000 blocks. Riehl's factory supplied the uranium in aluminum clad cylinders 32 mm diameter and 100 mm long, about 1.4 kg each, giving 36,000 uranium slugs, which were distributed in a matrix through the central portion of the graphite 'pile.' The engineer in charge of graphite purification at Moscow Electrode, G.K. Bannikov, was awarded a prize of 150,000 rubles for his success.[120]

7.16. Potsdam Conference Revisited

July was a busy month in 1945. In Chapter 6 we sent President Truman to the Potsdam Conference, which started on July 17, 1945. Potsdam is a suburb of Berlin. The conference was held there because the city of Berlin itself was destroyed. Delegates were housed in the vicinity of the conference site, Cecilianhof, which still stands. The 'Big Three,' Roosevelt, Churchill, and Stalin, had met at Yalta in the Crimea in February, 1945, while the war in Europe was still very much ongoing. The Red Army had not penetrated the eastern borders of Germany. Several topics first discussed at Yalta – one example being the post war disposition of Germany - were to be taken up again about three months after the cessation of hostilities. The time and place for that meeting –in July, at Potsdam – was set by Harry Hopkins, one of Roosevelt's aids, and by Stalin. There has been speculation that the date was chosen to coincide with the Trinity Test for Truman's political advantage, but there is no

---

[120] /Atomnyy_proekt_SSSR.T.2.Kn.3.(2002).pdf #75.

evidence that Truman, who had a lot on his mind, tinkered with the dates for the conference.[121]

The big three had changed between February and July, and was to change again during the Potsdam Conference. Franklin Roosevelt died in April, 1945, and was replaced by his vice president, Harry Truman. General elections in Britain ousted Churchill's government during the conference, and Clement Attlee, who had accompanied Churchill to Potsdam, took his seat at the table. The Big Three were then Truman, Attlee, and Stalin. Stalin commented: "Western democracy is a worthless system if it could exchange the great Churchill for the pathetic Attlee." [122]

The Trinity test of the first plutonium implosion bomb took place in the New Mexico desert the day before, July 16. Secretary of War Henry Stimson, who accompanied the president to Germany, received a coded message from Washington at 7:30 in the evening, dated July 16 which read:

"Operated on this morning. Diagnosis not yet complete, but results seem satisfactory and already exceed expectations. Local press release necessary as interest extends great distance. Dr. Groves pleased. He returns tomorrow. Will keep you posted."[123]

Thus did Truman learn of the success of the Trinity test. Of course, Groves was not a doctor, he was a major general, in charge of the Manhattan Project. It is interesting that the message mentions a press release. It came from the Alamogordo Air Base, where flights had been

---

[121] Bert Cochran, "Harry Truman and the crises presidency," Funk and Wagnalls, New York, 1973. p 156.

[122] Edvard Radzinsky, "Stalin Zhizn' I Smert'," AST, Moskva, 2010 p 656. This is my translation of Stalin's quote. Another rendition is in the English version, *ibid.*, p 509. Radzinsky is a popular Russian writer, and much of the flavor can be lost in translation. His biography of Stalin is very earthy, in keeping with the subject.

[123] Quoted by Harriman: W. Averell Harriman and Ellie Able, "Special Envoy to Churchill and Stalin," Random House, New York, 1975, p 489.

grounded on July 16, and was intended to calm members of the local population who may have seen a bright flash or heard a loud noise. It said:

> "Several inquiries have been received concerning a heavy explosion…A remotely located ammunition magazine containing a considerable amount of explosives and pyrotechnics exploded. There was no loss of life or injury to anyone…Weather conditions…may make it desirable for the Army to evacuate temporarily a few civilians…"[124]

The release got some circulation in the US. It did not fool chemists at the DuPont Company, who knew that pyrotechnics and high explosives would never be stored in the same place.

Stimson took the message at once to Truman in the Little White House on Kaiserstrasse, about 5 km from the meeting site. The next day Stimson told Churchill that the atomic bomb was a reality. The Prime Minister was very interested in the news, and was "strongly inclined" not to share it with Stalin.[125] A full report to Stimson, containing the results of some post mortem measurements of the energy of the explosion (approximately 20,000 tons of TNT), the extent of the blast damage, and the amount of radiation was received in Germany on July 21. Truman had a new weapon in his pocket, and felt energized in his dealings with the Soviet delegation regarding the disposition of Germany, whether the USSR was entitled to war reparations from the part of Germany under the control of the Western Allies, and many other matters pertaining to peace in Europe. Stalin was a skilled negotiator, and had a very large military force in Eastern Europe. President Truman was not optimistic that Soviet ambitions in that area could be restrained. There were also issues regarding the entry of the Soviet Union in the war against Japan. [126] By this time operations were well under way for the bombing of Hiroshima and Nagasaki, about two weeks hence, after the conference had ended. Although Stalin knew about the Trinity Test in advance, it was the destruction of Hiroshima, an event

---

[124] Leslie M. Groves, "Now It Can Be Told," *ibid.*, p 301.

[125] W. Averell Harriman and Ellie Abel, *ibid.*, p 490.

[126] Bert Cochran, "Harry Truman and the crisis presidency," Funk and Wagnalls, New York, 1973, Chapter 9, p 158-175.

for which espionage was not necessary, that convinced him of the reality of nuclear weapons.

Stalin did not like to fly. His son Vasily was an expert pilot, and an officer in the Red Army Air Force, but Vasily's father preferred ground transportation. So it was that a train carrying Stalin, Molotov, and their entourage left Moscow for Berlin at 5:30 pm on July 16, 1945. TASS (the Soviet news agency) reported on July 16 that Stalin was already there – no doubt a ruse to confuse possible assassins.[127] Fifteen hundred NKVD guards saturated every kilometer of the route, so that the train could pass through unscathed.[128] Another 2000 guarded the palace where the meeting took place. Before Stalin left, Beria received a note from Vsevolod N. Merkulov, People's Commissar for State Security, that reliable intelligence agents reported that the first explosion of an atomic bomb, made of ^{239}Pu, was expected around July 10.[129] It is natural to assume that Beria so notified Stalin. In fact in Beria's diary for July 13, Stalin is quoted asking:" The tenth has come and gone, what about the explosion?"[130] Beria said that maybe it misfired. Stalin said perhaps it was all a bluff, but Beria didn't think so. Then Stalin observed that three tons of explosives would be a truck load, but 5000 tons would be a train load! Something to think about.

The Potsdam Conference chugged along at its tasks. Germany was divided into zones of occupation, an arrangement that persisted until the mid 1950's. First there were three zones: British and American in the West, and Soviet in the East, with the Elbe River, where US and Soviet troops first met in 1945, serving as the dividing line in the North. Then France was given a zone, carved out of the West. This arrangement placed Berlin, the German capital, smack in the middle of the Soviet zone. Berlin was divided into four parts also, creating West Berlin accessible only through East German territory. This arrangement, a legacy of Potsdam, led

---

[127] L.P. Beria, "Secret Diaries 1937-1953," *ibid.*, p 421.
[128] Edvard Radzinsky, "Stalin," Doubleday, New York, 1996, p 508.
[129] /APSSSR/, *ibid.*, #371.
[130] L.P. Beria, "Secret Diaries 1937-1953," *ibid.*, p 421.

to many problems in East-West relations, and was a hot button throughout the cold war. In 1948 there was the Berlin blockade and the airlift, and in 1961 there was the Berlin wall. Berlin was one of the few places where US and Soviet troops confronted each other directly. On the other hand, viewed from the 21st century, impractical and divided Berlin played an important role in the cold war, in providing a window on the West, and giving hope to some of Berlin's population. The alternative would have been all East Berlin.

Truman decided to inform Stalin of a new weapon, but to do it in an informal way, as if making conversation to pass the time. On July 24, 1945 during a break in the conference Truman recollected: "I casually mentioned to Stalin that we had a new weapon of unusual destructive force. Stalin showed no special interest. All he said was that he was glad to hear it, and hoped we would make 'good use of it against the Japanese.'"[131] Vladimir N. Pavlov, Stalin's English-Russian translator, in his memoirs, said that Stalin said nothing in reply to Truman's news, and walked away.[132] On the other hand, Charles Bohlen, Truman's translator, is quoted by Harriman as saying: "So offhand was Stalin's response that there was some question in my mind whether the President's message had got through. I should have known better…"[133] Harriman then recalled a conversation with Molotov after they had returned to Moscow from Potsdam. Harriman said: "Molotov looked at me with something like a smirk on his face and said, 'You Americans can keep a secret when you want to.' The way he put it convinced me that it was no secret at all."[134]

7.17. Beria's Committee

After Hiroshima in August, the Soviet Atomic Project was placed in the fast lane. Beria notes in his diary on August 18, 1945, that Stalin has

---

[131] Truman Memoirs, Vol I, p 416, quoted by Averell Harriman and Elie Abel, *ibid.*, p 491.
[132] V.N. Pavlov, quoted by Graham Farmelo, "Churchill's Bomb," *ibid.*, p 302.
[133] Charles Bohlen, "Witness to History 1929-1969," Norton, New York, 1973, p 237, quoted by Averell Harriman and Elie Abel, *ibid.*, p 491.
[134] Averell Harriman and Elie Abel, *ibid.*, p 491.

put him (Beria) in charge of the 'Uranium Work,' and that things should get going quickly.[135] Accordingly, the State Defense Committee issued order #9887 on August 20, 1945, which set up a Special Committee on the Uranium Problem headed by Beria (who displaced V.M. Molotov in this job). Included on the committee were government ministers with military rank: M.G. Pervukhin (chemical industries), A.P. Zavenyagin (mining, deputy to Beria), and B.L. Vannikov (armaments). Beria insisted that Vasily A. Makhnev, one of the leaders of the Soviet Alsos mission to Germany, be a member of the committee. Scientists on the committee were Igor Kurchatov and Peter Kapitza. Two members of the Politburo, the top group in 'collective leadership' headed by Stalin, who also joined the committee were Nikolai A. Voznesensky and Georgy Malenkov. In the fall of 1945 the volume of work on the uranium problem significantly increased, and time became of the essence.[136] The ministers - Pervukhin, Zavenyagin, Vannikov, and Makhnev – already had substantial involvement in the project, as did Kurchatov. Makhnev became the secretary of the committee. He was one of Beria's deputies, but not really an NKVD intelligence operative. We have already commented on the skill and intelligence that is reflected in his written correspondence. Pavel Sudoplatov, as head of Section 'S' of the NKVD, responsible for intelligence in atomic research, was an adjunct member of the committee, and attended most of its meetings.[137] Beria's committee had complete authority over the resources of the country for the development of nuclear weapons.

A working group within Beria's committee was chaired by Vannikov, and given the generic title "Primary Management Committee," with no hint of its secret purpose. In Russian it was called "Pervoe Glavnoe Upravlenie," or PGU for short. The PGU had two working subcommittees, a Scientific and Technical Subcommittee, chaired by Vannikov, and an Industrial and Engineering Subcommittee chaired by Pervukhin.

---

[135] L.P. Beria, "Secret Diaries 1937-1953," *ibid.*, p 425.

[136] HISAP96, Vol 2, *ibid.*, p 102. Article by N.V. Knyazkaya, "I.V. Kurchatov and the first stages of solution of the atomic problem."

[137] Pavel Sudoplatov, "Razvedka I Kreml," *ibid.*, p 238.

Kurchatov was a member of the scientific subcommittee, but also a regular attendee of the industrial one. The committees met once a week in the early going.[138] In April, 1946, the two subcommittees were merged into one, chaired by Vannikov, with Kurchatov and Pervukhin as deputies. In December, 1949 Kurchatov became chair.[139] The PGU had complete power to orchestrate the entire project with all of its branches. It was located one level below Joseph Stalin. Beria's Special Committee met 84 times between its establishment and the test of the first nuclear weapon in 1949. More than 1000 orders were promulgated during that four-year time span.[140]

Kapitza was appointed because of his stature in the scientific community, his experimental skills, and his engineering talent. Kapitza had not worked on problems in nuclear physics since his early days with Rutherford in Cambridge. He was, however, an active member of the Soviet Academy of Sciences, and participated in discussions about the practicality of atomic energy. He headed a laboratory in Moscow devoted to low temperatures and material science. Landau worked at the same place. Kapitza had interacted with the Soviet government at the highest levels on several occasions in the process of setting up his laboratory, and continued to do so. As discussed in Chapter 2, he also developed an industrial scheme for liquefying air to obtain large amounts of oxygen, used in metallurgy and other applications. His technique was more efficient than procedures then in use. Engineers and factory managers with a stake in the old procedure formed a movement to undermine confidence in Kapitza and his ideas. However, in 1943 Kapitza became State commissar for industrial oxygen. In his new post Kapitza interacted with V.M. Molotov, Georgy Malenkov, and their deputies. In the spring of 1945 he was awarded a gold star as Hero of Socialist Labor.[141] This was the high water mark in Kapitza's influence. It is not clear who recommended Kapitza for the committee, but it was not Beria. Malenkov, with Stalin's

---

138 Atomnyy_proekt_SSSR.T.2.Kn.4.(2003).pdf, Number 2 and 29.
139 Atomnyy_proekt_SSSR.T.2.Kn.6.(2006).pdf, p 45.
140 Atomnyy_proekt_SSSR.T.2.Kn.6.(2006).pdf, p 42.
141 I.M. Khalatnikov, "Dau, Kentavr, I Drygie," *ibid.*, p 30.

approval, would have been a likely source. Malenkov was the only high ranking Politburo member with an engineering education, and he was impressed with Kapitza's ability in industrial organization.[142] As a sign of things to come, Beria had his secretary notify Kapitza that he should report to Beria's office for service on the committee, rather than calling him on the phone himself.

Beria assumed that Kapitza, now a member of Beria's committee, would be obliged to do whatever Beria requested of him.

Kapitza was a reluctant member of the committee from the beginning, but the early meetings were more or less harmonious. Acting on a suggestion by Beria, Sudoplatov gave Kapitza a hunting rifle as a friendly gesture. Kapitza mentioned that he had a worn out copy of a book written by his father in law, Academician Krilov, who was an expert in shipbuilding, and he would like to obtain the book in good condition for his library. Amazingly, Sudoplatov somehow got the old book reprinted on good quality paper, and gave Kapitza two copies. Kapitza in turn gave one copy to Stalin.[143]

Sudoplatov was basically a fly on the wall at committee meetings, and was an astute observer of the dynamics of the process. Not surprisingly, there was tension on the PGU regarding the disposition of the limited resources of Soviet industry in 1945. Given the scope of the Manhattan Project, which was well understood by members of the committee, and the devastated state of Soviet industry after WWII, it was inevitable that the Soviet Atomic Project and virtually all other needs of the Soviet economy were to compete for what remained of the pie. Nikolai Voznesensky, the politburo member of the committee, was an economist, and responsible for state economic planning. Sudoplatov writes:

> "The Special Committee on atomic problems had broad powers to mobilize any and all resources for the creation of an atomic bomb. This meant that when we started

---

[142] HISAP96, Vol 2, *ibid.*, "Kapitsa, Beria, I Bomba," P.E. Rubinin, p 260ff.
[143] Pavel Sudoplatov, "Razvedka I Kreml," *ibid.*, p 239.

to build factories in Siberia for refinement of uranium ore, we took electricity away from a large number of enterprises. I recall sharp arguments and uncensored abusive language between Pervukhin and Voznesensky regarding the issue – which factories would retain their full allotment of electricity. I was completely surprised that Pervukhin, in defending his life work – the Soviet chemical industry – would attack Voznesensky, his senior as a member of the Politburo.

"In the early years after the war the operation for gathering intelligence on nuclear weapons had high priority. In December, 1945, Beria left his post as people's commissar for internal affairs, moved out of Lubyanka, and occupied a new office in the Kremlin. He was now deputy chief of the Soviet of People's Commissars [Stalin was the chief]. Meetings of the Special Committee on atomic problems, which always occurred in Beria's office, moved from the NKVD to the Kremlin. As chief of the second bureau of the committee, a government worker, I acquired a permanent pass to enter the Kremlin at any time, day or night.[144] …In addition to sharp arguments regarding the distribution of electric power, Pervukhin quarreled with Voznesensky about funding for the chemical plants charged with the production of nuclear fuels. The vigor of quarrels among high ranking members of the government never ceases to amaze me. Beria usually was a calming influence, calling Pervukhin and Voznesensky to order. I saw for the first time that each member of this government committee considered himself to be the equal of everyone else, regardless of rank."[145]

Sudoplatov commented that much of the post war intelligence was gathered through the open literature: economic data, news sources, industrial journals and annual reports, and of course scientific publications. We have only a few first hand accounts of the dynamics of high-level government committees in the USSR during this period. Vigorous arguments, shouting, name calling, and colorful profanity were probably frequent phenomena. To a lesser extent this was true of Soviet scientific seminars. But the fear of immediate liquidation hanging over anyone in Beria's office must have been a fairly unique feeling. Physics seminars were not life and death situations. Failure in Beria's eyes most assuredly was. When he wanted to, however, Beria could be personable, if not friendly. Sudoplatov was impressed by Beria's wide range of interests. During the war Beria was responsible not only for state security, but also for the production of war materials, and the energy sector. He was

---

[144] In Stalin's time the Kremlin was closed to the public. Now it is an open park.
[145] Pavel Sudoplatov, "Razvedka I Kreml," *ibid.*, p 240-241.

especially interested in the production and refinement of oil. There was a model of an oil refinery in his office.[146]

7.18. Kapitza and Beria's Committee

Regarding Kapitza's participation on the committee in its early days, Sudoplatov writes:

"I observed a growing competition between Kapitza and Kurchatov. Kapitza was an outstanding personality, an excellent tactician and strategist, and a formidable scientific organizer. He often made humorous comments during serious scientific discussions. I remember that one meeting of the Special Committee in 1945 took place at the same time as a soccer match between British and Soviet teams in England. The match was broadcast in Russian from London. Everyone was shocked when Kapitza suggested that we interrupt the discussion and listen to the match. There was an awkward pause. Then Beria, to everyone's amazement, called a recess. The tension in the room disappeared, and we listened to the radio. In the end, the Soviet team won, and the mood of the participants improved.

"Kapitza, having played an important role in getting the work on the atomic problem started, and establishing contact with western physicists, Niels Bohr in particular via Terletsky [the Terletsky mission is covered in Section 8.20], naturally assumed an independent management position in the realization of the atomic project. But soon the relationship between Kapitza, Beria, and Voznesensky spoiled. Kapitza proposed that Kurchatov consult with him regarding the reliability of experimental results before presenting them to the committee. Pervukhin supported Kapitza, but Beria and Voznesensky did not agree. Beria had an alternative suggestion, namely that Kapitza repeat Kurchatov's measurements at Kapitza's Institute." [147]

The experimental group of Kapitza's Institute focused on low temperature physics, and was not set up to repeat the neutron experiments of Laboratory #2. Consequently, Kapitza was not enthusiastic about Beria's idea. While this incident does not seem particularly important, it was one example of stress on the committee.

---

[146] Pavel Sudoplatov, "Razvedka I Kreml," *ibid.*, p 241.
[147] Pavel Sudoplatov, "Razvedka I Kreml," *ibid.*, p 239-240.

Kapitza's adversaries in the oxygen business began a character assassination campaign. They claimed that his factories had a 'capitalist' character, and they questioned his loyalty to the system. He found Beria to be overbearing and difficult to work with. He wanted to build a Russian atomic bomb, developing new and more efficient techniques than those used by the Manhattan Project. He was not fully aware of the extent of knowledge of the Manhattan Project in hand from espionage. He did not want to work on a project simply to copy what had been already done – that seemed to him to be insulting and not interesting or challenging. Kapitza made the reasonable point that the Soviet Union would never catch up with the United States unless it developed its own, better techniques. But Stalin wanted the new weapon right away, the sooner the better, and Beria worked for Stalin.

Khalatnikov in his thoughtful and informative book discusses the question of Kapitza's loyalty to the Soviet Union.[148] He points out that Kapitza lived for 13 years in England, was elected to the Royal Society, and was awarded public support for his research. But he never renounced his citizenship in the Soviet Union, even though it restricted his travel, and eventually led to his confinement in Moscow. He loved his homeland, and recognized and valued the Soviet leadership, although he often thought that they did stupid things. He tried to straighten them out with frequent letters to the Kremlin. He supported and contributed to the broad program of industrialization of the USSR before WWII. He certainly supported Soviet physics. Like many other successful entrepreneurs, he was individualistic, liked to think for himself, and do as he pleased, putting him on a collision course with Beria and the Politburo. This is a fundamental problem in a totalitarian society, particularly a communist one, because in order for things to move forward, you have to have competent people, and the USSR had plenty of them. The Soviet system was basically unworkable, but very skilled people made it work for a time. The problem for the Kremlin was: How do you keep them in line? A recurring solution was terror.

---

[148] I.M. Khalatnikov "Dau, Kentavr, I Drugie," *ibid.*, p 32.

Sudoplatov describes the atmosphere at the meetings. He was impressed that the committee members were not inhibited by hierarchy – any member of the committee felt free to argue with any other member, regardless of rank. Sudoplatov continues:

"Beria played the role of arbitrator, trying to settle arguments and keep things moving forward. Beria was rough and cruel with subordinates, but he could be attentive, considerate, and supportive with those who were doing important work. He would protect them from repression by the NKVD and the Party. He always warned project leaders of their personal responsibility for strict fulfillment of the task, and he had a unique ability to instill the feeling of terror while inspiring the work ethic. Naturally, his personality reminded directors of industry of the NKVD. It seems to me that at first terror prevailed, but gradually, working with him for a few years, the fear diminished, and was replaced with the conviction that Beria would support those who successfully fulfilled the most important tasks for the national economy. [If the individual in question survived. LGP] Beria would often encourage freedom of action in order to solve some difficult problem. I think that he acquired these qualities from Stalin – the ability to establish an atmosphere of trust in addition to firm control, and extremely high demands. Those who successfully fulfilled the assigned task were guaranteed support."[149]

Sudoplatov might have been praising his boss too much.

There were meetings between Kurchatov and Stalin. Upon his return from Potsdam, where Truman mentioned in passing that the USA had a new weapon of extraordinary force, Stalin accused Kurchatov of a lack of diligence.[150] Kurchatov answered: "So much has been destroyed, so many people killed, and the country is starving. Things are not good." Stalin replied: "If the children don't cry, the mother does not know their needs. Give us your requirements. You will not be refused." Later, early in 1946, Kurchatov prepared a set of progress reports in advance of his meeting with Stalin. Kurchatov recorded his impression of that meeting: "In terms of the future development of the work, Stalin said that we should not waste time on small tasks, but develop the effort on a broad front, with grand Russian scale, and that the project would have the widest national support. Stalin also said that it was not necessary to save money, but rather work

---

[149] Pavel Sudoplatov, "Razvedka I Kreml," *ibid*, p 242.
[150] HISAP96, Vol 2, *ibid.*, p 102. Knyazkaya quotes A.P. Alexandrov, who was a witness.

as quickly as possible in as many directions as necessary." That is about as close to a blank check as one could ask.

Beria's request that Kapitza write an introductory letter to Bohr for Terletsky to take with him on his mission to Copenhagen in the fall of 1945 (described in detail in Chapter 8) was one of the early causes of tension between the two. The Terletsky visit was an NKVD spy operation – to find out what Bohr had learned from his travels to the US and Britain regarding the Manhattan Project. Kapitza did not want to mislead his friend Bohr with the impression that Terletsky was an innocent young nuclear physicist in Copenhagen to learn physics from a world expert. Kapitza also did not want to have anything to do with the NKVD. He did as he was asked, but he did not like the assignment.

It is interesting to compare and contrast Kapitza and Kurchatov. Either one could have managed the Soviet Atomic Project, but only Kurchatov stayed with it to the end. Both had solid scientific credentials, Kapitza perhaps more than Kurchatov. Both had organizational abilities, knew how to work within the Soviet system, and had industrial experience – again Kapitza perhaps more than Kurchatov. Both were absolutely fearless – a necessary qualification for success. Both were directors of large laboratories of the Soviet Academy of Sciences in the Moscow region. Was Kapitza more headstrong and less flexible than Kurchatov? Surely Kurchatov was as tough and determined as anybody. Perhaps Kurchatov was smoother in inter-personal relations. Kapitza admitted that he – Kapitza – had a sharp tongue, and did not suffer fools lightly. And as we all know, fools are liberally distributed all over. In any event, Kurchatov was the only one left standing at the end of 1945.

Kapitza was an independent force, with no obligations towards Beria whatsoever. Beria therefore may have seen Kapitza as a special challenge to his authority. Not one to shrink from challenges, perhaps Beria went after Kapitza, to probe his limits, and to determine his weaknesses, if any; to bring Kapitza under his thumb, as it were. There may have been Politburo rivalry between Beria and Malenkov for second place in Stalin's firmament. Kapitza was strongly supported by Malenkov, and may have been subject to punishment for that reason. In any event, work on the Special Committee became quickly intolerable for Kapitza.

Kapitza often wrote Stalin a letter about this or that, usually for 'educational' purposes. Stalin never replied, but Kapitza had reason to believe that his letters had been read. On October 3, 1945, Kapitza wrote another letter to Stalin. Beria had suggested that Kapitza become deputy to someone unnamed whom Kapitza did not like. Kapitza was becoming ever more frustrated with Beria and his committee, and in desperation requested that he be relieved of duty both in the liquid oxygen operations and in the Special Committee. No answer. Two months later, on November 25, 1945, came another letter Kapitza to Stalin, which left collaboration between Kapitza and Beria impossible. Kapitza repeated his call that the Soviet Union go its own way in bomb development – that it would be cheaper and faster than copying the Americans. Then he commented on Beria's style in managing the committee. P. E. Rubinin in his article in HISAP'96 observes that Kapitza makes a rather remarkable comparison of Beria to an orchestra conductor.[151] After all, Beria was a murderer, and in no way similar to a musician. However, Kapitza said in his letter to Stalin that Beria waved the baton like a good orchestra conductor, but that he did not know how to read music. In effect, Kapitza said that Beria was running a vast project that he, Beria, knew nothing about. In a post script to the letter, Kapitza asked Stalin to communicate the contents of the letter to Beria. Stalin obliged.

Beria was a top candidate to replace Stalin, as were at that time Malenkov and Molotov. Stalin had rewarded Beria with wide powers, but Stalin did not trust anyone, and may well have been collecting data to use against Beria should the occasion arise. Kapitza's letters accusing Beria of insensitivity in dealing with scientists, and general lack of any technical knowledge, may have been useful data for the anti-Beria files. Whether accidental or by design, the value of Kapitza's letters to Stalin may have saved his life. Stalin wanted to keep Kapitza around as a trump card for

---

[151] P.E. Rubinin, "Kapitza, Beria, and the Bomb," HISAP'96 Vol 2, *ibid.*, p 270.

Beria. Stalin, certainly a complex personality himself, may have even been sympathetic towards Kapitza.[152]

Beria called Kapitza on the phone, and said that he had read the letter to Stalin, and that the two of them should chat. So could Kapitza please come to Beria's office in the Kremlin? Kapitza replied that he was very busy, and had nothing in particular to say to Beria, and if Beria wanted to talk to him, he should come to the Institute for Physical Problems.[153] Igor Tamm told the story to his younger colleague, Andrey Sakharov. Tamm said that Beria was obviously much busier than Kapitza. Sakharov wondered if Tamm were criticizing Kapitza, but decided that he was being facetious. Kapitza interpreted the exchange to his advantage, having put Beria firmly in his place. In any event, Beria came to Kapitza's Institute, and brought a double barrel shot gun as a gift (not clear whether this is another gun, or the one referred to by Sudoplatov above). Kapitza admired the gift, but did not admire the giver. The rift between the two was irreparable. Beria decided that the Atomic Project could proceed without him, and Kapitza's work in the liquid oxygen industries was drawing to a close, so Kapitza could be dispensed with. Stalin intervened to prevent Kapitza's execution.

On December 21, 1945, Kapitza was relieved of all duties on the Special Committee, but was allowed to keep his post at the liquid oxygen industry, and his directorship of the Laboratory for Physical Problems in Moscow. However, the commission in charge of the liquid oxygen industry was 'strengthened' with some of Beria's followers. Then in August, 1946, the commission issued a memorandum stating that since Kapitza had spoiled the plans for industrial oxygen (!), he could no longer remain in charge. On August 17, Stalin signed an order relieving Kapitza from duties as chief of industrial oxygen, and additionally from the directorship of the Institute for Physical Problems. Kapitza was alive, but

---

[152] I.M. Khalatnikov, "Dau, Kentavr, I Drugie," *ibid.*, p 37. It is hard to believe that Stalin was ever sympathetic towards anyone.

[153] This would be funny if it weren't so serious. The story was so well known that it is repeated in Sakharov's "Memoirs," p 129.

had no occupation. A.P. Aleksandrov assumed the directorship of the Institute for Physical Problems.[154]

As was the case some ten years earlier, when he was told that he could not return to England, Kapitza at first was very depressed. He was alive, and still had his dacha, so he had a place to live. (He probably also had quarters on the Institute grounds, which he lost.) He gathered himself together, and started assembling a 'hut laboratory' at his dacha. At first he did experiments in hydrodynamics, and then high power electronics. While most of his friends avoided him, Landau and Lifshits from his old institute came to visit him. His wife and two sons helped get things set up. Eventually the president of the USSR Academy of Sciences quietly helped him out with some equipment, and the loan of his technician who had worked for him at the Institute. In retrospect, Kapitza thought that these were some of his best years. Within reason, trying circumstances can indeed bring out quality work. It is possible to be too comfortable. But surely Kapitza's 'hut laboratory' was a Spartan place.

Kapitza was not strictly under 'house arrest.' He taught at Moscow State University for two years 1947-49. 1949 was Joseph Stalin's 70th birthday, and many celebrations were held at MSU and the Academy of Sciences. Kapitza was still insulted by Stalin's actions, and boycotted the celebration. Kapitza's absence from any jubilee in honor of Stalin did not go unnoticed, and he was dismissed from his teaching post at MSU. He subsequently wrote another letter to Stalin explaining that he avoided crowded public events like the plague. The Academy of Sciences had to intervene to keep him from being thrown out of his dacha.

In 1955, nine years after his removal from office, and two years after Stalin and Beria's deaths, Peter Kapitza was rehabilitated, and restored to the directorship of the Institute for Physical Problems, which he had

---

[154] work.atomlandonmars.com/APSSSR/Atomnyy_proekt_SSSR.T.2.Kn.3.(2002).pdf #1 "Order #1815-782s on the production of oxygen by the method of Academician Kapitza."

founded.[155] He never worked on the Soviet Atomic Project. Many writers have wondered why, and whether he would have made a difference.

Khrushchev in his Memoirs quotes a conversation with Kapitza after his reinstatement.[156] "I asked him why he did not do weapons work, when his country needs weapons for defense?" He answered: "I am a scientist, and scientists like to work in the open, the same as artists or writers, and like to have their results known to others, and discussed. But weapons work is secret. If you do weapons work, you will be buried behind the walls of an institute, and your name will vanish from the open literature."

One reason for Kapitza's withdrawal was Beria, whom he disliked immensely. Another reason was the duplication of effort – he preferred an Atomic Project that struck out on its own and developed new and better techniques.[157] A third reason was perhaps the magnitude and scope of the enterprise, and the inherent difficulty of making a substantive, original contribution. And possibly a fourth reason was the secrecy. It is difficult to imagine Kapitza working at Arzamas-16, surrounded by barbed wire, with a prison camp next door to supply labor. Kapitza spent 13 years working in England, and, although he was patriotic, he was perhaps not the dedicated communist that many people were.

The Soviet Atomic Project proceeded without Kapitza, who together with Landau formed the greatest aura in Soviet physics in mid twentieth century. Landau did not like the idea of weapons work either, but was drawn into it through contacts at his Institute, as we shall learn.

7.19. Components for the Reactor Arrive

Around New Year's Day, 1946, Laboratory #2 began receiving purified graphite from the electrode plant. The first metallic uranium from Riehl's Electrostal plant started coming in at about the same time. The

---

[155] Aleksandrov went to Laboratory #2
[156] Quoted by P.E. Rubinin, HISAP'96, Vol 2, *ibid.*, p 278.
[157] The Project did manage innovations in parallel with the main task.

quality of the materials was confirmed by measurements in the prism rigs inside the tent, as described in Appendix B. The reactor was assembled in its own new building on the laboratory site. Construction of the graphite-uranium matrix occupied the calendar year 1946. On Christmas day, 1946, the final layer was installed, and the F-1 reactor, the first in Europe, went critical. The reactor was a spherical design, with a radius of 3000 mm. The actual assembly had a cylindrical base with a diameter of 7600 mm, and a hemispherical top 7300 mm above the floor. The 200 mm × 200 mm × 200 mm uranium and uranium oxide matrix was within a 6000 mm diameter sphere, and the outer shell of graphite 800 mm thick served as shielding. The final assembly had 35 metric tons of uranium metal, 13 metric tons of UO_2, and 420 metric tons of pure graphite.[158] After stacking the 62nd layer, it was clear that the reactor had achieved criticality – so, as was the case at Chicago four years earlier, the device worked sooner than the design had anticipated. Stalin was so impressed by the success of F-1 that he held a reception on January 9, 1947, in the Kremlin for the PGU committee, the Laboratory #2 scientists, and members of the politburo. This was the last time that Stalin met with scientists involved in the Soviet Atomic Project. Kurchatov had met Stalin once before, in January, 1946.

Like its ancestor at Chicago, F-1 had no provision for cooling, and relatively little shielding, so it was not safe to operate at high power levels for extended periods of time. There was a thermal column on top for a copious source of slow neutrons for measurements. As mentioned at the beginning of this chapter, F-1 still stands today, with its original charge of uranium, and will work if asked to.

The production of enough ^{239}Pu to study its chemistry and to measure its fission cross sections and secondary neutron yields was emphasized by Kurchatov as a prime mission of Laboratory #2 from its beginning. At first four kg of uranium were irradiated by neutrons from a Ra-Be source containing 2 gm of radium to produce a tiny sample of plutonium, about 5×10^{11} atoms. Then operating Nemenov's cyclotron for three months, irradiating 50 kg of uranium made about 2×10^{14} atoms (≈ 0.1 μ gm) of

---

[158] Atomnyy_proekt_SSSR.T.2.Kn.6.(2006).pdf, p 47.

plutonium, enough for chemical studies, decay lifetime measurements, and neutron cross sections.[159] Larger samples had to await the operation of the F-1 reactor.

A vertical column on the top of F-1 was used to allow thermal neutrons to escape from the reactor interior for experimental purposes – measuring various neutron cross sections of interest to the project. Micrograms of ^{239}Pu were produced inside the reactor by absorption of the neutron flux on the natural uranium slugs. The reaction forms ^{239}U, which β decays into ^{239}Np, which subsequently β decays into ^{239}Pu. The first decay half-life is 23 minutes, and the second one is 2.36 days, so it takes several days of irradiation for the plutonium to build up. The uranium slugs were then removed, and the tiny amounts of plutonium chemically extracted from the highly radioactive slugs.

Vladimir Mostovoy was a graduate student/post-doc from Kiev working at Laboratory #2 in 1947, after the F-1 reactor was in routine operation. He has described the techniques used to study the fission by thermal neutrons of ^{239}Pu and ^{233}U compared to ^{235}U as a standard. [160] They constructed a split ion chamber, with a high voltage cathode in the middle, and two grounded charge detector plates on the ends. A thin film of ^{235}U was deposited on the cathode on one side, and the test sample – a few micrograms of ^{239}Pu – was on the other side.[161] Uranium and Plutonium both decay by emitting α particles, which could be counted by their ionization within the isolated chamber. Fission was triggered by neutron bombardment (spontaneous fission was a small effect). Fission fragments gave more than ten times as much signal as the α's. The thermal column of the F-1 reactor was the neutron source. A cadmium sheet was used to measure thermal neutron cross sections – no cadmium sheet in front of the chamber gave total fast + thermal neutrons, and the cadmium sheet

---

[159] HISAP96, Vol 1, p 267.

[160] V.I. Mostovoy, in Raisa Kuznetsova, "Kurchatov v Zhizni," Rossiiskii Nauchnyi Tsentr 'Kurchatovskii Intsitut," Moskva, 2007, p 522.

[161] An ion chamber to observe fission is described in Appendix E – the paper by Petrzhak and Flerov.

removed the thermals, leaving only the fast neutrons. Thus the difference rate was due to thermals, and ratio of the counting rates on the two sides gave the fission cross section ratio ^{239}Pu/^{235}U.

The sample of ^{233}U deposited on a platinum foil used by Mostovoy was, unknown to him at the time, smuggled out of Chalk River by Alan Nunn May during the Manhattan Project, and turned over to the NKVD. The story was in May's confession after his arrest, and is described in Chapter 8. It is very unusual for specific espionage material to be traced from source to consumer in this way. Mostovoy said that he got the sample from Dzhelepov. Since Laboratory #2 had fully functional reactors at that time, it could have produced the ^{233}U itself by irradiating ^{232}Th with neutrons, so Mostovoy did not think about where it came from. Probably no one would have told him had he asked. In any event, the fission measurements were of interest only for nuclear physics, and not for weapons, which at that time did not employ the isotope ^{233}U.

The F-1 reactor became a test bed for designs of high power uranium-graphite reactors to be used for industrial production of ^{239}Pu.[162] Evidence that the reactor still exists is presented in Figure 7.3, where the author stands in front of it. The optimal arrangement of the lattice, the placement of the uranium blocks and aluminum channels, the aluminum cladding for the uranium, and other design questions were studied with F-1 in order to build the new reactor "A" in "Mayak," a new facility in the Chelyabinsk region 1500 km east of Moscow, on the eastern side of the Ural Mountains. Nuclear physics research with the F-1 reactor at Laboratory #2 in 1947 and thereafter followed by two or three years what was done in the Manhattan Project by Fermi and Libby using CP-1, that had been moved from Ellis Avenue in Chicago to the rural area that would become Argonne National Laboratory. Secondary neutrons from fission are fast – a few MeV. The spectra of fission neutrons were measured using photographic emulsions to record n-p recoils. This was the technique that Hugh Richards employed at Los Alamos to make the same measurements on the same transuranic element (^{239}Pu). As time went on, Laboratory #2 became more a general

---

[162] I.F. Zhezherun, "First Soviet Reactor F-1," HISAP96, Vol 2, *ibid.*, p 68.

purpose research lab, as contrast with a weapons lab.[163] It had research reactors, cyclotrons, hot chemistry labs, and could handle various problems in nuclear physics, nuclear chemistry, and biology. The weapons program outgrew Laboratory #2, and had to set up large industrial complexes in areas remote from major population centers, both for safety and security. Laboratory #2 continued to use its considerable experimental resources to do nuclear physics measurements pertinent to the weapons program. But it remained more a physics research laboratory, not a factory producing kilograms of fissile material.

Figure 7.3. The Author in front of the F-1 reactor, Europe's first, at the Kurchatov Institute in Moscow in September, 2014. Credit Artem Bekzhanov.

---

[163] V.I. Mostovoy in HISAP96 Vol 1, *ibid.*, p 264.

Chapter 8

Soviet Espionage and the Atomic Project

8.1. Introduction

Soviet penetration of the societies of their own western Allies, England, Canada, and the United States, was complete. There were spies in government, in industry, and in the Manhattan Project. They learned everything that was going on. They were really thieves – stealing information by the truck load. There were no heroes in the story – certainly not in MI5 and FBI counter intelligence, because they never caught anyone red handed, but not on the Soviet side either, because the job, although dangerous, was not heroic.

Espionage is a constant feature of our lives today, and has been around as long as recorded history. It is always useful to know what the other fellow is thinking. In wartime, there is a tactical advantage to be gained from daily intelligence. Even knowing the units and their commanders on the other side of the line can be an advantage. Personalities of commanders are scrutinized to figure out how they will respond to a particular situation. Captured soldiers are supposed to give only name, rank, and serial number, keeping the unit secret. Timetables for battle plans are particularly valuable, and this type of information is often obtained from POW's. This is intelligence on the personal level, usually without having to deal with codes. Field communications over wires are reasonably – but not always - secure, but that is not true for wireless signals. Anyone can pick up long distance wireless signals, which for most of WWII were transmitted in Morse code. To be secure, the information has to be encrypted. The unscrambling of encrypted information can be achieved with some sort of a code book, or an encryption device run backwards. Code books were often stolen, or captured from ships or embassies. Encryption devices, like the German Enigma, were obtained before the war and studied to attempt to break the code of intercepted messages. A more technical discussion of encryption is contained in Appendix G.

The Zimmerman telegram, described in a book by Barbara Tuchman,[1] was one of the sensational decodes of WWI. Arthur Zimmerman, the German Foreign Secretary, sent the coded telegram to the German Ambassador in Washington, Count Johann von Bernstorff, in January, 1917, before the United States had entered the war. He first said that in February, 1917, the German navy planned to resume unrestricted submarine warfare. Submarine activity was scaled back after the international uproar over the sinking of the Lusitania off the Irish coast in May, 1915. Zimmerman assumed that this provocation would move the United States to declare war on Germany, and this motivated the second part of his message. Part two authorized the German ambassador in Washington to contact Mexican authorities through the German Embassy in Mexico City, and offer Mexico the states of Texas, New Mexico, and Arizona in return for joining the conflict with Germany against the United States, after the inevitable US surrender! And it was just a Western Union telegram. The contact was made, and in the end the Mexican government decided not to take the Germans up on their offer.

The German Enigma machine, which was widely used by the German army and navy in WWII, was a electro-mechanical typewriter device that exchanged individual letters. It had a keyboard with 26 letters, and in its simplest form three wheels, each with 26 positions. Its workings are briefly explained in Appendix G. The British established a decoding factory at Bletchley Park where tactical enigma communications were decoded in real time, fast enough to affect battlefield operations. This was especially important for U-boat deployment. Enigma coded messages were decoded

---

[1] Barbara Tuchman, "The Zimmerman Telegram," Ballentine Books, New York, 1966.The actual telegram and its transcription are shown in the illustrations after p 148.

with success throughout the war.[2] One of the brains behind the operation was Alan Turing, famous for his early conceptualization of digital computers.

In WWII in the Pacific the Americans tried to learn Japanese intentions by the location and intensity of radio traffic, as well as the decoding of the traffic itself. Several listening posts were needed to locate a source by triangulation, and the vastness of the Pacific theater meant that simply knowing the location of the traffic was useful information. The Japanese spoken language can be written in Kana, which has 48 characters, most of which are consonant – vowel pairs. Many words in Kana have multiple meanings, and the correct meaning must be discerned from context. The Chinese characters, called Kanji in Japanese, have unique meaning. Japanese could be encoded like the Zimmerman telegram, where a kanji character is assigned a number, and decoding is done with a lookup table. Kana can be transmitted with a special type of Morse code, called Wabun Morse code, and can be scrambled by an Enigma type machine. The code breakers in Honolulu had to deal with all of these problems, and once it was decoded, it was in Japanese.[3]

Decoding is always a dicey business. Some of it is logic, and some is guesswork, and most intercepted communications are only partially decoded, if at all. There is always a paradox in handling decoded tactical information. If you act on it, the enemy will figure out that you are reading their mail, and will change the code. But if you don't act on it, it is useless.

8.2.　Soviet Espionage in the West

The Soviet spy network in the United States during the time of Stalin was a robust operation. It was well established before WWII, which enabled it to pursue the Manhattan Project with a vigor that would have been difficult if not impossible to achieve from a cold start, given the brief

---

[2]　Ronald Lewin, "Ultra goes to War," McGraw Hill, New York, 1978.

[3]　Edwin T. Layton, "And I Was There," William Morrow, New York, 1985. Lt. Commander Layton was intelligence officer for Admiral Nimitz in WWII.

three year time frame. The Soviets gave the Manhattan Project the code name ENORMOZ. The great economic depression of the 1930's, and the allure of the Soviet Union and the building of communism, caused many Americans to at least sympathize with the USSR. Many Russian émigrés lived in the United States, mostly refugees from the 1917 revolution, including Alexander Kerensky himself, who was at Stanford. For Soviet espionage the émigrés were a mixed bag. Some still loved the old country, and were willing to help it. Others were not sympathetic to Stalin's needs. There was, and still is, a US Communist Party, founded in 1919. The espionage, however, was controlled strictly by Moscow, and not by the local party. The organization was roughly as follows: 1.) the spy, who collected and transmitted intelligence information, sometimes in response to a request, and sometimes voluntarily. This individual was typically not a Soviet citizen. 2.) The handler, or courier, who met with the spy periodically to receive the stolen goods. In most cases, this individual was not a Soviet citizen either. And 3.) the station chief, who transmitted the materials to Moscow. This individual was a Soviet citizen, usually an NKVD operative, living in the US either as part of the embassy or a consulate, or a foreign correspondent for TASS, called a 'legal', or as an 'illegal' with some sort of cover job, like a shoe store. A foreign trade business in New York City called AMTORG served as cover for one of the Soviet spy centers, staffed mostly by illegals. AMTORG was founded in 1924, before the United States and the Soviet Union had diplomatic relations (which happened in 1933), to expedite import-export between the two countries. It played a big role in Lend-Lease. New York City was the principal receiver and transmitter of intelligence information. This purpose was served by the consulate, and various cover businesses. Access to overseas shipping was doubtless one of the reasons for the importance of New York. The embassy in Washington, DC, was in second place. The west coast site was San Francisco. Each location had an NKVD agent who handled communications with Moscow.

The espionage ring was a hub and spoke arrangement, with the spy at the rim, the courier being the spoke, and the resident station chief at the hub. The resident, a Soviet citizen, was subject to FBI surveillance.

Disconnecting the resident from the spy by using an American citizen as a courier decreased the danger of discovery. The preservation of internal security also dictated that there were no cross links after the hub. In that way, if part of the operation was compromised, the rest of it was unaffected, because the apprehended individuals have no knowledge of other branches. Of course they could compromise the resident, who may or may not have diplomatic immunity, but who would be post-haste recalled to Moscow in the event.

The most important player on the team was also the least reliable - #1, the spy. These people were volunteers, who freely led a double life. They earned a living at a day job, and gathered intelligence information as a side line. Many worked for zero pay from the Soviet Union. Consequently, the NKVD had little direct control over them, and was always concerned that, for some reason, they would cease to produce. This worry caused endless fussing over the character, stability, and personality of the spy by his/her handlers. It was a pressure cooker, wherein raw nerves could lead to carelessness and disaster. Psychological examinations of spies by their handlers made up a fair amount of the coded telegraph traffic between New York and Moscow. One might wonder why the spies did it? Money was almost universally declined. That leaves dedication to the Communist cause and admiration for the Soviet Union in its time of troubles. Theodore Hall, an important Los Alamos source, is quoted as having said: "There is no country except the Soviet Union that could be entrusted with such a terrible thing [an atomic bomb]."[4] In addition, the risk and danger of the activity may have been attractive to some.

There is one other individual in the picture, upstream of #1, the spy, and that is the scientist or engineer who did the work. Sometimes the two were the same person –Klaus Fuchs - but in most instances the worker was an unwitting source, unaware that secret results were being transmitted to a foreign country. It seems to be part of a code of ethics for espionage that the names of the scientific sources and the spies are not to be revealed. Sources of information are protected, like in journalism. Accordingly,

---

[4]   Allen Weinstein and Alexander Vassiliev, "The Haunted Wood," *ibid.*, p 196.

many names are redacted from declassified documents, and the operatives are referred to in secret communications by code names. Some of the code names involved in Manhattan Project espionage have been identified, but some have not. Many of the spies remain unknown today. The information gathered by the spy network, after sifting and winnowing in Moscow by Kurchatov and others, proved to be of enormous value. We have already noted that the highest levels of the Soviet government knew the scope of the Manhattan Project as a result of the espionage, and that fact alone ensured the commitment of the necessary resources. The scientific value of the information was also considerable, but here a word of caution is necessary. A map of Oak Ridge gives you an idea of the scale of the gaseous diffusion plant K-25, and of the electromagnetic separation plant Y-12, but it does not tell you how to build the plants and make them work. We have only seen hints of how garbled scientific information can be, but it is safe to assume that there was a lot of noise on the signal. There is no evidence of intentional deception, although Beria was continuously suspicious that the USA was tricking him. But mistakes can come from copying, from memory, and from language translation, to name a few.

Vladimir Barkovsky, one of the ranking members of the NKVD intelligence network, has said that it is important to choose your source of information carefully. Do not aim too high in an organization, because the leaders will be independent and strong minded, and hence difficult to manipulate. It is better to concentrate on sources lower down in the hierarchy, although of course not too low, because they do not know anything.[5] This proved to be a good working rule for the Manhattan Project, where the top leaders maintained tight security, despite the many pressures on them to do otherwise, but there were an unknown number of middle level sources who were very reliable.

Deciphering what actually went on to produce the staggering amount of information is not so easy. It was supposed to be kept secret, and much of it still is today. People had assumed names and code names, and this certainly adds to the confusion. Many code names remain unidentified.

---

[5] V.B. Barkovsky, Voprosy Istorii Estestvoznaniya I Technikhi (VIET), 1994, #4 p 122.

Anatoly Yatskov, one of the Soviet operatives, is referred to as Anatoly Yakovlev by the FBI. His agents knew him only as "John," or "Johnny." It is one of the curious features of this story that much of the known information from the American side comes from testimony and confessions in the course of the Rosenberg trial in 1950, well after the war. It is generally conceded that Julius Rosenberg, although a dedicated agent for Soviet espionage, was not an important source of information on the atomic bomb, and that his wife, Ethel, was only peripherally involved if at all.[6] Yet they were the only two people executed for espionage by the United States. However, the various grand juries and the actual trial generated information that did reveal much of what we now know – or think we know - about the espionage ring. Sworn testimony and confessions are not completely reliable, so everything has to be treated with some skepticism.

8.3. The Sources

There are memoirs from each side of the competition. For the Soviets NKVD agents Pavel Sudoplatov and Aleksandr Feklisov have each written books about their service, both in Russian[7] and English[8]. Sudoplatov's Russian version is later, and perhaps more reliable, although one must always keep in mind that authors of memoirs have motives to look their best. For Feklisov, the Russian version is more constrained – "The Man Behind the Rosenbergs," first published in 1999 in French, contains much more information about Feklisov's relations with Julius Rosenberg than the Russian book. The name Rosenberg is not in "Kennedi I Sovetskaya Agentura." He is referred to as "agent Stanley." The description of the activities of the Rosenberg network is sufficiently disguised as to be unrecognizable. The descendants of the NKVD in the

---

[6] The two Rosenberg sons requested exoneration of their mother by President Obama, reported in The Washington Post, December 1, 2016. Nothing happened.

[7] Pavel Sudoplatov, "Razvedka I Kreml," TOO Geya, Moskva, 1996, and Aleksandr Feklisov, "Kennedi I Sovetskaya Agentura," Algoritm, Moskva, 2011.

[8] Pavel and Anatoly Sudoplatov, "Special Tasks," Little Brown & Co, New York, 1994. Aleksandr Feklisov, "The Man Behind the Rosenbergs," Enigma Books, New York, 2001.

modern Russian Federation did not want Feklisov to say that Rosenberg was a Soviet spy.[9] He may have felt obliged to publish outside Russia for that reason.

The two men had different views of what went on. Sudoplatov was one of Beria's deputies, stationed in Moscow during the war, responsible for management of wide ranging NKVD operations, with the rank of major general. Feklisov was boots on the ground – a Soviet agent in New York City, and then after the war in Britain. Much later, he was to play an important role as an intermediary in the Cuban Missile Crisis. But during our story, Feklisov was an NKVD agent in a foreign land – the USA – for the entire four years of WWII. As an intelligence agent, Feklisov had a dangerous assignment, but not nearly as dangerous as the front lines back home. His good fortune, relative to his family and friends who remained in the Soviet Union, would be a lingering guilt for the rest of his life. In their memoirs Sudoplatov comes across as a tough, hard boiled pragmatist, dedicated to getting the job done, while Feklisov seems more flexible and reasonable, like a good neighbor. But they are both thieves.

Sudoplatov established his usefulness to Stalin's regime in clandestine operations before WWII. He was born in 1907 in Melitopol in southern Ukraine, near the Sea of Azov. He ran away from home at the tender age of 12 to join his older brother in the Red Army. The boy soldier Pavel served in communications. His unit fought the Ukranian independence forces led by Symon Petlyura and Evgeny Konovalets. Soon his transcription and coding skills merited a transfer to the Ukranian security organization that would become the NKVD. He worked in Kharkov and Kiev developing his skills in intelligence gathering. He soon gave up an attempt at law school. In the 1930's Sudoplatov studied German in preparation for a tour as an agent in Western Europe.

The resolution of the conflict between the Red Army and the Ukranian freedom fighters in 1923 granted amnesty to the leaders, and Petlyura and

---

9   Aleksandr Feklisov, "The Man Behind the Rosenbergs," *ibid.*, p 103. Klaus Fuchs was similarly disowned; Pravda, March 8, 1950, p 5.

Konovalets went into exile in Western Europe. Petlyura was assassinated in Paris in 1926. Sudoplatov first met Konovalets in Berlin in 1936. They became friends. Konovalets probably did not know exactly who Sudoplatov was – just a friendly fellow Ukranian. Stalin learned that Konovalets was supported by the Germans in his efforts for a free Ukraine.

Upon his return to Moscow Sudoplatov for the first time was summonsed to Stalin's office in the Kremlin. Stalin asked if Konovalets had any habits that could be exploited. Sudoplatov replied that he was very fond of chocolate candy. Stalin agreed that that could be used. NKVD headquarters in Lubyanka crafted a bomb in a chocolate box. The fuse was set to go off 30 minutes after the box was rotated from vertical to horizontal.

In May, 1938, Sudoplatov made arrangements to meet Konovalets at a restaurant in Rotterdam. He carefully carried the package vertically. He also had a pistol, for the purpose of killing himself if the operation did not go according to plan. Konovalets was waiting for him at a table in the restaurant. They exchanged greetings, and Sudoplatov gave him the now rotated horizontally chocolate box gift. Sudoplatov then excused himself, saying that he had another appointment, but that they could meet again later on. On his way down the street, Sudoplatov bought a new hat and overcoat to change his appearance, and headed for the train station. Konovalets was killed.[10]

Feklisov met with Molotov in October, 1940, before going overseas to America. Molotov asked why Feklisov should be trusted, since he was single, and therefore subject to meeting an American girl and defecting. After persuasive arguments favoring his loyalty, Molotov said," OK, go and work. But don't let us down." Molotov in his speech and mannerisms reminded Feklisov of Lenin, whom he had seen in movies.[11] Feklisov boarded a train at the Yaroslavsky Station at 4 pm on a cold day in January, 1941, for the first leg of his trip to the United States. His family was at the

---

[10] This account is from Sudoplatov, "Razvedka I Kreml", *ibid.*, Chapter 1.
[11] Aleksandr Feklisov, "Kennedi I sovetskaya Agentura," *ibid.*, p 15.

station to see him off. It was the last time he would see his father, who died in 1942. The trip to Vladivostok took 10 days. Feklisov played chess and dominoes, and read two books. He spent five days in Vladivostok at a hotel for foreign service personnel, before catching a Japanese steamer from Vladivostok to Tsuruga on Wakasa Bay, the west coast of Honshu, Japan. Another train ride took him to Yokohama, where he stayed at the Imperial Hotel for seven days, waiting for a ship to San Francisco. In Yokohama Feklisov was followed everywhere he went – a sign of things to come in the US. He was in a party of five Soviet diplomats and their families, all travelling together, and taking meals together when possible. They boarded a Japanese passenger ship, 'Yavota Maru,' to sail from Yokohama to San Francisco, with an intermediate stop in Honolulu, which took 12 days. (The Pearl Harbor attack was 11 months hence.) Then a cross country train from San Francisco to New York took another 5 days. The whole journey Moscow → New York via Vladivostok took 40 days total. The Soviet consulate in New York was a four-story house on 61st Street. The consulate also owned the house next door, where some Soviet diplomatic families lived. The canteen was located in the next door house, and the English language classes were held there. Feklisov rented a room on 87th Street.

Some materials from the Soviet NKVD archives have been released, and are available on-line.[12] Some of the intercepted Soviet messages from USA to Moscow and back have been decoded – VENONA transcripts – and are available on the National Security Agency website.[13] On rare occasion the VENONA transcript and the message released from NKVD archives match up. This does not happen very often, partly because the VENONA decoded messages are only about 1% of the total radio traffic. The fraction of relevant information released by Moscow is unknown.

---

[12] http://work.atomlandonmars.com/APSSSR/
[13] https://www.nsa.gov/public_info/declass/venona/

An account from the American side was written by an FBI agent at the time, Robert Lamphere.[14] He tells us what the FBI was doing for most of the war, why the penetration of Los Alamos was not detected, and how the FBI responded after the war when evidence began to break. He gives a detailed account of the testimonies and confessions around the Rosenberg trial. There are also on-line files from the FBI and the US Congress, particularly the Joint Committee on Atomic Energy, that present the results of congressional hearings into the espionage.[15]

8.4. Job Description

Some discussion of the mechanics of obtaining and transporting secret materials is in order. This was before the days of the Xerox machine, so everything had to be photo-copied, transcribed by hand, or read and memorized. Microfilm was a common medium. Most everything of interest was under lock and key. The spy had to remove it somehow from the site to get it copied, or find a way to copy it without removal, and put it back before it was missed. If several copies of a classified document existed in the files, they were all numbered, so one could not simply remove a copy and hope that it would not be noticed. It was perhaps more low key than James Bond, but still a very risky activity. Meetings between spy and handler were rather cloak and dagger, especially when there was a change in contact personnel, so that the two did not know each other. Wait outside a specific bar, carry a Physical Review in plain sight, and ask for directions to the post office – that sort of thing. The handler delivered the documents to the NKVD agent, who had the responsibility to get them to Moscow.

The material was usually in paper form – documents, manuals, blue prints, etc. Sometimes fairly lengthy documents were involved – several hundred pages. Terletsky (see below) noted in 1945 10,000 pages of

---

[14] Robert J. Lamphere and Tom Shachtman, "The FBI-KGB War," Berkeley Books, New York, 1986.

[15] https://archive.org/details/sovietatomicespi1951unit. This is a transcript of hearings by the JCAE on Soviet atomic espionage.

materials on the atomic project in the NKVD files in Lubyanka. Atomic espionage was, however, only a small part of the total amount of printed material shipped from the United States to the Soviet Union. Literally tons of stuff left the country during the Lend-Lease period, beginning in 1941. Information pertinent to industry – factory plans of all kinds, specifications for machines and equipment, armaments, etc. – was secret prior to Lend-Lease, and had to be obtained surreptitiously for the expansion of Soviet industry in the 1930's. After Lend-Lease however, much of the industrial material was freely available, and was sent to the Soviet Union by the carload. If tanks were shipped to the USSR, tank service manuals had to be shipped too. Obviously, there was way too much paper to encode and send by wireless, and as a result the VENONA files, to be discussed below, contain relatively little stolen information. Also, it was all in English, and had to be translated at the receiving end into Russian (an imperfect form of decoding).

Design specifications in the US are in feet and inches, but the USSR, and almost everyone else, used the metric system, and this difference alone could lead to foul ups. There are 2.54 cm to the inch, and one would think that this ratio is large enough to avoid mistakes, but it isn't. Working in millimeters is better – a factor of 25.4 is noticeable. Tolerances, important in machining specifications, are difficult to convert, as are screw sizes, threads, etc. So Kurchatov and his group had to exercise extreme caution in interpreting the data received. This they did. We do not have the original documents in English. We do know that the edited and organized form presented, for example, in the memorandum from Merkulov to Beria on October 18. 1945 which describes in detail layer by layer the Nagasaki implosion plutonium bomb (translated in Appendix H) is very accurate and polished, showing clear competence in handling the input information. The translation, verification, and synthesis of the original intelligence – a big job - was done in Moscow.

The stolen information, as received in English, sorted and translated into Russian, and reviewed by Igor Kurchatov, has not been released, and may not still exist.[16] All we have are Kurchatov's comments and questions on the material addressed to the NKVD in Moscow for transmission to centers in the USA, Canada, and Britain, and reports on the contents of the espionage prepared for Beria and Stalin. The lists of questions were typically transmitted by telegraph, and many were intercepted by US intelligence, as a part of the large volume of wireless traffic between the US and the Soviet Union. The Soviet diplomatic wireless communications were encoded using a 'One-time pad', which uses a table of random numbers to scramble the message.[17] In order to unscramble, you need the same table of random numbers. Obviously, a carload of documents cannot be practically transmitted in this way. So how did all of this information leave the USA and find its way to Moscow?

8.5. Major Jordan and Great Falls, Montana

Lend-Lease hardware was delivered mostly by ship from the East Coast of the USA to Murmansk on the Barents Sea, and diplomatic pouches could also go this way. However, a surprising amount of the mixed paper – both clandestine and legal – left the US by air. The Russians called this 'Super Lend-Lease.' Initially, the air link was Newark Airport in New Jersey. George Racey Jordan (he went by his middle name) was a captain in the US Army Air Force charged with assuring that the airlift lend-lease went smoothly. He learned early on that the USSR had priority over everyone else, including the US Army. Soviet Colonel Anatoly N. Kotikov was the coordinator for the Russians, stationed in the US. Jordan in his memoirs tells how he learned of the clout that Russian Lend-Lease had. One day in May, 1942, an American Airlines passenger plane scraped a warplane destined for shipment to the USSR on the tarmac. The damage was minor, and could be repaired in a few hours, but Kotikov went ballistic, and called the Soviet Embassy in Washington, who in turn called

---

[16] VIET 1992, #3, p 105 article by A.A. Yatskov: 'V arkhivax poka ne obnaruzheno kopi samix materialov, dobiytykh razvedkoi I peredannykh Kurchatovy.'

[17] users.telenet.be/d.rijmenants/en/onetimepad.htm

Harry Hopkins, head of Lend-Lease, in the White House.[18] In a few days an order appeared from the Civil Aeronautics Board closing Newark Airport to all commercial flights! Subsequently orders were cut giving Russian Lend-Lease priority over everything else.

President Roosevelt believed that he could charm Joseph Stalin. Frank Freidel, in his biography of Roosevelt, writes: "Roosevelt arrived at Tehran full of zest and optimism, feeling that he had spent his career reaching accommodations with men, and that Stalin could not be so different. Even if he could not persuade him to become a good democrat, Roosevelt thought they could come to a working agreement."[19] The Tehran conference was held in November, 1943. Nothing was firmly settled there. The Soviets wanted to retain their boundaries from 1939, which included the Baltic states and eastern Poland. Roosevelt's assumption that Stalin would be reasonable if he were treated fairly goes back to the beginnings of Lend-Lease, in June, 1941, just after the German invasion of the USSR. Stalin could be charming when needed, and Harry Hopkins left Moscow after early negotiations over Lend-Lease with a very favorable impression. This turned into a propaganda campaign in the US to sell the Lend-Lease idea to help out "good old Uncle Joe."[20] William C. Bullitt ambassador to Moscow from 1933 to 1936, was discussing Soviet policy with Roosevelt, and not agreeing with the president's views. Roosevelt said: "Bill, I don't dispute your facts. They are accurate. I don't dispute the logic of your reasoning. I just have a hunch that Stalin is not that kind of man. Harry (Hopkins) says he's not, and that he doesn't want anything but security for his country. And I think that if I give him everything that I can and ask nothing from him in return, *noblesse oblige*, he won't try to annex anything and will work with me for a world of peace and democracy."[21] There is some evidence that Stalin returned Roosevelt's

---

18 George Racey Jordan, "From Major Jordan's Diaries," Harcourt Brace, New York, 1952, p 21ff.

19 Frank Freidel, "Franklin D. Roosevelt," Little Brown and Company, Boston, 1990, p480.

20 Frank Freidel, "Franklin D. Roosevelt," *ibid.*, p 375.

21 George Racey Jordan, "From Major Jordan's Diaries," *ibid.*, p123-124. Quote from Life Magazine, June 30, 1949.

friendship. Andrei Gromyko, then Soviet ambassador to the United States, was also at Yalta. Stalin asked Gromyko of his opinion of the American President. "Stalin did not hide his fondness for FDR from Gromyko, which amazed the young diplomat, because his character was so harsh that he 'rarely bestowed his sympathy on anyone from another social system.'"[22] The Yalta conference, in January, 1945, after Roosevelt's election to a fourth term as president, settled the fate of Eastern Europe in favor of the Soviets. Roosevelt died in April of that year. Whether he would have been able to manage a post war Stalin is unknown.

As the Lend-Lease air traffic increased, a route was established to go west to Siberia. An airport in Great Falls, Montana, called Gore Field, was used to fly C-47 transports (twin engine Douglas DC-3's) west to Fairbanks, Alaska, then across the Bering Strait to Siberia, and on to Moscow.[23] US aircrews were replaced by Soviets at Fairbanks. The airlink was a daily affair. Jordan was promoted to major – on the insistence of Colonel Kotikov – and was transferred to Great Falls as the US officer in charge of Lend-Lease traffic at the airbase. Colonel Kotikov moved to Great Falls with his wife.[24] Jordan describes the procedure followed by the Soviets to send plane loads of documents and plans from the USA to the USSR in wartime. The materials were packed in black suitcases, wrapped in white cord, and sealed with red wax. The first load was six bags, that Jordan passed as personal luggage. They kept coming:

"But the units mounted to ten, twenty, and thirty and at last to standard batches of fifty, which weighed almost two tons and consumed the cargo allotment of an entire plane."[25]

Jordan bullied his way onto a plane to open the bags. He found a bit of everything, from road maps and train tables to catalogs and scientific

---

[22] Gromyko is quoted by Simon Sebag Montefiore, "Stalin, in the Court of the Red Tsar," Vintage Books, New York, 2005, p 482.

[23] Richard Rhodes, "Dark Sun," Simon and Schuster, New York, 1995, p 94ff.

[24] George Racey Jordan, with Richard L. Stokes, "From Major Jordan's Diaries," Harcourt, Brace, New York, 1952.

[25] George R. Jordan, *ibid.*, p 69.

journals, all perfectly legal. There were some State Department documents that looked as if they were classified. Then there was a large map of Oak Ridge, Tennessee, a place he had never heard of before, and unknown words like neutron, proton, and uranium 92.[26] He allowed the plane to take off with the cargo.

Jordan became concerned about the outflow of information, and sent red flags up the chain of command. He even went to Washington once to appeal to the State Department. But each time he was told to relax, that the shipments were all authorized. It was very likely, however, that his contact in the State Department knew nothing at all about the Manhattan Project. The matter boiled over in 1949, during the hearings of the House Committee on Un-American Activities.[27] While it is clear from Jordan's diaries that a substantial amount of contraband material moved out of the US via air from Great Falls, how much was atomic espionage is not so obvious. For example, he quoted 700 kg of uranium in various forms (1kg of uranium metal), and 500 gm of heavy water[28], which are consistent with requests approved by Groves for Russian Lend-Lease. Igor Kurchatov was very short of both uranium and heavy water until the German surrender in May, 1945. This shortage produced considerable fussing within the Soviet government regarding uranium prospecting, and getting hydroelectric plants up and running for heavy water. It was pointed out in Chapter 7 that Kurchatov and Pervukhin had requested these substances through Lend-Lease channels in 1943. It is not clear that any of this ever reached Kurchatov's Laboratory #2.

After the war, when the wheels of the Soviet espionage network began to fall off, and it was realized that the secrecy of the Manhattan Project had been compromised, the US Congress became interested in exactly how much material left the country. This led to the HUAC hearings referred to

---

[26] George R. Jordan, *ibid.*, p 83.
[27] George Racey Jordan, "From Major Jordan's Diaries," *ibid.*, p 234.
[28] George Racey Jordan, "From Major Jordan's Diaries," *ibid.*, p 142. Natural uranium presumably, although if we had enough depleted uranium, that would be better.

above, which were summarized in the JCAE report published in 1951.[29] Jordan tried to tell his story during and after the war, but after the successful Soviet atomic test in August, 1949, he became determined. A prominent radio newscaster, Fulton Lewis Jr, interviewed him in national broadcasts beginning December 2, 1949.[30] The programs attracted plenty of attention, all right, and Jordan's book, published later, is partly an attempt to answer his critics. One sore subject was his claim that he found in one of the black suitcases a note on White House stationary addressed to Mikoyan which contained the passage "… had a hell of a time getting these away from Groves," signed "H H."

Jordan identified Groves as the Manhattan Project chief, and H. H. as Harry Hopkins, Roosevelt's aide who was in charge of Lend-Lease. Well, this caused a big flap. Everyone at the Congressional hearing testified that Harry Hopkins, who had died in 1946, always signed H.L.H, and never H.H. Then Groves testified that he never met Harry Hopkins, didn't talk to him on the phone, nor write him any letters. Jordan could only point to notes he took at the time, inside a C-47 on the runway at Great Falls, about to fly to Siberia.

Jordan had no technical training, and only interrogated the contents of the black bags on occasion, so he really did not know how much Manhattan Project intelligence went through his airport. But he worried about it, kept a diary, and told his story after the war. And since the intelligence did flow out, no doubt Great Falls did its share. The JCAE chapter ends with the summary:

> "Review of the data examined and interviews conducted in connection with this inquiry indicate that two shipments totaling 1420 pounds (645 kg) of uranium salts and one shipment of 2.2 pounds (1kg) of uranium metal were made to Russia under Lend-Lease with the knowledge and consent of all agencies and parties concerned in 1943 and 1944. In addition, two shipments of heavy water, totaling 1,100 grams, were made to Russia during the war within existing legal provisions for such exports.

---

[29] Joint Committee on Atomic Energy, "Soviet Atomic Espionage," US Govt Printing Office, 1951, part II, Chapter 5 "Allegations Concerning Wartime Exports."

[30] George Racey Jordan, "From Major Jordan's Diaries," *ibid.*, p 240.

"The Joint Committee staff could find no indication that documents, maps, blue prints or classified papers other than those authorized in connection with the wartime agreements with Russia were shipped during the war through Great Falls, Montana. At the same time, the volume of material in transit and the existence of channels covered by diplomatic immunity through which unauthorized shipments might have reached Russia preclude any positive statement that nothing ever went to Russia without approval via Great Falls."[31]

8.6. The British Group

The Soviet Union had a vast world-wide espionage network in place when WWII began in 1939. Every embassy or consulate was a base for intelligence work. In Germany, there was the 'Red Orchestra', that was a communist cell for information about German plans until it was destroyed by the Gestapo in December, 1942.[32] Richard Sorge was the famous Soviet spy who posed as a German newspaper man in Tokyo until he was uncovered in 1944.

Yuri Modin was a code clerk and translator at NKVD headquarters in Moscow in the early 1940's, responsible for data coming from London, when the first word about British and American interest in an atomic bomb arrived. He has described some of the procedures.[33] In most instances Moscow Center directed the search for specific information, requested by the Kremlin, but this news about atomic weapons was not anticipated. Information from London came in the form of coded telegrams. Reports went directly to Molotov, Beria, and Stalin. Lower ranks rarely saw them. Once it started coming in, Molotov valued atomic energy intelligence very highly. Technical data were rarely transmitted by telegram.

---

[31] JCAE, *ibid.*, p 192.

[32] Anne Nelson, "Red Orchestra," Random House, New York, 2009. In the 1920's Arvid and Mildred Harnack, members of the group, lived in Madison, WI, in the same house where I am now writing this book. Small world. 'Mir tesen' in Russian.

[33] Yuri Modin, "My Five Cambridge Friends," Farrar Straus Giroux, New York, 1994, p 115.

Modin shared a small office with two others in Lubyanka. He worked at a desk with an English dictionary and a pile of papers.[34] He had to decide what was important enough for him to translate himself, and what could be sent to a pool of female translators for English to Russian. He also processed microfilms, which were developed and printed there. He then sifted through the translations, and picked out which ones should be sent to the Kremlin. He checked with his boss, who usually agreed with his assessment. An NKVD officer took the documents in a sealed packet over to the Kremlin. Each time Modin waited for a response, but none ever came.

One of Modin's 'friends' was John Cairncross, a young British Civil Servant, who worked in the Cabinet Office as private secretary to Lord Maurice Hankey, the Chancellor of the Duchy of Lancaster.[35] In this position Cairncross had access to many high level confidential documents. Cairncross had been a student of modern languages, particularly French, at Trinity College, Cambridge, in the early 1930's. He spent a year at the Sorbonne, and became an expert in Moliere and other French writers. He was attracted to Communism at Cambridge, and was recruited for Party work by James Klugman, the founder of the Communist cell there.[36] After graduation in 1936, Cairncross took a job in the Foreign Office. He became Lord Hankey's private secretary in 1941, and began to supply his handler, Anatoly Gorsky, with very valuable information. Cairncross' code name was 'Carelian.'[37] Hankey was chairman of the British Scientific Consultative Committee, and had access to documents regarding 'Tube Alloys." Accordingly, Cairncross alerted Moscow in 1941 to the fact that British and American governments had been considering a weapon made of ^{235}U since 1940. He continued to supply intelligence of military value to Moscow from Hankey's office until March, 1942, when he was transferred to Bletchley Park to work on the Enigma decryption problem.[38]

---

[34] Yuri Modin, "My Five Cambridge Friends," *ibid.*, p 121.
[35] en.wikipedia.org/wiki/john_Cairncross
[36] Yuri Modin, "My Five Cambridge Friends," *ibid.*, p 106.
[37] Code names are useful when reading NKVD documents, but are otherwise confusing, and will be used sparingly here.
[38] Yuri Modin, "My Five Cambridge Friends," *ibid.*, p 111.

Donald Maclean was another member of Klugman's communist cell in Cambridge. After graduation, he passed examinations to join the British Foreign Office in October, 1935, despite his admitted past record of communist party activities.[39] He was assigned to duties in the London Foreign Office, and was cultivated for NKVD espionage by Theodore Mally, the station chief in London.[40] Maclean was a capable diplomat, and was assigned to many important posts before, during, and after WWII. In each assignment he contributed substantially to Soviet knowledge about British and American foreign policy, war plans, Lend-Lease, the second front, and Tube Alloys. A report from Maclean via Gorsky, who had succeeded Mally as London station chief, to Moscow on September 25, 1941, began:

"Vadim [Gorsky] has relayed a report from Leaf [Maclean] about a meeting of the Uranium Committee of September 16, 1941. The meeting was chaired by Boss [Hankey]. The meeting was advised of the following.

The uranium bomb may conceivably be developed in the course of two years, particularly if the Imperial Chemical Industries company is obliged to manufacture it under a crash program.

A representative of the Woolwich arsenal, Fergusson, declared that the fuse for the bomb can be designed within several months. There is no need or possibility to obtain the minimal velocity of the relative movement of the mass of explosive at 6000 feet per second [1800 m/sec]. The explosion will be produced prematurely. But even in that case the force of the explosion will be infinitely greater than that of conventional explosives. [...]

Three months ago the Metropolitan Vickers company was issued an order for designing a 20 stage apparatus, but the permission for that was granted only recently. The execution of this order is set as a top priority. The Imperial Chemical Industries company holds a contract for the production of uranium hexafluoride, but it has not yet been started. Fairly recently a patent was issued in the United States for a simpler production process employing uranium nitrate. At the meeting it was also reported that data about the best type of diffusion membranes can be obtained in the USA.

---

[39] Yuri Modin, "My Five Cambridge Friends," *ibid.*, p 97.
[40] Allen Weinstein and Alexander Vassiliev, "The Haunted Wood," *ibid.* p 72.

> The Chiefs of Staffs Committee at their meeting on September 20, 1941, made a decision to immediately launch construction in Britain of a plant to manufacture uranium bombs. Vadim requests an appraisal of Leaf's materials on uranium.[41]"

This is a fairly detailed report of a secret meeting early on in preliminary development of the Tube Alloys project, shortly after the MAUD report. A muzzle velocity of 1800 m/sec is much higher than was actually used in the Hiroshima gun-type bomb (300 m/sec). The 20 stage apparatus refers to gaseous diffusion, and 20 stages are not enough to do much, but one has to start somewhere. Although bomb development did not proceed to completion in Britain, this document leaves no doubt that the British were serious about it, and that Moscow was well informed.

8.7. Klaus Fuchs

Klaus Fuchs enters the espionage story in Britain, long before he was photographed at Los Alamos, shown in Figure 8.1. He was born in Germany in 1911. He joined the Communist Party in Germany, and fled to England in 1933, when the Nazis rose to power. [42] He was not Jewish, but he feared persecution for his communist beliefs. He completed graduate studies in theoretical physics under the guidance of Nevill Mott at Bristol in 1937. His work in the electron theory of metals and other aspects of solid state physics attracted the attention of several theorists, including Rudolf Peierls.[43] After graduation Fuchs went to Edinburgh to study with Max Born - another refugee from Nazi Germany. Fuchs was interned as an enemy alien in 1940, during the Battle of Britain. He was first sent to the Isle of Man, and then to Quebec in Canada. Fuchs resented his incarceration, which may have affected his subsequent activities. In

---

[41] Pavel and Anatoly Sudoplatov, "Special Tasks," Little Brown and Company, Boston, 1994, p 437.

[42] Biography of Klaus_Fuchs on www.atomarchive.com.

[43] Rudolf Peierls, "Bird of Passage," Princeton University Press, Princeton, NJ, 1985, p 163.

any event, Max Born helped to get him reinstated, and he returned to Edinburgh early in 1941. Rudolf Peierls, co-author of the Frisch-Peierls Memorandum, and a member of the MAUD committee (discussed in Chapter 3), was at that time absorbed in theoretical problems involved in isotope separation and other matters of nuclear weapons production, and was looking for someone to help him out. Peierls offered Fuchs the opportunity to work on a project directed against the Nazis. Fuchs as a refugee from Nazi Germany should naturally be receptive to such a proposal. Peierls was well aware of Fuchs' communist past, but attributed it to student activities, not all that exceptional in the 1930's. Fuchs had to have a security clearance to work with Peierls. After some difficulties a clearance came through, and Fuchs joined Peierls' group at Birmingham in May, 1941. Peierls found him an efficient worker, and a pleasant person to have around. He was unmarried. Fuchs was the silent type, but could give clear responses when asked a question. For a time he was a lodger at the Peierls home. Peierls writes:

> "There were a few brief periods when he felt unwell. He did not go to work, stayed in bed or on a deck chair in the garden, and showed no interest in food. But this passed in a day of so. We realized the significance of these attacks only much later"[44]

These absences were connected with Fuchs' rendezvous with his NKVD handler. Fuchs became a British subject in 1942. As a British subject his loyalty was never questioned until 1950.

---

[44] Rudolf Peierls, "Bird of Passage," *ibid.*, p 163-164.

Figure 8.1. Los Alamos badge photo of Klaus Fuchs. Credit Los Alamos National Laboratory.

Fuchs was originally recruited by the GRU – Soviet Military Intelligence – which was a separate operation from Beria's NKVD, as early as August, 1941.[45] Since he was a collaborator of Rudolf Peierls, who was a key member of the British team, Fuchs was able to supply direct information on Tube Alloys. Roosevelt, Churchill, and Mackenzie King, the Prime Minister of Canada, met in Quebec in August, 1943, and produced what is known as the Quebec Agreement, that moved Tube Alloys across the Atlantic Ocean. Britain and the United States agreed to coordinate rather than duplicate, and since the resources in the US were already committed, a delegation of British scientists was dispatched to America. Fuchs accompanied James Chadwick, Rudolf Peierls, and others

---

[45] Allen Weinstein and Alexander Vassiliev, "The Haunted Wood," *ibid.*, p 185. See also Pavel Sudoplatov, "Special Tasks," *ibid.*, p 138.

to the United States in 1943. Moscow anticipated the departure of the British group, and was already organizing a liaison for Fuchs in America.[46]

8.8. The American Group

In North America there were many 'operatives', spread out, it would seem, almost everywhere. Gaik Ovakimyan headed the agency in New York City based in the AMTORG trading company. Perhaps his most important agent at the beginning of the war was Semen Semenov.[47] Semenov had advanced degrees in engineering from MIT, and spoke perfect English. Aleksandr Feklisov and Semenov were both New York agents before and during WWII. Feklisov described Semenov thus:

"He was short, with a duck bill nose, and big eyes which rolled around like radar antennas when you talked to him. He got along with people quickly and easily. He had a solid appearance, and on the whole could be taken for a middle level American business man." [48]

In April, 1941, the NKVD split its US intelligence network into two parts: a 'political' group and 'scientific-technical' group.[49] Pavel Fitin was in charge in Moscow, and many of the coded telegrams between New York and Moscow were addressed to him. In the US Semen Semenov assumed control of Harry Gold, who would become the courier in the United States for Klaus Fuchs.

8.9. Harry Gold

Harry Gold called himself a biochemist. He was born in Berne, Switzerland in 1910 of Russian Jewish emigre parents. The family moved

---

[46] Allen Weinstein and Alexander Vassiliev, "The Haunted Wood," *ibid.*, p 186.

[47] Pavel Sudoplatov, 'Razvedka I Kreml," *ibid.*, p 212. Some authors transliterate Semen Semenov as Semyon Semyonov, which is the correct pronunciation. There are two ways to pronounce the Russian letter 'e.' Sometimes even Russians are confused.

[48] Aleksandr Feklisov, "Kennedi I Sovetskaya Agentura," *ibid.*, p 60.

[49] Allen Weinstein and Alexander Vassiliev, "The Haunted Wood," *ibid.*, p 173.

to the US in 1914, and Harry grew up in Philadelphia. He was a good student, but his education was interrupted by the depression, and he took a job in a soap factory to earn some money for family support.[50] He joined the US communist party, and was recruited to obtain industrial information for the Five Year Plans of the USSR before WWII. Gold was a courier, reporting from 1940 to 1944 to Semen Semenov (Sam) in New York City. Semenov was recalled to Moscow in 1944, and replaced by Anatoly Yatskov (John). Abraham Brothman was an early source of industrial information for Gold. Brothman was a chemical engineer in industry sympathetic to the Soviet cause. After the war, Gold worked for Brothman's chemical company. This link, although peripheral to atomic espionage, is important because it helped the FBI in unraveling the conspiracy after the war.[51] The first courier for Brothman was Elisabeth Bentley an American citizen who became a double agent in 1945. Her testimony implicated Brothman, who in turn knew Gold. When Moscow learned that Klaus Fuchs would go with the British delegation to the United States in late 1943, the resident in New York assigned Harry Gold as his courier.

Fuchs and Peierls were working together in Britain on the design of a plant to separate the isotope ^{235}U from natural uranium by gaseous diffusion. The first test was to be a 20 stage plant, but it was never completed, and the British decided to combine forces with the Manhattan Project to build it at Oak Ridge – the huge K-25 plant that came into full operation in 1945. A contractor for the plant was the Kellex Corporation, with offices in New York City. Harold Urey at Columbia University was the chief scientist. And so it was that Fuchs and Peierls began their American expedition in New York.[52]

The meetings between Gold and Fuchs had been pre-arranged by Moscow. The vast Soviet espionage operation in the United States was carefully managed, with close attention to detail in every nook and cranny

---

[50] Richard Rhodes, "Dark Sun," *ibid.*, Chapter 4.
[51] Robert J. Lamphere and Tom Shachtman, "The FBI-KGB War," *ibid.*, p 37.
[52] Rudolf Peierls, "Bird of Passage," *ibid.*, p 184.

– and they were never caught during the war. Moscow knew that Fuchs had a married sister, Kristel Heinemann, who lived in Cambridge, Massachusetts. Kristel could, and did, serve as a link to Klaus when contact was broken. Fuchs' initial instructions were to come on the first and third Saturdays of every month at 4 pm to an address on the Lower East Side of Manhattan, starting in January, 1944. He was to stand holding a green book and a tennis ball. He would be approached by a man wearing gloves (not unreasonable, it being January in New York City), and holding a third glove. The man would ask Fuchs how to get to Chinatown, and the reply would be "I think Chinatown is closed at 5 pm."[53] In fact, meetings between Fuchs and Gold started in February, 1944, in New York City, and totaled five in all before Fuchs left for New Mexico. They met at pre-arranged locations in the various New York boroughs.[54] Fuchs passed on information about progress in the design of the Oak Ridge K-25 gaseous diffusion plant.[55] Since the plant was still under construction at that time, Fuchs did not know the final design. He lost contact with the project when he was transferred to Los Alamos.

Figure 8.2 is an attempt to simplify the 'hub and spoke' structure of the New York consulate, the couriers, and the sources in operation at Los Alamos. Alexandr Feklisov was omitted from the circle with Semenov and Yatskov, because this diagram shows contacts with Los Alamos. Feklisov was handed Julius Rosenberg in early 1944, when Semenov left the USA, but by the time David Greenglass was recruited, Yatskov was handling Rosenberg. Yatskov was indicted in absentia in the Rosenberg trial, but Feklisov was never named. Feklisov reappeared in England in 1946 as a courier for Fuchs after the war. Code names were: Semen Semenov (Sam); Anatoly Yatskov – the FBI calls him Yakovlev – (John); Saville Sax - Ted Hall's friend who acted initially as his courier – (Star); Lona Cohen (Leslie); Harry Gold (Goose, also Raymond); Julius Rosenberg (Liberal,

---

[53] Allen Weinstein and Alexander Vassiliev, "The Haunted Wood," *ibid.*, p 186. This description is so absurd that it deserves quoting in full.

[54] Robert J. Lamphere and Tom Shachtman, "The FBI-KGB War," *ibid.*, p 161.

[55] The plant began operation in 1945, and after the war became the sole source of enriched uranium.

although sometimes just Julius); Ted Hall (Mlad); Klaus Fuchs (Charles, Rest); David Greenglass (Caliber). Most of the code names seem arbitrary. Perhaps 'Sam' makes some sense for Semenov, and maybe 'Goose' for Harry Gold. 'Star' and 'Mlad' mean elder and youngster respectively in Russian. Hall was certainly a youngster – 19 years old in 1944 – and his schoolmate Sax was a bit older. 'Goose' and 'Rest' are the subjects of the VENONA decrypt of a cable from New York to Moscow on February 9, 1944, describing Fuchs' first meeting with Gold.[56]

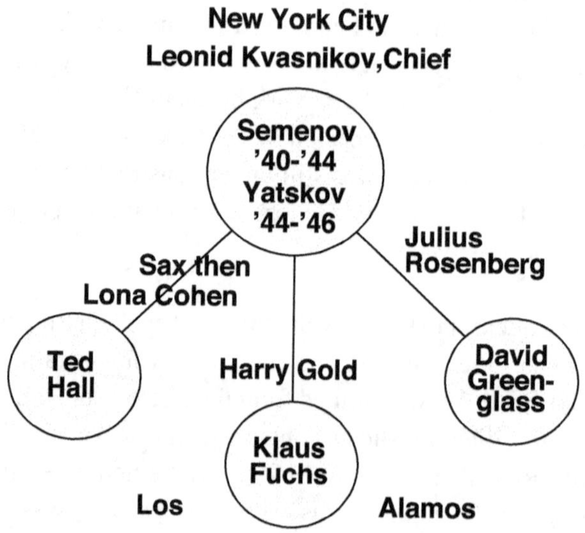

Figure 8.2. Schematic of the New York → Los Alamos hub and spoke. The spies at the rim were not supposed to know about other couriers or spies.

---

[56] www.pbs.org/wgbh/nova/venona/inte_19440209.html#cable

8.10. George Koval

While the network penetration of Los Alamos played a central role, other sites were [57] compromised as well. George Koval was born in Sioux City, Iowa of Russian émigré parents in 1913. In 1932 the family returned to a collective farm in the Vladivostok region - the far eastern Soviet Union. Young George's mechanical ability was soon recognized, and he was accepted for study at the Mendeleev Institute of Chemical Technology in Moscow.[58] After graduation in 1939 George was drafted into the Red Army, and promptly recruited by the GRU (Red Army Intelligence) for espionage. He was sent back to the United States, and became a deputy station chief for the GRU in New York City. An American citizen, he was drafted into the US Army in 1943, trained at City College of New York, and sent to Oak Ridge, where he was a Health Physics officer, with security clearance and access to the secret laboratories. His codename was "Delmar," and he passed intelligence regarding Oak Ridge operations to a Soviet courier known only as 'Faraday.' After Oak Ridge Koval was transferred to a chemical plant in Dayton, Ohio, where Monsanto was purifying polonium. The details of the bomb trigger – a polonium beryllium source mixed by the detonation, and called the 'urchin' by Manhattan Project scientists – were among the secrets reported to his courier by Koval. He was unusual in the Soviet spy network in that he was actually a trained agent himself, and directly obtained the information, rather than playing the usual role of coordinator, like Feklisov or Semenov. He was honorably discharged from the US army in 1946, completed his degree in electrical engineering at CCNY, and left for the USSR in October, 1948, ahead of any legal complications for his espionage work. He spent the rest of his life in the Soviet Union, and died in Moscow in 2006. George Koval was known as a Soviet spy by the FBI in the 1950's, but his identity was not revealed. Vladimir Putin, president of the Russian Federation, honored Koval posthumously in Moscow on November 2, 2007 for his courage in discharging special missions. This recognition was

---

[57] Owen N. Pagano, "The Spy who stole the Urchin," Senior Thesis, History Department, Washington University, class of 2014.

[58] www.atomicheritage.org/profile/george-koval

not a routine honor – intelligence operatives of any nation are rarely praised. As a result of this event his story was made public, and widely reported in the press.

8.11. Berkeley

The Soviet spy ring on the West Coast operated out of the San Francisco consulate. Grigory Heifitz was the station chief.[59] The agency was very active in unsuccessful attempts to enlist Robert Oppenheimer, his brother Frank, and his wife Kitty, all known to have had communist sympathies at one time or another, as sources of information on the Manhattan Project. These activities were exceptional, in that they attracted the attention of the FBI in real time. Access to the Radiation Laboratory at Berkeley was restricted, but compared to Los Alamos it was wide open, and relatively easy to penetrate. A few people working for Lawrence were released, but no one in any leadership role was compromised, nor was anyone arrested.

8.12. Los Alamos

In the summer of 1944 the British delegation, including the Peierls family and Klaus Fuchs, moved to Los Alamos. The Peierls family went from New York to Chicago by train, changed in Chicago to the California Limited, and got off in Lamy, New Mexico, the station closest to Los Alamos. They were housed in one of the new Army built apartment buildings on the ground floor. The Fermis soon moved into the second floor above them.[60] Fuchs, being single, was quartered in the 'big house', a building original to the Los Alamos boys' school, which was converted to a dormitory. Richard Feynman occupied an adjacent room to Fuchs.[61] Feynman was married, but his wife was ill in a hospital in Albuquerque, so he lived at Los Alamos as a bachelor.

---

[59] Strict transliteration would be 'Kheifitz." My adaptation conforms to the violinist.
[60] Rudolf Peierls, "Bird of Passage," *ibid.*, p 189.
[61] Rudolf Peierls, "Bird of Passage," *ibid.*, p 197.

Security at Los Alamos was very tight, in the sense that the town was located on an isolated mesa in the New Mexico mountains that could be fenced and guarded 24/7. It officially did not exist, was not on any map, and had no post office. The postal address was PO Box 1663, Santa Fe, New Mexico. There were almost no telephones. Half of the population was US Army. No one could leave the base without a pass. Because the Manhattan Project was under Army control, the national civilian security force, the FBI, had no jurisdiction. There was no local civilian police force. The guards were military police. General Groves had his own security operation, headed by Lt. Colonel John Landsdale, Jr.[62] Lt. Colonel Boris Pash also worked in security, and headed up Groves' 'Alsos' mission to Europe to learn about German activities in atomic energy during the war. Since the United States was at war with Germany and Japan, these nations were the principal focus of counter intelligence. Although Communists were watched, and to some extent controlled, particularly at Berkeley, where Lawrence was building his calutrons, the Soviet Union was an ally, and therefore viewed in a different light than Germany and Japan. Groves has stated that Klaus Fuchs joined Los Alamos as a British subject, a key member of the British mission, whose loyalty was confirmed by Britain. Groves did not wish to undertake separate, lengthy security clearances on members of the British team.[63]

Part of Groves' security was compartmentalization, and need to know. He kept the various sites, Oak Ridge, Chicago, Hanford, and Los Alamos, separate, so that only a few top people knew what was going on across the board. However, Oppenheimer convinced Groves to allow weekly scientific seminars at Los Alamos, where the scientific staff could learn of the Laboratory's progress. Since the other sites were created to feed Los Alamos, some knowledge of Oak Ridge and Hanford unavoidably came with the seminars. Klaus Fuchs profited by the relative openness of Los Alamos, and by his high rank in the British mission. What he learned he transmitted to Harry Gold.

---

[62] Richard Rhodes, "Making of the Atomic Bomb," *ibid.*, p 508.
[63] Leslie Groves, "Now it can be Told," *ibid.*, p 143.

By all accounts of those who knew him, Klaus Fuchs was a perfect spy. He was quiet and reserved, and dedicated to his work. He was single. He did not attract attention by way of his behavior. On the other hand, he was helpful and personable in social settings. Hans Bethe, chief of the Theory Division at Los Alamos, and Fuchs' boss, said after Fuchs' arrest: "We were very friendly together, but I didn't know anything about his real opinions. If he was a spy, he played his role absolutely perfectly."[64] He was very popular with the Peierls' children. In his own confession in 1950 Fuchs said that he maintained two separate and distinct personalities - one the theoretical physicist devoted to his work, and the other the informer, aiding the cause of communism in the Soviet Union.[65] He said that communist doctrine helped him keep the two worlds separate. At Los Alamos Fuchs had a car. Many scientists had cars. The Peierls arrived without one, but eventually bought a second hand 1927 Nash.[66] Fuchs often either drove Feynman to Albuquerque to visit his wife, or lent his car for Feynman to drive himself. Owning a car at Los Alamos was not entirely in keeping with Fuchs' quiet and unassuming image. It may have helped facilitate his clandestine meetings with Gold in Santa Fe. Fuchs was allowed one day per month off site for shopping. Of course, Gold never went to Los Alamos. He probably did not know how to get there.

Gold lived in Philadelphia. He lost contact with Fuchs in the summer of 1944, when Fuchs suddenly went to Los Alamos, but he had been told of a possible contact through his sister Kristel in Cambridge, Massachusetts. Gold visited Kristel, and was informed that Klaus would come to see her at Christmas, and that Gold could then resume contact.[67] The contact took place in January, 1945, at Kristel's apartment in Cambridge. Kristel was not at home at the time. They met for about 20 minutes, and Fuchs told Gold about Los Alamos. He said that his colleagues treated him very well, and that the uranium bomb would be ready soon, with the plutonium bomb not far behind. He furnished written information, design drawings, and mathematical calculations. Gold had

---

[64] JCAE "Soviet Atomic Espionage," *ibid.*, p 14.

[65] JCAE "Soviet Atomic Espionage," *ibid.*, p 16.

[66] Rudolf Peierls, "Bird of Passage," *ibid.*, p 196.

[67] Robert J. Lamphere and Tom Shachtman, "The FBI-KGB War," *ibid.*, p 161.

$1500 in an envelope for Fuchs, but he emphatically declined taking any money.[68] Gold travelled by train to New Mexico for two or three subsequent meetings. One was in June, 1945, before the Alamogordo test. After this meeting Gold went to Albuquerque to see David Greenglass – a break in the rules of operation described below. Their last meeting was September 19, 1945, after WWII was over, and Los Alamos was no longer a secret place, but surveillance and questioning of curios people had increased, as Lona Cohen found out at the Albuquerque train station, also described below.

8.13. Theodore Hall

Moscow had more than one source of information at Los Alamos. All escaped detection at the time, and some have never been identified. One important source not linked to the Fuchs – Rosenberg case, and not mentioned at all in the trial testimony was Theodore Hall. He does not appear in the JCAE 1951 report on Soviet espionage, and he is barely mentioned in passing by Lamphere (p183). His name surfaced in the decoded VENONA files, but he was never prosecuted. Hall's Los Alamos badge photograph is shown in Figure 8.3.

---

[68] Aleksandr Feklisov, "Kennedi I Sovietskaya Agentura," *ibid.*, p 163.

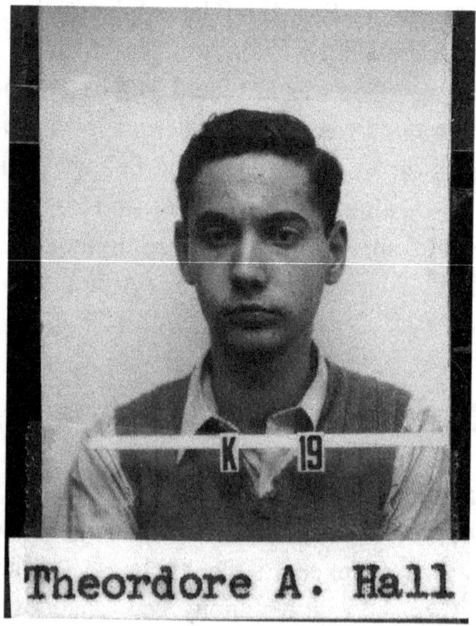

Figure 8.3. Los Alamos badge photo of Theodore Hall with his name misspelled. Credit Los Alamos National Laboratory.

Theodore Hall entered Harvard at age 16 in the fall of 1942, after two years of college elsewhere. He was an outstanding undergraduate physics student, receiving a bachelors' degree in 1944. He met Saville Sax at Harvard. Sax was a dedicated communist and became Hall's best friend.[69] John H. Van Vleck, a distinguished Harvard professor and Nobel Prize winner (1977), recommended Hall to work on the Manhattan Project. He was one of the youngest scientists recruited to work at Los Alamos.[70] He began work on the design of the implosion bomb. He was then drafted into the US Army, but remained at Los Alamos, doing what he had been, only in uniform, and for less pay. The Manhattan Project had special hooks out for skilled scientists and technicians who were drafted. On leave in New York City in mid October, 1944, Hall linked up with Saville Sax, his old

---

[69] Bulletin of the Atomic Scientists, Jan/Feb 1998, p 47.
[70] en.wikipedia.org/wiki/Theodore_Hall.

Harvard pal, and decided to find an appropriate contact in New York to give intelligence information from Los Alamos to the Soviet Union.[71] Hall had with him some materials on Los Alamos to use as evidence of his sincerity. Getting the attention of an appropriate NKVD agent was not as easy as one might imagine. The Soviet consulate was reluctant to embrace someone who walked in off the street with a package. But Hall and Sax persisted, and soon found Sergei Kurnakov, a Soviet author and journalist writing for communist newspapers in New York. Kurnakov proved to be the man Hall had been looking for. The resulting contact was duly reported in a November 12, 1944 cable from New York to Moscow, intercepted, decoded, and translated by VENONA.[72] The cable, addressed to Pavel Fitin, was as follows:

> "BEK [Kurnakov] visited Theodore Hall, 19 years old, the son of a furrier. He is a graduate of Harvard University. As a talented physicist he was taken on for government work. He was a GYMNAST [communist youth league], and conducted work in the Steel Founders' Union. According to BEK's account, Hall has an exceptionally keen mind and a broad outlook, and is politically developed. At the present time H. is in charge of a group at CAMP-2 [Los Alamos]. H. handed over to BEK a report about the CAMP and named the key personnel employed on ENORMOZ. He decided to do this on the advice of his colleague Saville Sax, a GYMNAST living in TYRE [New York City]. Sax's mother is a FELLOWCOUNTRYMAN, and works for RUSSIAN WAR RELIEF. With the aim of hastening a meeting with a competent person, H. on the following day sent a copy of the report by S. to the PLANT [New York consulate]. ALEKSEJ [Anatoly Yatskov] received S. H. had to leave for CAMP-2 in two days' time. He – Yatskov – was compelled to make a decision quickly. Jointly with MAY [Stepan Apresyan] he gave BEK consent to feel out H., to assure him that everything was in order, and to arrange liaison with him. H. left his photograph and came to an understanding with BEK about a place for meeting him. BEK met S. [1 group garbled] our automobile. We consider it expedient to maintain liaison with H. [1 group unidentified] through S., and not to bring in anybody else. MAY has no objection to this. We shall send the details by post."

This message says it all. It gives Hall and Sax good working class backgrounds – son of a furrier and a Russian War Relief worker;

---

[71] Joseph Albright and Marcia Kunstel, "Mlad and Star," Bulletin of the Atomic Scientists, Jan/Feb 1998, p 48ff.

[72] www.pbs.org/wgbh/nova/venona/inte_19441112.html#cable

emphasizes their dedication to the cause, and sets up the contact machinery for subsequent intelligence. Saville Sax was the courier for Ted Hall in the beginning. He made at least one trip to New Mexico by bus. He was interrogated by immigration agents in Albuquerque, and had to explain that the reason for his trip was to possibly enroll in the local university.[73] Sax obtained information that Hall smuggled from Los Alamos, and returned to New York. After this trip he was replaced by Lona Cohen. Yatskov and Apresyan were both vice-consuls in New York at that time.

8.14. Lona Cohen

Lona Cohen, and her husband Morris Cohen, were US citizens. Morris had fought in the Abraham Lincoln Brigade in Spain, and was a guard at the Soviet Pavilion in the 1939 New York World's Fair. There he met Lona. They were married in 1941.[74] Semen Semenov recruited the couple for intelligence work in the United States.[75] In 1942 Morris was drafted into the US Army, and sent overseas. Morris Cohen served honorably in the army, first on Alcan highway construction, and later in combat in France after the Normandy Invasion. He was discharged in November, 1945.[76] Lona assumed responsibility for Morris' contacts when he was drafted. Lona worked in a defense plant, and transmitted useful information to Semenov. In 1944 Semenov was recalled to Moscow, and replaced by Anatoly Yatskov. Yatskov assigned Ted Hall to Lona Cohen.

Lona went to New Mexico using a visit for treatment at a tuberculosis sanatorium as a cover. In 1992 Yatskov would remember Lona as a very attractive young woman.[77] Lona made her final trip to Albuquerque, New Mexico in August, 1945, after the bombing of Hiroshima and Nagasaki. There she obtained very important documents regarding the nuclear

---

[73] Allen Weinstein and Alexander Vassilev, "The Haunted Wood," *ibid.*, p 208.

[74] Robert J. Lamphere and Tom Shachtman, "The FBI-KGB War," *ibid.*, p290.

[75] Pavel Sudoplatov, "Razvedka I Kreml'," *ibid.*, p 226-227.

[76] There is a recent biography of the Cohens: Barnes Carr, "Operation Whisper," University Press of New England, Lebanon, NH, 2016.

[77] Pavel Sudoplatov, "Razvedka I Kreml'," *ibid.* Lona subsequently appeared on a Russian Federation postage stamp.

weapons designs from Ted Hall, and went to the railroad station to return to New York. Because the existence and purpose of Los Alamos was now known, security at the Albuquerque railroad station was similar to that experienced by air travelers at US airports today. Suitcases and carry-on bags were searched. Lona pretended to be flustered looking for her tickets, and she handed a tote bag to one of the security guards as she fumbled around. Under a Kleenex box in the tote bag were the secret documents. She got on the train without the bag, but it was soon returned to her by the guard, unmolested. When Lona got back to New York, she told Yatskov that the police held the materials in their own hands, but cheerfully returned them to her.[78]

Lona and Morris were evacuated to Moscow during the Rosenberg trial in 1950. A few years later they turned up in Britain under the names of Peter and Helen Kroger, where they continued espionage using an elaborate shortwave radio set in the attic of their British bungalow. That scheme unraveled in 1961 through the arrest of Gordon Lonsdale – a pseudonym for a Soviet citizen named Konon Molody. After six years in prison the Cohens were released, and lived the rest of their lives in Moscow.

8.15. The Rosenbergs and the Greenglasses

Julius Rosenberg was born in New York City of Russian émigré parents in 1918. He grew up on the Manhattan lower east side, and graduated from City College (CCNY) in 1939 with a degree in electrical engineering. He married Ethel Greenglass in June of that year. He had participated in communist activities in college, and attracted the attention of Soviet operatives after taking a job with the Army Signal Corps in 1940. He was first assigned to Semen Semenov, who served as his contact until early 1944, when Semenov was recalled to the Soviet Union. He was then handed over to Aleksandr Feklisov, and the two worked together until

---

[78] The story of Lona and the Kleenex box has been told many times. This version comes from Sudoplatov.

1946. Rosenberg was in the electronics and radar business, and virtually all of the information that he passed to Feklisov was on these subjects.

The development of radar was a very big undertaking in WWII – almost rivaling the Manhattan Project. Radar was also an early example of British – American collaboration. The practical cavity magnetron oscillator, an early compact source of radar waves, and still used in modern microwave ovens, was a British invention.[79] A working model was brought to the United States as part of the Tizard Mission in 1940.[80] Many physicists were employed at the 'other' Radiation Laboratory at MIT for radar development. The Tizard Mission also brought over an early version of the radar proximity fuse. Rosenberg had access to the output of the electronics/radar industries. Feklisov tells the story of his Christmas present from Rosenberg in 1944. Somehow Rosenberg succeeded in smuggling a working proximity fuse out of a factory in New York where they were being made. Proximity fuses were difficult to manufacture, and there were many rejects. Rosenberg managed to squirrel away enough fuses that failed the quality control to assemble a working device. He delivered the fuse to Feklisov at a Horn & Hardart Automat[81] on West 38th Street on December 24, 1944.[82] Feklisov had some gifts for the Rosenberg family, and they exchanged packages. The fuse weighed about 15 pounds. Feklisov later observed that it was a proximity fuse that brought down Francis Gary Powers' U2 spyplane over the Soviet Union in 1960.

Julius Rosenberg was an effective recruiter of agents willing to steal technical data for transmission to the USSR. Feklisov was the handler for Rosenberg's network of electronics spies, but he did not handle the Manhattan Project. Less than 10% of Rosenberg's espionage activities

---

[79] A cavity magnetron works on a resonance principle between an electron current and shaped copper cavities, that look like a revolver cartridge cylinder. When asked how it works, an expert replied – like a whistle. OK, how does a whistle work?

[80] https://en.wikipedia.org/wiki/Tizard_Mission

[81] These early 'fast food' restaurants dispensed dishes through glass doors when nickels were inserted in a coin slot. They fed New Yorkers for almost 100 years.

[82] Aleksandr Feklisov, "The Man behind the Rosenbergs," *ibid.*, p 124.

were involved with David Greenglass and Los Alamos. Feklisov and Rosenberg became friends during their two year collaboration.

The brother of Ethel Rosenberg, David Greenglass, was a machinist, drafted in the US army in April, 1943, assigned to Oak Ridge in July, 1944, and then transferred to Los Alamos in August. He worked in one of the on-site machine shops, building various parts for weapons development. He had no scientific training, and only minimal understanding of the mission of the laboratory, but he had the opportunity to talk to people, and met with scientists when they brought in sketches of objects that they needed to have made. Julius Rosenberg realized that his brother-in-law may have access to secret information. He pressed Ethel to recruit her brother. Ethel communicated with David through his wife, Ruth. Ruth found David resistant to the idea, but eventually she managed to convince him to go along. Ruth Greenglass was to serve as the courier for David.

This brings us to the Jello box story. While all of this was being set up in the Rosenberg home in New York, the question arose of how to identify a new courier, should it be necessary to employ one? Julius Rosenberg brought out from the kitchen the side of a Jello box, and cut it in two. One part was given to Ruth, and Julius kept the other.[83] Should it become necessary to send a new courier, Julius would ensure that the missing box part would accompany him/her for identification.

Ruth Greenglass moved to Albuquerque to be closer to David. Apparently, several families of service men working at Los Alamos lived in Albuquerque at that time. David would be granted leave to see his family, and go to Albuquerque with whatever information he could extract from his contacts at the lab. In most cases the specific information transferred by one of the sources to a courier, and subsequently to New York and Moscow, is not known. However, David Greenglass in his testimony at the Rosenberg trial does say explicitly that he was asked to fabricate molds for the shaping of explosive lenses, and that plans for these molds were transmitted. They were two dimensional compression lenses,

---

[83] JCAE "Soviet Atomic Espionage," *ibid.*, p 76ff.

shaped like a four leaf clover, to surround and compress a pipe. The explosive was poured into the mold in liquid form, and allowed to set. These were tests made preceding the three dimensional spherical version ultimately used. So this communiqué was a small detail, but accurate, and hence useful.

In June, 1945, Harry Gold planned to make one of his periodic trips to Santa Fe to meet Klaus Fuchs. Anatoly Yatskov, with the permission of Moscow Center, asked Gold to make a detour to Albuquerque after Santa Fe, to pick up a package from David Greenglass. Yatskov gave Gold the half Jello box top, and told him that the introduction was 'I come from Julius.' Gold at first objected to the arrangement – it was a cardinal rule of Soviet intelligence never to allow one spoke to cross over into the territory of another. But he agreed. Later on this act would cause Sudoplatov to be called on the carpet by the Kremlin, as we shall see.

So it was that Harry Gold, after making contact with Fuchs in Santa Fe, took the bus over to Albuquerque and knocked on the door of Ruth Greenglass' apartment on a Sunday morning in June, 1945. Ruth opened it, was greeted by 'I come from Julius', and matched the two parts of the Jello box – one from her purse, and the other from Gold. David did not have the package prepared, but it would be ready later in the afternoon. Gold then returned to New York, and delivered the materials from both Fuchs and Greenglass to Yatskov. This was the last contact with David Greenglass.

8.16. Things Fall Apart

The Soviet espionage network began to fall apart in 1945, but the damage was already done. There were perhaps three events that caused the downfall. There were two defections: one by Igor Gouzenko on September 5, 1945, and the second by Elizabeth Bentley in early November of the same year.[84] The third factor was the VENONA decryptions, which continued for more than a decade, and proved to be very valuable in identifying the spies, but was initially classified, and not available in open court. Bentley was not directly involved in atomic espionage, but she knew a lot about the structure of the system, and caused its virtual shutdown almost overnight. As mentioned earlier, Bentley knew a chemist named Abe Brothman, and he knew another chemist named Harry Gold, and it is through this chain that Gold attracted the attention of the FBI. However, after questioning Gold the FBI let him go. The actual link was not Gold to Fuchs, but Fuchs to Gold. In response to the demise of the spy network, Yatskov was recalled to Moscow in 1946.

Gouzenko was a code clerk in Ottawa, Canada, for the Red Army Intelligence, the GRU. He came to Canada in 1943 with the intention of defecting as soon as possible, and threats from Moscow to return him to the Soviet Union in 1945 convinced him to act.[85] He took with him some documents incriminating Alan Nunn May in atomic espionage. At first the Canadians could not believe that the Soviet Union was spying on them, but after a day or two of suspense, Gouzenko was taken into custody by the Royal Canadian Mounted Police. According to Lamphere, one of the GRU agents in Canada uncovered by the Gouzenko affair was a certain Israel Halperin. An address book containing the name 'Klaus Fuchs' was found in his apartment.[86]

A front page story in the New York Times, February 3, 1947, began "Bernard M. Baruch was reliably reported to have told the Joint Congressional Committee on Atomic Energy that Russia evidently had tapped United States atomic-bomb secrets."[87] It took three more years before Klaus Fuchs was arrested.

---

[84] Allen Weinstein and Alexander Vassiliev, "The Haunted Wood," *ibid.*, p 104. Strict transliteration would be 'Guzenko.' The version 'Gouzenko' is intended to aid pronunciation – like 'goose' rather than 'gun.'

[85] Robert J. Lamphere and Tom Shachtman, "The FBI-KGB War," *ibid.*, p 32ff.

[86] Robert J. Lamphere and Tom Shachtman, "The FBI-KGB War," *ibid.*, p 139. Lamphere was puzzled, because GRU and NKVD usually did not overlap.

[87] JCAE "Soviet Atomic Espionage," *ibid.*, p 13.

Lamphere told the story of Fuchs' arrest in an oral history published by PBS.[88] In 1948 FBI headquarters obtained some fragmentary KGB messages which had been deciphered by the Army. This was early work of the VENONA project.[89] More messages were sent over, and one of them indicated that a member of the British mission to the Manhattan Project had supplied information to Moscow regarding isotope separation by gaseous diffusion. Lamphere quickly zeroed in on Klaus Fuchs, partly because his name had appeared in Halperin's address book, partly because of his communist activities in Germany before the war (which were known through captured Gestapo files), and partly by the process of elimination. The FBI notified British Intelligence MI5, as Fuchs was then in Britain. Indeed, he was head of the Theory Division at Harwell, the British atomic energy laboratory.

The British arrested Klaus Fuchs, and he soon confessed to about 10 years of atomic espionage for the Soviet Union. Soon after his sentence to prison, TASS denied in Pravda that the Soviet Union had ever heard of him.[90] He knew his US courier only as 'Raymond.' The FBI sent Lamphere to London to question Fuchs, armed with several pictures of Gold, whom they suspected to be Fuchs' courier. In the meantime, FBI in Philadelphia were questioning Gold and searching his belongings, where they found a map of Santa Fe, New Mexico.[91] This was sufficient evidence to arrest Gold. Fuchs signed the back of a photograph of Gold, stating that he was indeed Raymond. Harry Gold confessed, and pleaded guilty to espionage.

Then the consequence of the courier cross-over in 1945 came back to bite the network. Gold identified a young soldier whom he met in

---

[88] www.pbs.org/redfiles/kgb/deep/interv/k_int_robert_lamphere.htm

[89] Feklisov, p 181, writing in 1990, did not believe that the Soviet diplomatic code had been compromised by US intelligence.

[90] Pravda, March 8, 1950, top of page 5. Most issues have four pages. This one is longer because of the award of Stalin Prizes, including to Dmitry Shoshtokovich.

[91] Robert J. Lamphere and Tom Shachtman, "The FBI-KGB War," *ibid.*, p 157. Feklisov, p 180, adds that fingerprints of Gold and Fuchs, together with a marked rendezvous, were on the map.

Albuquerque as another source at Los Alamos. The FBI had no trouble identifying David Greenglass, who in turn identified the Rosenbergs. Thus in 1950 began the Rosenberg trials. The testimonies of Gold and Greenglass at the Rosenberg trials are valuable sources of information regarding atomic espionage in WWII. The Rosenbergs themselves never admitted to anything.

Alan Nunn May was arrested in London in February, 1946.[92] In his deposition to the police, Nunn May said:

"I carefully considered the question should the development of atomic energy be the sole prerogative of the USA? I came to the painful conclusion that there should be general transparency regarding the atomic question. I accepted the proposition of an individual who visited me in my apartment, and whose name shall remain secret. I met with him several times in Canada. At one of these meetings I gave him two samples of rare isotopes of uranium $-\,^{235}U$ and ^{233}U. A milligram sample of enriched ^{235}U was in a glass tube in the form of UO_2. One tenth of a milligram of ^{233}U was deposited as a thin film on a platinum foil, all covered by a paper wrapper. I also gave my contact a report containing all that I knew about the atomic problem. He gave me a package containing money, that I accepted in spite of myself."[93]

This 0.1 milligram sample of ^{233}U deposited on a platinum foil was used by Mostovoy in his fission cross section measurements at Laboratory #2 in 1947 discussed in Chapter 7. It is one of the rare documented examples where an item obtained by espionage actually found its way into the experimental physics labs of the Soviet Atomic Project. Nunn May does not say, but presumably he smuggled these samples out from the Chalk River reactor in Canada. Did the guards at Chalk River not do body searches for radioactive substances?

---

92 Stanislav Pestov, "Stranitsy Istorii Tainy Atomnoi Bomby" in 'Argumenty I Fakty' #41, October 14, 1989.

93 Translated by the author from Pestov's Russian account. Thus it suffers from English-Russian-English confusion, but should be approximately what May said.

8.17. Sudoplatov in Trouble

Pavel Sudoplatov was not perturbed by the Rosenberg arrests, as he considered them to be minor players in the scheme of things, as indeed they were. He was surprised, however, about one year later, when the new minister of state security, Ignatev, ordered him to turn over all materials relating to the collapse of Soviet intelligence operations in the USA and England connected with the Rosenberg case.[94] Sudoplatov was called before the Central Committee of the Communist Party of the Soviet Union to answer the question: who authorized the ill-fated meeting between Gold and Greenglass in Albuquerque? A subcommittee was formed to look into it, and concluded that it was due to a lack of training. Ovakimyan was relieved of duty in the NKVD as a result.

The case was renewed the next year, when a deputy of Malenkov named Kiselev ordered Sudoplatov back to the Kremlin and accused him of deceiving the Central Committee into thinking that the Rosenbergs were minor players in the spy operation, when in fact they were the opposite. Kiselev said that the Central Committee had received an anonymous letter claiming that the Rosenbergs were major players in the enterprise. Sudoplatov repeated that they were minor players. Besides, intelligence operations in the USA were shut down in 1946, and switched to England. The Rosenbergs were dedicated to the cause of the Soviet Union, and would fulfill any task asked of them, but they were not significant sources of atomic secrets. Kiselev in an official tone of voice said that he would convey Sudoplatov's response to the Central Committee, and personally to Comrade Malenkov, and the Committee for Party Control would establish who was guilty of the demise of the intelligence operation in the USA. The Rosenberg trial was played for propaganda purposes by both sides of the Cold War. The fuss blew over when the Kremlin learned that the Rosenbergs went to their deaths without revealing anything. So the cross-linking within the network had actually done no damage, except to the Rosenbergs themselves.

---

[94] Pavel Sudoplatov, "Razvedka I Kreml'," *ibid.*, p 255.

8.18. Fuchs in England Before His Arrest

We return to Feklisov's account of his dealings as a courier with Fuchs after 1946. In the summer of 1946 Aleksandr Feklisov was recalled to Moscow after more than five years in the United States. In the interim he had married, and had a young daughter. He was soon shipped out again, this time to London. Living conditions in London for a few years after the war were severe. Parts of the city were still destroyed after the German bombings. The Feklisovs were on their own in finding a place to live, and tried out several possibilities before settling on an apartment recently vacated by a Soviet diplomat who was returning home. Feklisov's salary was £90 per month, and the rent was £10 per month, so they could afford it. Hot water was supplied by feeding shillings into a meter on the hot water heater, not an unusual arrangement in England at that time. His wife got a job, and their daughter went to British kindergarten, where she became expert in the English language. The Feklisovs spoke only English at home. Food was still short. There were ration cards for everything – even sugar and tea. Russians like more sugar and not so much tea, and Brits the other way around, so the Feklisovs could trade with their neighbors. They seemed like ordinary neighbors, but they were there for unfriendly activities.

A visit to the Irish playwright George Bernard Shaw was one of Feklisov's first assignments by the Soviet Embassy in London. Shaw was a dedicated socialist, and well known to the Soviet Embassy. He visited there for an open house, met Feklisov, and asked about an honorarium that he considered his royalty for publications and performances of his works in the Soviet Union. There were no international copyright agreements between the USSR and western countries, so there was no formal arrangement to pay for the use of copyrighted materials. The American playwright Arthur Miller had a similar experience, as his play "The Death of a Salesman" was viewed as a prequel to the demise of capitalism, and frequently performed in Moscow.[95]

Shaw's request was forwarded to Moscow. £20,000 was authorized to be given to Shaw in recompense, and Feklisov was chosen as the courier to deliver the news. Feklisov had no hesitation in driving a British Vauxhall on the left hand side of the road to Shaw's country home in Hertfordshire. The mission was of course harmless, but had the roles been reversed, and Feklisov was a British diplomat driving out from Moscow, he would have been for sure followed. The Soviets, and others, have successfully exploited the difference in surveillance of foreigners.

Shaw was in his 90's – he would die in 1950 at age 94 – but still irascible, having terrorized British arts and letters for over 50 years. Feklisov found the house, and parked the car under close supervision by Shaw. Shaw had an assistant with him. They discussed the honorarium, and Shaw observed that, should he accept it, the British government, which he despised, would take most of it in taxes. In the end Shaw declined the £20,000, saying that the USSR needed it more than he did. Feklisov drove back to London, having met one of the 20th century's more interesting people.

Counter-intelligence activity by MI5 and MI6 in Britain had increased after WWII, compared to the experience of the NKVD in the USA during the war, when the FBI was relatively quiescent towards Soviet activities, and the CIA did not yet exist. The cold war had settled in, the USA had a nuclear force monopoly, and NATO was gearing up to contain perceived Soviet aggression. As a result, the Soviet Embassy in London was under constant surveillance. For the first time the NKVD was feeling pressure. Feklisov established a pattern of travelling around London daily by bus, tube, taxi, auto, and on foot, in order to convince the British counter-

---

[95] In Miller's case, he was reimbursed by an all expenses trip to the Soviet Union, recounted in the book "In Russia," by Inge Morath and Arthur Miller, Viking Press, New York, 1969.

intelligence that this was part of his normal routine, and nothing to be concerned about.

Feklisov was occupied trying to learn NATO's plans for European defense. He thought that there was British sentiment in favor of the Soviet Union, and a general dislike of the United States. This may have been true, for had Hitler not invaded the Soviet Union in 1941, he might well have invaded Britain. The US was hard to love, because it had replaced Great Britain on the world stage. The Soviet Embassy in London continued to be effective in learning about British foreign policy, partly through MacLean and Burgess, members of Yury Modin's five Cambridge friends.

Feklisov's next assignment was to serve as courier for Klaus Fuchs. This was an unorthodox arrangement, in that couriers – the spokes in Figure 8.2 - were not usually Soviet agents, because of the danger of compromise, as previously discussed. This departure from normal procedures did not cost the NKVD, however. Fuchs himself was compromised in 1950 by evidence supplied from the FBI, as previously discussed, and not by the work of British counter intelligence. Fuchs did not implicate Feklisov, and he was never exposed. The assignment may have been based on the extremely high value Moscow placed on Fuchs' information, and their confidence that Feklisov, the experienced agent, would not jeopardize the operation.

Feklisov had no knowledge of nuclear weapons, and was subjected to a crash course in classified material given by an unnamed scientist in order for him to be able to interact with Fuchs in a useful manner.

At the last meeting between Klaus Fuchs and Harry Gold in Santa Fe in September, 1945, Gold, in anticipation of Fuchs's return to England, was instructed to tell Fuchs how to restore contact with Soviet agents in London. As mentioned above, Fuchs was highly regarded by the British atomic energy establishment, and was appointed to head the Theory Division of Harwell upon his return. In that capacity, he conducted his own research, directed the work of others, handled personnel issues, hired and fired workers, etc. Harwell, near Oxford, was started in 1946 as the center for both civilian and military applications of atomic energy. The close collaboration between the United States and Briton that existed during the war was terminated by the McMahon Act passed by the US Congress in August, 1946, leaving Briton to develop atomic weapons on its own. Work had restarted in early 1947.

8.19. Aleksandr Feklisov in England

Feklisov first met Fuchs in 1947, at a bar outside of central London. He gave a detailed description of the first meeting, which seems to have come from a grade B spy movie.[96] Feklisov arrived about 15 minutes early, to make sure that he had not been followed, and that there were no occupied parked cars snooping around. Satisfied that everything was safe, he went inside the pub and saw a man seated at the bar, reading a specified newspaper. A half full glass of beer sat in front of him.

The man fit the description of Klaus Fuchs. Feklisov sat down at the bar, and ordered a beer. He took out a specific magazine for Fuchs to see, and placed it on the bar. Feklisov kept one eye on the entrance, to make sure that no 'tail' came in after them. Two men came in, but were obviously locals, frequent customers of the pub, and were greeted by the bartender. A poster displaying British heavyweight boxing champions was hanging on the wall. By prearrangement, Fuchs carried his beer glass over to the poster, where other customers were discussing the merits of the boxers. Feklisov joined them. Fuchs said, "Bruce Woodcock was the best boxer of all time," to which Feklisov responded, "Tommy Farr was significantly better."[97] After a brief argument over the two fighters, Fuchs paid his bar bill and left the pub. A few minutes later Feklisov also left. The two met outside, and chatted as they walked along the street.

---

[96] Aleksandr Feklisov, "Kennedi I Sovietskaya Agentura," *ibid*, p 171ff.
[97] Both were British heavyweight champions, Farr in 1937, and Woodcock in 1945-1950.

Someone had to have set all of this up. London is full of pubs, but this one was chosen, probably not at random.[98] The Soviet Embassy knew that there was a poster of heavyweight boxers next to the bar. The dialog had to be written, and sent out to the two parties, Fuchs and Feklisov. Bruce Woodcock was the reigning British heavyweight champion, so most pub goers would know the name, but Tommy Farr was ten years earlier. Who came up with him? We will never know. Was Feklisov followed from the Embassy, and did he lose them en route to the pub? We are not told. Fuchs was at that time above suspicion, and not subject to being followed.

The materials supplied by Fuchs naturally pertained to the British atomic project. One of the first decisions was to construct nuclear reactors patterned after the ones at Hanford to produce plutonium for fissile material. Two graphite reactors were designed for this purpose to be built at Windscale on the northwest coast of England. The first Windscale reactor went critical in October, 1950, so Fuchs could only discuss the plans in 1947.[99] The British successfully tested a plutonium implosion bomb of the Nagasaki type in October, 1952, in the Montebello Islands off the northwest coast of Australia.[100]

Fuchs continued to work in the Theory Division at Harwell, and to meet with Feklisov, for two more years, until 1949. They always made arrangements for the next meeting in advance. Fuchs responded to questions passed on from Kurchatov, as well as supplying independent information. Some materials pertained to the hydrogen bomb, which Fuchs had worked on before he left Los Alamos, but on this subject, as we shall see, the Soviets did not really need any help.

Fuchs and Feklisov became friends in the course of their mutual espionage activities. Feklisov asked Fuchs why he did not marry and settle

---

[98] The 'Nags Head' pub is pictured in "The Man behind the Rosenbergs."

[99] On October 10, 1957, Windscale Unit 1 caught fire – the worst nuclear accident in Britain. See https://en.wikipedia.org/wiki/Windscale_fire.

[100] www.tiki-toki.com/timeline/entry/59244/The-History-of-the-UK-Nuclear-Weapons-Programme/

down. Fuchs answered that he had thought about it, but decided that, as long as he was busy helping the Soviet Union with secret information, his life was too precarious. One little slip and he would be in prison. He was willing to take the risk personally, but did not want to subject a family to that sort of threat.

After his confession and conviction in 1950, Fuchs served nine years of a 14 year sentence in prison. He then moved to Dresden in East Germany, and did marry and settle down. The couple had known each other before the war in communist activities, and she spent the war in the Soviet Union. Fuchs became deputy to Heinz Barwich, the director of a nuclear physics research laboratory at Rossendorf. Later on, after Khrushchev's de-Stalinization, Feklisov attempted unsuccessfully to obtain recognition by the USSR for Fuchs' contributions to national defense. Mstislav Keldysh, then president of the Soviet Academy of Sciences, said that electing Fuchs to the Academy would cheapen the accomplishments of Soviet scientists. Fuchs died in 1988.

The next year, 1989, an East German film maker decided to make a movie of Klaus Fuchs, and invited Feklisov to come to Dresden to be interviewed for the film. Feklisov agreed, and combined the trip with a visit to Fuchs' widow. She asked him where he had been – Klaus waited 25 years for any recognition by his Soviet colleagues.

Moscow ordered the return of Feklisov after the arrest of Klaus Fuchs in January, 1950. Boris Rodin, the London station chief, urged that Feklisov be allowed to stay a few months, until things blew over, arguing that his immediate departure would implicate him in the Fuchs affair, and that Fuchs under interrogation may well not identify Feklisov. Moscow agreed, and Feklisov remained in London until April, 1950. His clandestine activities ceased at once when Fuchs was arrested, and Feklisov was never identified as Fuchs' courier.

Aleksandr Feklisov worked in Moscow for a time training NKVD agents. He accompanied Khrushchev on an official visit to the United

States in September, 1959.[101] Feklisov (under the cover name of Aleksandr Fomin) was again in Washington D.C. in October, 1962, for the drama of the Cuban Missile Crisis. He met several times with an ABC correspondent named John Scali, who was then reporting to President Kennedy's brother, the attorney general Robert Kennedy. Feklisov and Scali are credited by some historians with the agreement that ended the crisis – withdrawal of Soviet missiles for a guarantee that the United States would not invade Cuba. There is a plaque on the wall of the Occidental Restaurant in Washington, D. C. that reads: "At this table during the tense moments of the Cuban missile crisis a Russian offer to withdraw missiles from Cuba was passed by the mysterious Russian 'Mr. X.' to ABC-TV correspondent John Scali. On the basis of this meeting the threat of a possible nuclear war was avoided."[102]

Aleksandr Feklisov died in Moscow in 2007, aged 94 years. His life long espionage activities were to some extent exonerated by his diplomacy in 1962.

8.20. Terletsky Visits Bohr

Yakov Terletsky was sent on a mission to visit Niels Bohr in November, 1945. In Chapter 6 we described the Heisenberg-Bohr meeting in 1941, and all of the confusion it generated. There are some similarities, in that Bohr was still Bohr, but Terletsky was not Heisenberg. Heisenberg's trip from Germany to Denmark was officially sanctioned for his participation in a conference on astrophysics at a German cultural establishment in Copenhagen. He was not sent on an intelligence gathering mission. Terletsky's mission was an NKVD espionage operation from start to finish, although it was hardly typical. For one thing, we have a detailed account by Terletsky himself. Usually for spying activities the less said the better. For another thing, the mission went according to plan, but the Soviet Union did not learn much that they did not already know. It was

---

[101] On this occasion Khrushchev banged his shoe on the lectern at the United Nations in New York.

[102] "Mr. X' was Feklisov. The plaque is quoted on the Occidental Restaurant web site.

a success – everyone got back alive. It was also a failure – Bohr did not tell them anything new. Terletsky's description of how they got to Copenhagen from Moscow just after the war, and what that part of western Europe looked like to Soviet eyes, give some insight into living conditions. Terletsky's experiences working with Sudoplatov in the NKVD Lubyanka headquarters, and his meetings with Beria also tell us something about daily life in the NKVD. Sudoplatov also described the trip in his memoirs.[103] His claims regarding the intelligence gained on the Manhattan Project are exaggerated to make the operation, which was his responsibility, look good. This might be called irrational exuberance. This was to lead to further confusion, to be described below.

Ya.P. Terletsky visited Niels Bohr in Copenhagen in November, 1945.[104] How this came about is a story in itself. We learned in Chapter 7 that the Soviet Atomic Project was put in the fast lane after the Hiroshima bomb. A special committee was set up, with Beria in charge, to get going quickly. There were two scientists on the committee – Igor Kurchatov and Peter Kapitza. Beria acquired the services of Kapitza as a member of his committee, meaning that, from Beria's point of view, Kapitza was obliged to do whatever he asked.

Beria decided that it would be useful to find out what Niels Bohr knew about the Manhattan Project. Bohr was back in Copenhagen, and had earlier visited Los Alamos. (See Chapter 6). Yakov P. Terletsky was trained in theoretical physics at Moscow State University. He later became a physics professor at the University for World Friendship named Patrice Lumumba, and wrote several textbooks on electrodynamics and statistical physics. The friction between MSU and the Academy of Sciences has been discussed before, and may have had something to do with Terletsky's career. While he received the Stalin and Lenin prizes for his research, he was never elected to the Academy. He was a convinced, dedicated

---

[103] Pavel Sudoplatov, "Razvedka I Kreml" Geya, Moskva, 1996 p 244ff.

[104] Memorial biography of Yakov Petrovich Terletsky, Uspekhi Fizicheskikh Nauk, Vol 164, #2 February, 1994. His NKVD work is not mentioned. His own memoir of the mission to Bohr is: Ya.P. Terletsky, "Copenhagen Operation of Soviet Intelligence" Voprosi Istorii Ectectvoznania I Tekhniki, (VIET) #2, pp 18-44, 1994.

communist. His reaction to the Hiroshima bomb in August, 1945 was visceral: "The Americans have left us behind, and can now dictate to us their own conditions. Our glorious victory over fascism has been cheapened. It will not be possible to dream about freeing the peoples of Europe and Asia from capitalism any time soon."[105] Terletsky soon learned that Beria was put in charge of the uranium project, a move that convinced him that the government was determined to go ahead vigorously. He wanted to help. He received an invitation to visit NKVD headquarters in Lubyanka prison in Moscow.

Such 'invitations' were not usually received with much enthusiasm, but Terletsky gathered up his courage, and went for an interview at the office of Pavel Sudoplatov, Beria's go-to guy. He was asked what he knew about the uranium project, which he answered vaguely. Then he was asked if he wanted to join the effort. He replied that he had previously talked to Kurchatov, who offered him a job, but that he deferred in order to finish graduate school. Now, however, he was ready to jump in. Joining Laboratory #2 was not what Sudoplatov had in mind. When informed that he was to join Section 'S' of the NKVD ('S' for Sudoplatov), Terletsky said that he preferred to continue doing scientific work, rather than work in the Ministry. Sudaplatov answered that he could continue his research at MSU, but that his job in the NKVD would take precedence. Terletsky was sent home to think about it. He was not comfortable with the proposed arrangement. He worried that, despite the perquisites he would have from a position in the Ministry, he would be shut off from physics research and teaching, which could be detrimental to his career development. In the end he decided to accept the job, believing that, in that way, he could significantly contribute to the national effort.[106] Perquisites followed in the form of housing, a higher rank ration card, and permission to shop in special stores reserved for the elite. His new job was to edit reports of overseas activities in nuclear weapons obtained by espionage. The

---

[105] VIET, *ibid.*, p 22.

[106] Ya.P. Terletsky, *ibid*, p 24. He stayed with the NKVD until 1950, and worked with Beria's committee that managed the Project. He then he wrote a letter to Stalin requesting permission to return to teaching, which was granted.

documents were obtained in English, then translated into Russian by NKVD translators. Terletsky had to make sure that the translated documents made scientific and technical sense. He later learned that there were about 10,000 pages of information regarding the uranium problem in the safes of Section S of NKVD headquarters, covering experimental measurements of nuclear parameters, designs of nuclear reactors, isotope separation techniques, etc. It was all there.

Zoya Zarubina was one of the translators in Section S. Her father, Vasily Zarubin, was the NKVD New York station chief until recalled to Moscow in 1944.[107] Zoya was a language expert. She had served as interpreter and organizer for the Yalta, Teheran, and Postdam conferences of the Allied Powers during WWII. She was then assigned to NKVD headquarters under Sudoplatov, given a dictionary, and a stack of intelligence material in English on the Manhattan Project. She had no technical training, and, while she could translate the words, the resulting sentences did not make sense.[108] She met more than once with Kurchatov, who encouraged her to learn some physics and do a better job of technical translation. The sources and contents of her translations remained top secret. She never discussed the materials even with her father, and he in turn never told her anything about his work as the New York station chief. Later on she translated the Smyth Report into Russian. Sudoplatov confirmed that Zoya was one of a group of expert translators.[109]

Kurchatov also had an office in Lubyanka for viewing these reports, but apparently Terletsky saw them first, after Beria and Sudoplatov. Beria liked to maintain competition among his subordinates, including the scientists, for the purpose of keeping people off-balance, uncomfortable, and insecure. He did not trust anyone, and some of the intelligence was kept even from Kurchatov, in order to check up on the authenticity of his progress reports. Intelligence information did not move past Kurchatov to

---

[107] Inez Cope Jeffery, "Inside Russia, the life and times of Zoya Zarubina," Eakin Press, Austin, Texas, 1999.

[108] Inez Cope Jeffery, "Inside Russia, the life and times of Zoya Zarubina," *ibid.*, p 16-17.

[109] Pavel Sudoplatov, "Razvedka I Kreml," *ibid.*, p 219.

other scientists and engineers in the project, with some specific exceptions – Kikoin obtained information on isotope separation.

Terletsky's first day at work was October 11, 1945. Sudoplatov gave him an office previously occupied by an NKVD colonel, who did not particularly like being kicked out in favor of a very young physicist. Sudoplatov warned him to show no interest either in the origin of the information, or in who wrote or transmitted the documents. That was none of his business. His job was to verify the technical accuracy of the translations.

As described in Chapter 6, Niels Bohr visited Los Alamos after leaving Denmark and landing in Scotland in October, 1943. After his successful escape, Kapitza sent him a letter, inviting Bohr and his family to come to the Soviet Union. This letter was not delivered to Bohr until April, 1944, when he was back in England, after visiting Los Alamos off and on for about a year. British Intelligence was nervous about Bohr's invitation to the USSR, but the letter from Kapitza was innocent enough, having been written before Bohr went to Los Alamos. Subsequently, on May 5, 1945, after the liberation of Denmark from German occupation, Kapitza sent a congratulatory telegram to Margrethe Bohr. On August 25, 1945, after the Hiroshima and Nagasaki bombs, Bohr returned to Denmark.

October 7, 1945 was Bohr's 60th birthday, and the Danish government planned a lavish reception on the anniversary. Congratulations flowed in from all over the world, except the USSR. Neither Kapitza, nor Landau, nor any other of Bohr's friends, nor the Academy of Sciences as a whole, sent any word of praise on the occasion. Then on October 20 Kapitza sent Bohr a telegram wishing the Bohr family well after the liberation of Denmark. This all seems strange. Why not observe Bohr's birthday? Why congratulate the family in October on reuniting after the liberation, when it all happened in August? This strange behavior is attributed to Beria. Before the formation of the special committee on uranium in August, 1945, headed by Beria, Kapitza was an independent operator. In fact, he was perhaps too independent for his environment, as we shall soon see. In any event, Beria had no direct control over Kapitza until he was placed on the committee. After that, whatever he did had to be satisfactory to Beria. Some of the friction between the two has been described in Chapter 7, taken from the memoir of Pavel Sudoplatov. There is no clear reason for the USSR's neglect of Bohr's anniversary, but the neglect itself demonstrates the rigorous control of Stalin's society.[110] By late October, 1945, Beria had his reasons to renew contact between Kapitza and Bohr.

According to Sudoplatov, in the fall of 1945 it became important to learn more about the nuclear reactors of the Manhattan Project. The first idea was to send a group of physicists to the USA for a secret meeting with Oppenheimer in Chicago, which was not adopted for several reasons. The second idea was to send Kapitza to meet Bohr. This did not work either, because Beria and Voznesensky denied Kapitza permission to leave the USSR.[111]

Beria decided to send someone else on a mission to Copenhagen to contact Bohr, and learn what he knew about the Manhattan Project. There were several reasons for this. Bohr was a world famous physicist, and whatever he said would be above suspicion. He had made seminal contributions to nuclear physics, and understood the problems of nuclear weapons. Bohr had visited the United States, and spent time at Los Alamos. He knew all of the important scientists in the Manhattan Project. And, perhaps most important of all, he had expressed strong desires for global dissemination of nuclear weapons technology, in the interest of world peace. Beria was anxious to demonstrate to Stalin that the NKVD was doing its utmost to obtain useful information from the capitalist countries.

Accordingly, Kapitza was told to write and introductory letter to Bohr on behalf of Terletsky, and Terletsky, who knew virtually nothing about the uranium problem, was to be dispatched to Copenhagen. Terletsky, a mere post doc, does not in retrospect seem to be a good match to interview

---

[110] HISAP 96, Vol 1, *ibid.*, p 488.
[111] Pavel Sudoplatov, "Razvedka I Kreml," *ibid.*, p 243.

the great Niels Bohr. If Kapitza was not a suitable candidate, why not send someone like, say, Kikoin? The counter argument was that Kikoin knew too much, and if he were questioned, could reveal the true status of the Soviet Atomic Project. Even in casual conversation with Bohr, an individual like Kikoin could unintentionally reveal more than he should. Terletsky on the other hand, was a blank slate, and therefore risk free. Even if hung by his thumbs he couldn't tell them anything of importance. Lev Vasilevsky, an NKVD colonel and Sudoplatov's deputy in Section S, would accompany Terletsky on the trip. Vasilevsky had worked in Mexico during the Trotsky affair.

Beria's request was one that Kapitza could not refuse, but he had no enthusiasm for the undertaking. Kapitza understood that the objective of the Terletsky mission was to find out what Bohr knew about Manhattan Project, and he did not want any part of it. [112] Landau was present when Terletsky visited Kapitza in his office at the Institute for Physical Problems in Moscow for an interview on October 22, 1945. In addition to Landau's presence, the office door was left open, all in violation of the confidentiality required by the NKVD for discussion of an intelligence mission. Sudaplatov was very upset when he learned of Kapitza's disregard for security.[113] Terletsky left Kapitza's office with the letter to Bohr and small gifts for the Bohr family. Kapitza omitted some standard phrases from the recommendation letter to Bohr, in hopes that Bohr could read between the lines, and would have some suspicions and reservations regarding Terletsky's visit. The correspondences between Kapitza and Bohr, and subsequently between Terletsky and Bohr after Terletsky's arrival in Copenhagen are preserved in the Niels Bohr Archive.[114]

Terletsky was to meet with Beria to get instructions and encouragement for the mission, and he needed a new outfit to wear, both

---

[112] I.M. Khalatnikov, "Dau, Kentavr, I Drugie," *ibid.*, p 35.

[113] I.M. Khalatnikov, *ibid.*, p 35.

[114] Kapitza-Bohr 10/22/45; Terletsky-Bohr 11/13/45; Bohr-Terletsky 11/13/45; Bohr – Kapitza 11/17/45; Bohr-Terletsky 11/17/45; //Niels Bohr Private Correspondence Niels Bohr Archive, Copenhagen (in the mm/dd/yy format).

for Beria and for the trip to Denmark. He had only wartime uniforms that did not fit too well. He was thin, and the suits at the NKVD store and at the store for people going overseas – there was such a thing – were all too big for him. A suitable outfit was finally obtained-everything from the skin out- from a clothing store in Moscow that was available for spies only!

Terletsky and Sudoplatov were waiting on October 24 for a phone call from Beria's office. The phone finally rang, and they went along a system of corridors to the office of the People's Commissar for State Security. They passed through rooms full of armed officers who looked at them attentively. We quote some of Terletsky's description of Beria's office from his memoir:

"We then waited in Beria's outer office. There besides members of Section S – Sudoplatov, Sazykin, and Vasilevsky, were the Kobulov brothers, Bogdan and Amayak, who were close assistants to Beria. Bogdan was a 200 kilogram egg shaped fat man whose repulsive appearance did not match that of a trusted communist party member, an assistant to the all powerful People's Commissar. Waiting time was passed telling jokes and talking about women. Bogdan and Amayak remembered Stockholm, through which we will go on our way to Copenhagen, and the conversation turned to Stockholm's 'women'…Such a person was the chief assistant to Beria. I later learned that he was a killer and a sadist, personally participating in the torture of prisoners. Yes! I had not imagined that such people surrounded the chief guardian of the revolutionary laws.

"Finally we were admitted to Beria's office. When we entered, Beria stood behind a writing desk in the middle of an enormous room. He came over to a large conference table in the center. I was introduced to the People's Commissar. He was medium height, growing older. His head was round on top, and his facial features were severe, without any warmth or smile. Beria did not give the expected impression from his portraits (young, energetic, intelligent with pince-nez glasses). We all sat at the conference table. A large white marble ash tray in the form of a polar bear with ruby eyes was in the center of the table. This was the only object on the very long table. Everyone looked at it. It was obvious that it had never been used as an ash tray. After a few biographical questions addressed to me, Beria started discussing the operation. Turning to me and Sudoplatov, he asked how we thought the task would unfold. I had no idea what I would ask Bohr, and Sudopaltov was no help. My knowledge of English was not adequate for a personal conversation with Bohr, and Vasilevsky only knows French. It was therefore decided that a translator working for Mikoyan named Arutyunov, who had just returned home from a long assignment in the USA, would accompany us. It was also decided that a group of scientists would compose a list of questions for Bohr, and would also bring me up to speed on the atomic project. Beria ordered the scientists, together with Vannikov and Zavenyagin, to report to his office. In 30 or 40 minutes they all came in the room. The intervening time was taken up by Beria questioning me."

Beria cleared the transfer of Arutyunov with his fellow commissar, Mikoyan. Terletsky was briefed before departure in Sudoplatov's office in Lubyanka. He was given a list of 22 questions to ask Bohr. Kurchatov, Artsimovich, Khariton, and Kikoin participated in the briefing. The subject of nuclear reactors was of particular interest. Terletsky was to play the role of a young physicist working on the Soviet Atomic Project, and anxious to learn more about the subject from the great master. He was to memorize the questions, so that he did not have to refer to notes during the interviews. The Wilson Center Digital Archive has listed Terletsky's questions, and Bohr's replies.[115] Terletsky was very impressed with the knowledge of the scientists. It was only later that he realized that a substantial portion of that knowledge was the result of espionage.

After the briefing in Sudaplatov's office the group went to Beria's office for a send off. On the way there Khariton urged him not to go. It would be difficult and dangerous for a young scientist. Terletsky in his memoir agreed with Khariton. The trip did him no particular good professionally, and may have hurt his reputation among other physicists. But he could hardly back out on the way to Beria's office![116] As usual, the group had to wait outside for a time before going in. While waiting, Boris Vannikov told him how to get the most information out of an intelligence mission – be observant. He said that before the war he was with an official delegation to the Krupp armaments works. Vannikov poked his finger in the barrel of a gun in a factory, and got enough data to determine the caliber of the weapon. After entering Beria's office they went over the list of questions one more time, and there were some alterations.

---

[115] digitalarchive.wilsoncenter.org/document/111839
[116] Ya.P. Terletsky, VIET, *ibid.*, p 30.

The journey began on October 25, with a flight from Moscow to Leningrad that did not go because of fog. So the group decided to go overnight by train. On the train Vasilivsky asked Terletsky what he thought of Beria, eliciting a response like 'very impressive.' Terletsky simply said that Beria was not what he expected, without any additional comments. They spent the next day in Leningrad, which was a city in desperate need of repair. They went to the war museum. Terletsky met one of his former teachers, who asked what he was doing in Leningrad – a question that he could not honestly answer. That evening they boarded a train for Helsinki. They crossed the Finnish border at dawn, going through border guards and customs. The trip through the Finnish countryside, which not long ago was enemy territory, took the rest of the day. They reached Helsinki in the evening. At the hotel in Helsinki, Terletsky got his first glimpse of prosperous bourgeous western life. There were weddings going on in the evening in the hotel, with well-dressed people and lots of food and wine. So close to the USSR, and yet so far away. The next morning they boarded a ship for passage to Stockholm. Terletsky enjoyed the trip along the Finnish coast, with rocky islands covered by pine trees. He had some difficulty with the food on board the ship. Finns on the ship were friendly, but reminded him that they were enemies not long ago. The USSR invaded Finland in 1939, and Finland subsequently joined Germany in the fight against the Soviet Union.

On October 29, a bright sunny morning, they landed in Stockholm. It was a livelier place than Helsinki. The Soviet embassy sent a car to pick them up and take them to their hotel. They enjoyed a stroll around the city, and ate at a stylish restaurant. At the restaurant Vasilevsky spotted American military intelligence, which served as a reminder that they were not on holiday. People were well dressed, the stores were full, and there was plenty of food. Terletsky thought to himself: "Is all of this at the expense of our country? If we had not fought fascism, and won, would neutral Sweden be able to keep its material and human resources?"[117]

---

[117] Ya.P. Terletsky, VIET, *ibid.*, p 33.

From Stockholm, they took a train to Copenhagen. The train was loaded on a ferry boat to cross over the Oresund from Malmo to Copenhagen. They cleared customs and took a taxi to the Soviet embassy. The embassy building was the official residence of the Soviet ambassador and his wife who lived on the second floor. Formal receptions were held on the first floor. Terletsky's small group were housed on the top floor. The ambassador was very cordial. Vasilevsky said that the ambassador was the local representative of the USSR, and had to be treated with appropriate respect, but Terletsky behaved as if the two of them were on equal terms, and the ambassador and his wife did not seem to mind. They called Terletsky 'professor', a title which he was hardly accustomed to, having received his advanced degree only a few months before.

They spent the next couple of days preparing to meet Bohr. Terletsky filled the time studying the prepared questionnaire and going over his notes regarding the Soviet project which he had learned from Kurchatov, et al. He was supposed to learn all of the questions by heart, and recite them like a parrot. He would speak in Russian, and Arutyunov would translate into English. Then Bohr's responses in English had to be carefully recorded. The questionnaire was top secret, and carried in a diplomatic pouch that Vasilevsky always kept in his possession. Vasilevsky established contact with a left wing Danish deputy, who could serve as a link to Bohr.

On the weekend, the ambassador took them on an outing in the countryside. They drove north to Helsingor, across the water from the Swedish town of Helsingborg, and passed Hamlet's castle on the way. The next day they went south from Copenhagen to Roskilde, where Danish royalty are buried. The height of each foreign monarch who had visited the cemetery was recorded on a pole. The tallest visitor was the Russian Tsar Peter the Great, who was in Denmark in 1716. Walking around Copenhagen they admired the ceremony of changing of the guard at the royal palace. They went to the movies, and Terletsky was allowed to walk alone to do some shopping, but there were not many gifts available in Copenhagen at that time.

The meeting with Bohr was postponed. November 6 was the 28th anniversary of the Russian revolution. Molotov gave a speech to the Moscow Soviet that was broadcast on the radio, and received in Copenhagen. Molotov said that the secrets of the atomic bomb were known to the USSR.[118] Workers in the embassy must have known something about Terletsky's training and perhaps even his mission, because they bombarded him with questions, but he was not allowed to say anything other than general physical principles.

The embassy held a reception the next evening. The ambassador and his wife stood in the receiving line for hours, as dignitaries were announced upon entry. This was difficult for the ambassador's wife, who was six months pregnant. There was a stand up buffet dinner, with caviar, all types of hors d'oeuvres, and limitless vodka. An American military attaché had too much vodka, and had difficulty standing up, but most of the guests managed to look sober despite the flow of alcohol.

Then Terletsky suddenly recognized Niels Bohr across the room. He remembered seeing Bohr in Moscow before the war, when Bohr lectured on his debates with Einstein on the interpretation of quantum mechanics and Heisenberg's uncertainty principle. That was followed by a popular lecture on the physics of the atomic nucleus. Had Terletsky been able to muster enough English, he would have approached Bohr at the reception. He could not find Arutyunov in the crowd for a helpful English lesson. So the chance slipped away, and Bohr left the embassy without an introduction to Terletsky.

The next few days were spent waiting for an appointment to meet Bohr at his institute. The weather was warm and damp, just right for a fall overcoat, which he did not bring with him. One of the embassy staff showed Terletsky a file on Niels Bohr. It said that Bohr from his Institute, responding to requests from British intelligence, analyzed information regarding secret German work on the atomic bomb. Terletsky does not say how Bohr obtained this information, or how useful it was. We have seen

---

[118] Pravda, November 7, 1945.

in Chapter 6 that, as a result of Heisenberg's 1941 visit, Bohr knew that a German atomic project existed, but he did not know the extent of its accomplishments, nor did he have access to updating information after 1941. In fact the Manhattan Project knew very little about the German atomic bomb – that was a major motivation for them to keep working as fast as possible. Only in 1944, after Alsos penetrated Italy, did some hard intelligence begin to come in. Paul Rosbaud, the editor at Springer Verlag mentioned in Chapter 7, was a source for British intelligence through the Norwegian underground and other neutral channels, but it is not known how much was learned regarding the Uranverein through his efforts. In any case, Groves and the Manhattan Project knew that the German effort existed, but did not know much about its achievements before Alsos.

Terletsky said that Bohr left Denmark in 1943 because of suspicion of espionage by the Gestapo. Most sources say that the Bohr family departure was motivated by the anti-semitism activities, and the fact that Niels' mother was jewish. Terletsky said that the British intelligence service is setup to guard the bourgeois, and is therefore the class enemy of the USSR.

On November 11 they took another excursion, this time across the water to the peninsula of Jutland. Terletsky noted that Denmark has almost no forests, and the countryside is divided into small farms. He also observed that the rural people and the city people dressed about the same, in contrast to his country, where the rural population looks very different from the urban one.

On November 13 they received at the embassy an invitation from Bohr to visit him at his institute on the morning of the next day. So the next morning the three of them – Terletsky, Vasilevsky, and Arutyunov, went by embassy automobile to the Bohr institute, which was not far from the embassy. It was so close that they got there before the appointed time, and had to drive around the block a few times. Vasilevsky was the chauffeur, and he stayed in the car while Terletsky and Arutyunov, who were very nervous, went inside to meet Bohr. His secretary led them in to Bohr's office. Bohr greeted them with restraint, and introduced his son Aage, who was in the room. Niels Bohr did not speak Russian, and said that it was too late for him to learn it, but Aage was studying it, and could understand some conversation. The senior Bohr came across as a modest person. He spoke softly in English with a Danish accent. Even if English was your mother tongue, Bohr could be difficult to follow, as many of Bohr's collaborators have testified. He had a tendency to mumble and swallow words. Terletsky noted that his hands shook slightly. Terletsky apologized for his inability to speak English, which was becoming a universal language. (He later studied English in a special school.) Bohr accepted the presence of Arutyunov after it was clear that he was a professional interpreter, and nothing more.

Terletsky gave Bohr the letter and gifts from Kapitza. Bohr asked about Kapitza's family and how Landau was doing. Bohr had great fondness and respect for Landau, dating back to the days before the war when Landau was a frequent visitor at Bohr's institute in Copenhagen. Later on Bohr was to play an important role in the award of the 1962 Nobel Prize in physics to L. D. Landau.[119] Another gift was a picture album about the Soviet Union, which had Stalin and Molotov on almost every page. Bohr observed that physicists should in general stay away from politics, but they had a responsibility to protect mankind, whatever the politicians did. Bohr said that all nations, in particular the Soviet Union, should have the atomic bomb. The only way to guarantee safety is to distribute the weapon to everyone. Happily, the bomb is no longer a secret, and qualified physicists everywhere understand the basic processes of atomic energy. The Americans have successfully assembled such weapons, and this should shorten the development work for everyone else.

Niels Bohr's basic idea: "Give the atomic bomb to a large number of nations - this will guarantee peace in the future."[120]

---

[119] As an aside, it is interesting to note that Bohr had a good batting average for his Nobel Prize nominations. See Abraham Pais "Niels Bohr's Times," Oxford University Press, New York, 1991, p 216. The Nobel committee released the private correspondence up to 50 years ago, so Pais lists Bohr's nominees 1920-1939.

[120] Ya.P. Terletsky, VIET, *ibid.*, p 37.

Then Terletsky and Arutyunov got a tour of Bohr's institute. The laboratory was well supplied with the latest equipment – a cyclotron, an electrostatic accelerator, and a mass spectrometer. They saw photographs of tracks of fission products from uranium. They then returned to Bohr's office. Niels' son Aage was still present. Before the meeting ended, Terletsky started asking from his list of questions, and Arutyunov listened carefully to Bohr's answers. Bohr talked softly, and in very general terms. Terletsky did not get through all of the questions that afternoon. Bohr had to go to a meeting, but he agreed to schedule a second interview in two days, on November 16.

All the next day Terletsky and Arutyunov worked on transcribing Bohr's responses from memory. Arutyunov had a phenomenal memory, but he did not know many of the technical terms. Terletsky's English was not all that great, and Bohr was legendary for being difficult to understand. So the transcription process was slow and painful. On November 16 they returned to Bohr's institute and the interview.

As before, Aage Bohr was present in the room, and Niels Bohr's answers were committed to memory. After the question period, Bohr relaxed and there was a discussion of Soviet physics, in which he expressed interest in inviting young post docs to visit his institute. He produced a stack of reprints and preprints to deliver to Kapitza. Aage gave Terletsky a copy of the Smyth Report to take back to the USSR. Terletsky believed that this was the first time that anyone from the USSR had seen the Smyth Report, although shortly after his return to Moscow other copies appeared from USA via NKVD channels.[121] The Report was a declassified description of the Manhattan Project, published in August, 1945.[122] It was translated into Russian and devoured by Kurchatov and the staff of Laboratory #2.

---

[121] Ya.P. Terletsky, VIET, *ibid.*, p 39.

[122] Henry D. Smyth, "Atomic Energy for Military Purposes," Princeton University Press, Princeton, N.J. 1945.

Bohr accompanied his visitors out the door of the institute, and gave them a cordial sendoff. They returned to the embassy. Vasilevsky had sent a preliminary report to Moscow two days earlier, and the final report of a completed operation was sent on November 16.

The next day they crossed over to Malmo by steamer, and then by train to Stockholm. They felt that they had company on the train, but all of the material was carried in a diplomatic pouch, which Vasilevsky kept in his hands at all times – under his pillow while he slept. At the borders they had diplomatic immunity, and were not subject to search. In Stockholm, they learned that on November 20 there would be a Soviet military plane in Stockholm that would take them back to Moscow by air. The next day they went shopping. Terletsky was troubled, thinking of the poverty at home, and communist ideology ingrained in his generation. But he became reeducated, for this was the NKVD way of rewarding foreign travel. He was given ample money to spend.

The next morning, they boarded an airplane configured to transport army paratroops. It had benches along the sides, and a cable on the ceiling for parachute releases. The crew were army pilots. It was Terletsky's first airplane flight. They landed at Moscow's Vnukovo airport. After some confusion Sudoplatov appeared with a car, and took them home.

The next day they all convened in Sudoplatov's office in Lubyanka. They were to prepare for the presentation to Beria of the results of the mission. Terletsky already knew that strict adherence to the questionnaire was required. On the trip out, Vasilevsky had told him a story about an NKVD general who decided to disregard orders and do things on his own. Beria asked: "So, you don't want to work with us? Maybe you would like to trade in vegetables?" Beria called up Mikoyan, then People's Commissar for agriculture, and said: "Anastas Ivanovich, do you have an opening for a vegetable manager? I have someone to send you." The general was dispatched to vegetables, and considered himself lucky to still be alive.

With this sobering incident in mind, Terletsky, Vasilevsky, Arutyunov, and Sudoplatov went over the questions and answers, and settled on their presentation to Beria. Beria received them as if he had seen them before, and listened to Terletsky's report, interrupting him occasionally with abusive comments directed at Bohr and at the Americans. At the end Beria said: "Bohr was used by the Americans and then dropped, but we need to support him."

Examples of the Terletsky-Bohr mission questions and answers, taken from the Wilson Center Digital Archive, are as follows:

Question 1.
Terletsky: By what practical method was uranium 235 obtained in large quantities, and what method is considered to be the most promising (diffusion, magnetic, or some other)?

Bohr: The theory of obtaining uranium 235 is well known to scientists of all countries; they were developed before the war and present no secret. The war did not introduce anything basically new into the theory of this problem. Yet I have to point out that the issue of the uranium pile and the problem of plutonium resulting from this are issues which were solved during the war, but these issues are not new in principle either. Their solution was found as the result of practical implementation. The main thing is separation of the uranium 235 isotope from the natural mixture of isotopes. If there is a sufficient amount of uranium 235, realizing an atomic bomb does not present any theoretical difficulty. For separation of uranium 235 the well known diffusion method is used, and also the mass spectrographic method. No new method is applied. The American succeeded by realizing in practice installations, basically well-known to physicists, in unimaginably big proportions. I must warn you that, while in the USA, I did not take part in the construction of these apparatuses, and, moreover, I have never seen a single installation. During my stay in the USA I did not visit a single plant. While I was there I took part in all the theoretical meetings and discussions on this problem which took place.

I can assure you that the Americans use both diffusion and mass spectrographic installations.

Comment: This question was proposed by Kikoin, the man responsible for isotope separation in Laboratory #2. Bohr answered the question in a rambling fashion. Both gaseous diffusion and thermal diffusion were actually used, as was electromagnetic separation (the calutrons). This information was available in the Smyth Report.

Question 2:

Terletsky: How can the space charge of the ionic beam in a mass spectrograph be compensated for?

Bohr: If the gas from the vacuum chamber is pumped out completely, we will have to think about a way to compensate for the volume charge of the ionic beam. But if the gas from the chamber is not pumped out completely, it is not necessary to worry about compensating for the volume charge. Or, in fact, compensation of the volume charge of the ionic beam is accomplished by means of the incomplete pumping of gas from the vacuum chamber.

Comment: This answer is correct, but the phenomenon was already known to Artsimovich, Mesheryakov and others working on electromagnetic separation.

....

Question 10:

Terletsky: What is the number of spontaneous fissions per unit time for all the above-mentioned substances (uranium 235, uranium 238, plutonium 239, and plutonium 240)?

Bohr: Few spontaneous disintegrations take place, and in calculations it is not necessary to take them into consideration. The period of spontaneous fission is approximately 7,000 years. I cannot give you the precise numbers, but you understand that with such a period of spontaneous disintegration, there is no reason to expect it to influence the process significantly.

Comment: Bohr's answer is baloney. Spontaneous fission is discussed in Appendix E. The highest rate is ^{240}Pu, which has a half-life of 6569 years, and a branching fraction for spontaneous fission of $(^{240}$Pu$\rightarrow$fission$)/(^{240}$Pu$\rightarrow$all$) = 5\times10^{-8}$.

...

Question 16:
Terletsky: How often is plutonium removed from the machine, and how are the terms for the removal determined?

Bohr: I do not know for sure. By unconfirmed hearsay, the removal of the rods takes place once a week.

Comment: This could be useful information, although once a week may be too short. The isotope ^{240}Pu, which because of its high rate of spontaneous fission and hence neutron emission is a problem for weapons – but not for reactors – results from neutron capture on ^{239}Pu, which is created from ^{239}U after two sequential β decays (see Appendix C). The longer the plutonium stays in the reactor, the higher the content of ^{240}Pu. So to obtain weapons grade ^{239}Pu, the material must be removed from the reactor periodically. The cycle time depends on the operating power level of the reactor, and the ratio of plutonium isotopes desired. ^{239}Pu grows linearly with time in the reactor, while ^{240}Pu grows quadratically because its creation depends on the presence of the lighter isotope.

...

Question 19:
Terletsky: From what materials were atomic bombs made?

Bohr: I do not know of which substance the bombs dropped on Japan were made. I think no theoretician will answer this question for you. Only the military can give you an answer to this question. Personally I as a scientist, can say that these bombs were evidently made of plutonium or uranium 235.

Comment: After confusing remarks about theorists and the military, he gets it right in the last sentence: uranium for Hiroshima, plutonium for Nagasaki.

…

Question 22:

Terletsky: Is the phenomenon of overcompression of the compound under the influence of the explosion used in the course of the bomb explosion?

Bohr: There is no need for this. The point is that during the explosion uranium particles move at the same speed equal to the speed of the neutrons' movement. If this were not so the bomb would have given a clap and disintegrated as the body broke apart. Now precisely due to this equal speed the fissile process of the uranium continues even after the explosion.

Comment: Neither the question nor the answer makes any sense. A free translation of the original Russian question is: "Is the compression of the material taken into account in describing the explosion?" [123] In this form the question is reasonable, and has a clear answer-"yes." Bohr declined to give an answer, for whatever reason.

All 22 questions can be read on the Wilson Center website. These quoted are a fair representation of the substance of the questioning. There were several questions regarding nuclear reactors – moderators, cooling, control rods, etc – which Bohr answered in very general terms. In some instances Bohr was confusing, and in others he was correct. It is difficult to assess just how valuable this information was for the Soviet Atomic Project. Much of the useful information – but nothing about ^{240}Pu – was already available in the Smyth report. These questions were obtained from a memo to Stalin dated November 28, 1945, that was declassified from the KGB archives. The memo to Stalin was in Russian (reprinted in VIET 1994, #4 p 121). The questions started out in Russian, and Stalin probably

---

[123] VIET 1994, #4, p 121 gives the Russian version of the question:
Используется ли при взрыва бомба явление переуплотнения вещества под действием взрыва? The reader can take a crack at it him/herself.

got the original versions, but Bohr's answers were originally in English, which was not the native language of anyone involved. The dialog presented here is in English, so Bohr's responses have been translated twice English→Russian→English, which may explain their awkward tone. The old joke comes to mind: English "The spirit is willing but the flesh is weak"; then to Russian, and back to English "The vodka is good, but the meat is bad." A clue to the quality of the translation comes from the fact that the article about Terletsky's visit in Scientific American prints questions one, ten, nineteen, and twenty two, and the English is different from the Wilson Center website, although the meaning is more or less the same. Hans Bethe couldn't make any sense out of question 22 either.[124]

Terletsky himself stated that the information obtained from the Bohr mission was not substantively new. What was not already contained in the NKVD intelligence files could be extracted from the Smyth Report.[125] Question 10 about spontaneous fission and question 16 about recycling fuel rods from a reactor are related, in that it is the high spontaneous fission rate of ^{240}Pu that made it an undesirable component of a plutonium bomb, and hence the fuel rods were removed periodically to keep the amount of ^{240}Pu small. Kurchatov and Nemenov had made some ^{239}Pu by bombardment of ^{238}U with neutrons from the cyclotron deuteron beam, but such a method made only a very small amount of the isotope ^{240}Pu. Kurchatov did not have a reactor, with its high flux density of slow neutrons, needed to make measurable amounts of the 240 isotope. Laboratory #2 would have obtained valuable information from Bohr, if the questions had been answered clearly, and if knowledge of the ^{240}Pu problem had not already been obtained from other intelligence sources.

Sudoplatov was responsible to Beria, and therefore to Stalin, for the operation. Consequently, he was more upbeat in his evaluation, stating that what Bohr said was important in that it confirmed the previously obtained intelligence material, and that Bohr verbally said more about atomic

---

[124] Scientific American, May, 1995, p 85. "Did Bohr Share Nuclear Secrets?" by Hans Bethe, Kurt Gottfried, and Roald Sagdeev.
[125] Ya.P. Terletsky, VIET, *ibid.*, p 41.

weapons than was recorded in the documents. Sudoplatov pointed out that Bohr reported the meeting with Terletsky to British intelligence, and stated that he had given Terletsky a copy of the Smyth Report, but was silent on the content of the questions. He concluded that, until the confessions of Klaus Fuchs, British intelligence was not aware of the extent of Soviet knowledge regarding the Manhattan Project.[126]

The Wilson Center also has Kurchatov's response to the Bohr questionnaire, dated December, 1945:

EVALUATION

Of the answers given by Professor Niels Bohr to the questions on the atomic problem

Niels Bohr was asked two groups of questions:

1. Concerning the main directions of the work.
2. Those containing concrete physical data and constants.

Definite answers were given by Bohr to the first group of questions.

Bohr gave a categorical answer to the question about the use of methods for obtaining uranium 235 in the USA, [question 1] which completely satisfied the correspondent member of the Academy of Sciences Prof Kikoin, who put this question.

Niels Bohr made an important remark dealing with the effectiveness of using uranium in the atomic bomb. This remark must undergo a theoretical analysis, which should be the task of Professors Landau, Migdal, and Pomeranchuk.[127]

---

[126] Pavel Sudoplatov, "Razvedka I Kreml," *ibid.*, p 245. An earlier version of this memoir in English is different in its treatment of the Terletsky visit and other matters. Pavel and Anatoli Sudoplatov, Jerrold and Leona Schecter. "Special Tasks: The Memoirs of an Unwanted Witness," Little Brown, Boston, 1994.

[127] digitalarchive.wilsoncenter.org/document/111838.

Academician Kurchatov.

Kurchatov's first comment supports Sudoplatov's assertion that the information was useful for confirming the accuracy of intelligence data. A careful reading of Bohr's responses to the 22 questions does not reveal which one Kurchatov refers to in his second comment. Kurchatov's comments were restrained, in contrast to his usual evaluation of intelligence material, which was enthusiastic, even exuberant. This may have been a reflection of his low opinion of the whole operation.[128]

So this concludes our treatment of the Terletsky-Bohr visit. If it had not been a trip to see Bohr, it probably would not have received much attention, for its yield of useful information was modest at best. But it was an actual clandestine NKVD intelligence operation in a foreign country, and we are privileged to know all of the principal actors in the drama. Terletsky has given us a detailed day by day diary. This kind of information is sadly lacking for most of the widespread NKVD spy network. Sudoplatov and Beria had to put the best possible spin on it, because it was a considerable expense for an economy that had little hard currency spending money, and was devastated by the war.

Barkovsky made some interesting observations on the Terletsky mission after the publication of Terletsky's memoir in VIET.[129] He wrote that the operation was a publicity stunt from the start, with no anticipated useful intelligence information. The questions covered well known subjects, and the answers were not particularly revealing. Vasilevsky had visited Bohr before the Terletsky mission to ask Bohr if he would consider joining the Soviet physicists on the bomb project. Bohr declined to visit the USSR, citing commitments at home, but Vasilevsky's visit alerted Bohr before his conversations with Terletsky to the fact that the NKVD had an interest in him. Vasilevsky had hoped to meet Bohr again, but was

---

[128] Yu.V. Gaponov, F. Ozerud, and P.E. Rubinin, HISAP96, Vol 1, p 504.

[129] V.B. Barkovsky, Letter to the editor regarding Terletsky's visit, VIET 1994 #4 p 122. Barkovsky was an NKVD operative in London beginning in 1941. He was named a hero of Russia in 1996 for his intelligence work.

required to stay in the car while Terletsky and Arutyunov talked to Bohr. Barkovsky went on to say that Sudoplatov's Section "S" did not succeed in recruiting many German nuclear scientists, and the Bohr-Terletsky mission was one way to save face. It did not really work, however, since Section "S" ceased to exist in October, 1946.

Sudoplatov recalled that Beria in 1943 ordered him to describe to Ioffe, Kurchatov, Kikoin, and Alikhanov information on the Manhattan Project obtained through espionage. He was not to reveal the source of the information, nor the physicists involved in the work. He first described the Chicago chain reacting pile, and Kikoin said that must be Fermi, no one else could do such a thing! Then Sudoplatov covered the page with his hand, concealing the sources of the information, but the scientists said that he was wasting everyone's time, they knew the field of nuclear physics, and could identify who did what. After this Sudoplatov obtained permission from Beria to reveal the sources of the work (but not the informants).[130]

Sudoplatov (p221) related an incident involving ill advised statements by the younger brother of Kikoin, and Beria's reaction. Sudoplatov was instructed to meet with Kikoin, and urge him to talk to his brother, to straighten him out. This is an example of important scientific members of the Soviet Atomic Project and their families obtaining leniency. Then on p 222 Sudoplatov says that the scientists had perks as per Beria's orders, like those assigned to the German visitors, and discussed in Chapter 4: special ration cards, special stores, good housing, and health care.

8.21. Bruno Pontecorvo

No treatment of Soviet espionage would be complete without saying something about Bruno Pontecorvo. He came from a large Jewish Italian family, and grew up in Pisa. Pontecorvo was a graduate student in Fermi's group in Rome during the neutron activation experiments that won

---

[130] Pavel Sudoplatov, "Razvedka I Kreml," *ibid.*, p 217-218.

Fermi the Nobel Prize. According to Sudoplatov, Grigory Heifitz, before going to San Francisco, was in Rome, and befriended the young Pontecorvo.[131] Heifitz suggested that Pontecorvo might escape Mussolini's anti-semitism by going to Paris to work with Frederic Joliot Curie. Sudoplatov implied that Heifitz could furnish an introduction for Pontecorvo, since Joliot-Curie was a dedicated communist, and had contacts with the Soviet Union. Fermi also could have recommended Pontecorvo on scientific, rather than political, grounds. Bruno's communist sympathy grew stronger in Paris, but could have already been present in Rome, because some of his brothers and sisters also had communist connections. Bruno met and married his Swedish wife Marianne, and their first son Gil was born in Paris. Many people who knew him in his youth have commented that Bruno had a striking presence, and could have been taken for a movie star. In 1940 the family escaped the German invasion of France, and landed in Tulsa, Oklahoma. An oil prospecting firm in Tulsa had offered Bruno a job, based on his experience with neutrons in Rome. Bruno developed techniques for searching for oil by using neutrons and gamma rays to document the strata of the well below the earth's surface. He did not invent the technique, but made substantial improvements in it. It is still in use today. In neutron well logging a neutron source and a BF_3 neutron detector (shielded from the source) are inserted into the well hole. The flux of neutrons scattered back into the counter is a measure of the hydrogen content of the rock. Pontecorvo was particularly skilled in handling sources and building counters and electronics.

In 1942 Pontecorvo went to Canada. He worked on developing the NRX nuclear reactor at Chalk River – a natural uranium reactor with a heavy water moderator. His skill at radiation detection was again applied to uranium ore prospecting in Canada. [132] He visited Chicago, and met with Fermi a time or two. He did some experiments with muons from cosmic

---

[131] Pavel Sudoplatov, "Razvedka I Kreml," *ibid.*, p 98.
[132] Simone Turchetti, "The Pontecorvo Affair," University of Chicago Press, Chicago, 2012, p 58.

rays – elementary particle physics with no military applications. After the war he returned to England.

Bruno, his wife Marianne, and their three children mysteriously disappeared while on vacation in late August, 1950. The family lived near Harwell, the British nuclear physics laboratory, where Bruno worked. Klaus Fuchs had also worked at Harwell – in the same section as Pontecorvo - and his confession to espionage the previous January had caused a big stir, not only at Harwell, but all over England, and across the Atlantic in the USA. Harwell become more security conscious, and since Pontecorvo had communist sympathies, he was eased out of the national laboratory, and landed a new job at the University of Liverpool, where he planned to go at the end of the year. Because of the increased scrutiny he perhaps feared for his safety. In any event, in July, 1950, the Pontecorvos loaded up their automobile and headed for the continent of Europe. They stopped first in Switzerland, then went to Italy. On September 1, 1950, the Pontecorvos flew to Stockholm, stayed overnight near the airport, and flew the next morning to Helsinki. From Finland, they travelled either by train or by ship to Leningrad.[133] The Pontecorvo exit had to be pre-arranged by the NKVD, starting in Italy. The Kremlin was expecting him when he arrived.

The Pontecorvo disappearance was noticed in England after he failed to return to work at Harwell. Given the spy ring angst of that time, the Fuchs case, the arrests in connection with the Rosenbergs, the HUAC hearings in Washington, etc, the Pontecorvos were immediately suspected of having defected to the Soviet Union (albeit without any hard evidence). The perception of another loss of an atomic scientist only fueled the fires of anti-Soviet feelings in the West.

In his interesting article in HISAP 96 B.L. Ioffe of ITEP Moscow (a younger nuclear physicist, not to be confused with A. F. Ioffe of LFTI in Leningrad) tells us a little about Pontecorvo's arrival behind the iron

---

[133] Simone Turchetti, "The Pontecorvo Affair," *ibid.*, p 116.

curtain.[134] He says that the Pontecorvo family sailed from Finland on the Soviet ship "Beloostrov." The departure from Helsinki was illegal, and the arrival of the Pontecorvo family in the USSR was not announced in the press or anywhere else. They went to the new laboratory at Dubna, that had only been in existence for a few months, and officially was not on any map. The Dubna laboratory was created 120 km north of Moscow on the Volga River to house the new synchrocyclotron that Kurchatov wanted to build. The Soviet government – Lavrenty Beria - constructed the laboratory with prison camp labor. It was called the 'Hydro-technical Laboratory of the USSR Academy of Sciences,' presumably as a cover, exploiting its proximity to the Volga River. No one was supposed to know it existed. Although Dubna never did weapons work, all research in nuclear physics, regardless of its true aim, was classified.[135] This policy persisted until the mid 1950's. The Pontecorvo family was cooped up in Dubna for five years, before their presence in the USSR was acknowledged.

Physicists in the Moscow region visiting Dubna were forbidden to report the presence of Pontecorvo to the outside world – that is, outside Dubna. Pomeranchuk often visited Dubna, and upon his return to Moscow would say that he discussed such and such with the 'professor', and the 'professor' said so and so. The 'professor' was Pontecorvo. Sometime in 1950 Ioffe's colleague in reactor work, A.D. Galanin, was summoned to the Kremlin. This caused some consternation, because scientists were often ordered to go somewhere or other, but not to the Kremlin. Some years later Galanin told the story that he and some other physicists were called to the Kremlin to meet with the newly arrived Bruno Pontecorvo, for the purpose of obtaining any useful information that Pontecorvo had regarding design of nuclear reactors. They learned that, while Pontecorvo understood the principles well, he had no knowledge of the details of reactor design, and therefore they did not get any useful information. Ioffe does not say in whose office in the Kremlin the meeting took place.

---

[134] Boris L. Ioffe, "'Truba', Pochemu ona ne Proshla," HISAP96, Vol 2, *ibid.*, p 224-225.
[135] V.P. Dzhelepov, "Kogda Dubny ne bylo na Karte," HISAP96 Vol 1, *ibid.*, p 284.

Ioffe went on to say that Pontecorvo's life in the Soviet Union was an unenviable one. He was not allowed to leave Dubna, which in the early years was a closed town requiring special permission to visit. Pontecorvo therefore had very limited contact with the scientific world, even as represented by the Soviet Union. Contact with the West was nil. To quote Ioffe:

"For example, he could not interact with Landau – Landau did not go to Dubna. Pontecorvo could not publish any scientific articles – for five years his name disappeared from science. None the less, he did not change his communist convictions."[136]

The JCAE in its 1951 report on Soviet Espionage writes:

"While Pontecorvo had no direct contact with weapons work, it is possible to speculate that he may have betrayed reactor data from 1943 onward, supplementing the bomb details and the U-235 processing information divulged by Fuchs and thereby furnishing Russia with a particularly well rounded picture. In any event, as of September, 1950, the Soviets acquired in Pontecorvo not only a human storehouse of knowledge about the Anglo-American-Canadian atomic projects, but also a first class scientific brain."

Ioffe's account belies the claim that Pontecorvo had detailed knowledge that the Soviets lacked. After all, his arrival in the USSR was after the successful detonation of the first Soviet nuclear weapon in August, 1949, so they did not need his talents in that regard. Pontecorvo was, however, a first class physicist, who had many original ideas, and would have been even more productive had he remained in the west. Among his contributions to elementary particle physics was the idea of neutrino oscillations, which has been experimentally confirmed only after his death.

Pontecorvo was on Sudoplatov's radar screen, but not all of the claims that Sudoplatov makes are plausible. He mistakenly attributes the code name 'Mlad' to Pontecorvo, thereby identifying him as an agent during

---

[136] B.L. Ioffe, HISAP96 Vol 2, *ibid.*, p 225.

the war years. It is now generally accepted, based primarily on VENONA transcripts, that 'Mlad' was Theodore Hall.[137] There is no evidence that Pontecorvo was a spy for the Soviet Union during the Manhattan Project. In fact, Pontecorvo had little, if any, contact with the Manhattan Project.

Sudoplatov credits Lev Vasilevsky, his deputy who accompanied Terletsky on the mission to interrogate Bohr, with making arrangements for Pontecorvo's defection, and implementing them in September, 1950.[138] According to Sudoplatov, Vasilevsky and Pontecorvo were friends, and continued to see each other now and then at lunch at the writers' club in Moscow. (This must have been after Pontecorvo was allowed to leave Dubna.)

Pontecorvo reappeared in the Soviet Union by way of an article in Pravda, signed by him, on March 1, 1955.[139] He professed his dedication to peaceful uses of atomic energy. He said that one reason for his defection was the desire to avoid all weapons work. He also mentioned his concern regarding the increased security at Harwell prior to his departure. He implied, but did not say in so many words, that his defection was, if not planned, at least considered before the family went on holiday. In no instance, in this or later interviews, did he ever reveal the contacts and arrangements necessary for his extraction from the West.[140] He expressed appreciation for the hospitality that he received at the Institute for Physical Problems, and the opportunities to do experiments with the synchro-cyclotron (Dubna). He sniped at I.I. Rabi for saying to him, Pontecorvo, in 1949 that the USSR did not have the industrial capacity to build nuclear power reactors. He expressed the conviction that the people and

---

[137] Pavel Sudoplatov, "Razvedka I Kreml," *ibid.*, p 237; John Earl Haynes & Harvey Klehr, "Venona," *ibid.*, p 107 and p 1624 – 'mlad' means 'youngster' in Russian.

[138] Pavel Sudoplatov, "Razvedka I Kreml," *ibid.*, p 253.

[139] Pravda, 1 March 1955, page 3. "Zapretit' Atomnoe Oruzhie" – a message from scientist Pontecorvo to scientists of the world. If written by Pontecorvo in Italian, it was translated by someone else, since Pontecorvo's Russian was not that good.

[140] There is no mention of Pontecorvo in Beria's Diaries, so arrangements for his departure probably did not get that high up. Someone in the Academy of Sciences must have been involved.

government of the Soviet Union were doing all they could to maintain peace world-wide. He ended with another call for peaceful uses of atomic energy.

In 1961 Heinz and Elfi Barwich arrived in Dubna from Dresden in East Germany. Heinz assumed the post of deputy director of the Joint Institute for Nuclear Resrearch, JINR. The director of JINR at that time was Dmitri Blokhintsev. Being an international laboratory, the scientific staff of JINR had representatives from member states, like East Germany.[141] As deputy director of the laboratory, Barwich was accorded one of the private houses at Dubna, next door to Bruno Pontecorvo. Elfi was very impressed with her next door neighbor. She said that he was a fascinating man, of medium build, with a tan complexion and beaming, intelligent eyes. Pontecorvo had been at Dubna for 11 years, but had not lost any of his captivating presence. Elfi had worked as an interpreter before her marriage, and was fluent in Russian. Pontecorvo did not speak German, so they communicated in Russian. Elfi noted that Bruno's Russian came with a soft English accent, but it was apparently serviceable.

Travel restrictions eased for Pontecorvo during the Khrushchev era, allowing him to travel within the Soviet Union, and in 1978, 28 years after his departure, he was allowed to visit Italy.[142] The occasion was Eduardo Amaldi's seventieth birthday, and Emilio Segre was also there from Berkeley – a reunion in Rome of three members of Fermi's old group: Amaldi, Pontecorvo, and Segre. Fermi had died in 1954 in Chicago. Travel in the west remained problematic, but Pontecorvo attended international conferences in the Soviet Union for the remainder of his career. Late in life Pontecorvo suffered from Parkinson's disease. He died in Dubna in 1993. He took to the grave many secrets about his life.

---

[141] Heinz and Elfi Barwich, "Das Rote Atom," Scherz Verlag, Munich, 1967, p 193ff.
[142] Simone Turchetti, "The Pontecorvo Affair," *ibid.*, p 181.

8.22. Summary

An operation as vast as Soviet espionage from 1941 to 1945 in England and the United States is difficult to summarize in a few sentences. That it was well organized and managed is without question. It was also remarkably successful. What impact it had on the Soviet atomic project is still debated. Yuly Khariton, chief at Arzamas-16, and Igor Kurchatov's deputy, has said that they gained at least a year from the intelligence information. From reading high level memos between NKVD and the Politburo in 1944-45, one gets the impression that Soviet ministers with no technical training knew a lot about the scope of the Manhattan Project, and what sort of commitment it required to reproduce the effort in the USSR. This was of inestimable value in marshalling the industrial might of the Soviet Union behind Kurchatov and his compatriots. Comparisons of the relative efforts on the national economy show that the USSR spent ten times as much of its gross national product on the atomic bomb as did the United States.[143]

Of course in the end it was Kurchatov's physicists, engineers, chemists, and technicians who did the job. Anatoly Yatskov, who should know what he is talking about regarding intelligence, had this to say in 1992:

"Komsomolskaya Pravda for November 18, 1990, under the title 'The Bombs of General Filatov,' published an interview by military journalist S. Bogdanov with the editor of 'Military History Journal', General V.I Filatov. Excerpts of that interview follow:

Question: It seems that there are many sensational claims in your journal. Could you comment on some of them?

Answer: Consider the material in 'The Bomb for Sakharov.' They permitted us to uncover one of the spies who delivered drawings and other scientific documents of American factories where the super-secret bombs were made. All of this was correctly analyzed by our 'luminaries', who

---

[143] HISAP'96, Vol 1, *ibid.*, p 291 "Soviet Atomic Project and Mobilization of the Economy," V.V. Alekseev and B.V. Litvinov; Vol 2 p 320, "USSR Atom Project – Material Resources 1945-1950," R.B. Kotel'nikov and V.A. Tumbakov.

became famous afterwards. The Intelligence Service wanted to publish this history, but Keldysh [president of the USSR Academy of Sciences] and others protested – it would undermine their authority.

Question: In other words, you wish to say that without the KGB we would not have so many geniuses?

Answer: These scientists would never have become famous. Let them kiss the hands of our spies!

On November 26, I [Anatoly Yatskov] sent a letter to the editors of Komsomolskaya Pravda. Here it is, although it was never published:

'To the editors of Komsomolskaya Pravda,

I am greatly disappointed by the remarks of General Filatov regarding the atomic bomb, scientists, and spies, where he says 'These scientists would never have become famous. Let them kiss the hands of our spies!' This absurd statement could give the impression that the intelligence service claims a predominant role in the making of the Soviet atomic bomb. But espionage did not make the bomb. It was made by scientists and technicians, exploiting the scientific-industrial capacity of the country.

All Soviet citizens, including espionage agents and General Filatov, should bow to Igor Kurchatov and his cohorts, who in unbelievably difficult circumstances, in no way comparable to those of the United States, knew how to make atomic weapons in a timely manner, thereby preventing an unpredictable development of events which could have had fateful consequences for our country. The capabilities of our scientist were the equal of those in the USA, where they assembled the best in the world.

Espionage played a significant role in the fabrication of Soviet atomic weapons, and this should be recognized. But it does not follow that spies and scientists should confront one another – they live according to their own laws, each does his own thing, and it is not necessary to kiss hands. And Andrei Sakharov should be left alone – let him rest.' [144]

---

[144] Voprosy Istorii Estestvoznaniya I Tekhniki, 1992, #3, p 106-107.

So let the matter rest here. Even the Russians cannot agree on the importance of the espionage. But it remains a vital part of the story. The 'unpredictable development of events' mentioned by Yatskov refers to the fear present in the USSR between 1945 and 1950 that the USA, with its nuclear supremacy, would attack the Soviet Union.

Chapter 9

Players in the Drama
Stalin, Beria, and Kurchatov

9.1. Joseph Stalin

Figure 9.1. Formal portrait of Stalin as leader of the Communist Party. Courtesy of the Library of Congress.

Since virtually nothing happened in the USSR (apart from the German invasion) that was not approved, if not actually initiated, by Joseph Stalin, shown in Figure 9.1, some attempt at understanding his personality is in order. This is hardly a first try – there are library shelves full of biographies of Stalin. Acquaintances, associates, historians, journalists, and virtually everybody else, including his own daughter, have tried to figure out the

360

behavior of this complex individual.[1] He died without writing his memoirs, so apart from official documents we do not have his viewpoint on contemporary events.

The history of Stalin's rise to power has been told many times. He was born in Georgia in the little town of Gori, not too far from the capital of Tbilisi, in 1878. His father, Vissarion Djugashvili, was a shoemaker who drank too much and threatened his family. His mother Ekaterina was a laundress. Stalin grew up in a poor, harsh environment. At his mother's insistence, he attended a Russian orthodox seminary for the priesthood in the Georgian capital, Tbilisi. He became a revolutionary while a student in the early 1900's. He learned about Lenin, who was the leader of the Russian Marxist party, and became dedicated to the fate of Russia. He adopted the revolutionary name 'Stalin' in place of his Georgian name Djugashvili.[2] He spent many years before WWI in exile or in Tsarist prisons. After the 1917 revolution, Stalin was a member of Lenin's inner circle, and after Lenin's death in 1924 – Lenin was physically incapacitated for a year or so before he died – Stalin maneuvered to achieve supreme power in the Soviet Union. Perhaps his most able opponent for power was Leon Trotsky, minister of defense during the civil war, and a very talented and determined revolutionary. Stalin wanted to build communism in Russia, while Trotsky wanted to pursue world revolution. Stalin managed to exile Trotsky in 1928.

9.1.1. *Family Life*

Stalin had a family. He had two wives. He married a Georgian woman, Ekaterina Svanidze, in 1905. They had a son, Yakov, born in 1907. Ekaterina died of typhus that same year. The Svanidze family did not fare well later on, in Stalin's purges. Stalin's second wife, Nadezhda

---

[1] A few sources: Edvard Radzinsky, "Stalin, Zhizn' I Smert,'" AST, Moskva, 2010; Simon Sebag Montefiore, "Stalin, the Court of the Red Tsar," Vintage Books, New York, 2003; Robert C. Tucker, "Stalin in Power," W.W. Norton, New York, 1990; Svetlana Alliluyeva, "Twenty Letters to a Friend," Avon Books, New York, 1967. A Russian version "Doch' Stalina," Algoritm, Moskva, 2013, includes an interview.

[2] This seems to be the accepted spelling. Transliteration should be Dzhugashvili.

Alliluyeva, was born in 1901 to Russian revolutionary parents. Her father, Sergei Alliluyev, was a railway worker and an ardent communist, who sheltered Stalin in exile in the Caucasus. He became acquainted with Nadezhda as a child, 23 years younger than he. She grew up a dedicated Bolshevik, and worked for Lenin in Moscow. Stalin and Nadezhda were married in 1919. They had two children, Vasily, born in 1921, and Svetlana, born in 1926.[3] Yakov's stepmother, Nadezhda, was only six years older than he was. Stalin had a habit of publicly mocking and criticizing his young wife. After one such episode at a dinner in 1932, Nadezhda committed suicide.

Kremlin children, like children of famous people everywhere, had mixed fortunes as adults. Stalin's elder son, Yakov, retained the family name Djugashvili, while the younger son was Vasily Stalin. Svetlana's memoir, "Twenty Letters to a Friend," was translated into English and many other languages, and has enjoyed wide circulation. In addition, at least one of Stalin's grandchildren has written a family history.[4] Yakov grew up with the Svanidze family in Tbilisi, Georgia, and only came to Moscow in 1921 at age 14. His relationship with his father was never smooth, but Yakov was a good companion to his much younger half brother and sister. Svetlana was very fond of him.[5] Vasily was a difficult child, and Yakov sometimes acted as disciplinarian. Yakov paid little mind to papa Stalin, and vice versa. He attended engineering school, and graduated from Frunze Military Academy just in time to go to the front as a lieutenant when the Germans invaded. He was soon captured by the Germans, and became a prisoner of war. The Germans tried to exploit the fact that Stalin's son was a POW. German aircraft dropped leaflets behind the lines claiming that Yakov was well treated, as were all Soviet POW's, and encouraging other Red Army troops to surrender. After Field Marshal Friedrich Paulus surrendered at Stalingrad on February 2, 1943, and became a prisoner of the Soviet Union, the Germans offered to trade

---

[3]   en.wikipedia.org/wiki/Nadezhda_Alliluyeva

[4]   Galina Djugashvili Stalina, "Taina cem'i Vozhdya," AST Zebra, Moskva, 2007. Galina
      was Yakov's daughter. Several grandchildren of Stalin are still alive.

[5]   Svetlana Alliluyeva, "Doch' Stalina," Algoritm, Moskva, 2013, p 24. See also "Twenty
      Letters to a Friend," Chapter 14.

Yakov for Paulus. Stalin refused, saying that he would never exchange a field marshal for an ordinary soldier. This may seem a cold response, considering that Yakov was his own son, but it was probably the correct thing to do politically. There were over a million Russian POW's in German hands, and Stalin wanted to avoid any sign of favoritism. Yakov was not among the Soviet POW's liberated when the Red Army invaded Germany. His fate remains unknown. Did he starve to death in one of the camps, or was he executed?

The younger son, Vasily, was a problem for Stalin from an early age. He was a mischief-maker in school, exploiting the fact that he was the son of the 'father of the country' to justify outrageous behavior, like riding a motorcycle through the halls of the school building. Stalin was obliged to discipline Vasily from time to time, and to try to smooth things over with his teachers, by encouraging them to remain calm and firm, and not to fear retribution for appropriate control. Papa terrified Vasily, but that did not suppress his activities. He went to flight school, and became a competent military pilot. He advanced in the officer corps of the Soviet Air Force partly because of his skill, and partly because of his family name. He had a good record during the war. He was married twice, and had children. Eventually he became an alcoholic, and enough of an embarrassment that he was removed from command by Stalin's order.[6] After Stalin's death in 1953, Vasily became completely ostracized. He died in 1962.

Svetlana was Stalin's youngest child, and only daughter. Svetlana died in Richland Center, Wisconsin in 2011, and was buried there as Lana Peters. How all of this happened is a story worthy of a book all by itself. Stalin was very fond of Svetlana when she was a child and teen ager. She was personable and intelligent – probably the best of the lot of Stalin's children. She was married four times, the last time to Wesley Peters, an architect with Frank Lloyd Wright. Svetlana's third husband was a native of India, who died in Moscow. Svetlana received permission to take her

---

6 Svetlana Alliluyeva, "Doch' Stalina," *ibid.*, p 29. There is a Russian made for television movie series called "Syn Ottsa Narodov" (Son of the father of the country) about Vasily and his exploits, broadcast in 2013. Some artistic license, but probably largely accurate.

husband's ashes to India in 1967. In New Delhi, she requested asylum in the United States, and came to New York City. Svetlana and Wesley met through Wright's widow, Olgivanna, who was herself Russian, and invited Svetlana to visit Taliesin, the Wright architecture complex near Spring Green, Wisconsin.[7] They had a daughter, Olga, who still lives in the United States. Svetlana's impressions of her father are of interest to us.

Svetlana remembered fondly her relationship with her father while her mother was still alive. Six year old Svetlana sensed a change in her father's behavior after her mother died. He seemed more remote, less interactive. He spent more time at his dacha, away from the family in Moscow, and was in constant contact only with members of the Politburo. Stalin exchanged letters with his young daughter, asking how she was doing in school, but they saw each other only rarely. Svetlana describes a visit she made to her grandmother in Georgia – her grandmother was also a freckle faced red head in her youth, and the family resemblance was said to be striking.[8] Svetlana was confused and upset by the purges, that seemed to affect normal, innocent people. She blamed her father, but also other members of the politburo, believing that they all shared responsibility. She emphasized that Stalin was of modest appearance – he dressed neatly but without lavish decorations, and he lived a simple and dedicated life for someone with so much power. He did not acquire the trappings of a tsar, although he certainly had all of the muscle.

Although Stalin himself remained in Moscow for the entire war, his family was evacuated to Kuibyshev (now called Samara) in the fall of 1941, along with many foreign embassies and much of the Soviet government. Svetlana went along, and stayed there for one year. She was observant of local conditions, and realized that, as Stalin's daughter, she was treated better than the indigenous population, who were dislocated by the influx of Soviet bureaucrats.[9]

---

[7] Obituary for Lana Peters, Milwaukee Journal Sentinel, November 28, 2011.
[8] Svetlana Alliluyeva, "Doch' Stalina," *ibid.*, p 37.
[9] Svetlana Alliluyeva, "Twenty Letters to a Friend," *ibid.*, Chapter 15.

Svetlana believed that her father was at his very best during the war. This could be said of the Soviet Union as a whole – the war was its finest hour, to borrow a phrase from Churchill. In August, 1942, Winston Churchill and Averell Harriman flew to Moscow to meet with Stalin. It was a tense time during the Allies war effort against Nazi Germany. The Wermacht was advancing on Stalingrad, and Stalin was anxious for a second front to relieve the pressure on the Red Army. That front was not to be for another two years, and Churchill had to carry the bad news that in 1942, at least, a second front was out of the question. Churchill said that it was like carrying ice to the North Pole. It is fair to say that no leader of the twentieth century hated communism more than Churchill. So one could not look forward to a chummy relationship. Nevertheless, Churchill and Stalin both behaved diplomatically, and seemed to get along well.[10] Towards the end of their meeting Stalin introduced his daughter Svetlana, a red head. Churchill said that when he was younger he also had red hair. Svetlana was anxious to practice her English, which she had been studying, but the conversation was between Stalin and Churchill, mostly about military matters, and she was soon released from the company. This incident points towards Stalin's efforts to appear as a normal parent, proud of his 16 year old daughter.[11] After the war Stalin's health declined, and he lost some of his focus, although he remained formidable enough to discourage any movement to replace him. He became even more suspicious, and the political repressions of the 1930's started up again. The country was easier to manage during the war, because of a looming common danger, and afterwards perhaps the strains of running things were beginning to have an effect on Stalin. This atmosphere was bound to impact the Soviet Atomic Project.

We are interested in Stalin for the last decade of his life, 1943-1953. In the middle of this time span, in 1948, the Soviet Union celebrated Stalin's 70th birthday. Stalin did not achieve absolute power until a few years after Lenin died in 1924, but he was certainly among the top communist

---

10 W. Averell Harriman and Elie Abel, "Special Envoy to Churchill and Stalin," *ibid*, p 149ff.

11 Svetlana Alliluyeva, "Doch' Stalina," *ibid.*, p 41-42.

leadership since that time. So he had been in power for almost 20 years by 1943, and if not comfortable, was confident in his role as supreme leader. After WWII Stalin was beginning to age, and lost some of his phenomenal energy and memory, but no one questioned his authority. Instead of mellowing, Stalin became more suspicious and cruel in his last 10 years. His years in power were characterized by continual chaos, oppression, and hardship, and the last decade of his life was no exception. We have already noted that many Soviet citizens hoped that after victory in WWII the political repression would ease, and life would become more like the West, but it was not to be. It is almost impossible for a modern scholar, in the internet age, to grasp the environment of those times. The considerable demands made on the economy by the secret Soviet Atomic Project contributed to the national hardship in the early post-war period.

9.1.2. *Zhukov on Stalin*

The impressions of people who had regular dealings with Stalin during this period should be useful in understanding his character. Marshal Georgy Zhukov interacted with Stalin almost every day during the war years 1941-1945. Here is what Zhukov had to say:

"Stalin made a strong impression in conversation despite his unremarkable appearance, and short stature. He captivated his audience by the simplicity of his approach, rather than the grandeur of his presence. A free form dialog, the ability to sharply formulate his thoughts, a naturally analytic mind, broad knowledge, and an uncommon memory forced even the most sophisticated and experienced people to be on the alert when in conversation with Stalin.

"Stalin did not like to sit during conversations, and typically walked slowly around the room, stopping now and then to approach his subject closely and look him in the eye. He had a sharp and penetrating stare. He spoke softly, but distinctly, separating one phrase from the next, and rarely gesticulating. There was usually an unlit tobacco pipe in his hand, and he liked to rub the stem across his moustache. He spoke fluent Russian with a Georgian accent, and he liked to use colorful comparisons, examples from literature, and metaphors.

"Stalin rarely laughed, and when he did, he laughed softly, as if to himself. But he had a sense of humor, and enjoyed a good joke. He had good eyesight, and could read without his glasses any time of the day or night. As a rule, Stalin wrote everything by hand. He read widely, and retained information on a broad range of subjects. He had a remarkable ability for work, and could quickly grasp a chain of logic, allowing him to master a quantity of varied material that would overwhelm an ordinary person.

"It is hard to say precisely which characteristics were dominant. While he had many talents, Stalin was certainly not even tempered. He had a very strong will, and a secretive and violent character. While usually quiet and reasonable, he could on occasion erupt into violence. When this happened his face became pale, and acquired a severe, cruel appearance. I have not known many brave souls who could withstand Stalin's wrath and deflect his blows.

"Stalin's daily life was unconventional. He usually arose around noon, and worked mostly at night. The Central Committee of the Communist Party had to adjust to Stalin's routine, as did all of the people's commissars, and the basic government and planning organs. This exhausted everyone.

"It was difficult for me to evaluate before the war the depth of Stalin's knowledge of military science, the arts of tactics and strategy. Subjects of discussion with the Politburo or with Stalin himself before the war concentrated on organization, mobilization, and supply questions, as I have already mentioned.

"I can only say that even before the war Stalin was occupied with many questions regarding armaments and war industry. He would frequently call the industrial managers of weapons manufacture to the Kremlin, and ask them detailed questions regarding the manufacture of war materials at home and abroad. I have to give him credit for grasping the qualities of new types of weapons.

"Stalin knew the industrial managers personally, and he required that the types of airplanes, tanks, artillery, and other important armaments be of the highest quality – not only on the same level as overseas, but better.

"Not one type of armament either started or finished production without Stalin's approval. This limited the authority of the people's commissar for defense and his deputy, responsible for supplying the Red Army. On the other hand, Stalin's interest in many instances hastened the production of new forms of weapons technology.

"I am often asked, was Stalin actually a great thinker in the field of armed forces and questions of operational strategy?...

"Many political, military, and general governmental questions were discussed and decided, not at official sessions of the Politburo, but at dinner in the evening in either Stalin's apartment or his dacha, where the members of the Politburo closest to Stalin were present. Included among these were Molotov, Beria, Malenkov, Zhdanov, Mikoyan, and Voroshilov. Stalin assigned tasks to the individual ministers at dinner. Sometimes the chief of the General Staff (Zhukov) was invited to attend with the minister of defense.

"Stalin was a strong willed person, and in no way a coward. I only saw him break down once. This was at dawn on June 22, 1941, when fascist Germany invaded our country. He was unable to gain control of himself for that first day. The shock was so great that the sound of his voice dropped to a whisper, and many of his orders did not correspond to the reality of the situation.

"However, after June 22 Stalin firmly managed the country, the armed conflict, and international relations. Even at the moment of deadly danger, when the enemy was only 25-30 kilometers from Moscow, Stalin did not leave his post in the Kremlin, and conducted himself like a true commander in chief."[12]

Salient features of Zhukov's description, repeated by others, are Stalin's remarkable memory, his attention to detail, his broad interests, his political instincts. Stalin was a people person – he liked company, and worked with others around him. He rarely left the confines of power – either the Kremlin, or one of his dachas near Moscow, or on vacation in Sochi. He enjoyed gardens and walking in the woods, although he himself did not do much gardening.[13] Affairs of state continued as long as Stalin was awake – in the office, at meals, walking in the woods, etc. His nocturnal habits were hard on everyone else who had to keep regular hours because of children and other commitments, and who also had to be at Stalin's beck and call at 3 am. As has been noted earlier, the nocturnal habit was a serious problem during the war, when all of the fighting tended to take place during the daylight hours, when the boss was asleep. Physics

---

[12] Georgy K. Zhukov, "Vospominaniya I Razmyshleniya," Vol 1 *ibid.*, p 339ff.
[13] Svetlana Alilueva, "Twenty Letters to a Friend," *ibid.*, p 38.

experiments, on the other hand, often work best at night, so that Stalin's schedule was not a problem for Laboratory #2.[14]

He did not travel around to see what was going on in the countryside (or the battlefield during the war). He was afraid of flying, and either rode in an automobile or on a train when he did leave Moscow. He gained information from reports, both written and oral. His orders were either verbal, or penciled onto various state documents that he received. There was a log of every person that visited his office, day or night, their names, arrival times, and departure times, but not a record of what business transpired.

It is remarkable that no one seriously tried to kill him, at least not after he had achieved supreme power. Assassinations have been attempted on Russian Tsars and US presidents for centuries, and sometimes they have been successful. But there is no record of anyone since the early 1930's even trying to kill Stalin. He lived with body guards, of course, and as we said he did not move around a whole lot, but still, he seems like such a good target.

Either a botched assassination attempt or a careless use of firearms occurred in September, 1933, while Stalin was on a vacation fishing trip on the Black Sea. According to Montefiore, Stalin and crew were in a fishing boat off the coast, when machine-gun fire opened up from land aimed at the boat. No one was hurt; Beria was suspected of having initiated the attack, but his efficient clean up actions left him above suspicion.[15] Sergo Beria, Lavrenty's son, tells a slightly different story of the event.[16] Stalin's entourage was enjoying itself on the shore, when Mikeladze (local security chief in Abkhazia) spotted a coast guard cutter cruising along, and decided to hail it so that the vacationers could go for a ride. Mikeladze started shooting into the air (not aimed at the boat) to attract the attention

---

[14] Electricity usage is minimal at night, leading to stability. Also, the apparatus works better, because fewer experts are messing with it.

[15] Simon Sebag Montefiore, "Stalin," *ibid*, p 125 and photographs following page 162.

[16] Sergo Beria, "Moi Otets Narkom Beria," Moskva, Algoritm, 2013, p 31.

of the cutter. But Mikeladze did not know that Stalin was on board the cutter…

9.1.3. *Mikoyan on Stalin*

Anastas Mikoyan was another source close to Stalin. Mikoyan was born in 1895, and was an Armenian communist revolutionary, who joined the politburo near its beginning, while Lenin was still alive, and who survived through Stalin and Khrushchev, to retire in the 1970's during the leadership of Leonid Brezhnev. He was truly a "Teflon" minister, having survived almost 50 years at the top of the communist leadership of the Soviet Union. He had a reputation, deserved or not, of being 'reasonable' with foreigners.

Mikoyan tells an interesting story that follows up on Zhukov's observation that Stalin was incapacitated by the German invasion:

"A day or two after the invasion, about 4 o'clock in the afternoon, while Voznesensky was in my office, Molotov called on the phone and asked us to join him in his office. There we found Malenkov, Voroshilov, and Beria in conversation. Beria said that it was necessary to establish a State Defense Committee (GKO), which would have full powers over the country. We should hand over to this committee all government functions, the Supreme Soviet and the Central Committee of the Party. Voznesensky and I agreed to this.

"We then agreed that Stalin should be the head of the GKO. We did not consider anyone else. We thought that Stalin's presence would have such an influence on the consciousness, feeling, and trust of the people, that it would expedite the mobilization and management of all military activities. We decided to go visit him at his close-by dacha (in the Kuntsevo district, about 15 km west of the Kremlin).

"Molotov said that Stalin had been completely dysfunctional for the last two days; nothing interested him, he took no initiative, and was generally in a bad mood. Then Voznesensky said indignantly: "Vyacheslav (Molotov), you go forward, and we will follow." He had in mind that should Stalin remain in his funk, then Molotov should be the leader of the GKO. Other members of the Politburo paid no attention to Voznesensky's suggestion. We were confident that we could organize the defense, and conduct the war properly. We knew it would not be easy, but we were not depressed. Only Voznesensky was particularly excited.

"We arrived at Stalin's dacha, and found him sitting in an armchair next to a small table. Seeing us, Stalin got up from his chair and looked at us with surprise. He asked: "Why have you come?" His appearance was suspicious, somewhat strange. His question was at least as strange as his face. Indeed, given the situation, he should have summoned us himself. I had no doubt – he thought that we had come to arrest him.

Molotov spoke for us. He said that it was necessary to concentrate all power in order to put the country on its feet. For this we should create a State Defense Committee. Stalin asked: "Who will head the committee?" Molotov answered that he, Stalin, would be the head. Stalin looked somewhat surprised, but said nothing other than "OK." "[17]

After agreeing to the proposal, Stalin recovered his normal demeanor and assumed full control of the country. This is a revealing story, because it shows that the "collective leadership" could actually act when necessary, and that Mikoyan thought that Stalin expected to be arrested.

Stalin was well aware of serious mistakes made by the Soviet leadership in the early stage of the war. At a reception in the Kremlin on May 24, 1945, after the war in Europe was over, Stalin proposed a toast to the ordinary Russian people, who had supported the Soviet government in the war effort despite the incompetence in 1941-42. He said that other people might well have dismissed the government, but the Russians did not.[18] Sudoplatov was present at the Kremlin reception, and greatly admired Stalin. Sudoplatov thought that the triumph over Hitler justified the purges and tragedies of Stalin's pre-war period. The repression was to continue after the war, partly because of the burden placed on the economy by the Soviet Atomic Project. Terletsky (Chap 8) was not alone in thinking that the American atomic monopoly had cheapened the great Soviet victory, and hobbled any moves to extend communism to Europe.

---

[17] A.I. Mikoyan, "Tak Bylo," Vagrius, Moskva, 1999, p 390-391.

[18] The date of the reception is quoted by Simon Sebag Montefiore, "Stalin," *ibid.*, p 492. The reception is described by Sudoplatov, "Special Tasks," p 171. The made for TV movie "Zhukov" stages Stalin's speech at the reception following the parade on June 24, 1945.

Leaders who think themselves infallible periodically get crazy ideas that must be suppressed by nervous associates. Mikoyan describes some instances of this phenomenon with Stalin. One was a fixation on growing wheat instead of other crops like barley, corn, or rye in the realm of Soviet agriculture. The conversion of some fields in the Moscow region that had previously grown rye or barley into wheat was done, with some difficulty. Mikoyan argued that rye bread is essential for Russian survival. Eventually the idea faded. Before the German invasion Stalin was worrying about the safety of Soviet infrastructure from aerial bombardment. He proposed that new electricity generating stations be small and dispersed, rather than large and localized, to minimize damage. This idea too was mostly ignored. Stalin was fascinated by natural gas, and industrial gasification of coal. He proposed converting coal to gas directly in the mine, rather than extracting the coal first. Some engineers were set to working on this idea, but the discovery of natural gas from petroleum wells took away the motivation to pursue it. Then Stalin decided that of all products of petroleum, fuel oil was the most valuable, and should be conserved as a national resource. Stalin also often moved personnel around, to the dismay of the ministers. Many transfers took place, but often did not end well.[19] These instances seem relatively benign, and serve to illustrate the breadth and detail of Stalin's interest in everything going on.

A humorous incident given by Mikoyan would be called "bananagate" in modern speak. A bit of background information would be helpful in setting the stage for this event. Stalin loved bananas, and Mikoyan, as head of foreign trade, was responsible for importing them, in addition to a huge list of perishable and non-perishable goods. Bananas were introduced in the major cities, Moscow, Leningrad, Kiev, etc. Stalin regularly had a delivery. Mikoyan personally did not partake. In 1949 Mikoyan had been minister of foreign trade for ten years, and Stalin asked him, as well as the other ministers, to come up with a short list of possible replacements from their own staffs, should such a replacement become necessary. Mikoyan complied. In February of that year Mikoyan was posted to China for talks

---

[19]　A.I. Mikoyan, "Tak Bylo," *ibid.*, p 523.

with Mao Zedong and Chou en Lai. Upon his return to Moscow, Stalin asked him if it were not a good time for him to step aside as minister? What could Mikoyan say? Then Stalin asked who should be his successor? Mikoyan nominated Mikhail Alekseevich Menshikov, who was duly appointed minister of foreign trade on March 4, 1949.[20]

In the summer of 1951, Mikoyan was on vacation with his family in Sukhumi, and they all went to Stalin's place for supper. Suppers with Stalin tended to last a long time, and about 4 am a plate of bananas appeared on the table. They were too green. Stalin bit into one, and said to Mikoyan: "Try a banana, and tell me how you like it." Mikoyan found the banana rather similar to a raw potato – inedible. "So, how come the bananas are inedible?" asked Stalin. "Hard to say," answered Mikoyan, "could be they were too green when they were shipped." "Well", said Stalin, "when you were minister of foreign trade, the bananas were always perfect, and now we have a new minister, who is doing a lousy job." Mikoyan interceded for Menshikov, saying that he could not be held responsible for everything imported into the Soviet Union. Stalin wanted to get to the bottom of the problem – find out who is responsible. Mikoyan agreed.

Mikoyan got back to his vacation dacha about 5 am, and decided that it would be best to get some sleep, and call Moscow in the morning to initiate an investigation into the banana affair. He got up about noon, and called one of his deputies in the ministry, a certain Vlasov. Vlasov said that he already knew all about the banana flap, that Beria had called him earlier that morning. Mikoyan called Beria, asking him how come he, and not Mikoyan, was following the banana crisis? Beria said that Stalin had called him about 6 am, and asked him, Beria, to solve the problem, but he was happy to turn the responsibility completely over to Mikoyan. Mikoyan on the other hand was happy to let the matter rest, hoping that it would all go away.

---

[20] A.I. Mikoyan, "Tak Bylo," *ibid.*, p 529.

A few days later it was time for Mikoyan and Malenkov to return to Moscow. They went together to say goodbye to Stalin. Mikoyan acted as if the banana problem had never existed, but Stalin suddenly said, out of the blue, that Menshikov was obviously incompetent as minister of foreign trade. OK, the banana affair was not all that serious, but it was symptomatic of a deeper problem at the ministry. So Mikoyan upon his return to Moscow should relieve Menshikov of his duties, and replace him with another of Mikoyan's former deputies, Pavel Nikolaevich Kumykin. Mikoyan objected, saying that Menshikov was doing a good job, and that Kumykin was not ready to run a ministry. Stalin was not able to convince Mikoyan to make the replacement, so he turned to Malenkov, and told him, upon his return to Moscow, to relieve Menshikov, and install Kumykin.

After they both returned to Moscow, Malenkov asked Mikoyan to do the dirty work of firing Menshikov and installing Kumykin. Mikoyan first talked to Kumykin, who was terrified at the prospect of running the ministry of foreign trade. Mikoyan tried to calm him with assurances that his friends would help him out. Then he talked to Menshikov, who was stoical about the whole thing, and accepted his fate. Kumykin stayed in office until 1953, after Stalin's death. Mikoyan managed to hide Menshikov as customs inspector in Amur, on the Chinese border. He later became ambassador to the USA (1958-1962). Such was the fallout of green bananas.

This is a funny story, but the players are all deadly serious. Any boss could have a tantrum over green bananas, and Mikoyan may have told the story to show Stalin's human side.

Other Mikoyan anecdotes exposed Stalin's dark side, if only in a presentable format. Stalin liked movies, and one of his favorites was a British film about a pirate who murdered all of his accomplices in order to avoid sharing the loot. Stalin admired his resolve. Stalin also admired Tsar Ivan IV, Ivan the terrible, for his ruthlessness. But Stalin said that Ivan would have done better had he murdered all of the Boyars (Russian nobles) rather than just some of them. That way he would have been able to move more freely. Before the German invasion, Stalin admired the tactics of Hitler, particularly the Roehm purge, in which Hitler exterminated many critics of his regime in what has been called the 'night of the long knives' in 1934.[21] Generally speaking, Stalin admired killing for political purposes.

9.1.4. *Boris Vannikov – Prisoner to Minister*

Boris Vannikov told a story about Stalin's capricious behavior.[22] It was mentioned in Chapter 1 that Vannikov, People's Commissar for Armaments, was arrested on June 7, 1941, released and reinstated just after the invasion on June 22, and the arrested again in December of that year. Vannikov recalled:

"In January, 1942, after returning from work on the way to the camp barracks, I was stopped and taken to the commandant. He said that I was to leave the camp at once, without any explanation. My request to return to the barracks to get some personal affects and say goodbye, was denied. Therefore, I left the camp with the clothes I had on. A special plane from the nearest airport took me to Moscow. I was given no explanation. However, as we drove past the Kremlin, the car from the airport turned into the Borovitskie Gates, and I knew that I was going to see Stalin, for only Stalin could call me to the Kremlin. [23] I stopped at the entrance to Stalin's office. Stalin was walking along the table, looking at his feet. Finally, he spoke softly about the problems at the front, and the supply and material difficulties of the armed forces. I then realized that I would not be punished further, and I relaxed a bit. I listened carefully, trying to understand what Stalin wanted me to do. Stalin stopped walking, and said: 'We have decided to reorganize the People's Armaments Ministry, and we want you to head it.' Pointing to my prison clothes, I asked how I was to head a ministry as a common prisoner? Stalin looked at me attentively, as if seeing my clothes for the first time, and waved his hand, saying that I should go to the Hotel Moscow, where I would have a room, get some rest, and return to the Kremlin in three days with an organizational outline for the Armaments Ministry so that the army would be assured of the necessary supplies in a timely fashion."

---

21 A.I. Mikoyan, "Tak Bylo," *ibid.*, p 534.
22 G.A. Sosnin, "Brief about the founders of atomic industry," HISAP96 Vol 1, *ibid.*, p 189.
23 The gates are at the southwest corner of the Kremlin, near the riverbank.

Vannikov went on to say that Stalin was a very dangerous individual. He could listen patiently to your point of view, and it was possible to argue with him, but once he had made up his mind, you had better fulfill your assignment to perfection, or woe be unto you. There was no legal justice. Your fate depended more upon Stalin's whim than on the law codes. Even though Stalin imprisoned Vannikov twice, he awarded him great responsibility, and Vannikov performed, without showing any grudge towards Stalin.

9.1.5. *Milovan Djilas on Stalin*

Milovan Djilas was one of Tito's deputies in the Communist Party of Yugoslavia. In WWII the Germans occupied Yugoslavia, and Tito went underground with a communist partisan army, hoping to gain control of the country after the Germans were expelled. The Soviet Union supported Tito fighting the Germans, and in late 1944 the Red Army entered Yugoslavia and chased the Germans out. Over the course of these events Djilas had the opportunity to visit Stalin in the Kremlin on three occasions.[24] The first mission was in the spring of 1944, before the withdrawal of the German army. Djilas was sent to Moscow to seek aid for Yugoslav communists. The Yugoslav delegation flew in to Moscow via Teheran. They were warmly received in the Soviet capital, but sat around for a few weeks waiting to see anyone in the Kremlin. While waiting, Djilas and his colleagues were taken for a visit to Marshal Konev's headquarters on the front lines.[25] One evening after his return to Moscow Djilas and his companions were loaded into an automobile and whisked to the Kremlin. Molotov and Stalin awaited them in a conference room. Djilas writes:

"Stalin was in a marshal's uniform and soft boots, without any medals except a golden star – the Order of Hero of the Soviet Union, on the left side of his breast... This was not the majestic Stalin of the photographs or the newsreels- with the stiff, deliberate gait and posture. He was not quiet for a moment. He toyed with his pipe,

---

[24] Milovan Djilas, "Conversations with Stalin," Harcourt Brace & Company, New York, 1962 (renewed in 1990).

[25] Milovan Djilas, "Conversations with Stalin," *ibid.*, p 52.

which bore the white dot of the English firm Dunhill, or drew circles with a blue pencil around words indicating the main subjects for discussion, which he then crossed out with slanting lines as each part of the discussion was nearing an end...

"I was also surprised at something else: he was of very small stature and ungainly build. His torso was short and narrow, while his legs and arms were too long. His left arm and shoulder seemed rather stiff. He had a quite large paunch, and his hair was sparse, though his scalp was not completely bald. His face was white, with ruddy cheeks. Later I learned that this coloration, so characteristic of those who sit long in offices, was known as the 'Kremlin complexion' in high Soviet circles. His teeth were black and irregular, turned inward. Not even his mustache was thick or firm. Still the head was not a bad one; it had something of the folk, the peasantry, the paterfamilias about it – with those yellow eyes and a mixture of sternness and roguishness.

I was also surprised at his accent. One could tell that he was not a Russian. Nevertheless his Russian vocabulary was rich, and his manner of expression very vivid and plastic, and replete with Russian proverbs and sayings. As I later became convinced, Stalin was well acquainted with Russian literature – though only Russian – but the only real knowledge he had outside of Russian limits was his knowledge of political history."[26]

Stalin got down to business, and agreed to help the Yugoslav communists with arms and supplies. The recognition of Tito's government as legitimate in Yugoslavia, replacing the pre-war king of Serbia, was postponed pending negotiations with the British. Stalin displayed his usual firmness and close attention to detail. They exchanged gifts upon departure.

Djilas' second visit with Stalin was in the summer of 1944, just before the Allied invasion of Normandy. On that occasion Stalin entertained him with dinner that lasted six hours from 10 pm until 4 am. The importance of these dinners for matters of state has already been mentioned, as has been the nocturnal behavior of the Politburo in response to Stalin's habits.

---

[26] Milovan Djilas, "Conversations with Stalin," *ibid.*, p 60-61. This is a translation from Serbo-Croatian, and may lack some color thereby.

Djilas was very impressed with Molotov, who was clearly Stalin's right hand man.[27] Djilas writes regarding dinner table conversation:

"An uninstructed visitor might hardly have detected any difference between Stalin and the rest. Yet it existed. His opinion was carefully noted. No one opposed him very hard. It all rather resembled a patriarchal family with a crotchety head whose foibles always caused home folks to be apprehensive."[28]

A dispatch from Churchill arrived during the dinner, informing Stalin of the anticipated Normandy landings the next day (June 6). Stalin made fun of the dispatch, saying that the landings might take place if there was no fog, and if they did not meet any Germans.

The next trip to Moscow was in the winter of 1944-1945, after the Red Army had liberated Yugoslavia, and driven out the Germans. The behavior of the Red Army towards the civilian population of Yugoslavia gave the Yugoslav communists, who were trying to win popular support, serious problems. Tito's followers idolized the Red Army, and that made coping with criminal behavior even more difficult. To make matters worse, the Soviet generals were not at all sympathetic to Yugoslav complaints.[29] Instances of rape, murder, looting, and assault were recorded.[30] Djilas found the Soviet leaders to be indifferent to allegations of rape, which he thought especially heinous. His objections caused friction within the Yugoslav Communist Party, because it threatened to jeopardize Soviet support that they desperately needed.

A sizeable delegation of Yugoslav communists journeyed to Moscow. There was another dinner at the Kremlin. Stalin singled out Djilas for criticism:

"And such an army (the Red Army) was insulted by no one else but Djilas! Djilas, of whom I could least have expected such a thing, a man whom I received so well!

---

[27] Djilas hardly mentions Beria.
[28] Milovan Djilas, "Conversations with Stalin," *ibid.*, p 77.
[29] Milovan Djilas, "Conversations with Stalin," *ibid.*, p 88.
[30] Similar complaints were made in Austria.

And an army which did not spare its blood for you! Does Djilas, who is himself a writer, not know what human suffering and the human heart are? Can't he understand it if a soldier who has crossed thousands of kilometers through blood and fire and death has fun with a woman or takes some trifle?"[31]

This is a translation from Serbo-Croation of Djilas' recollection of what Stalin said in Russian. With all those caveats, it remains an outrageous statement by a head of state. It gives us a glimpse of the true nature of the leader that Igor Kurchatov and his colleagues had to deal with.

Mikoyan bemoaned the fact that most writers of memoirs who had some exposure to Stalin only talked about his good qualities. Zhukov, quoted above, is an example. Foreign ambassadors were virtually uniformly impressed with Stalin. He was very smooth at the Yalta and Potsdam conferences, and even asked Churchill for advice on how to handle Charles de Gaulle, who would soon be visiting Moscow. Stalin was crude, coarse, uneducated, and brutal, yet he always managed a dominating presence, and could be charming if needed. He had the ability to command deference in any discussion, and to be heard respectfully, in the presence of foreign leaders, politburo colleagues, and military marshals. He sometimes compromised in reaching consensus decisions, but without seeming to yield any ground to opposing views.

But no one talks about the murders. Stalin killed a lot of people, in his purges, in his prisons, and through his policies. Mikoyan was present during the excesses of Stalin's rule, both during the purges of 1937-38, and the terror after the war. His son Sergo wrote in an introduction to "Tak Bylo" that his father was completely blinded by his admiration of Stalin, and, although Mikoyan pere did not agree with the repression, he felt powerless to do anything to stop it.[32] One documented example of Mikoyan's agreement to mass execution is the NKVD memorandum from Beria to Stalin in 1940 calling for the murder of 12,000 Polish officers –

---

[31] Milovan Djilas, "Conversations with Stalin," *ibid.*, p 95.
[32] A.I. Mikoyan, "Tak Bylo," *ibid.*, Introduction by Sergo Mikoyan, p 5ff.

the Katyn massacre. The memorandum was signed by Stalin, Voroshilov, Mikoyan, and Molotov.[33] No one had clean hands, but none of the excesses would have taken place without Stalin's explicit approval.

A national sentiment in the Soviet Union towards a more open society after the victory over Germany in WWII was mentioned in Chapter 1 of this book. Mikoyan devoted Chapter 41 of his memoirs to the subject "What I expected after the War."[34] Mikoyan believed, as did many others, that the pre-war political repressions would not be repeated. The war exposed the Soviet population to western ways, and created new thoughts. Soldiers returning from the front lines had seen the living standards of the west, and had acquired expanded horizons and new demands. The army, despite wartime discipline, had acquired a certain comradely democracy. To quote Mikoyan:

"Wartime tactics and combined operations at the front and in the rear had a positive effect in bringing people closer together. Considering this, and the evolution in Stalin's character and behavior, I supposed that a process of democratization in the party and the country at large would begin, so that at a minimum we would return to the form of democracy that we had in 1929, and then go forward. I was convinced of this, and had a joyous feeling. I also felt a trusting and friendly relationship with Stalin. Stalin had entrusted me with many tasks during the war, that I had successfully fulfilled. We had almost never had any arguments or disputes."[35]

Alas, greater freedoms were not forthcoming. We have already noted the expansion of the prison population, partly German POW's not repatriated, and partly the imprisonment of Red Army soldiers and officers by the NKVD after liberation from German POW camps. Anyone who had contact with the west as a result of the German conquest was considered unreliable. Many tribal peoples within the Soviet Union were deported to exile for their lack of enthusiasm for the Soviet state.[36] Rather than

---

[33] katyn.ru/index.php?go=Pages&in=view&id=6 reproduces the document in full in Russian, and shows the signatures on the first page.

[34] A.I. Mikoyan, "Tak Bylo," *ibid.*, p 513ff.

[35] A.I. Mikoyan, "Tak Bylo," *ibid.*, p 513.

[36] As noted elsewhere, these displaced groups were a significant factor in the disintegration of the USSR 45 years later.

becoming more tolerant with age, Stalin became more suspicious. The repression extended to the intelligentsia, and the scientific community. While during the war initiative and self-determination on the part of those with technical training was encouraged as a means towards victory, such an atmosphere was not to be tolerated in peace time. There was to be no Soviet spring.

9.1.6. *Modern Physics vs Marxist Philosophy*

In Chapter 2 we learned that some Marxist thinkers did not believe that relativity and quantum mechanics were consistent with Marxist philosophy, and that that was one of the reasons for the defection of George Gamow and his wife to the United States. Then A.A. Maksimov, a philosophy professor at Moscow State University, started a campaign against Ioffe and his journal 'Uspekhi Fizicheskikh Nauk' in 1937 – see Chapter 4. Maksimov questioned the relevance of 'western physics' (relativity and quantum mechanics) to Marxist thought. The fuss led to a meeting of the Soviet Academy of Sciences, wherein Ioffe successfully defended the relevance of his activities. There were also members of the Moscow State Physics Department who rejected the new thinking. This was before the war, at a time when not all research universities world-wide offered classes in quantum mechanics, and there were certainly department members who thought the whole thing was baloney. However, a few nuts should not stop progress, and although the anti-physics movement was a nuisance, it should not have been a real threat.

But it did not go away. The first organized threat was in bio-agriculture. It would be too much of a digression to go into the Lysenko affair, but in essence Trofim Lysenko was a pseudo-geneticist, who rejected Mendel's theory of heredity. He won Stalin's support, partly because he was influential in coaxing the peasants to resume agriculture after the great collectivization of the 1930's. In August, 1948, the All-Union Academy of Agricultural Sciences named V.I. Lenin – with the jaw-breaking Russian acronym of VASKhNIL – had a meeting in Moscow, organized by Lysenko and his associates. Lysenko was alarmed by a shift in the

Soviet Academy of Sciences under its new president, the physicist S.I Vavilov, towards cooperation with the 'bourgeois' western countries. Sergey Vavilov's elder brother Nikolai had been a prominent geneticist, and enemy of Lysenko. Nikolai was arrested, and died in prison in 1943. Therefore, the president of the Academy of Sciences was not sympathetic to the aims of the VASKhNIL meeting. The discussion was vigorous, and a member of the audience got up and poured a glass of water over one of the speakers.[37] Despite the opposition of many respectable geneticists, the result of the meeting was a victory for dielectical materialism – a victory of Marx over Mendel. This diversion in Soviet biology would persist until after Stalin's death.

Emboldened by the success in biology, the opponents of 'bourgeois' physics gathered together for another attack. Times had changed since the 1930's. Relativity and quantum mechanics were well accepted by the end of WWII. Most physics departments had professors who were competent to teach the subjects. The atomic bomb was known to work. The energy gained by fission comes from $E = mc^2$ (special relativity)[38], and the neutron capture cross sections require quantum mechanics. So the existence of the bomb meant that 'bourgeois' physics was just sound engineering principles, not crazy at all. Nevertheless, A.A. Maksimov, the same philosopher from Moscow State University who hassled Ioffe before the war, started to organize a physics meeting based on the model of VASKhNIL.[39]

The opponents of 'bourgeois' physics had essentially the same objections in 1948 as were expressed ten years earlier. The physics taught to Soviet students was western physics, without sufficient respect for Marxist theory. Concepts of time, energy, mass, etc. in textbooks were inconsistent with Marxist teachings. All of this had to be changed, so that students learned the right stuff. The plaintiffs convinced Voroshilov, a

---

[37] ru.m.wikipedia.org.VASKhNIL (1948).

[38] Serber in "The Los Alamos Primer" asserts that the kinetic energy comes from the coulomb repulsion of the fission products. He is right, but he is nit-picking.

[39] A.S. Sonin, "The Meeting that did not take place," Priroda 1990, #3 p 97, #4 p 91, and #5 p 93.

member of the Politburo, to direct the Ministry of Higher Education and the Soviet Academy of Sciences to call a meeting in Moscow for January, 1949.

If such a meeting deflected and crippled Soviet physics, as VASKhNIL did to biology, the result would be disastrous. Most members of the Academy of Sciences, including S.E. Vavilov, the president, scientists working on the Atomic Project, laboratory workers and teachers recognized the potential danger, and mobilized to derail the conference. Plans, programs, and disputes went on for several months. The meeting was postponed to late March, 1949, and then canceled altogether. Before the end, however, the matter reached the highest levels of government. Beria asked Kurchatov if relativity and quantum mechanics were minor subjects that Soviet physics could happily do without. Kurchatov answered that the atomic bomb depended on relativity and quantum mechanics, and if it were necessary to dispose of these subjects, then we should stop working on the bomb. Beria replied that he would have to take the question to Stalin.

Stalin gave a memorable answer that stopped the objectionable conference in its tracks: "Leave them in peace. We can always shoot them later."[40]

Accordingly, Beria called president Vavilov. A witness to one side of the conversation in Vavilov's office heard the following:

Telephone rings. The president picks up the phone.

"Hello, Lavrenty Pavlovich"

"Yes, Lavrenty Pavlovich"

"Of course, Lavrenty Pavlovich"

"Good bye, Lavrenty Pavlovich"

---

[40] I. Zorich, Priroda 1990 #9 p 106. The Russian is "оставь их в покое. расстрелять их мы всегда успеем.

Vavilov then said that there would be no discussion of the merits of modern physics. The meeting was cancelled, and the whole enterprise was finished.[41] Soviet physics was saved from having to defend itself against the barbarians.

So this was the character of the leader of the Soviet Union during the Soviet Atomic Project. Stalin was ruthless, crude, even barbaric at times, and all powerful. He could be a smooth diplomat and a formidable negotiator. He became more suspicious and vengeful as he aged. His private life was unpretentious, and he devoted all of his energy to his work. His rule was characterized by a 'cult of personality' that was projected nation-wide, and to some extent world-wide. Whether this cult was initiated by Stalin, or simply tolerated by him, it was certainly a major component of the government while he was in office. He was determined that the Soviet Union have nuclear weapons, in order to level the playing field with the western capitalistic countries. That the atomic bomb was a western creation did not matter. He would defend Soviet physics in the face of Marxist doctrine to achieve that goal.

---

[41] B.C. Gorobets, "Sovietskie Fiziki Shutyat," Knizhny Dom Librokom, Moskva, 2012, p 251.

9.2. Lavrenty Pavlovich Beria

Figure 9.2. Lavrenty Beria, Stalin's deputy and leader of the Soviet Atomic Project. Courtesy of the Library of Congress, New York World Telegram and Sun.

Lavrenty Beria is shown in Figure 9.2. His son Sergo has written that Stalin was a bad influence on his father. On the other hand, Stalin's daughter Svetlana has written that Beria was a bad influence on her father! Both were probably right.[42] After he came to Moscow from Georgia to assume direction of the NKVD in 1939, Beria was very closely associated with Stalin. Molotov was more visible, particularly in diplomatic settings where foreign ministers were present, but Beria was more powerful. He enjoyed his relative obscurity.

9.2.1. *Early Life*

Lavrenty Beria was born in the Georgian village of Merkheuli, near Sukhumi on the Black Sea coast in 1899, 20 years after Stalin. He was a member of the Mingrellian tribe – one of many separate small tribes in the region. There was an unwritten Mingrellian language, but every school child learned Georgian. He was too young to participate in revolutionary activities before WWI, and was only 18 years old when the Bolshevik

---

[42] Svetlana Alliluyeva, "Twenty Letters to a Friend,' *ibid.*, p 149. Sergo Beria, "Moi Otets Narkom Beria," *ibid.*, p 53.

revolution took place in 1917, so he did not qualify as one of the 'old Bolsheviks' who experienced tsarist prisons and underground operations. But he made up for lost time with his vigor for expediting the Bolshevik cause through the secret police in his native Georgia.[43] One of his first tasks was to calm the Mingrellians, which he did with distinction. Stalin visited Georgia, and noticed Beria. Beria was soon promoted to Moscow, and at the end of the 1937-38 purges Beria replaced Ezhov as head of the NKVD. Beria did some house cleaning, but he assumed responsibility for the prisons and the gulag camp system that remained largely intact.

Beria's early contact with the physics community has been described in Chapter 4. In response to Kapitza's letter of appeal on behalf of Lev Landau, Beria released Landau after one year in prison in 1939. During the war Beria was busy with ongoing intelligence work, partisan operations, and the prison camps. He was also responsible for much of Soviet wartime industry. Through his managerial skills he therefore gained control of the economy except for agriculture and the military – industry plus the gulag. The story of the Soviet 'Alsos' mission to Germany just after the war, to collect scientists, engineers, equipment, and raw materials for the Soviet Union has been told in Chapter 7. This was one of Beria's operations, as was the Terletsky-Bohr mission in Chapter 8. The description of the espionage operations in Chapter 8 – all under Beria's control - makes it clear that the Soviet Union had substantial detailed information on the Manhattan Project by the end of the war.

Thus Beria was a skilled organizer who had extensive knowledge, without technical training, of the atomic project when he was appointed by Stalin to head the Special Committee on Uranium after Hiroshima in August, 1945. He was also a ruthless tyrant and a murderer.

---

[43] Amy Knight, "Beria, Stalin's First Lieutenant," Princeton University Press, Princeton, NJ, 1993, p 11ff.

9.2.2. *Recollections of Contemporaries*

We may use the same technique for Beria as for Stalin, that is, obtain impressions of the man from those who had contact with him. The antagonism between Kapitza and Beria has been described in Chapter 7, and clearly Kapitza had nothing good to say about Lavrenty Pavlovich.

Igor Golovin, Kurchatov's deputy and biographer, was generally positive:

"Beria was an exceptional organizer – energetic and corrosive. For example, if he took home documents overnight, by morning they returned with pithy remarks and useful suggestions. He was a good judge of people, verifying each one personally, and it was quite impossible to hide any mistakes from him..."

Golovin then tells a story of a meeting at Laboratory #2: "Once Beria came to the Kurchatov laboratory and I (Golovin) showed him one of our measurement setups under construction. Beria gave me a strong look after the presentation, and from his expression I understood, that he understood, that I understood that he did not understand. And I was scared."[44]

A.P. Aleksandrov became the director of Kapitza's Institute after Kapitza was removed in 1946. He was active in the Soviet Atomic Project, and later developed industrial and marine nuclear reactors at the Kurchatov Institute (Laboratory #2). He made several observations regarding Beria.

Because of its importance to national security, the Soviet Atomic Project was held in total secrecy from the Soviet people, and the rest of the world. Beria guaranteed that the scientists, engineers, and their families would not be touched by the NKVD during the course of the project, and he kept to his promise. One of the reasons for the hands-off policy was that interrogation was one of the standard features of an NKVD arrest, and the arresting officers would ask the occupation of the suspect. The NKVD

---

[44] B.S. Gorobets, "Sovetskie Fiziki Shutyat," *ibid.*, p 257

was not permitted to know anything about the occupations of the Atomic Project workers. Several years later Aleksandrov observed:

"Stalin's word generally determined the fate of the project. Beria with one gesture could send any one of us to oblivion. But Kurchatov was at the top of the pyramid, and he was our savior, competent, responsible, and strong."[45]

K.I. Shchelkin, an expert on shock wave detonation and one of the physicists at the Semipalatinsk firing range for the bomb test in 1949, recounted a Beria episode. Beria was there for the test. He called Shchelkin over and asked:

"Do you need the engineer Ivanov?"

Shchelkin: "Lavrenty Pavlovich, I do not take to the firing range anyone whom I do not need."

Beria: "Then take a seat and listen."

A young rosy cheeked NKVD general with a thin brief case walked up.

Beria: "Give your report."

Rosy cheeked general: "It is necessary to remove from the firing range and arrest engineer Ivanov."

Shchelkin: "On what evidence?"

RCG: "In 1941 he repeatedly expressed his dissatisfaction with the retreat of the Red Army."

Shchelkin: "And you were satisfied with the retreat?"

Shchelkin said that he had never suspected that the complexion on the face of an individual could change so quickly. Beria said:

---

[45] Scientific American "V Mire Nauki," special issue 2013, p 17.

"Go, we will talk later. Leave Ivanov alone."

Shchelkin recollected that for the eight years that Beria was responsible for the Soviet Atomic Project, from 1945 to 1953, not a single scientist was arrested.[46]

One of the industrial reactor types under study at Laboratory #2 was natural uranium and heavy water. To make such a reactor required several tons of heavy water, and the Soviet Union had zero. Aleksandrov proposed a cryogenic method for isotope separation in liquid hydrogen, using catalytic conversion of ordinary hydrogen from ortho to para.[47] A test setup to study the technique in Dneprodzerzhinsk exploded. Post mortem showed that some oxygen contaminate froze solid, and under certain conditions could cause an explosion. Aleksandrov proposed to build an industrial sized plant.[48]

A meeting was called in Beria's office in the Kremlin to discuss the proposal. Beria sat at the head of the table, with Makhnev on his left, and Aleksandrov on his right. Makhnev started the discussion. "Comrade Aleksandrov proposes to build a plant for production of heavy water." Beria had the proposal in front of him. Beria asked Makhnev (ignoring Aleksandrov, who was sitting there) "Does Comrade Aleksandrov know that the experimental plant in Dzerzhinsk blew up?" Makhnev answered "He does." Then Beria again asked Makhnev, "He has not removed his signature from the proposal?" Makhnev, "No." Beria: "Does he know that if this plant explodes he will be sent far away?" Aleksandrov spoke up to say that he knew the risk. Beria finally addressed Aleksandrov himself: "Do you still sign the proposal as principal investigator?" Aleksandrov: "I do." Beria: "Build the plant," and he initialed the document "Approved, LB." The plant cost about 100 million rubles, worked well for years, and

---

[46] B.S. Gorobets, "Sovetskie Fiziki Shutyat," *ibid.*, p 255.

[47] The two proton spins are parallel in molecular orthohydrogen, and anti-parallel in parahydrogen. The ortho to para conversion proceeds slowly in the liquid. Heat is liberated by the conversion, and some of the liquid boils off.

[48] HISAP96, *ibid.*, Vol 2, p 290.

never exploded.[49] Many years later, in the 1970's, when Aleksandrov was himself president of the Soviet Academy of Sciences, he was asked what the Soviet Union had to do in order to reproduce Silicon Valley in their country. Aleksandrov replied: "Dig up Beria."

Another example of Beria's decisiveness involved the site selection for the laboratory at Dubna. As mentioned in Chapter 8 in connection with Bruno Pontecorvo, Dubna was built in 1949, 120 km north of Moscow on the Volga River to house a large synchrocyclotron that Kurchatov wanted for nuclear research. In the beginning Dubna was a branch of Laboratory #2. A committee was formed in the tradition of such things to select the best site for the new laboratory. The committee report favored a site on the north side, but nearer the city of Moscow. There were three sites on the list. The committee gave an oral presentation of its findings in Beria's office.

"Having heard the committee report, Beria asked for a map, and pointed with his finger to the location of the future Dubna (this location was not among those recommended by the committee), and said: 'We will build it here.'

'But Lavrenty Pavlovich, there is a swamp there – not good for support of a heavy accelerator.'

'We will drain it.'

'But there are no roads.'

'We will build them.'

---

[49] B.S. Gorobets, "Sovetskie Fiziki Shutyat," *ibid.*, p 256-257. Beria quoted a Russian proverb "go there where Makar did not drive his cattle," which I translated as 'far away.'

'But there are few villages in the area. It will be difficult to find construction workers.'

'We will find them.' Said Beria."[50]

One of the reasons Beria gave for his choice was a nearby prison camp, which could furnish the labor to build the laboratory and the town. I am told that the camp was across the Volga River from the present town of Dubna. The camp is no longer there. Another reason for the location was its remoteness, which facilitated secrecy (although Dubna never did weapons work), and also, according to Beria, kept the scientists busy with the task at hand, not going back and forth to distractions in the big city. A hydroelectric plant was already located only 5 km away, so that power was not a problem, and there was plenty of cooling water from the Volga River.[51] It has been said that the area was known to Beria as a pleasant place to go. As part of its disguise the Dubna laboratory was initially called the "Technical Hydrology Laboratory of the USSR Academy of Sciences."

There are not very many recorded examples of the deadly threat atmosphere that pervaded high level meetings of the Uranium Committee in Beria's office. Sudoplatov's observations were discussed in Chapter 7. He recalled that there were sharp arguments and abusive language among the members, and that Beria often served as a moderator. He also claimed that those who worked directly for Beria – like Sudoplatov himself - became inured to his threats, and were relatively safe as long as they did their jobs. Still, working with the man who ran the prison camp system must have been an emotional strain, and the principal scientists had to contend with it on a daily basis.[52]

---

[50] B.L. Ioffe in HISAP96, Vol 2, *ibid.*, p 226.
[51] V.P. Dzhelepov in HISAP96, Vol 1, *ibid.*, p 285.
[52] Technically Beria was not head of NKVD after 1946. He was replaced by Sergei Kruglov, and his office moved from NKVD headquarters in Lubyanka to the Kremlin. But Beria still had considerable influence over the NKVD.

9.2.3. *Sakharov and Beria*

Andrei Sakharov visited Beria's office several times. Sakharov is world famous for his human rights campaign against Soviet oppression, but he was also a very respectable theoretical physicist. Born in 1921, Sakharov was college age during WWII. Unlike most young men of his generation, he did not serve in the Red Army, but rather worked on technical problems in defense plants. He graduated from Moscow State University after the war, and joined the Soviet Atomic Project at Arzamas 16 (see Chapter 10) in 1950. He worked at Arzamas 16 – which he called the 'object' - for 18 years, until 1968. Like Los Alamos during the war years, Arzamas 16 was a secret place, not on any map.[53] Sakharov's young family lived in Moscow, and he lived at the laboratory during the week, making weekend visits home. He made important contributions to the design of a practical, deliverable hydrogen bomb. He was also interested in controlled nuclear fusion, and worked with Igor Kurchatov on the design of various thermonuclear reactors. After fifty odd years of work around the world, a practical controlled thermonuclear device to generate electric power has not been invented. But Sakharov's work on early designs was the motivation for his first visit to Beria. He was a technical expert rather than a laboratory big shot.

Fusion power is based on the deuterium – tritium reaction $^2H + {}^3H \rightarrow {}^4He + n$, resulting in 14 MeV neutrons. The energy balance for this reaction is described in Chapter 3. The reaction proceeds with a deuteron beam from a cyclotron and a tritium target. But for a large energy release, many reactions are necessary per second, and this requires dense, high temperature gases. In a hydrogen bomb these conditions are supplied by a fission bomb trigger, but in a power reactor there must be control, ability to turn it on and off, vary the power level, etc. A high temperature ionized gas is called a plasma. Plasma heating and containment are the names of the game.

---

[53] The laboratory is located at the town of Sarov, which was once a monastery. It is still closed to foreign visitors.

A young sailor in the Soviet Pacific Fleet named Oleg Lavrentev sent a letter on plasma confinement by electric fields to Beria, who forwarded it to Sakharov and Igor Tamm at Arzamas 16 for critique and evaluation.[54] Sakharov and Tamm did not think that the suggestion was practical, but it started them thinking about confinement by magnetic fields instead. This led to the development of the tokomak, which is the design of choice for many modern experimental fusion confinement devices.[55] It also led to an invitation for Sakharov to visit Beria in the Kremlin.

Sakharov writes:

"Each time I was called to Moscow [from Arzamas 16] I had to wait idly for at least a week before I was taken to the Kremlin. When the call from above finally arrived, we were all taken inside the Kremlin walls, where each of us was given a personal pass that we had to show several times as we went along. One officer verified the passport and Kremlin pass, while another looked me in the eye to assure no suspicious expression. To be sure, they asked if we carried any weapons, but they did not do a body search.[56]

"This time I went alone. I saw Oleg Lavrentev in Beria's reception area. They had called him to Moscow from the fleet. We were admitted to Beria's office together. Beria sat at the head of a long table, wearing his signature pince-nez glasses and a light cape over his shoulders. His constant companion Makhnev, formerly the boss at the Kolyma prison camp, sat by his side. After Beria was removed from office Makhnev became chief of information for the Ministry of Medium Industry [code name for the Atomic Project, referred to as MSM]. In general, MSM served as a "national park" for former cohorts of Beria.

"Beria somewhat condescendingly asked me what I thought of Lavrentev's proposal. I repeated the evaluation that I had written. Beria then asked some questions, and dismissed Lavrentev. I never saw him again. I learned that he later studied physics or radio engineering at an institute in Ukraine, and after graduation took a job at LIPAN [Laboratory #2]. After a few months he got in trouble with his coworkers, and he returned to Ukraine. In the 70's I received a letter from him asking me to send whatever documents I had pertaining to his 1950 proposal and

---

54 Andrei Sakharov, "Vospominaniya," Alfa-Kniga, Moskva, 2011, p 143.

55 ITER, the international fusion reactor under construction at Cadarache, France is a tokomak. See Science, Vol **350**, p 1011, 27 November, 2015.

56 It is harder to get on a plane now than it was to enter Kremlin sanctuaries in 1950.

my response thereto. He wanted to establish priority for his invention. I did not have anything on paper, but I sent him what I could from memory, registering my letter with the office of the Physics Institute of the Academy of Sciences. The first letter for some reason did not arrive, and he asked me to re send it, which I did. I never heard from him again.

"After Lavrentev left Beria asked me how the work on magnetic fusion confinement was going under Kurchatov's supervision. I answered him. He then stood up, signaling the end of our interview, but suddenly asked: 'Do you have any questions for me?'

"I was completely surprised by the question. Without thinking, I asked: 'Why do our development works go so slowly? Why are we always behind the USA and other countries, losing the technical competition?'

"I did not know what kind of answer to expect. Over a 20 year time span we had discussed this subject among ourselves, and came up with the answers – the authoritarian structure of management, insufficient exchange of information, and limits on intellectual freedom. But I hardly expected Beria to say this.

"Beria answered: 'Because we have no broad industrial base. Everything here depends on one factory, but the Americans have hundreds of firms and a strong base.'

"(That answer was of course of no interest to me.) He gave me his hand. It was chubby, somewhat damp, and deathly cold. It was then that I realized that I was talking face to face with a dangerous individual."[57]

This story has some interesting facts. Sakharov asserts that the Soviet Atomic Project was a dumping ground for old NKVD folks, which could not have made life any easier for the workers. Beria's close attention to detail, and his ability to judge the character of individuals are demonstrated by this interview. Beria responded to a letter from a sailor in the Pacific Fleet – he must have received hundreds of letters a day – and sent it to Sakharov, who was a rapidly rising star, but still a very junior researcher. Both were invited to a personal interview, not with one of

---

[57] Andrei Sakharov, "Vospominanya," *ibid.*, p 150-151. Translated by the author.

Beria's assistants, but with Beria himself. Beria had control over millions of people, and a substantial part of the economy of the Soviet Union, and yet he spent an hour or so with a young sailor and a promising young physicist. It is interesting that Sakharov and his fellow students discussed the problems associated with Soviet R&D, and came up with their own reasons for them. Sakharov knew very well whom he was dealing with, but it was only in parting that he fully realized that he was in the presence of supreme evil.

Beria's companion, Vasily A. Makhnev, was also present during the episode with Aleksandrov. We have met Makhnev before, in Chapters 7 and 8, where he was involved in the Russian 'Alsos' mission into Germany to look for signs of the German atomic project, and as a minister closely following the progress of Laboratory #2. Sakharov exposed Makhnev's sinister side – prison camp commandant – but like many other people in this history he was also intelligent and resourceful.

Milovan Djilas also recorded his impressions of Beria. Like Svetlana, Djilas found Beria offensive. Djilas first met Beria at one of Stalin's long, late dinners, where Beria drank too much vodka. At the second meeting, also a dinner, Djilas paid closer attention to Beria. He writes:

"Beria was also a rather short man – in Stalin's Politburo there was hardly anyone taller than himself. He, too, was somewhat plump, greenish pale, and with soft damp hands. With his square-cut mouth and bulging eyes behind his pince-nez, he suddenly reminded me of Vujkovic, one of the chiefs of the Belgrade Royal Police who specialized in torturing Communists. It took an effort to dispel the unpleasant comparison, which was all the more nagging because the similarity extended even to his expression – that of a certain satisfaction and irony mingled with a clerk's obsequiousness and solicitude. Beria was a Georgian, like Stalin, but one could not tell this at all by the looks of him. Georgians are generally bony and dark."[58]

Djilas, like Sakharov, comments on Beria's hands. The greenish pale complexion matches the description of the 'Kremlin complexion' mentioned by Djilas earlier. It is not unusual for executioners to look like

---

[58] Milovan Djilas, "Conversations with Stalin," *ibid.*, p 108.

school teachers, or low level bureaucrats. Djilas did not attend a meeting in Beria's office, so his exposure was limited.

9.2.4. *Riehl and Beria*

The German chemist Nikolaus Riehl attended several high level meetings during his ten years in the Soviet Union from 1945-1955. Many of these meetings were at his own laboratory at Elektrostal.[59] Riehl's direct contact was one of Beria's deputies, Avraami Zavenyagin. One day in January, 1946, when the uranium processing at Elektrostal was full of difficulties, Zavenyagin paid a visit. Sudoplatov wrote about abusive language in the Soviet Atomic Project committee meetings. Zavenyagin's inquest into the failures of Elektrostal to deliver uranium was filled with the most abusive and insulting forms of invective. To quote Riehl:

> "The quality of the disciplinary criticisms we received was beyond Western standards as Zavenyagin directed his words, one by one, to the entire group: the director of the factory, the head engineer, the NKVD general in charge of the construction workers, the leader of the group directing the program from Moscow, the division leader in charge of procurement, and so forth. It was clear that I [Riehl] was to be included in all of this. The tone of the criticism was unbelievable. Zavenyagin addressed the individuals in the most vulgar manner of which the Russian language is capable, irrespective of rank or age. The mildest of these curses would have been judged a major insult in any Western democratic country."[60]

According to Riehl the Russian slang for this sort of treatment is a "mordoboy," which could be translated 'in your face.' Since the Soviet Atomic Project had many setbacks and difficulties along the way, mordoboys must have been frequent, although we do not have many descriptions of them. Riehl spares us the expressions used, but the Soviets were more accustomed to this type of discourse, and consequently would not have been as upset. Subsequently, when too high a concentration of boron appeared in the uranium that Elektrostal was routinely delivering to Laboratory #2, Zavenyagin appeared again, and threatened the chief

---

[59] Nikolaus Riehl and Frederick Seitz, "Stalin's Captive," *ibid.*, p 105ff.
[60] Nikolaus Riehl and Frederick Seitz, "Stalin's Captive," *ibid.*, p 107.

engineer with imprisonment if the situation did not improve quickly. This is reminiscent of Beria's threat to pack Aleksandrov off to God knows where if his factory exploded. Such was standard procedure.

Riehl had two encounters with Beria. One was when he and other German scientists had just arrived in the Soviet Union, and Beria invited them to his office to get acquainted. Igor Kurchatov was there. The meeting was cordial. The second time Beria came out to Elektrostal to visit the factory after uranium production had become a reality. Beria arrived in a motorcade of about 15 cars from Moscow. After some presentations regarding the status of production, Beria asked if the German group had any complaints or questions. Riehl grumbled about the restrictions on travel and the ever present security, without any positive response from Beria, and afterwards the meeting disbanded.[61]

Beria had a reputation during his lifetime for womanizing. He was accused of using his position of power to seduce young women, with particular attention to female athletes. There was enough of a buzz about this around town that there must have been some truth to it.[62] Stalin, who had a reputation for disinterest in such activities, must have known of Beria's dalliances, but chose to look the other way. In July, 1953, after Beria's arrest, evidence for his infidelity was presented at a meeting of the Communist Party Central Committee, based on testimony by one of Beria's body guards, a certain Sarkisov.[63] Beria's supporters claimed that the accusations were overblown, that Beria was too busy to have a lot of time to fool around. On the other hand, there was proof of at least one indiscretion. Before his death, Beria confessed to his son Sergo that there was a half-sister born out of wedlock.[64] Sergo subsequently became acquainted with her. She married twice, and had two children. Her life was unremarkable, and apparently not scarred by her birthright. Such events are not uncommon – many famous people in history were born out of

---

[61] NIkolaus Riehl and Frederick Seitz, "Stalin's Captive," *ibid.*, p 115.
[62] Amy Knight, "Beria," *ibid.*, p 97.
[63] Sergo Beria, "Moi Otets Narkom Beria," *ibid.*, p 24-25.
[64] Sergo Beria, "Moi Otets Narkom Beria," *ibid.*, p 28.

wedlock. On the other hand, Beria had enormous power, and the use of power for such activities is not acceptable.

In summary, Stalin and Beria had much in common. Both were Georgians, had sharp minds and good memories, paid close attention to detail, were completely ruthless, and were indispensable to the Soviet Atomic Project. Without the close attention of both of them, particularly Beria, the Project would certainly have dragged out much longer than it did, and may not have succeeded at all. Stalin and Beria were men of their times – hard to love, but impossible to do without.

9.3. Igor Kurchatov

Figure 9.3. Igor Kurchatov in the garden of his home at Laboratory #2. Courtesy AIP Emilio Segre Visual Archives, Ioffe Physico-Technical Institute.

9.3.1. *Work Habits*

Igor Kurchatov's early life and education have been described in earlier chapters. Kurchatov had to make it all work. He had to procure materials, build factories, attend an infinite number of contentious meetings, and somehow keep moving forward on an immense and complex project, full of all kinds of dangers. He is shown later in life in the garden of his dacha in Figure 9.3. He was a skilled experimental physicist, but no genius. His pre-war and wartime work demonstrated competence, and contributed to his reputation for reliability and resourcefulness. Kurchatov's supervision of the work of Flerov and Petrzhak on the spontaneous fission of uranium was a good example of his thoroughness in experimental procedure – careful cross checks, background studies, etc. He maintained these standards, and an interest in the phenomena of nuclear physics – little puzzles and big puzzles - throughout the Soviet Atomic Project. He also gained valuable experience in working with Soviet industries and ministries. While all positive, his resume hardly served as proof of the administrative skills necessary to run the Soviet Atomic Project. In this sense, he and Oppenheimer had a lot in common – someone in authority gambled that they were the right people for the job.

Like most of his contemporaries with management responsibilities in Stalin's Soviet Union, Igor Kurchatov was a workaholic, who rarely took a day off, did not get enough exercise, smoked too many cigarettes (but drank alcohol sparingly), and died relatively young (57). He was constantly on the phone, in meetings, or on the road to inspect progress at the plants and laboratories of his widely scattered project. He hardly had time to think, but surprisingly he seems to have done so – keeping up with developments in physics and technology, and having ideas for directions for further R&D. As the Atomic Project began to achieve its initial goals, Kurchatov became interested in controlled fusion power. Controlled fusion got started at Laboratory #2 in the early 1950's. All work at any of the laboratories was classified secret. Kurchatov managed to convince Khrushchev that there was no need to classify controlled fusion, and when the two went to England in 1956, he successfully convinced the British scientists to declassify their work as well. Controlled fusion work in the USA was also secret, and subsequently opened. This is an example of Kurchatov's skill in manipulation of the bureaucracy, and his world view of free exchange of scientific information for peaceful uses.

Kurchatov and Igor Tamm became interested in biology through the biological effects of radiation, and he was excited about the double helix of Crick and Watson when it was announced in 1953. He encouraged biological work at Laboratory #2, and was firmly opposed to 'Lysenkoism.' Sakharov thought that the Kurchatov biology initiative at Laboratory #2 was a good example of his stature in the government – most people would not have been able to oppose Lysenko at that time.[65]

Kurchatov was patient and persistent. He questioned progress reports in detail. He did not suffer fools lightly, and quickly squelched any idle chatter. His interest in biology exposed him to the Lysenko affair, and he had absolutely no use for Trofim Lysenko. We have already discussed the Moscow State University opposition to 'western physics' – relativity and quantum mechanics. By the 1940's these theories were well established experimentally, and not subject to philosophical debate. Whether the members of this opposition were stupid or merely misinformed, Kurchatov had no patience with them whatsoever.

Yakov Zeldovich, an important theorist in the Soviet Atomic Project, who later made seminal contributions to astrophysics, commented on Kurchatov's interest in theoretical physics.[66] Zeldovich placed experimentalists in two distinct categories with respect to their attitudes regarding theory. "One group," says Zeldovich, "follows the old saying: 'Pay theorists the same attention as you would your wife. Ask for advice, and then do the opposite.'"[67] These people share the point of view that too much attention to theoretical speculation could deprive experimentalists of their own originality of thought. They especially value experiments that disprove current theory. Important discoveries are often unanticipated. On the other hand, there are experimenters at the other extreme, who ask what are the effects that are predicted by theory, and should be verified by experiment. Experiments of this nature are easier to justify. Kurchatov managed to avoid either extreme. He showed an interest in theory, and shared the satisfaction of experimental confirmation of theoretical calculations. One example given by Zeldovich was the design of the F-1

---

[65] Andrei Sakharov, "Vospominaniya," Alfa-Kniga, Moskva, 2011, p 118.

[66] Ya. B. Zeldovich in Raisa Kusnetsova, editor "Kurchatov v Zhizni," Rossiiskii Nauchny Tsentr 'Kurchatovskii Institut," Moskva, 2007, p 466.

[67] The author is quoting Zeldovich, and does not necessarily endorse this advice.

reactor, where the volume and purity of graphite moderator, the placement of uranium slugs in the lattice, and the amount and purity of uranium in each slug all were calculated by complicated numerical methods, using as input cross sections measured in the laboratory, before the reactor was built. Successful operation of the reactor as designed confirmed the theoretical analysis, to the delight of all concerned, including Kurchatov.

Associates referred to Kurchatov as 'the Beard,' a nickname in common use during the Atomic Project. Kurchatov grew a beard when he was ill in Kazan during the war, and did not shave it off. He said that he would shave after the successful test of the Soviet atomic bomb in 1949, but did not do so, saying that if he did, he would no longer be 'the Beard.' Aleksandrov gave him a new razor and a bowl full of shaving cream, to no avail. The razor is part of the exhibit of personal effects at the Kurchatov home on the Laboratory grounds. Beards were not that common among men in the USSR, in contrast to tsarist Russia.

9.3.2. *Kurchatov and a Post-doc, Mostovoy*

V.I. Mostovoy, whom we have already met in Chapter 7, was a Red Army veteran and Hero of the Soviet Union. He started graduate school in nuclear physics in Kiev in 1946, working under Leipunsky.[68] Leipunsky recommended in 1947 that Mostovoy go to Moscow for a month or so of practical work in Kurchatov's Laboratory #2. Mostovoy was given the phone number of a contact at the Laboratory named Pevzner. Similar to Djilas' experience in trying to see Stalin, Mostovoy was delayed, and sat around for about a month before he acquired security clearance and got permission to enter Laboratory grounds. After the Laboratory gates he went through several guard stations before he arrived on the third floor of the main building, where he was warmly greeted by Pevzner. Pevzner's group worked in radiochemistry, headed by Igor Kurchatov's brother, Boris. The F-1 uranium graphite reactor was operating, supplying external thermal neutrons for research, and bombarding the natural uranium in the reactor to produce tiny amounts of ^{239}Pu for chemical extraction studies.

---

[68] V.I. Mostovoy in Raisa Kuznetsova, editor "Kurchatov v Zhizni," *ibid.*, p 522.

Pevzner's group measured the amount of plutonium by observing its decay $-$ ^{239}Pu $\rightarrow$ ^{235}U $+\ \alpha$ (half-life 24,100 years). After three days in Pevzner's lab Mostovoy was invited to meet the Laboratory director, Igor Kurchatov, in his office. Mostovoy described his thoughts prior to the meeting:

"The closer we got to the time of the meeting, the more anxious I became. What would he ask me, and how would I answer? I already knew who Igor Kurchatov was. He was the front commander. Yes, the front – unseen and unheard, but still the front. The battle after our triumphal victory in WWII was proceeding in Kurchatov's laboratory. The battle with the secrets of the atomic nucleus, that had to be grasped and put to work in the defense of our country, blackmailed by Truman and his military advisors with the atomic bomb. Igor Vasilevich commanded the front lines! How would a meeting unfold with a former Red Army soldier, who has only briefly been on the scientific front? I had never before met such a famous person."

Pevzner and Mostovoy did not have far to go. Kurchatov's office was also on the third floor of the main building. They went into the reception room, and were told that Kurchatov was expecting them. Mostovoy described his impressions:

"A powerful looking man was standing behind a writing desk, a man like Ivan Susanin (a 17th century Russian hero). He came straight towards us, smiling happily. It was not only an external smile, with his face, but also internal, with his eyes. His striking eyes, happy and penetrating, made an immediate impression. We met in the middle of the room. Igor Vasilevich greeted us and we shook hands. He invited us to sit down in soft armchairs next to his desk. My anxiety disappeared, and I completely relaxed. Kurchatov asked me when I would graduate, and what I had been doing. Then he started describing the work with the uranium-graphite reactor, and the production and extraction of micrograms of ^{239}Pu, and the measurements that had to be made – fission of ^{239}Pu and ^{233}U by thermal neutrons, and comparison to ^{235}U. We briefly discussed how to do this. Then Kurchatov briskly stood up, and just as briskly said goodbye. Pevzner and I left the office. The meeting lasted all of 15-20 minutes, but I will remember it all my life. My scientific fate was determined. The personality of Igor Vasilevich seemed somewhat fantastic, embodying the strength and intelligence of the Russian people."

This is a nice description of Kurchatov's daily interactions as seen by a raw recruit in the army of the Soviet Atomic Project. Igor Vasilevich's people skills spanned the spectrum of everybody, from Stalin and Beria to the chauffeur who drove his car, and the bodyguard who was with him constantly.

9.3.3. *Kurchatov the Peace Maker*

The Soviet Atomic Project must have had many internal conflicts. Choices had to be made on a daily basis, resulting in winners and losers. The losers were often necessary for the project, and had to be comforted and restored to duty. There were people who thought they had good ideas that were not given the attention and support they deserved by the management. Such conflicts, if important enough, inevitably led to some sort of peace keeping effort on the part of Kurchatov. One example is described by Georgy Flerov.[69] A theorist whom Flerov called 'N' proposed a cheap and quick means for isotope separation. Flerov and Artsimovich did some calculations to show that it would not work. In the meeting with 'N' Kurchatov said that it was a terrific idea, and that it should be developed into a specific proposal for construction of industrial plants to carry out the separation. Would 'N' kindly furnish such a proposal in two weeks? After two weeks 'N' said that he was working on it, but he would need six more months. After six months 'N' had faded away. By such manipulations at least some 'N's were shunted off to oblivion, but that technique did not work for everyone – especially for those whose skills were needed. It was obviously advantageous if at all possible to resolve technical conflicts before presenting anything to Beria, and ultimately this was Kurchatov's responsibility.

Beria had assured the scientists working on the Soviet Atomic Project that they would be protected from harassment by the NKVD. He kept his word, but he had to be reminded of it on a regular basis. Kurchatov was usually the one to do the reminding. Occasionally Kurchatov appealed to Stalin. If the residence of a scientist had a telephone, it was bugged. Everyone was watched, and informers were everywhere. This situation was generally accepted as normal. Travel was forbidden without special permission, and no one associated with secret work was allowed abroad.

---

[69] G.N. Flerov in Raisa Kuznetsova, editor, "Kurchatov v Zhizni," *ibid.*, p 471.

Khariton was reminded of the surveillance at a birthday party for Kurchatov at his home. The 'sword of Damocles' was always hanging over his head. Kurchatov's wife was aware of the NKVD's presence. It is probably apocryphal, but it has been said that Beria took with him to Semipalatinsk for the first Soviet atomic test in 1949 two sets of orders. One set, in the event of success, was a collection of medals and awards for all concerned. The other set, in the event of failure, was a group of arrest warrants. Beria is quoted as saying to Kurchatov after the successful test: "If it had not succeeded, there would have been terrible misfortune."[70] It did succeed, and medals and awards were the result. Had it failed, there would have been consequences, but one wonders how many people would have been arrested, since a successful atomic weapon was still essential, and no one else could do it.

In the Soviet Union, as in many western countries, people of Jewish extraction occupied a number of high positions disproportionate to their fraction of the population. The Soviet Atomic Project was no exception. Indeed, many of the theorists of the Landau School, including Landau himself, were Jewish. There was no official policy of discrimination against them, but from time to time they were singled out as a class for being 'unreliable.' Igor Golovin described such instances in the scientific personnel of Laboratory #2. In 1949, after the successful bomb test, two workers in Artsimovich's division, a brother and sister named Braverman, were arrested as 'spies.' The laboratory director, Kurchatov, intervened, and they were released. Then in 1951 Beria revoked the security clearance for two future members of the Academy of Sciences, A.B. Migdal and A.G. Budker.[71] Without security clearances the two could not work. Kurchatov and Golovin, his deputy, wrote a letter to Beria pointing out that without these scientific experts the development work on controlled

---

[70] Raisa Kuznetsova, editor, "Kurchatov v Zhizni," *ibid.*, p 42.

[71] The XX Party Congress in 1956 created 'Academy Town' in Novosibirsk to help develop the economy of Siberia. Budker took the job of director of the Physics Laboratory of the Siberian Division of the Academy of Sciences. The author met him there in 1976. Budker said jokingly that he was the only Jew in Siberia not in a prison camp.

nuclear fusion, approved by Stalin, would not proceed. Their rights to classified materials were restored.

One of Kurchatov's many duties was to keep his team together against periodic invasions by the NKVD.

9.3.4. *Kurchatov the Trouble Shooter*

Beria for the most part stayed in Moscow, only making site visits occasionally. Not so Kurchatov, who was constantly on the move, and who made extended visits to remote sites when the situation warranted. Kurchatov was an experimental nuclear physicist, and his training showed in his response to difficult technical problems that arose in the course of the project.

B.V. Brokhovich, the chief reactor engineer at Chelyabinsk-40 (see Chapter 12), recounts problems with the start-up of reactor 'A,' the first plutonium production reactor of the Project.[72] Chelyabinsk-40, also known as Combine #817, is now called 'Ozersk,' and is a town east of the Ural Mountains, about half way between Sverdlovsk in the north, and Chelyabinsk in the south. (See Map #3). Chelyabinsk is 1500 km east of Moscow. Sverdlovsk is famous for the execution of the family of Tsar Nikolas II. The choice of location for the reactor complex was made partly for its remoteness, and to begin with there was nothing there – not even a railroad line. As was true of many of the Atomic Project facilities, the construction was done by prison labor and Red Army recruits. The infrastructure was still primitive in 1947, when assembly began on reactor 'A,' a graphite moderator natural uranium power reactor, with channels to insert and extract uranium slugs for processing after neutron irradiation to obtain ^{239}Pu. Boris Vannikov recounted some of the living conditions, shared by himself and Kurchatov, in his oral history.[73] The reactors were located 17 km from the town, and Vannikov did not want to drive 34 km back and forth every day, so he had a branch of the railroad line extended

---

[72] Raisa Kuznetsova, editor, "Kurchatov v Zhizni," *ibid.*, p 499.
[73] B.L. Vannikov in Raisa Kuznetsova, editor, "Kurchatov v Zhizni," *ibid.*, p 476ff.

as far as the first reactor, and lived there in a railroad car. Kurchatov could have lived in town, but he chose to join Vannikov in the rail car, where the amenities were limited, and there was not a whole lot of heat. In the winter it routinely dropped below freezing in the car over night, but Kurchatov was not phased in least bit. Later on they set up two 'Finnish houses' in the woods nearby.[74]

Reactors similar to those at Ozersk had been successfully operated at Hanford, Washington in the Manhattan Project. Kurchatov was on hand when reactor 'A' was commissioned in the summer of 1948. There were problems from the start with the cooling water, and over-heating of the uranium slugs. Since uranium can burn – as demonstrated by the Windscale accident in Britain,[75] and the tent that burned down at Laboratory #2 – this was potentially a dangerous situation, and Kurchatov played a hands-on role in working around the problems, and starting plutonium production. He was on site for several months, under substantial strain to get the reactor up and running. Kurchatov and Vannikov played card games late into the night in order to relax. Kurchatov often took a nap after lunch. But he stuck with it until the crisis was past, showing good crisis management style - he was there in person.

The Soviet Atomic Project employed thousands of technically trained people. Laboratory #2 was in Moscow, but all of the other centers, Arzamas 16, Chelyabinsk 40, Kyshtym, Semipalatinsk, etc were remote, for reasons of safety and security. Scientists at these sites were effectively confined to quarters, unable to travel either internally or abroad without special authorization, and many remained in remote laboratory towns after retirement. They spent their lives there. One of Kurchatov's skills was in arm twisting – to get people that he deemed essential to give up what they were doing and join his operation at some god-forsaken place in the north woods. Yefim Slavsky, the metallurgist in aluminum refining who worked

---

[74] 'Finnish houses' are prefabricated wood frame houses of various designs, still available today. See current models displayed on www.finskydom.com. They probably were not that attractive in the 1940's. German 'visitors' had similar housing.

[75] See footnote 92 Chapter 8.

with the Moscow Electrode Factory, was one example of someone recruited by Kurchatov against his better judgment. But there were many others.

Many of Kurchatov's colleagues have commented on the use of prison labor. Boris Vannikov: "All of these highly secret buildings and equipment were built by prisoners. Only prisoners with the longest sentences were sent to the most secret places. Civilians were not used, since they could leave at any time, and carry off state secrets."[76] Yefim Slavsky: "Prisoners and soldiers for the most part built the complex. Civilians were minimal. The fate of all who worked there was planned in advance. That is why I am saying that people were afraid to work for the project. After the successful completion of our assigned tasks we were required to remain in the town that we had built in such a hurry. The whole team would have to live and work there without permission to leave. The prisoners would be sent north to live and work in the camps. The predetermined fate of the work force was not known to us until after Beria's execution."[77] They do not say, but other evidence points towards use of German POW's in addition to Soviet political prisoners and criminals to do construction labor. Prison camps of both kinds were widely distributed in the post-war Soviet Union.

Pavel Sudoplatov, Beria's go-to guy in the NKVD, knew Kurchatov well.[78] Sudoplatov observed that Kurchatov was a new version of Soviet scientist, different from Kapitza or Ioffe. Kurchatov was willing to give up scientific freedom and independence for the benefit of the state. Sudoplatov noted that it would not be possible for a scientist to be independent while working on a project of such national importance, and that all of the technical workers on the Atomic Project had to submit to severe restrictions by the government. As scientific director of the Project, Kurchatov had to have the administrative skills of a government minister, in addition to his scientific training. It was only in the late 1950's,

---

[76] Boris Vannikov, quoted in Raisa Kuznetsova, "Kurchatov v Zhizni," *ibid.* p 478.
[77] Yefim Slavsky, quoted in Raisa Kuznetsova, "Kurchatov v Zhizni," *ibid.,* p 488.
[78] Pavel Sudoplatov, "Razvedka I Kreml," *ibid.,* p 248-249.

however, that an extensive scientific bureaucracy developed to make life even more complicated. So there were some advantages to the beginning.

Igor Kurchatov was a hero of his time. He was a nuclear physicist, a project organizer, skilled in using the Communist system to his advantage, able to motivate and support his scientific and technical personnel, as well as keep them out of jail, able to tolerate the abuse and personal threats inherent in doing business with the Kremlin, and in the end to emerge victorious. He must have had some faults, everyone does, but tributes to his memory are completely devoid of any personal criticism. Unlike other high ranks in the Politburo, he had no responsibility for mass killings or incarcerations. He had to have known, however, that prison slave labor was extensively used in construction of his project, and in some instances for dangerous work in the factories. He absorbed this as a necessary evil – a resource indispensable for success.

Albert Speer made use of prison slave labor in German armament factories during WWII. He was Minister of Armaments and War Production for the Nazi Third Reich. He managed to maintain production despite the bombing, and the occupation of German territory until the bitter end. Use of slave labor was one of the charges against Speer at the Nuremburg war crimes trials in 1945.[79] While the Geneva Convention for the treatment of prisoners of war permits the use of prison labor, it explicitly forbids work in any way connected with armaments manufacture, or other activities necessary for the conduct of hostilities.[80] His sentence was mitigated by his efforts to improve living conditions for the prisoners. Speer was sentenced to 20 years in prison, and served the full time.

---

[79] Albert Speer, "Inside the Third Reich," *ibid.*, "Introduction," p xi.

[80] International Committee of the Red Cross, "Geneva Convention for the treatment of prisoners of war, 29 July 1929," Articles 27-34. The Soviet Union made extensive use of German POW labor, during and after WWII, often alongside Red Army recruits, or Soviet citizen prisoners – circumstances that could be consistent with the Geneva accords.

Igor Kurchatov suffered no such fate. He was honored by his government and people for heroic service. The Moscow laboratory that he founded has been named after him. The dacha built on the laboratory grounds for him has been preserved as a museum.

Chapter 10

Industrial Plants Move
to the Urals

10.1. The Soviet Project Outgrows the Moscow Region

Laboratory #2 was the center for personnel, and consequently also for research equipment and experimental and theoretical activity on the Soviet Atomic Project during its rapid expansion from 1945-1950. The analog for the Manhattan Project was perhaps the Metallurgical Laboratory at the University of Chicago. The Soviet leadership in the PGU (Beria's Primary Management Committee that had full powers in the Atomic Project) soon came to the same conclusion as the Manhattan Project that very large and expensive plants and factories were required to produce the necessary quantities of fissile material, either the isotope ^{235}U separated from natural uranium, or the new element ^{239}Pu made by neutron capture on ^{238}U. Several schemes for isotope separation were under test at Laboratory #2, using industrial facilities in the Moscow and Leningrad regions. Isaak Kikoin led the gaseous diffusion effort; Lev Artsimovich led the electromagnetic separation effort; a Soviet physicist of German origin Fritz Lange had been working for several years on centrifuges for isotope separation. German guest scientists at Sukhumi would contribute to each of these tasks. Igor Kurchatov himself would lead the construction of high power reactors for plutonium production. It was recognized from the early days that isotope separation might be much more difficult than plutonium production, and for that reason a plutonium bomb could be ready first.[1]

---

[1] Atomnyy_proekt_SSSR.T.1.Ch.2.(2002).pdf #233 Kurchatov to Pervukhin, May 18, 1944.

Mining and purifying tons of uranium, and manufacture of heavy water for power reactors were both constant problems in the background as the new plants were built.

The industrial plants were to be constructed along the line east of the Ural Mountains near Sverdlovsk in the north and Chelyabinsk in the south – isotope separation around Sverdlovsk, and power reactors near Chelyabinsk. This region had been heavily industrialized for armaments production during WWII, and had good electric power facilities and rail links, in addition to the remoteness required by the secrecy of the project. The names of the plants were ambiguous for security reasons. Sverdlovsk-44 included the gaseous diffusion plant #813, and the electromagnetic separation plant #814. Chelyabinsk-40 contained the 'A' power reactor and chemical plant for Combine #817.

10.2. Isaak Kikoin and Isotope Separation

Isaak Kikoin was born in 1908 in a village near Vilnius, the capital of Lithuania. His parents were school teachers. He enrolled in the Leningrad Polytechnic Institute in 1925, and in 1930 began doctoral work with A.F. Ioffe at the Leningrad Physico-Technical Institute. He studied the Hall effect in ferrous metals, and discovered an anomalous effect that is named after him.[2] In 1935 Kikoin moved to Sverdlovsk, where he became acquainted with Igor Kurchatov, and with the scientific and industrial capabilities of the Urals. He worked at the Urals Physico-Technical Institute (another UFTI, not to be confused with the one in Karkhov). During the war, he worked on land mines that would detonate when a tank or other iron object disturbed the local earth magnetic field. Thus Kurchatov and Kikoin both worked on magnetic mines – Kurchatov demagnitized ships so that the underwater mines would not sense their presence, and Kikoin designed land mines that would require demagnitization of German tanks. When Laboratory #2 got underway in

---

[2]   An external magnetic field perpendicular to a current carrying wire creates an electric field perpendicular to both the current flow and the magnetic field. The voltage is called the 'Hall voltage,' and is widely used to measure magnetic fields.

1943, Kurchatov chose Kikoin as his deputy, one of the first members of the staff, based on the Leningrad connection. Whether Kikoin chose isotope separation as his task in the project, or it chose him, is not clear, but in either case the two were joined for the duration.

Isotope separation was one of the major initial tasks of Laboratory #2 and the Project. Natural uranium contains only 0.7% of the isotope ^{235}U, and obtaining kilogram amounts of the fissile isotope was, and still is, a technological challenge. This is fortunate, for otherwise the proliferation of nuclear weapons would be worse than it is. Weapons grade uranium is about 90% pure ^{235}U, and the fractional difference in mass – the only handle on the separation – is only 0.013, so that achieving 60 kg of ^{235}U for a bomb is not easy.

The German scientists who joined the Project in 1945 preferred to help out in support tasks rather than work directly on nuclear weapons. The groups under Manfred von Ardenne and Gustav Hertz[3] worked on isotope separation at their respective laboratories in Sukhumi. Hertz had worked on isotope separation at Siemens in Berlin during the war, so he was continuing his earlier experiments. Gaseous diffusion is a colossal industrial undertaking, and doubtless attracted German technical skills for that reason. A deputy to Gustav Hertz named Heinz Barwich wrote a book about isotope separation in the Soviet Atomic Project from the perspective of a German visitor.[4] Like most of the German scientists, he spent 10 years from 1945-1955 in the Soviet Union. Barwich had been one of Hertz's graduate students and his doctoral thesis was on gaseous diffusion, so he was a natural choice to join the German delegation.

Kikoin was in charge of the problem at Laboratory #2. Many aspects of the Soviet Atomic Project stretched the capabilities of the Soviet heavy industry and economy, but probably no part of the enterprise was more

---

[3] Von Ardenne was a distinguished German inventor. Hertz won the Nobel Prize in physics in 1925. He worked for Siemens in Berlin until 1945.

[4] Heinz and Elfi Barwich, "Das Rote Atom," Scherz Verlag, Munich, 1967. Heinz Barwich was later deputy director of JINR at Dubna.

taxing than isotope separation. Kurchatov and Kikoin initially followed the lead of the Manhattan Project, and developed both gaseous diffusion and electromagnetic separation. These techniques are discussed in Appendix C, where it is shown that gaseous diffusion requires more than one thousand stages with compressors and membranes, all handling a chemically corrosive gas, UF_6. The first job was to make the uranium hexafluoride from uranium oxide. The second problem was to design and build the compressors, membranes, and plumbing in the required quantity, and build a factory. Any surface in the plant that came in contact with UF_6 had to be scrupulously clean to avoid chemical reactions. The Manhattan Project criterion was 'as inert as stainless steel in air.'[5] Scientific equipment was moved from Germany to the new labs at Sukhumi, but much German industry remained at home, and contributed technology such as special alloys and sophisticated chemical processes to the Soviet Atomic Project. German industrial equipment was acquired through war reparations.[6]

10.3. The Working Gas

Isotope separation by gaseous diffusion requires a uranium gas, and UF_6 or uranium hexafluoride, was the gas of choice. At room temperature and atmospheric pressure (STP) uranium hexafluoride is a solid. At one atmosphere pressure it sublimes to the gas phase above 56.5 ^{0}C. The triple point is at a pressure of 1.5 atmospheres, and a temperature of 64.05 ^{0}C. So temperature and pressure of the gas have to be carefully monitored to retain the gas phase. The feed gas pressure is below atmospheric, and after passage through the barrier the enriched gas pressure is even lower, so the entire system is a partial vacuum. Uranium hexafluoride is toxic, corrosive, and very heavy in addition to being radioactive and expensive – a thoroughly unattractive substance.

---

[5]   P.C. Keith, "Chemical and Metallurgical Engineering," February, 1946, p 112.

[6]   A.G. Plotkina and E.M. Voinov, "Academician Isaak Kikoin," HISAP96, Vol **2**, *ibid.*, p 195.

10.4. Measurement of ^{235}U Content

There are several ways to measure isotope enrichment. Both isotopes α decay to thorium. The transition energies are similar, but an ion chamber with good resolution could distinguish the two. The α from ^{235}U is about 10% more energetic.[7] Mass spectrometry could be used – an assay application of the electromagnetic separation technique. Von Ardenne's group at Sukhumi developed instruments for each of these measurements: a 1% precision proportional counter to measure α particle energy, and a stable precision mass spectrograph for heavy elements.[8] These methods were used in the early days of enrichment, before things became routine.

The standard method to measure isotope enrichment of choice, adopted both by the Soviets and the Manhattan Project, was to observe a 186 keV γ ray that accompanies the α decay to ^{231}Th about half of the time. The decay rate for ^{235}U $\rightarrow$ ^{231}Th $+\alpha + \gamma = 4.3 \times 10^4$/sec-gm, while the total decay rate $= 8 \times 10^4$/sec-gm. This relatively soft γ ray can be observed in situ, giving a measure of the purity of the sample in real time.[9]

10.5. Plant Design

Early in 1944 Kikoin and Alikhanov wrote a memorandum on isotope separation to Pervukhin.[10] They understood from their own work, and the information gained by intelligence, that isotope separation by gaseous diffusion was an industrial-technical problem, rather than a job for physicists and mathematicians. They listed three technical problems:

1. Design of the diffusion machinery.
2. Development of appropriate barrier on an industrial scale.

---

[7] Decays are: ^{238}U $\rightarrow$ ^{234}Th + 4.3 MeV α; ^{235}U $\rightarrow$ ^{231}Th + 4.7 MeV α. The lifetimes are $\tau = 6.5 \times 10^9$ and 1.0×10^9 years respectively, and the count rates dn/dt $= n/\tau$.

[8] Atomnyy_proekt_SSSR.T.2.Kn.4.(2003).pdf # 239

[9] nrc.gov/regulatory guide 5.21 "non destructive ^{235}U assay by γ ray spectrometry"

[10] work.atomlandonmars.com/APSSSR/Atomnyy_proekt_SSSR.T.1.Ch2.(2002).pdf #200.

3. Chemistry problems associated with production of enough UF_6 from uranium ore, and development of gaskets, grease, and surfaces resistant to corrosion.

The first job had been assigned to the Central Aerodynamic Institute in 1943. Research at the Institute was directed by Academician S.A. Khristianovich, who was an expert in aerodynamics. However, the project did not receive the attention of the Institute's A-team, probably due to heavy commitments for national defense, and got nowhere. Kikoin and Alikhanov wanted to get Khristianovich on board. In addition, they realized that ultimately the project would of necessity come under the complete control of an engineer. Looking over prospects in chemical engineering, they requested for that position a prominent hydro-electric power engineer and corresponding member of the Academy of Sciences, I.N. Voznesensky. Voznesensky was recruited, became one of Kurchatov's deputies, and would be with the project for many years. But it was still necessary to get started on an industrial design for the components of a separation plant – compressors, plumbing, etc. – with anticipation of many problems along the way. A test setup with only a few stages had to be implemented. The test setup had been assigned to the Central Aerodynamic Institute, and Kikoin and Alikhanov were agreeable to leave them the task, provided that progress be made in a timely manner. UF_6 reacts strongly with water vapor[11], and inert materials had to be found as seals and gaskets. Satisfactory performance of a small setup would be a necessary, but not sufficient condition for a large one. Big systems can have their own special problems.

10.6. The Barrier

Manufacture of a suitable barrier, with holes ~ 10 nanometers in diameter for the $^{235}UF_6$ to flow through, leaving the (slightly) more sluggish $^{238}UF_6$ behind, was another major technical problem. The barrier would come in thin sheets of some suitable material, inert to UF_6, covered

---

[11] The reaction is: $UF_6 + 2\times(H_2O) \rightarrow 4\times(HF) + UO_2F_2$. Hydroflouric acid is very disagreeable.

with tiny holes. The separation plant would require about one square kilometer of the stuff. Clearly this was a daunting task. Kikoin's own lab at Laboratory #2 was working on the development. Two methods were mentioned: sintering of metallic powders, and etching an alloy with an acid that preferentially removes one of the elements. Kikoin said that he was on the lookout for anyone who could help out in this endeavor. It is interesting to note that even today, presumably as a deterrent to nuclear weapons proliferation, the design of these barriers in both Russia and the USA remains classified.

In his note to Pervukhin, Kikoin stated that work on the centrifuge method of isotope separation was in progress at his laboratory in Sverdlovsk, and they would soon begin studies with UF_6. He also thanked Pervukhin for assigning competent chemists to the problems of handling the uranium gas.

Voznesinsky was well acquainted with the industrial capability of the Leningrad factories, and in March, 1944, a branch of Laboratory #2 was organized in Leningrad under Kikoin, with Voznesinsky his deputy. The city blockade had only been lifted two months earlier, in January, and conditions must have been very grave, but some research was going on. The Radium Institute had at the end of 1943 prepared 100 gm of UF_6 for Kikoin's studies of the physical and chemical properties of the gas.[12] Chemical plants in the Kirov region supplied industrial quantities of the gas in the late 1940's.[13]

The Leningrad branch was housed at first in the building of the Institute for Chemical Physics (Zeldovich and Khariton's old institute, that had not yet moved back from Kazan). They started work on compressor design, gas flow calculations, and questions of stability and control of the plant. KIkoin remained based in Moscow, but traveled to Leningrad frequently.

---

[12] A.G. Plotkina and E.M. Voinov, "Academician Isaak Kikoin" HISAP96 Vol **2** *ibid.*, p 199. Not clear whether this was done in Leningrad or Kazan, since the Institute was in both places.

[13] Kirov is 896 km northwest of Moscow, on the Vyatka River.

Early in 1945, however, when Beria assumed control of the project, the gaseous diffusion effort was concentrated in Moscow. Kikoin was not enthusiastic about the move, but his group transferred from Leningrad to Section 2 (Kikoin's operation) at Laboratory #2. Voznesinsky's group remained in Leningrad for the time being. The Elektrosila plant in Leningrad had a 220 ton electromagnet, suitable for ion source and collector studies for electromagnetic separation. This facility was useful for Lev Artsimovich and Manfred von Ardenne.

At that time Section 2 in Moscow had almost 100 people, counting scientists, engineers, technicians, and office workers. In an interview much later, Kikoin recalled the environment:

> "The main laboratory building was completely refurbished in the course of a few months. The best equipment available at the time, partly Soviet made, and partly from USA Lend-lease, was installed in the work areas. A large experimental hall was built attached to the machine shop. We acquired a group of highly qualified workers from other industries – the aircraft industry in particular. We also had some lab equipment as war reparations from the Kaiser Wilhelm Institute in Germany."

Work began on barrier design at Section 2 at the end of 1945. The first attempt was made by poking small holes in a copper foil with a bundle of needles. These holes were too large to support the necessary pressure differential, but could be used in early tests. Design help was sought from various sources: Moscow Combine for ferrous alloys; the Elektrostal plant (Riehl's factory); and the German scientists at Sukhumi (Manfred von Ardenne, Peter Thiessen, and Gustav Hertz). The Moscow Combine produced the first sample, a flat sheet 120×70 mm. These filters were used in the first diffusion plant built, D1, at Sverdlovsk-44, but the gas throughput was limited because the sheets could only stand a pressure of about 3% of an atmosphere. Peter Thiessen produced better filters that could stand higher pressures in the form of a 15 mm diameter cylinder with gas flow stabilizers.[14] The diffusion chamber was an array of these

---

[14] It is possible that the cylindrical design came from the Manhattan Project via intelligence.

small cylinders stacked in parallel to increase the total area of the barrier.[15] A sketch of the bundle of tubes is shown in Figure 10.1. The array contained several hundred 2 meter long tubes. Uranium hexafluoride gas at higher pressure entered the barrier cylinders in parallel, and lower pressure slightly enriched gas flowed out, to be repressurized and sent to the next stage. Pressures as high as 13% of an atmosphere could be held across the barrier. Each kg of natural uranium contains seven gm of ^{235}U, so as the enrichment process continued, the amount of gas to be handled could decrease (some of the ^{238}U had been discarded), and the scale of the apparatus shrank accordingly. However, the startup was slow, so a large number of full scale stages were needed – in the thousands. Kikoin's early planar filters were gradually swapped out for Thiessen's cylindrical arrays as other machinery in the plant was replaced, and gaseous diffusion became fully operational.

---

[15] V.N. Prusakov and A.A. Sazykin, "Uranium enrichment in the USSR," HISAP96, Vol 1, *ibid.*, p 166 Fig 2 is a photograph of the parallel array of barrier tubes.

DIFFUSION BARRIER

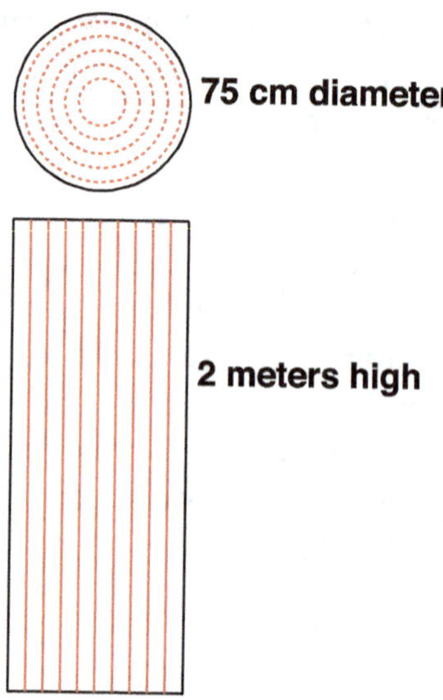

Figure 10.1. Schematic of a bundle of diffusion barrier tubes arranged in a two meter high cylindrical array. Groups of tubes are placed at each radius, giving an area of several square meters of barrier in each station.

Development of cylindrical barriers proceeded on several fronts. In the early 1950's double layer cylindrical filters were supplied to Sverdlovsk-44 that could support a pressure of 26% of an atmosphere. Work on resistance to corrosion by UF_6 was ongoing.

10.7. Manhattan Project Solutions

Kikoin was aware of the espionage information flowing into NKVD headquarters from abroad. Chapter 8 is devoted to this subject, but some repetition is useful here. Rudolf Peierls and Klaus Fuchs were working on the theory of gaseous diffusion in England, and when the British group moved to the USA in 1943, Fuchs worked in New York City with members of the Kellex Corporation who were responsible for the K-25 gaseous diffusion plant at Oak Ridge. NKVD headquarters in Moscow knew everything that Fuchs did. Fuchs was soon transferred to Los Alamos, before all of the design wrinkles in K-25 had been worked out (the plant achieved full operation in mid 1945), and he was cut off from further progress. George Koval was also at Oak Ridge, but it seems that crucial details of the K-25 plant remained unknown to the Soviet Atomic Project. However, Kikoin did not stop trying to learn more.

Kikoin commented on recent intelligence regarding isotope separation in February, 1945.[16] The volume of information is as interesting as its content. He was responding to 79 pages of printed text and 29 pages of photocopies, all acquired on September 16, 1944. There were three topics: 1. Sources and consequences of fluctuations in the output of a diffusion plant; 2. Influence of the degree of branching on the performance of the plant (how many in and out pipes for each stage); and 3. Some general remarks on the possibility of using thermal diffusion for a slight enrichment before gaseous diffusion, manufacture of the separation barriers, and a scheme for storing products of varying purity in the factory. Kikoin, like Kurchatov, often posed further questions to be forwarded across the Atlantic as assignments for the spies.

After the Hiroshima bomb in August, 1945, Stalin placed Beria in charge of the Soviet Atomic Project, and things started to happen quickly. The Smyth report appeared in the fall, and contained useful declassified information on isotope separation. Since the more detailed espionage data were known to only a small number of senior Soviet physicists, and to none of the Germans, the Smyth report was eagerly devoured by all concerned when it appeared first in English and then in Russian translation. The description of the scope of the Oak Ridge plants must have been daunting to those contemplating similar undertakings. Perhaps a large oil refinery could serve as an example of engineering complexity.

---

[16] Atomnyy_proekt_SSSR.T.1 Ch.2(2002).pdf #315. Kikoin was probably not aware of all of the intelligence information – only that which pertained to his own task.

However, there was virtually no technical detail in the Smyth report, so it must have been frustrating to those wanting to learn more information.

The Oak Ridge K-25 gaseous diffusion plant was a sufficiently significant achievement in industrial and chemical engineering that it was described in late 1945 and early 1946 in a series of articles that were available to the Soviets in the American journal "Chemical and Metallurgical Engineering."[17] These articles were in the spirit of the Smyth report in that they revealed no classified information, but gave broader descriptions of the plant, its design and development, and its operating problems. John Hogerton was chief engineer for the Kellex Corporation that built the Oak Ridge gaseous diffusion plant. He presented a reasonably complete qualitative description, paraphrased as follows:

"The feed gas pressure is less than one atmosphere, so that the system has to be vacuum tight; recompression of the gas heats it up, so that it must be cooled, but any contact with water is disastrous, so the coolant is a special liquid; the entire building and all of its contents have to be clean; uranium hexafluoride is poisonous, so before any part is worked on the gas has to be completely removed; the gas is also highly corrosive, requiring special handling in cleaning, in valve design, packings and O rings, etc."

Hogerton continued:

"A feature of the diffusion process encountered in few other processes is the so-called equilibrium time. When the cascade is first started up it must be allowed to run undisturbed until the proper concentration gradient has been established at each stage. Then and only then can product be taken off. Establishing equilibrium throughout the system involves a great deal of 'separative work' and, depending on the size of the cascade, may take anywhere from a day to several months. Since every cubic foot of internal volume adds to the equilibrium time, it is important to design the cascade with as little volume as practicable. For example, the stage arrangement should be compact in order to minimize the volume of interstage piping. Similarly, all of the process equipment should be specially designed with a view toward minimum hold-up. In the case of the pumps this means as high speed machine as possible."

---

[17] John F. Hogerton, "Chemical and Metallurgical Engineering," December, 1945, p 98; and P.C. Keith, *ibid.*, February, 1946, p 112.

The value of this information for the early designers of the Soviet plant is obvious. Another Kellex engineer Percival C. Keith emphasized corrosion and control problems in the operating plant. In his article published in February, 1946, on corrosion he wrote:

"The extreme reactivity or corrosive nature of uranium hexafluoride was particularly serious because of the vast internal surfaces of the plant and because the size of the holes in the barrier had to be maintained constant. In the first instance a corrosion rate well below tolerances for indefinite retention of mechanical strength could still be sufficient to convert all the 235 isotope into non-volatile compounds and thereby completely stop production. In the second instance opening up or closing of the holes of the barrier could reduce the separation to only a fraction of that required. As a result it was necessary to treat chemically all internal surfaces of the diffusion plant until the average corrosion rate was of the same order as stainless steel in air."

On controls he wrote:

"The most important of these was control of pressures on each of the stages of the plant. Without adequate control, pressure waves would have developed which would have overloaded pumps and mixed partially separated isotopes, thus nullifying some of the separation performed by the cascade. Because of the large number of stages this problem was without precedent in the history of process control...

"An adequate solution to the control problem was found only after the most exhaustive theoretical analysis of the cascade by thirty odd mathematicians experienced in the application of difference-differential equations to the solution of concrete problems. This time the engineer was faced with the problem of designing a valve with an automatic response defined by a mathematical equation which he could not fully comprehend; a valve that would operate dependably and could be manufactured by the thousand in short order. In the face of these conditions such additional problems as vacuum tightness and corrosion resistance seemed like child's play."

And finally on valves and piping:

"In terms of numbers and variety, the valve problem exceeded all others. Approximately ½ million valves are required to operate the diffusion plant. They vary in nominal size from 1/8 inch to 36 inches, and in type from conventional globe valves for water and air service to special units designed to the most rigorous specifications of cleanliness, corrosion resistance, and tightness. Two developments typical of many were the pressure control valves used to control pressure, and indirectly flow in the diffusion cascade, and process block valves used to isolate diffusional equipment."

Kikoin knew some technical detail, but not everything, from the spies. Everyone else had access to anything published in the open literature, and from the above quotes it is clear that this information, while qualitative, was very useful. They knew some of the difficulties they would face, if not the specific solutions.

In November, 1945, Terletsky (Chapter 8) was sent to Bohr in Copenhagen with 22 questions, and #1 was written by Kikoin, asking which methods were actually used to obtain enough ^{235}U for a weapon. Bohr's answer, also given in Chapter 8, was correct – both gaseous diffusion and magnetic separation were used – but not particularly helpful regarding any details. So Kikoin, Voznesinsky, and company were on their own.

10.8. Components for the Gaseous Diffusion Plant

Design parameters for the compressors were sketched by Kikoin and S.L. Sobolev. Each compressor was driven by an electric motor. The compressor could be axial or centrifugal, the former being more efficient, but the required compression factor of four to five was difficult to achieve in a single stage, so the less efficient centrifugal design was chosen. The compressor rotor looked like a jet engine.[18] It was powered by an electric motor that operated by induction – the rotor was placed inside the compressor housing and isolated from the stator outside in order to avoid having to seal the rotating axis. There was a meeting in Moscow in June,

---

[18] V.N. Prusakov and A.A. Sazykin, "Uranium enrichment in the USSR," HISAP96, vol1, *ibid.*, p 167 Figure 3.

1945, to design the first experimental machinery.[19] Voznesinsky prepared working drawings, and the construction job was given to a factory in Gorky,[20] with a completion date only one month hence, to be delivered to Moscow where a suitable test stand was built. The prototype compressor worked to specifications, and operated successfully at supersonic speed. At the operating temperature, the speed of sound in UF_6 is 85 m/sec. Kikoin sent the good news to the Kremlin in time for the celebration of the Revolution in November, 1945. The order to manufacture the compressors was divided between the Gorky plant, and the Kirov factory in Leningrad.

In August, 1946, these two factories were given the green light each to prepare 20 complete stages of a gaseous diffusion separation plant, with compressors, barriers, plumbing, controls, etc., to be delivered to Laboratory #2 not later than 15 November, 1946. The compressors were the largest and most numerous in the array, called 'T-56,' with a gas throughput of 25 kg/sec, consuming 369 kilowatts of electricity.[21] One thousand compressors would burn 369 Megawatts, indicating the need for a robust electric power source nearby. The amount of gas handled by each stage decreased slowly through the cascade, so that the output enriched gas was only a tiny fraction of the input. The Soviet design had three different compressor sizes; T-56 was the largest, and OK-6 was the smallest, with a throughput of 8 g/sec, a factor of 3125 smaller.

10.9. German Group Assignments

The German groups at Sukhumi were working on their assignments that originated in a meeting in Beria's office in the fall of 1945. Von Ardenne had originally proposed to continue his own research program, but Beria said that he must work on the Soviet Atomic Project, so von

---

[19] A.G. Plotkina and E.M. Voinov, "Academician Isaak Kikoin," HISAP96, Vol **2** *ibid.*, p 201.

[20] Now restored to its traditional Russian name, Nizhny Novgorod, on the Volga, about 350 km east of Moscow, near the closed city of Sarov (Chapter 11). Gorky was the city of exile of Andrei Sakharov.

[21] About 500 horsepower, which is a whole lot for one pump.

Ardenne's Laboratory 'A' at Sinop and Hertz's Laboratory 'G' (Russian alphabet has no 'H') at Agudzery, former sanatoria in Sukhumi, would both be involved in isotope separation. Laboratory 'A' would contribute von Ardenne's sophisticated ion source to electromagnetic separation, and Peter Thiessen's tube shaped barriers to gaseous diffusion. Heinz Barwich working in Laboratory 'G' was busy with the control problem mentioned above by P.C. Keith. Barwich was well aware of Keith's article.[22] Barwich was an experimentalist by training, but sat down to learn how to set up and solve the 'difference-differential equations' referred to by Keith to analyze the problems of control and stability in the huge plant. He did not have Keith's '30 odd mathematicians,' but he did have some students who could operate mechanical calculators just as was done at Los Alamos, and a very able Soviet mathematician named Sobolev in Kikoin's group. Partly based on the 'Chemical and Metallurgical Engineering' articles Barwich became acutely aware of corrosion problems as well.

10.10.   Moscow 20 Stage Test

Beria had set February 5, 1947, as the deadline for successful operation of the test stages. The test assembly had lots of problems, and Kikoin was obliged to write to Beria requesting an extension of the deadline date by an extra 12 days. The tone of Kikoin's request reflected the fact that he was not happy working under Beria's 'patronage,' with the constant threat to hang you out to dry. Like many other scientists in the Project, Kikoin had 'secretaries' 24/7, who were to insure his safety, but who also reported his every move to Beria. Both 20 stage cascades were working by February 23, due to an around the clock effort by the crew.

In his report to Beria regarding the successful tests, Kikoin wrote: "There is no danger of significant corrosion of either the compressors or the filters by UF_6 gas. We may therefore proceed with industrial production of these machines. I believe that it would be expedient to begin staged output by the Gorky plant, and to speed up the construction of the

---

[22] Heinz and Elfi Barwich, "Das Rote Atom," *ibid.*, p 85.

buildings to house the gaseous diffusion plant D-1(also called plant #813) in Sverdlovsk."

Kikoin had spent some time in Sverdlovsk, knew the area well, and appreciated the fact that the Ural region roughly between Sverdlovsk in the North and Chelyabinsk in the South had some of the necessary infrastructure like electric plants and railroads. The plant was to be located 70 km north of Sverdlovsk at a village called 'Verkh-Neivinsky' at the north end of a small lake that could supply cooling water. The town was called Sverdlovsk-44, a postal code for a place near Sverdlovsk, and the plant was 'Base 5' among the many sites of the project. It was, and still is, a closed town with no visitors allowed without special permission. Necessary construction of the town and plant were done by prison labor, as was the practice for most of the construction work for the Soviet Atomic Project. The reactor complex to produce ^{239}Pu was 'Base 10,' of Combine #817 located at Kyshtym between Chelyabinsk and Sverdlovsk – also called Mayak. The electromagnetic separation plant was 'Base 9' (plant #814), located 40 km north of Nizhny Tagil shown on Map #3.

Following the pattern of other sites for the Project, factories in the Urals were built mostly by prison labor. A letter written by Vannikov, Serov, Chernyshov, and Zavenyagin to Beria on September 25, 1948, stated that the construction work force for the PGU, composed of prisoners, former prisoners, repatriates, German immigrants, and Red Army soldiers, numbered 181,500 people.[23] Most of these workers were considered to be unreliable, and therefore for security reasons were not useable on other construction projects. Combine #817 ('A' reactor for plutonium production), Factory #813 (D-1 gaseous diffusion plant), #814 (SU-20 electromagnetic plant), and KB-11 (Arzamas-16) had a total of 27,000 Red Army construction battalion officers and soldiers. For obvious reasons the soldiers were considered more reliable, and more subject to military discipline than the prisoners, and the Project could always use more of them. On the other hand, industrious prisoners were rewarded with

---

[23] Atomnyy_proekt_SSSR.T.2.Kn.4.(2003).pdf #200.

extra rations and shortened sentences.[24] The disposal of prisoners after their sentence was a vexing problem for security – they could not be allowed to return to the general population because they knew too much. Many were rewarded with longer sentences in remote camps, belying the promise of shorter incarceration for labor on the Project. For the German guest scientists, a two year cooling off period was deemed sufficient – any knowledge that they took with them would be stale.

10.11.   Full Scale Gaseous Diffusion Plant Approved

Beria approved the construction of the first full scale gaseous diffusion plant, D-1, based on Kikoin's request. Training sessions started for plant technicians and operators, first at Laboratory #2 and Leningrad, and later at the Sverdlovsk site itself. Everyone expected the new plant to work. On May 22, 1948, the Soviet Ministry authorized commissioning of the first line. The startup was very difficult. Kilogram amounts of uranium were lost somewhere in the system. Kikoin complained of quality control problems in the machinery supplied by Soviet industry. Glass manometers for pressure measurement were of poor quality and leaked. The compressors came with all sorts of foreign materials – rubber gloves, tape, etc. Helium leak detectors were needed, together with helium gas, to help with vacuum sealing.[25] Zavenyagin and Vannikov came to the site, and there can be no doubt that recriminations of the sort delivered by Zavenyagin at the Electrostal plant and recounted by Riehl were repeated, although we have no record of such a meeting. Management was shaken up, factory #92 in Leningrad was scolded for the extraneous materials in the compressors, and the Germans were called to the rescue.

---

[24] Atomnyy_proekt_SSSR. *ibid.*, #48 and #67.
[25] Atomnyy_proekt_SSSR.T.2.Kn.4.(2003).pdf # 182.

10.12. Sukhumi Group Goes to D-1

In October, 1948 Hertz, Thiessen, and Barwich were summonsed to Moscow from Sukhumi with instructions to bring warm clothes.[26] The next morning they were on a two hour flight to Sverdlovsk, where a special train with darkened windows awaited them for the short trip to Sverdlovsk-44. They got off the train into a town that they estimated could have about 10,000 people, but they noticed very little activity – perhaps because of the cold weather. They were taken to the local hotel 'Ural' for supper and the night. They enjoyed a full Russian breakfast the next morning, with sosisky, pirozhky, and kefir.[27] If the town had a name the Germans were not told what it was, so Hertz named it 'Kefirstadt,' and it remained Kefirstadt for their several month duration there. They had an escorted tour of town and surroundings. There was the usual housing mix in town – dormitories, apartments, identical three- story houses, Finnish houses, and a few one-story dachas for the leadership, all clearly constructed on top of an old village. The local village church stood on a small hill, and from the hill the lake could be seen, partially frozen already in October. They did not see any shops. There was a movie theater, and a drama theater under construction for possible entertainment from Moscow. Contact with the local population was not permitted, the reason given was that the German delegation might not be well received.

The next day they were taken to the plant. While they knew from descriptions of Oak Ridge that the plant would be huge, they were nevertheless impressed with the real thing. Kilometers of pipes and pumps were arranged in a U-shaped building like the K-25 plant. The D-1 plant was slightly smaller than its Oak Ridge predecessor, but still pretty formidable. Vannikov, Kikoin, and the plant manager greeted Hertz, Barwich, and Thiessen in the control room. Vannikov had temporarily relieved Kikoin because of the incident. Barwich was well aware of the

---

[26] Heinz and Elfi Barwich, "Das Rote Atom," *ibid.*, p 103ff.

[27] Sosisky are sausages, pirozhky are stuffed pastries, and kefir is a sour milk drink that is available in the USA. Coffee and tea were also available. Russian breakfasts are quite good.

time frame in which all of the equipment had been built and assembled, and he congratulated Vannikov on the achievement. The plant was ready to go, once the loss problem was solved. The German delegation was not sure that they could be of any help. They had things to do in Sukhumi, and they may have been concerned that more involvement would lead to a longer stay away from Germany. Vannikov said that they were shipped to Sverdlovsk because they would be useful. They really had no choice.

The local group had been searching for the lost uranium. They took some of the machinery apart, and used chemical analysis to look for uranium in the diaphragms, nickel surfaces, compressors, armatures, grease and packings of the rotors and axes, all to no avail. The rest of the plant was operating, although not producing enriched uranium output in sufficient quantities. Barwich used the functioning machinery to measure input and output pressures and temperatures, and compared the numbers to the solution to the stability problem that he had worked out. The search for the missing uranium continued.

Increased presence of military police in town was a sign of an impending visit by the Kremlin, and shortly Beria arrived for an inspection tour. Beria's behavior in the presence of the Germans was restrained. He asked what could be done to make their lives better. The answer was more freedom to move about, and more information from the outside world. Beria encouraged them to keep working, and promised less restrictions. Nikolaus Riehl's experience with Beria was similar. Any criticism of the Soviet lab management for the plant's problems was not witnessed by the German visitors.

10.13.   D-1 Plant Requires New Compressors

A Soviet physicist named Voskoboinik found the missing uranium by using an ion chamber to measure residual radioactivity rather than applying chemical techniques. He found uranium on the laminated iron of the rotors that drove the compressors inside the compressor housing. There were serious leaks of UF_6 gas through the packings on the compressor shafts. The hexafluoride broke down chemically, and was deposited on the drive motors. The motor-compressor assemblies had to be redesigned and new ones built, so the D-1 plant in its original configuration was only a first try, and not able to produce kilogram quantities of weapons grade ^{235}U. The test stands assembled by Kikoin in Moscow might have been able to detect this problem given sufficient running time, but pressure from Beria's committee kept things moving at such a pace that decisions regarding construction of the plant had to be made in a timely fashion.

Barwich wanted to know what enrichment if any had been achieved in the output stage of the cascade – an OK-6 compressor. The standard technique of detecting the soft γ ray from the α decay of ^{235}U mentioned above had not been developed, so Werner Schutze's mass spectrograph from Sukhumi was used. It had been disassembled and shipped to Sverdlovsk-44, accompanied by Schutze himself. There was some concern about assembling too much ^{235}U in one place – it might exceed the critical mass![28] So the mass spectrograph worked with small samples. Igor Kurchatov showed up for the measurements. The results were only vaguely encouraging.[29] The assembly in one place of enough fissile material, either ^{235}U or ^{239}Pu, to approach a critical mass is dangerous, and is one of the many hazards in handling 'industrial quantities' of these substances.

In February, 1949, the German delegation returned by train to Moscow, their mission to Sverdlovsk-44 completed. But the task of reworking the entire system to perform up to specifications was far from over. Kikoin and his crew essentially had to start over again with the new compressors with better seals. No one was sent to prison, although there was certainly plenty of blame to go around. The new equipment came on line in June, 1949. There were new compressors in four sizes rather than three, and the initial output was 200 g of 40% enriched ^{235}U per day[30] that was further purified by Artsimovich's electromagnetic separators at Laboratory #2 –

---

[28] Barwich attributed this concern to ignorance of the amount required for a critical mass.

[29] Heinz and Elfi Barwich, "Das Rote Atom," *ibid.*, p 128.

[30] Atomnyy_proekt_SSSR.T.2.Kn.4.(2003).pdf #121.

described in Section 10.14 below. By 1950 factory #813 was producing 75% enriched ^{235}U, and the SU-20 plant at neighboring factory #814 was functional, so that it could be purified to weapons grade there. By 1950 the upgraded D-1 plant was producing weapons grade ^{235}U, and SU-20 was switched over to separating stable isotopes. Eventually in the 1950's several gaseous diffusion plants were built in the Urals, and in Siberia. The gaseous diffusion effort required the participation of industries throughout the Soviet Union. The first two Soviet nuclear weapons tested in August, 1949, and September, 1951 were both plutonium implosion bombs. Uranium combined with plutonium was used in later tests.

An amusing incident occurred in July, 1949, when Marshal Georgy Zhukov was commander of the Ural Military District, headquartered in Sverdlovsk. A certain I.M. Tkachenko from Combine #817 sent a telegram to Beria stating that Marshal Zhukov had ordered military training exercises in the Kyshtym region right on top of the Atomic Project's installations. The railroad had been ordered to transport 9 regiments to Chelyabinsk and 10 to Sverdlovsk. Zhukov declined to alter the plans, stating that authority came from the ministry of defense (Bulganin). Tkachenko requested that Beria ask Zhukov to change his plans. Beria called Bulganin on the phone, and told him to relocate the exercises away from the area. Bulganin in turn called Zhukov, and the training was moved to another location.[31]

This incident raises the question of how much the Red Army high command knew about the Atomic Project. Defense Minister Bulganin was a member of Stalin's Council of Ministers, and must have known something about the Project, if not the details, but he may not have told anyone else. Boris Vannikov in his oral history said that the Presidium of the Communist Party Central Committee was not informed of progress.[32] Zhukov was the most famous military officer in the country, Hero of the Soviet Union, and Marshal of Victory, and as commander of the Ural

---

[31] Atomnyy_proekt_SSSR.T.2.Kn.4.(2003).pdf #261. Ivan Tkachenko was a representative of the Politburo at Combine #817.

[32] Raisa Kuznetsova, "Kurchatov v Zhizni," *ibid.*, p 477.

Military District he had to have known that something was going on in the neighborhood, even if he did not know what it was. Literally thousands of construction workers, some of them Red Army soldiers, were employed in building vast industrial complexes in the region, but officers in charge did not have to know the purpose of the new buildings. Ivan Serov gave a counter example of strict 'need to know' secrecy.[33] In December, 1948, Serov was touring the labor camps in the Urals to see how things were going – some of the same camps that were building the nuclear plants. He visited the plutonium production facility Combine #817, and got a complete tour and a crash course on plutonium by none other than Igor Kurchatov himself. Thus Serov, a close associate of Zhukov when they were in Germany in 1945, knew quite a bit about what was going on near Chelyabinsk, but perhaps he kept this information to himself.

Another bit of evidence belies Zhukov's ignorance of the development of nuclear weapons, at least to the extent of progress made in Germany during the war. In the fall of 1945 Zhukov was commander of all forces in the Soviet zone of Germany. On October 2, 1945, Zhukov sent a detailed account of the findings of the Soviet Alsos mission regarding institutions and personnel involved in the Uranverein, research equipment like cyclotrons and electrostatic accelerators, and raw materials like uranium and heavy water. The document was quite sophisticated, and although it could have been written by someone else over Zhukov's signature, still, he must have read it himself.[34]

10.14. Electromagnetic Isotope Separation

The Manhattan Project developed the K-25 gaseous diffusion plant and the Y-12 electromagnetic separation plant in parallel, and both contributed significantly to the first uranium atomic bomb. The Soviet Project began with gaseous diffusion (#813) and electromagnetic separation (#814) on equal footing, but in the end gaseous diffusion won the race, and

---

[33] Ivan Serov, "Zapiski iz Chemodana," *ibid.*, p 376.
[34] Atomnyy_proekt_SSSR.T.2.Kn.6.(2006).pdf, #3 Zhukov to Stalin.

electromagnetic separation did not play an important role in obtaining a critical mass of ^{235}U.

The notion that mass spectrometer techniques could be applied in principle to uranium isotope separation had been around for some time. Arthur J. Dempster, one of the physicists who established the mass spectrometer as a practical tool for isotope separation, used the device in 1935 to discover the isotope ^{235}U in the first place.[35] Alfred O. Nier at the University of Minnesota produced small samples of ^{235}U and ^{238}U for the Columbia University group's fission studies in February, 1940.[36] Early cyclotron builders were attracted to this technique by the structural similarities between mass spectrometers and cyclotrons. Indeed, Ernest Lawrence the cyclotron inventor, was the leader of the mass spectrometers for isotope separation at the Oak Ridge Y-12 calutron complex. In his memorandum on the necessity of constructing new, more powerful cyclotrons in the USSR in 1945, Igor Kurchatov made the point that such a machine could be used to study uranium isotope separation methods.[37] M.G. Mesheryakov of the Leningrad Radium Institute where the first Soviet cyclotron was built and operated, began work on the problem during WWII. He was well aware of the promise and the difficulties of the technique. Promise because a simple calculation shows that the spatial separation of the two isotopes ^{238}U and ^{235}U in a practical mass spectrometer is about 1 cm – large enough for two clearly defined receivers. Difficult because 1 mA of singly charged uranium ions = 2.5 µg/sec, dominated by the heavy isotope. To get 8 kg of ^{235}U would require running for 14,300 years! A million machines would cut the total time to

---

[35] "Arthur J. Dempster" by S.K. Allison, Biographies of Members of the National Academy of Sciences. (1952). For constant magnetic field and accelerating voltage, the orbit radius ~ $(M/q)^{1/2}$, so ^{238}U^{++} and ^{119}Sn^{+} should have the same orbit. Dempster showed that tin lies inside uranium, indicating that energy would be released in fission. Using his mass spectrograph, Dempster established that nuclei near the middle of the periodic table are more tightly bound.

[36] A.O. Nier, E.T. Booth, J.R. Dunning, and A.V. Grosse, *Phys Rev* **57**, 546 (1940). Fission cross sections were measured using about 0.5 µg of ^{238}U and 0.003 µg of ^{235}U. They came close to the modern number of σ(slow n + ^{235}U → fission) = 585 barns, which is impressive.

[37] Atomnyy_proekt_SSSR.T.1.Kn.2.(2002).pdf #303.

about five days. Since one really needs more than 10 kg of ^{235}U to make a bomb, the practical solution had to be somewhere in between: ion currents greater than 1 mA; and many machines operating simultaneously. Y-12 had about 1000 machines. Important components of the device were the ion source, the accelerating voltage, the constant magnetic field, focusing of the ions, suppression of beam-beam interactions, receivers for the isotopes, and techniques for chemical extraction of all of the uranium that entered the vacuum chamber. Some of the design problems are discussed in Appendix C, with a diagram of a working device in Figure C.3.

In January, 1944, Lev Artsimovich was listed as head of Section #5 of Laboratory #2 to develop the ionic method of isotope separation.[38] As the Project grew the electromagnetic method became centered at Laboratory #2 in Moscow, although Leningrad and Mesheryakov continued to play a role. The Elektrosila factory in Leningrad had built magnets for cyclotrons, and had on hand the 220 ton magnet mentioned earlier for testing ion sources, charged particle optics, and receivers for the separated isotopes. In a letter to Kurchatov in April, 1946, Artsimovich reported from Leningrad that the Elektrosila plant was prepared to furnish the magnet in place with appropriate power supplies to pursue isotope separation studies.[39] When von Ardenne joined the effort with his lab in Sukhumi, he worked on ion source development, and traveled to Leningrad periodically to use the large magnet. Artsimovich was also involved in ion source development. One question was which of the many uranium compounds was optimum for an ion source? Von Ardenne preferred uranium metal, while Artsimovich worked first with the gas UF_6, then with the salt UCl_4.[40] An important requirement was high ionization efficiency independent of the choice of compound. Also, to be kept in mind was the necessity to retrieve all of the feed material from the chamber of the mass spectrometer – whether ionized or not. This requirement was absolutely mandatory for input feed that had had any enrichment, because of the expense involved in the enrichment process, either by gaseous

---

38 Atomnyy_proekt_SSSR.T.1.Kn.2.(2002).pdf #203.
39 Atomnyy_proekt_SSSR.T.1.Kn.2.(2002).pdf #303 footnote 7.
40 The Oak Ridge calutrons used UCl_4.

diffusion or electromagnetic separation in an earlier stage. The Soviet sources achieved good ionization efficiency with resonating electrons that moved back and forth through the uranium gas. Von Ardenne obtained high ionization efficiency with his duoplasmatron source that is discussed in Chapter 13.

There were at least two potentially harmful effects of a very large positive ion current density, both due to the electric fields created by the charge. One was in the source, and the other in the particle orbits outside the source in the magnetic field. The output current per unit length from a cylindrical source could be a limited amount of mA/cm, so that one technique for increasing the current was making the source longer parallel to the magnetic field. That required a large magnet gap to accommodate the source. For example, if a 20 cm long source was capable of 1mA/cm, then the total uranium mass flow would be 50 μg/sec, or about 5 g/day for one unit. Divide by 143 for the amount of ^{235}U so one thousand units would give 35g/day of weapons grade material – a realistic yield if actually achievable. Numbers like these kept people working on the development of a practical device.

The ion source was not the only important part of the electromagnetic mass separator. Large gap stable magnets capable of containing orbits of the order of 1 m radius had to be built, mounted, powered, and regulated. The fractional difference in square roots of the masses – the handle one has in EM isotope separation (and gaseous diffusion as well, only the centrifuge works with mass differences without taking the square root) – is only 0.0063. This quantity sets the scale for stability requirements on the magnetic field and the accelerating voltage – Artsimovich specified 0.03% stability. Then there are problems of handling and focusing the high ion currents, avoiding beam-beam interactions that could destroy the mass separation, and finally extraction of the enriched product from the receiver.

Kurchatov had Information about the Oak Ridge Y-12 plant obtained by the NKVD spy network. On April 7, 1945, he commented in a memorandum on the intelligence material. He noted that the source molecules were UCl₄, and that the magnetic field was inhomogeneous (without any details regarding the inhomogeneity). He said that the general approach was similar to the one being developed at Laboratory #2 in 1945. As was often the case in critiques of the espionage data, Kurchatov ended with four questions to be relayed across the Atlantic to the spies. The questions were as follows: 1. What was the ionization state of the UCl₄ molecules? 2. A dimensional drawing of the layout would be useful. 3. Was a light gas introduced into the vacuum chamber? And 4. What were the particle trajectories inside the chamber?[41]

The two subcommittees of Beria's PGU met every week in the early going. The second meeting of the Technical Subcommittee on 10 September, 1945, reported that while the electromagnetic method of isotope separation was able in principle to give 100% pure ^{235}U, the R&D for the instruments at Laboratory #2 was not sufficiently advanced to allow the design of industrial facilities.[42] A subsequent meeting on October 8 reported that von Ardenne's Laboratory 'A' in Sukhumi was joining the effort to study the application of electrons to cancel the space charge – Kurchatov's question about the introduction of a light gas into the chamber. Two days later Lev Artsimovich gave a tutorial on the electromagnetic isotope separation technique, with nice figures and many of the numbers presented in paragraphs above, and concluding that much work had yet to be done to build practical devices.[43]

In the fall of 1946 Laboratory #2 succeeded in obtaining small quantities of separated uranium isotopes with the electromagnetic method. In August, 1948, Artsimovich and Efremov gave a status report to the PGU. Ion source development was still in progress both at Laboratory #2 and at Electrosila. Adequate ion currents had been obtained, but the sources had lifetimes of only a few hours before they burned out – lifetimes measured in months were necessary for mass production.

---

[41] Atomnyy_proekt_SSSR.T.1.Kn.2.(2002).pdf #329. No information regarding answers to his questions
[42] Atomnyy_proekt_SSSR.T.2.Kn.4.(2003).pdf #2.
[43] Atomnyy_proekt_SSSR.T.2.Kn.4.(2003).pdf #146.

Industrial production of the vacuum chambers, pumps, and valves was not up to specifications. Criticism of Soviet industry was rare, but not unheard of in correspondence regarding the Project. Communism was supposed to be able to surmount any obstacle. Construction work on the SU-20 plant, factory #814, was proceeding as planned in the village of Nizhnyaya Tura (see Map #3 of the Sverdlovsk region), where 12000 prisoners and soldiers were busy on the job. SU-20 was to have 20 vacuum chambers for isotope separation, sandwiched between 3000 T of iron magnets, with a product goal of 150 g/day from input feed enriched to 40% ^{235}U by gaseous diffusion.[44] Artsimovich did not break this number down into currents required by the units, but from the arithmetic above, 150g/day of ^{235}U from a uranium feed that was 2/5 ^{235}U would correspond to 1.5 amperes total current, or 75 mA per unit. This modest current demonstrates the advantage of using the two isotope separation schemes in series. The current per unit would be 4.3 amperes starting with natural uranium feed.

In December, 1948, Vasily Makhnev, a member of the PGU, sent a critical evaluation of the status of the electromagnetic isotope separation effort to his boss Beria. Makhnev did not have any good news. Artsimovich's Section #5 of Laboratory #2 was performing poorly. Design specifications for the Electromagnetic Plant SU-20 at factory #814 were supposed to be done by July 1, 1948, and were not yet finished. The magnets were not designed properly, and will have to be re-done. High current buss bars to the magnets designed for 5500 amperes failed at 4500 amperes, resulting in a lower magnetic field (0.45 T rather than 0.5 T design). Design work on the ion sources was also not up to specifications. Two groups were working on it – one using uranium salts and the other metallic uranium – without sufficient coordination to converge on a working model. Makhnev claimed faulty management. SU-20 was still not operational in late 1949.[45]

---

[44] Atomnyy_proekt_SSSR.T.2.Kn.4.(2003).pdf #187. The Oak Ridge alpha calutrons were in a donut shaped 'race track' with about 50 chambers.

[45] Atomnyy_proekt_SSSR.T.2.Kn.4.(2003).pdf #221.

A report on German group progress of February 28, 1949, listed the properties of von Ardenne's source:

1. Stable arc length 20 cm.
2. Luminous area 5 cm^2.
3. Ion current 55 mA.
4. Accelerating voltage 32 kV.
5. 75%-80% singly ionized uranium
6. Cathode lifetime more than 100 hours
7. Oven lifetime up to 14 hours.

The obvious problem was the oven lifetime. Von Ardenne used uranium metal, that required an oven temperature of 2000^0 C. The performance of the source looked promising.[46] By June, 1949, von Ardenne's pure U sources were approved for use in the SU-20 isotope separation complex, together with the UCl$_4$ sources of Artsimovich. Principal centers for R&D remained at Laboratory #2, and at Elektrosila in Leningrad. By September, 1949, Artsimovich was producing several grams of high purity ^{235}U at his facility in Laboratory #2, and was authorized to use the same facility for processing partially enriched uranium obtained from the gaseous diffusion plant at Factory #813.[47] The SU-20 plant came on line the next year, and functioned satisfactorily. A larger plant with 80 units was planned, but apparently never built.

Unlike the Manhattan Project, where the Hiroshima (uranium) and Nagasaki (plutonium) bombs were ready simultaneously, sufficient Soviet ^{235}U for a weapon lagged the ^{239}Pu by a year or two. The Semipalatinsk test in August, 1949, was a plutonium weapon. Beria's committee had decided to emphasize plutonium production from the beginning, but isotope separation was not neglected, nor was it denied resources. It just turned out that way.[48]

---

[46] Atomnyy_proekt_SSSR.T.2.Kn.4.(2003).pdf #239.

[47] Atomnyy_proekt_SSSR.T.2.Kn.4.(2003).pdf #281 and #282.

[48] This account is based on the two articles in HISAP96 noted above. In Vol **1** by Prusakov and Sazykin, and in Vol **2** by Plotkina and Voinov.

While Laboratory #2 was building its first reactor, and beginning isotope separation studies in 1946, the Academy of Sciences set up a new laboratory for bomb development further east of Moscow at the town of Sarov, which had housed a munitions plant during the war.

Chapter 11

The Soviet Union Creates Arzamas-16

11.1. Site Selection

Another laboratory site - a branch of Laboratory #2 - was needed. Laboratory #2 was too close to Moscow, and did not have enough territory around it for safety and security in weapons development. The initial assignments of the new laboratory were the construction of two atomic bombs, called RDS-1 and RDS-2. RDS stood for 'Reaktivny Dvigatel Stalina,' or 'Stalin's rocket engine' in Russian, a code to hide the true mission of the lab.[1] RDS-1 was to be a plutonium implosion bomb based on the Manhattan Project design tested at Alamogordo, and RDS-2 a gun-type uranium bomb based on the one used at Hiroshima. The fissile materials for the bombs were to be supplied by other factories to be built elsewhere as part of the expanded Project, in close analogy with Los Alamos. The weapons had to be deliverable by aircraft. So, the new laboratory had a big job ahead.

The gun-type ^{235}U weapon RDS-2 was never built. The designation was later absorbed in a series of models after RDS-1, called -2, -3, -4, and -5, that were all implosion bombs of varying design, using a gap between the compression shell and the fissile material, and mixtures of ^{235}U and ^{239}Pu.[2] RDS-6 was the first hydrogen bomb test in 1953.

It may seem odd that nuclear weapons development requires substantial experimentation with conventional explosives, but that is

---

[1]   The prefix 'RD' is still on Russian rocket engines used by the US space program.

[2]   Atomnyy_proekt_SSSR.T.2.Kn.6(2006).pdf, p 51.

441

indeed the case. That meant an isolated place, but not too far from Moscow, since Laboratory #2 was still the center of nuclear R&D. The new site was named a 'construction bureau' of Laboratory #2, and given the number eleven, thus becoming KB-11 as a Russian abbreviation. Two people were appointed to manage the site. A lieutenant general in the Army, Pavel M. Zernov, was named overall commander, and Yuli B. Khariton, a nuclear theorist and early member of Kurchatov's enterprise, was named chief builder. During the war Khariton had been developing conventional explosives. Zernov, trained as a mechanical engineer, was responsible for industrial tank production during the war. The first task was to find a suitable location for the new town/laboratory. An official portrait of Yuli Khariton as laboratory director is shown in Figure 11.1.

Figure 11.1. Yuli Khariton, first scientific director of Arazamas-16. Credit VNIIEF Sarov.

Aside from remoteness, isolation, small indigenous population, and not too far from Moscow, it was also desirable that the new town had some technical infrastructure, and of course electric power.[3] A three-month search produced no candidates. Then Boris Vannikov, the former armaments minister who was now a deputy to Beria in charge of the Soviet Atomic Project, recalled a dormant munitions factory near the old

---

[3] L.P. Goleusova, "Arzamas-16," Voprosy Istorii Estestvoznaniia i Tekhniki 1994, #4, p 89.

monastery at the village of Sarov, near the city of Gorky (now Nizhny Novgorod), about 370 km east of Moscow. The Moscow region Map #2 shows the locations north to south: Gorky, Arzamas, and Sarov. Vannikov's expertise in conventional explosives fit well into the new thrust planned for KB-11. The Soviet government had closed the monastery in 1927. In the early 1930's the facilities at Sarov were managed by the NKVD, first as an institution for homeless children, and later as a sequence of work-correction camps for juveniles and adults. In 1938 Sarov acquired a factory for national defense, as part of the effort to disperse munitions plants to the east of the Moscow region. Factory #550 made shell casings for Katyusha rockets during the war. It was shut down in September, 1945, but not given any further orders regarding what to do next. The prison camp remained, not only to serve as construction labor for the new laboratory, but also to be a source of trouble.

Already in January, 1946, Khariton was worried about the power, water, and transport utilities of the Sarov site. Factory #550 was served by a narrow gauge railroad, and Khariton accordingly requested more narrow gauge locomotives and rolling stock.[4] He also wanted an airstrip. Vannikov requested of Marshal Bulganin, the defense minister, enough steel plates for a runway – an early involvement of the Red Army in the operations of the new laboratory.[5] Since its mission was weapons development, more military participation would follow.

Zernov and Khariton visited Sarov in April, 1946, and selected it for KB-11. Over the course of its existence Sarov had various other code names, Arzamas-16 being the most frequently used. Arzamas is an old Russian town in the same region as Sarov, only a bit to the north. The number 16 was a post office box – reminiscent of Los Alamos at PO Box 1663, Santa Fe, NM.

---

4   Atomnyy_proekt_SSSR._T.2.Kn.6.(2006).pdf, #16 Khariton to Vannikov.
5   Atomnyy_proekt_SSSR._T.2.Kn.6.(2006).pdf, #51 Vannikov to Bulganin June 4, 1946.

11.2. Prison Construction Crew

On 19 April, 1946, Vannikov ordered that factory #550, with all of its equipment, buildings, and personnel be turned over to the supervision of Construction Battalion #880, NKVD USSR, especially formed for the construction of the nuclear center, using the labor of prisoners of the Gulag.[6] Stalin signed the order, and on 21 June, 1946, construction began, with the first stage to be completed by October, and the second stage by 1 May, 1947, with an initial budget of 30 million rubles. A game preserve forest had to be cleared, and 500 local inhabitants deemed unnecessary for the nuclear laboratory had to be relocated. Local sources for food and provisions were not adequate, and that was to lead to significant difficulties for the project workers. The isolation and secrecy requirements were incompatible with supplying the place with all the necessary equipment, food, power, etc. Prisoners from other camps began to arrive to add to the Construction Battalion, and in early 1947 there were 9737 workers, including 1818 women.[7]

There were two types of prison labor: criminals, and political prisoners (so called article 58's of the criminal code). The political prisoners were better educated, and relatively benign and cooperative, although the system had the ability to toughen them up. The criminals formed their own group within the prison population, and had their own argot for communication. Paradoxically, the criminals were better trusted by the authorities, and often were given lighter work assignments, even supervisory roles over others.[8] The existence of KB-11 had to remain secret to the general population. As a result, prisoners whose sentences were fulfilled were released at KB-11 to become free volunteer workers, but were forbidden to leave the site.[9] This resulted in a criminal element in the population, and laboratory workers and their families were afraid to

---

⁶ L.P. Goleusova, "Arzamas-16," VIET 1994, #4 p 89.

⁷ L.P. Goleusova, VIET 1994 #4, p89ff.

⁸ Meyer Galler and Harlan Marquess, "Soviet Prison Camp Speech," University of Wisconsin Press, Madison, 1972, p 31.

⁹ At some secret sites freed prisoners were simply sent back to another, more remote camp in order that they not inform the outside world.

go out after dark. Zernov appealed to Beria in June, 1948, about the need to remove from the site 200 former criminal prisoners. He received no response, and the problem continued to get worse, partly from periodic amnesties that turned more prisoners loose. There was no accommodation for them outside the camp, and many of them did not continue work in construction, but rather hung about on the streets. A weapons laboratory of the Soviet Academy of Sciences obviously had no place for a loose criminal element. The solution to this problem developed over the years; released prisoners were sent to other camps farther to the east.

The war had stimulated an interest among Soviet scientists in conventional explosives. Yuli Khariton was born in Saint Petersburg in 1904, and educated in the Polytechnic Institute in that city – renamed Leningrad. He was a theoretical physicist at the Leningrad Physico-Technical Institute, now named after A.F. Ioffe, before spending two years at the Cavendish Laboratory in Cambridge from 1926-1928. He returned to Leningrad and joined the Institute for Chemical Physics, where he worked on explosives. During the war, the Institute was evacuated to Kazan, together with many institutions from blockaded Leningrad.

11.3. Yakov Zeldovich

Yakov Zeldovich was born in Minsk in 1914, and grew up in Leningrad. He joined the Institute for Chemical Physics in 1938, and received a doctor's degree in physics and math from Leningrad Polytechnic Institute in 1939. Zeldovich and Khariton wrote a paper on chain reactions in uranium in 1940, but they also worked together on conventional explosives. He was evacuated to Kazan with other members of various scientific organizations in Leningrad in 1941, returning in 1943 to Moscow. Zeldovich was one of the more versatile and remarkable Soviet physicists of his time. He seemed capable of understanding anything. He was a theoretical chemist; he made seminal contributions to the Soviet Atomic Project, particularly in the design of thermonuclear weapons, as head of the theory group at KB-11; he was a nuclear theorist, who was famous for his work in elementary particle physics; and last but not least, after his classified work on nuclear weapons, Zeldovich branched in a completely different direction, and became a celebrated theorist in astrophysics and cosmology. When branching out into a new field, Zeldovich would not only read the pertinent literature, but also talk to younger colleagues who were working in that field. Gershtein, in his article about Zeldovich, observed that the physics of stars combined his various fields of interest – fusion of light nuclei, weak interactions and neutrinos, and combustion and explosions.[10] Such a broad list of accomplishments boggles the mind. It is hard to believe that the same man did so many different things, and did them well. A young Yakov Zeldovich is shown in Figure 11.2.

Figure 11.2. Yakov Zeldovich looking very intelligent, which he was. Credit AIP Emilio Segre Visual Archives, Physics Today Collection.

The administrative setup of KB-11 in 1946 was as follows:

- Laboratory #1 M. Ya Vasilev (NII-6) Study and construction of focusing systems.
- Laboratory #2 A.F. Belyaev (Institute for Chemical Physics) Study of chemical explosives.

---

[10] S.S. Gershtein in "Under the Spell of Landau," M. Shifman, ed., World Scientific, Singapore, 2013, p 141.

- Laboratory #3 V.A. Tsukerman Development of high speed X-ray technology.
- Laboratory #4 L.V. Altshuler (Institute of Machines of the Academy of Sciences) Determination of states of matter under extreme pressure.
- Laboratory #5 K.I. Shchelkin (deputy to Khariton) Study of nuclear explosives, and development of special apparatus for measurement of the relevant processes.
- Laboratory #6 E.K. Zavoisky (University of Kazan) Measurement of the compression of the central component.
- Laboratory #7 A.Ya Apin (Institute for Chemical Physics) Development of a neutron initiator.
- Laboratory #8 N.V. Ageev (Institute for general and organic chemistry) Study of uranium and plutonium and the application of these elements into atomic weapons.
- Laboratory #9 G.N. Flerov measurement of critical masses, study of reactions of fast neutrons with heavy elements and fission. Creation of suitable instruments.
- Laboratory #10 A.P. Protopov Measurements with neutrons.

In the middle of 1948 the construction divisions were reorganized into two sectors:

- Sector #1 under N.L. Dukhov
- Sector #2 under V.I. Alferov

In 1950 the two sectors were merged under N.L. Dukhov

- Development of the ballistic part of the bomb
- Development of the nuclear charge
- Development of the automation controls
- Development of an external accelerator for a neutron initiator
- Development of remote control systems

In February, 1948, the Theory Division was established at KB-11, headed by Ya. B. Zeldovich.[11]

---

[11] This list comes from militaryrussia.ru

In 1949 KB-11 had 4507 employees, of whom 848 were scientific-technical personnel, the rest filling various support functions. A construction battalion of 8200 prisoners and soldiers was also on site.[12]

By the time that Andrei Sakharov arrived at KB-11, in 1950, Zeldovich was settled in with his wife Varvara Pavlovna and their children. Six children are listed in his biography.[13] At least two children were born to different mothers other than his wife, but were accepted and cared for by him. The story of one daughter, Anna Shiryaeva, was told by Sakharov in his memoirs,[14] and by Anna herself.[15] According to Sakharov, KB-11 was like an overgrown village, where everyone knew what everyone else was up to (excluding of course the secret work), and he knew of Zeldovich's romance with Olga Shiryaeva, a released prisoner from the local camp. She was an artist and architect, convicted of distributing 'slanderous fabrications' against the government. Her husband left her after her arrest – a fairly common occurrence. At the Sarov camp she distinguished herself by artistic decorations on the buildings, and was released from captivity. She met Zeldovich on a tennis court, and mutual attraction led to romance. Soon thereafter she was ordered to be deported to Magadan, in the far eastern Soviet Union, on the sea of Okhotsk, where at that time there were several prison camps. This was part of the policy of excluding former prisoners who had had any exposure to the Atomic Project from any contact with the main population of the USSR. Zeldovich awakened Sakharov in the middle of the night, and asked him for a loan. Sakharov had just picked up his pay, and had some cash on hand, which he gave to Zeldovich. Zeldovich was neither able to prevent her departure, nor to follow her out of KB-11. All he could do was give her money before she left. Their daughter Anna was born the following January in a Kolyma camp near Magadan, where the temperature reached -40^0 C. The child saved the mother from hard labor, and Zeldovich did his utmost to obtain

---

[12] Atomnyy_proekt_SSSR. T.2.Kn.4.(2003).pdf #276.
[13] ru.m.wikipedia.org/YakovZeldovich
[14] Andrei Sakharov, "Vospominaniya," Alfa-Kniga, Moskva, 2011, p 139.
[15] Anna Shiryaeva in "Under the Spell of Landau," M. Shifman, ed. *ibid.*, p 154ff.

their release. Twenty years later, at a meeting in Kiev, Sakharov met Anna with her father. Sakharov said that she looked just like Zeldovich's other daughter, born to his wife Varvara. Anna told of meeting a half-brother when both were no longer children. The whole family was reunited at Zeldovich's funeral in 1987. This story illustrates the fact that, while for many scientists and technical workers at KB-11 the prisoners were an indifferent fact of daily life, that was not true for everyone. Some people became deeply involved.

11.4. Early Experiments

Veniamin Tsukerman and Lev Altshuler were experimentalists developing X-ray high speed photography to study the time evolution of explosions at an aeronautics laboratory in Moscow. A 'Munroe jet,' mentioned briefly in Appendix H in connection with the design of the plutonium bomb initiator, was a good application of X-ray time evolution. German armor piercing shells were obtained for Tsukerman to analyze during the war in 1942.[16] A 76 mm shell could penetrate 100 mm thick armor plate, and it was not understood by the Soviets how this could be possible. The trick is the 'jet' formed by a metallic coating on the inside of a cone shaped explosive – the Munroe jet – that burns its way through the armor. This process was confirmed by Tsukerman's time sequence X-ray photographs. Khariton showed the pictures to Boris Vannikov, and Tsukerman's reputation in X-rays was secured. Khariton invited Tsukerman and Altshuler to join the laboratory soon after development at the KB-11 site had begun.

11.5. Oral Interview with Lev Altshuler

Lev Altshuler has left his impressions of life at KB-11 in an oral interview.[17] Altshuler was born in Moscow in 1913, and lived 90 years.

---

[16] N.G. Makeev, "Birth of impulse x-ray technology," HISAP96, Vol 2, *ibid.*, p 463ff.

[17] Oral interview with Lev Altshuler by Oleg Moroz, Literaturnaya Gazeta, June 6, 1990, p 13.

He studied X-ray technology in night school, and obtained a degree in physics from MGU. After brief service in the Red Army during the war, Altshuler in 1943 was released by request of the Soviet Academy of Sciences, and returned to application of X-ray techniques to weapons research. He was therefore a part of Khariton's group to begin work at KB-11. Khariton said that the job would take 1.5 – 2 years at most. Altshuler worked at KB-11 for 22 years. Everything done at KB-11, or Arzamas-16, was secret. As time went on, however, Altshuler and co-workers did publish compressibility studies in the open literature - ZhETF (Journal of Experimental and Theoretical Physics). But since Arzamas-16 did not exist, there were no sponsoring institutions listed in the articles, nor was there any mention of laboratories in the reports themselves. It was as if the work was miraculously done without equipment in someone's garage. Zeldovich also published articles that were indirectly related to weapons research – on radiation cooling in superheated air, for example – but he could hide behind the Institute for Chemical Physics.

Khariton offered Altshuler the opportunity to work in his scientific field under ideal conditions of facilities and support. Altshuler knew that the goal of the project was the development of nuclear weapons, and he agreed with the prevailing sentiment after the Hiroshima and Nagasaki bombs that it was of prime importance to establish equilibrium with the United States as soon as possible. Living and working conditions were excellent compared to the war-torn country outside the fence. Daily life even in Moscow was grim. The orders signed by Stalin creating Beria's committee, and the principal management structure (PGU in Russian), had created in Altshuler's view a 'white archipelago' of laboratory sites scattered around the country, in contrast with the 'Gulag archipelago' of prison camps described by Solzhenitsyn. Altshuler said:

"Thousands of highly qualified scientists, engineers, and builders who survived the war and repressions were assembled under Boris Vannikov and Igor Kurchatov. Our site – KB-11 – was at the very center of activity. My first visit was in December, 1946. The place of my future work was located 'in spite of arrogant neighbors' more than ten kilometers from the nearest railway station. We traveled from the station on a bus. We were thoughtfully supplied with fur coats to keep from freezing [it was December]. Through the windows of the bus we saw passing villages reminiscent of Russia before the time of Peter the Great. At the entrance to the site we saw the monastery, with its cathedrals and town houses in a forest. 'Finnish' houses, a small factory, and the inevitable remains of a prison camp were sprinkled among the trees…Local folklore tells of hundreds of pilgrims flocking to the monastery, where they enjoyed free meals from the monks; of a visit by the tsar, and of an uprising of the local prisoners, who escaped into the forest, led by an army pilot.[18] Usually a column of prisoners went through town to work in the morning, and trudged back to camp in the evening, looking beaten down…

"Social life at KB-11 was stratified by rank and occupation. Even the local prisoners lived better here than in the more remote camps. There were no political prisoners; most of them were 'ukazniki' [petty criminals, also discussed by Sakharov below]. The technical and scientific staff lived very well. Compared to half-starved Moscow, we were in paradise. Group leaders were well paid for the times. Our families did not experience any shortages. Any and all material needs were promptly met.

"In the beginning, we lived in 'Finnish' houses or in log houses. Later two story two family cottages were built for the scientific staff [these cottages still exist in Sarov, see Figure 11.4 below]. Sakharov lived there with his wife and family. Zeldovich lived alone – his wife stayed in Moscow to work. There were kindergartens, schools, and two movie houses in the village. There was also a legitimate theater, accommodating visiting actors. A branch of the Moscow Engineering Physics Institute opened in town later, giving us a technical college. In the early years, we had a lot of trouble with travel restrictions."

Travel restrictions for all residents, including scientists, and the official non-existence of the site were in conflict with Altshuler's claims of generous and speedy supply of all needs. Someone had to drive the trucks into town loaded with equipment, let alone all the cabbages, turnips and vodka required to keep everyone in good working condition. They had to know that the town was there, and that something was going on. They were no doubt sworn to secrecy on pain of death, but still, it must have been difficult to maintain security. Altshuler made a few remarks about security, and about political correctness on the part of the scientific staff.

"Severe consequences, even death, unavoidably awaited those who lost secret materials, not only at the site, but anywhere involved in the project. A co-worker at

---

[18] This uprising preceded the one recounted by Sakharov below.

one of the Moscow institutes carelessly put secret material in the wrong folder. He then forgot where he had put it, and could not explain the loss. He committed suicide. A senior scientist at KB-11, Dimitry Stelmakhovich, found himself in an untenable situation with respect to the security police, and also committed suicide.

"I would awaken from a recurrent dream in a cold sweat. I would be in Moscow, walking along Kropatkin Street, carrying a brief case full of secret and top secret documents, and I sensed that I would die, because I could not explain why I was carrying them.

"A little humor helped to assuage the tension. There was a sign board from the machine shop hanging on my office wall: 'Attention! Do not discuss project work here.

"The most important senior scientists had body guards who accompanied them everywhere (later only Yuli Khariton, the scientific director, had body guards). This naturally did not escape the notice of the local humorists. Andrei Sakharov wrote a poem:

There once was a lad named Adya

They gave him an uncle Dyadya

Dyadya took care of him day and night

Always behind him

Stopping him from entering a whorehouse[19]

Upholding trustworthiness"

When asked about the fact that Andrei Sakharov was a super star hero in the Atomic Project, and yet one of the leaders of the protest movement in the 1980's, Altshuler responded that Sakharov overestimated his influence over the ranking politicians in the Soviet government, and consequently said and wrote publicly more than was wise.

---

[19] The Russian slang is 'blyad' written only as b... in the poem. Hats off to Olga for interpretation. The Russian word for uncle is 'Dyadya.'

KB-11 was full of informers, and was periodically visited by travelling truth committees who checked up on everyone's reliability. Altshuler explains:

"The government approved all appointments. If a given worker turned out to be unreliable, the minister would be in trouble. If I spoke as an official somewhere, the government minister responsible for my appointment could react to what I said, and I have no problem with that. A government commission visited the lab site in 1951 to determine the effectiveness of local political education. When they asked me how I felt about the local indoctrination program, I responded that I go to the sessions, and I should have stopped there. But the devil made me add: 'Of course I do not agree with everything they say.' The visitors all sat up in unison, and asked what it was that I did not agree with. I answered that I believed that the classical geneticists were correct, and not Lysenko. Lysenko's statements make no sense. About a week later I was told that I had to leave the project, without any further instructions as to where I was supposed to go. Fortunately, Boris Vannikov's deputy, Avraami Zavenyagin, happened to be at KB-11 at that time. My colleagues, including Andrei Sakharov, appealed directly to Zavenyagin, saying that my removal from the project would cause irreparable harm. My departure was canceled.

"About a year later, without any prior warning, I was again told that I had to leave the site. This time Khariton called Beria directly. Beria asked if I was necessary to the project, and Khariton affirmed that I was. That solved the problem.

"Brief attempts to punish me for thinking freely occurred repeatedly, but were all unsuccessful. They got back at me only in 1969, when I was denied corresponding membership in the Academy of Sciences."

Altshuler was never elected to the Academy of Sciences, despite having received many awards for his work. The 1951 political correctness committee also interviewed Andrei Sakharov, who made the same response as Altshuler regarding Lysenko's pseudo-genetics, but Sakharov was not threatened. He was higher in the pecking order than Altshuler.

11.6. Andrei Sakharov at Arzamas-16

Andrei Sakharov spent 18 years of his life at KB-11, or Arzamas-16 as it was later called. Sakharov was a member of the younger generation working on the Soviet Atomic Project. Born in 1921, Sakharov was college age when the Germans invaded Russia. He was too young to participate in pre-war research. Unlike Georgy Flerov, who was 8 years older, Sakharov did not serve in the Red Army, but instead worked in a defense plant during the war. The plant was located in Ulyanovsk, on the Volga River, about 600 km east of Moscow, and considerably north of Stalingrad, where the fighting took place. Ulyanovsk was a relatively quiet town in wartime Russia. He attracted notice in the defense plant by developing new and more efficient manufacturing processes for anti-tank weapons. Sakharov thus merited good recommendations for continuing his education.

After the war, he returned to Moscow and became a graduate student under the direction of Igor Tamm at the Physical Institute of the Academy of Sciences (FIAN), now called the P.N. Lebedev Physical Institute, located at 53 Leninsky Prospect in modern Moscow. He married – his first wife was named Klava – and had a baby daughter. Living in post-war Moscow was very difficult. He wrote a dissertation in nuclear theory, and became a post-doc in Tamm's group in 1947. He was invited to give a seminar at Laboratory #2, with Igor Kurchatov in the audience. After the talk, he accompanied Kurchatov to his office. Sakharov said that Kurchatov's office was enormous, large enough to hold a conference. It had piles of journals on a large writing desk, and a nest of telephones of various colors.[20] Bookshelves loaded with scientific and reference works were hanging on the walls. Sakharov wrote:

> "Kurchatov sat behind the writing table and looked at me with his very expressive eyes. He now and then stroked his thick dark beard while talking to me. He was a very captivating individual. There was an original oil portrait of Stalin in front of the Kremlin hanging on the wall behind him. This was a symbol of the high position in the government hierarchy of the occupant of this office (the portrait hung there for a time even after the XX Party Congress).[21] Kurchatov invited me to join his Institute to work on theoretical nuclear physics."

---

[20] Andrei Sakharov, "Vospominaniya," Alfa-Kniga, Moskva, 2011, p 99.
[21] In 1956. Khrushchev denounced Stalin in a secret speech.

Kurchatov was always on the lookout for young talent, but Sakharov declined the invitation in order to stay at FIAN and work with Igor Tamm. Tamm's group became interested in nuclear fusion – both as a weapon (the hydrogen bomb), and as a controlled source of energy. Sakharov made seminal contributions to both subjects. The office in FIAN where he worked was locked and guarded, and the work was classified secret. He thought about other physics problems that should not have been classified. Thus, his original work on muon catalyzed fusion – an interesting but impractical source of 'cold fusion' - was not published in the open literature. The phenomenon was later observed experimentally in 1957 by Louis Alvarez.[22] Figure 11.3 shows Andrei Sakharov as a youth.

Figure 11.3. Andrei Sakharov as a young man. Credit AIP Emilio Segre Visual Archives, Physics Today Collection.

---

[22] L.W. Alvarez, et al., Phys Rev **105**, 1127 (1957). This letter refers to a later paper by Zeldovich, but not to Sakharov's unpublished work. A muon is 200 times heavier than an electron, and can bind a proton and a deuteron close enough to fuse, giving ^{3}He plus an energetic free muon. The process can be repeated in liquid hydrogen (with a naturally occurring deuterium contamination).

It was not until 1950, after the first successful Soviet bomb test, that Sakharov joined Kurchatov's theory group, but at Arzamas-16 rather than Laboratory #2. Sakharov's impressions of Arzamas-16 therefore pertained to a more developed and refined place than KB-11 in its early days, in 1947-48, but they are nevertheless interesting in describing the place that he saw.[23]

In the summer of 1949 Sakharov was at a shared dacha outside Moscow with his family when a black automobile drove up. An unnamed assistant to Boris Vannikov got out of the car, and said that the two of them were to travel together to Arzamas-16. (Sakharov referred to the place as the 'object,' and never identified it by its correct name.) The man gave him a Moscow address, and sent him off to get instructions. Sakharov went to the address, which was a warehouse with a sign 'Fruits and Vegetables.' He entered, and walked past people sitting around, some playing dominoes. A pale, nervous man sat behind a desk in the back room. When told that Sakharov was to go to the 'place,' and that he had never been there before, the man gave him a pass, and told him which train to take to get there. Eventually Sakharov obtained sufficient stature that he could arrange passage by telephone, but until then he had to go to the same 'Fruits and Vegetables' shop every time for authorization.[24] This must have been a nuisance, since he would often leave his family in Moscow, and come home on weekends.

Vannikov had his own railway car, so for the first visit at least Sakharov traveled in relative luxury. They boarded the train in the evening for an overnight trip. Vannikov invited Sakharov to join him in his suite in the car. M.G. Mesheryakov, the experimentalist working on isotope separation by mass spectrometry, and one of the founders of Dubna, was also present. Sakharov did not know Mesheryakov, and was mystified by their conversation, which was in some sort of secret code. He later learned

---

[23] Andrei Sakharov, "Vospominaniya," *ibid.*, Chapter 7, p 112ff.
[24] Sakharov did not say where the train went – perhaps Moscow to Gorky.

that 'Borodin' was in fact Kurchatov.[25] Sakharov did not sleep – he thought about controlled nuclear fusion. He did not say how long it took to get there – probably six or seven hours. He later learned that Arzamas-16 had an airstrip, and that it was possible to fly there from Moscow, which he did with his family – again not revealing the length of the flight.[26] Since Arzamas-16 did not officially exist, the time required to travel there had no meaning.

Vannikov and Sakharov were met at the train station, and driven to the site in a two car motorcade at breakneck speed over country roads full of pot holes. It was sunrise. The passing countryside was dotted with thatched roofed mud huts, and a few thin, raggedy looking cattle. A car passing in front of them killed a chicken crossing the road. Finally, they came up to the closed gates of the 'zone,' i.e. the prison camp attached to the town. The town itself, a considerable area,[27] was surrounded by two layers of barbed wire fencing, with a raked sand strip in between. The prison camp had its own fence inside the town. Sakharov became accustomed to seeing the prisoners under guard marching off to work every morning. The presence of the prison camp right in the middle of the laboratory made a vivid impression on Sakharov. Prisoners were an ever present feature of the Soviet Atomic Project. The Manhattan Project had no analog. Los Alamos was an army base, and the construction was done by army Corps of Engineers, who were mostly draftees, but not prisoners. Sakharov told of an incident that occurred in 1948, before his arrival at KB-11, and while the site was still more primitive.

---

25 Uranium was called 'tin,' and neutrons were 'zero points.' Manhattan Project called ^{235}U=25, and ^{239}Pu=49. Enrico Fermi was 'Henry Farmer,' which probably did not fool anyone.

26 The airstrip is only used for official business today. Tourists cannot fly in.

27 The area of modern Sarov is about 250 km², or 63 km of fencing if square. Goleusova quoted the original perimeter of 56.4 km – close to the present size.

"A small group of prisoners was digging a ditch. Among the prisoners was a former colonel (he could have been a member of the Russian Liberation Army[28]). One of the zeks (a short form for prisoner) bent over a wheel of the truck that had brought them to the work site, as if verifying something or other. One of the guards also bent over to look at the wheel. At that moment, another zek hit the guard on the head with a shovel, and the colonel grabbed his automatic weapon, shouting 'Boys, follow me!' The prisoners extracted the driver from the truck, and one of them sat behind the wheel. The colonel stood on the truck bed and shot at an approaching truck that was loaded with police. The zeks obtained more weapons from the police truck, and were armed to the teeth. They returned to the camp, disarming and shooting the guards. The colonel and about 50 others drove through the camp and out the gate of KB-11, hoping to get far enough away to be able to vanish into the woods and surrounding villages. At that time three divisions of NKVD troops, responding to an alarm, assembled. (This is what I was told. Probably no one knows exactly how many troops there were.) They formed a large ring around the escaped prisoners with trucks, tanks, and airplanes, and started to contract it, trapping the rebels inside. The final act of the tragedy was carried out in accordance with the articles of war. Every escaped prisoner perished as a result of massive artillery and mortar fire, even including fighter aircraft. It is quite possible that other prisoners, not part of the fugitive group, were also killed in the melee… After the insurrection, the composition of the prison population at KB-11 was changed. All prisoners with long sentences – who had nothing to lose – were sent away, and replaced with 'ukazniki,' that is those sentenced to shorter terms under the Order [ukaz in Russian] of the Presidium of the Supreme Soviet; typical sentences were 1 to 5 years: petty theft, stealing corn from a harvested field, hooliganism, voluntarily quitting work, like in a mine, especially if done repeatedly, unauthorized stopping of a train by pulling the emergency cord, and similar offences. There were no more insurrections."[29]

A prisoner insurrection that had produced a war zone must have been a considerable distraction for the workers at the laboratory. Sakharov did not mention political prisoners explicitly here. He did say that the release of prisoners who had knowledge of the secret location posed a problem that the government solved by sending them far away to other camps after their release. Sakharov then made a few personal observations.

"From 1950-1953 we lived right next to the prison camp. Every day in the morning a long grey column of people dressed in padded prison clothes and accompanied by

---

[28] A group of predominantly Russian soldiers that fought on the German side during WWII. The colonel was lucky he was not executed.

[29] Andrei Sakharov, "Vospominaniya," *ibid.*, p 119-120.

guards marched past our curtained windows. One could comfort oneself with the thought that they did not die of hunger, and that at other places like logging forests and uranium mines conditions were much worse. Occasionally one could help an individual some, with old clothes, a little money, some food. Once our neighbor's housekeeper cooked 12 chickens for the prisoners – that was really something! In 1953, after the amnesty [after Stalin's death], the prisoners disappeared, and were replaced by army recruits, who assumed the construction duties. Army recruits were not volunteers either, but at least they were not prisoners."

And so, eventually, Arzamas-16 became more like Los Alamos, where the army Corps of Engineers did all of the construction from the beginning. The two towns are now 'sister cities.'

The locals naturally wondered what was going on inside all that fence. The theory popular with those outside the fence was that the government was building 'trial communism,' and presumably if they got it to work it would be exported for everyone to enjoy. Los Alamos had the same sort of problem – the locals in Santa Fe wondered what was going on. The Laboratory attempted to sell the idea that the site was for rocket development, without much success. The official name for the first Soviet nuclear weapon was RDS-1, which referred to rocket engines in Russian, so the two projects independently used rocket development as a cover. Humorists said that RDS stood for 'Rossiya delaet sama,' or 'Russia did it herself.'

We now return to Sakharov's first visit. Upon arrival of the motorcade two officers appeared from the guardhouse, checked the papers of the first car, saluted, and let them in. Vannikov and Sakharov were in the second car, and Vannikov was in no mood to mess with the guards, so he ordered the driver to proceed without stopping. He could get away with it. Soon Sakharov found himself in the guest hotel, above the executive dining hall. The walls of the dining hall were decorated with stars that had been painted by a female prisoner.

Sakharov shaved and washed up after a sleepless train ride, and started down the hall only to meet Igor Kurchatov with two of his bodyguards ('secretaries') coming towards him. At that time only Kurchatov and

Khariton had bodyguards. Eventually, from 1954-57, Sakharov had them himself. They were special, high quality members of the NKVD, and Kurchatov treated them like family. Special treatment of his bodyguards may have been because of their accommodating nature, but could have also been one of many signs of Kurchatov's skills with different types of people. Fermi had a bodyguard too – a lawyer and army draftee named John Baudino, who was with him all the time, and also became a member of the family.[30] Khalatnikov told some stories about the 'secretaries.'[31] In 1953, after Stalin's death and Beria's arrest, Khrushchev and Malenkov assigned a bodyguard to Landau. Landau had a fit. He wrote a letter to Khrushchev saying 'A bird in a cage will not sing.' The idea was dropped. But many other top scientists suffered with the constant companionship of bodyguards. Artsimovich's bodyguard, accompanying him to his bedroom, said that Artsimovich was lucky, he could go to sleep, while his 'secretary' had to write up the report of the day's activities. Zeldovich prevented the entrance of his bodyguard into his quarters. Zeldovich lived in a modest three-story building on the grounds of the Institute for Chemical Physics in Moscow, and when Zeldovich was home, the bodyguard had to sit on the staircase. The building halls were not well heated in the winter. Bodyguards were a hindrance to his work, and Zeldovich eventually got rid of them altogether. But there were instances of harmonious relations, similar to Kurchatov. Academician N.N. Semenov (Nobel Prize in Chemistry in 1956), a physical chemist and contemporary of Kapitza, became so attached to his bodyguard NKVD Lieutenant P.S. Kostikov, that, when Semenov was dismissed from the Project, he hired Kostikov to work in an administrative position in his institute. He was an exceptionally skilled expediter for travel arrangements, and eventually wound up at Khalatnikov's Institute for Physical Problems, where he had a reputation for getting things done within the Soviet system. Kostikov was the exception rather than the rule, but he shows that the bodyguards were not all thugs.

---

[30] Leona Marshall Libby, "The Uranium People," *ibid.*, p 161.
[31] I.M. Khalatnikov, "Dau, Kentavr, I Drugie," *ibid.*, p 48.

Andrei Sakharov's house, which he occupied as one of the important people at Arzamas-16, is shown in Figure 11.4.

Figure 11.4. Duplex at Arzamas-16 that housed Andrei Sakharov's family in the 1960's. They lived in the right-hand half of the house. Credit AIP Emilio Segre Visual Archives, Physics Today Collection.

Kurchatov greeted Sakharov: "A Muscovite has arrived! Greetings!" He then departed in his staff car. Soon Zeldovich arrived to take Sakharov to meet the theory group. Zeldovich said that everything going on was secret, and the less one knew about it the better. He said that there were seminars all the time, and that Sakharov should not be upset if he were not invited. Zeldovich was not always invited himself! Khariton knew everything that went on, but he had broad shoulders to carry the burden. The Manhattan Project had been compartmentalized for security reasons also, but at Los Alamos Oppenheimer made a special effort that all scientific staff be kept informed of the progress. Arzamas-16 had more restrictive policies, and, as we have learned, Los Alamos' openness aided the atomic spies.

Arzamas-16 and the other laboratories must have been inconvenienced by the lack of a broad industrial base and consumer goods. The shortcomings were not fatal – the program succeeded in a timely fashion. However, little things like paper goods, tissues, aluminum foil, glues, Scotch tape, etc were either non-existent or in short supply – such stuff is useful around the lab.

Arzamas-16 had a supply of metallic fissile materials, either ^{235}U or ^{239}Pu, acquired from the remote plants that made it, and stockpiled on site. Zeldovich remarked that the metal looked perfectly ordinary, but that he could not help but think of the destructive power of each gram of the stuff – both in wear and tear on those involved in production, from mining all the way to product, and on the victims of its use. Zeldovich's theory group had three young men and one young woman, who had been his graduate students. More senior theorists, like Pomeranchuk, shown in Figure 11.5, came and went, and were not in residence at the time. The young woman not only did theoretical calculations, but also had the onerous task of keeping hand written notes of everything that went on. It was forbidden to let ordinary typists have access to secret materials. This policy was consistent with the use of code for the work – tin for uranium, etc.

Figure 11.5. Isaak Pomeranchuk at an international physics conference in the early 1960's. He made important contributions to high energy physics after his work on the Soviet Atomic Project. Credit AIP Emilio Segre Visual Archives, Physics Today Collection.

There was a conference in June, 1947 at Shelter Island, New York that stimulated feverish activity in quantum electrodynamics, and led to papers from the USA by Schwinger, Feynman, and Dyson, and from Japan by Tomonaga that introduced modern methods for calculation and greatly simplified the theory.[32] These papers were published in the open literature, and were eagerly read by Soviet theorists. However, Sakharov was otherwise occupied with nuclear fusion, and did not put in the time necessary to come up to speed in the new techniques. This neglect came back to bite him. Zeldovich asked him to give a lecture on the subject to the theory group, in order to keep everyone abreast of current progress. Unfortunately, Sakharov had only studied pre-war texts on the subject by Heitler and Wentzel, and was obliged to discuss rather outdated theories.[33]

---

[32] Feynman, Schwinger, and Tomonaga shared the 1965 Nobel Prize in physics. Dyson missed out on the three-person rule of the Nobel Committee.

[33] Gregor Wentzel was one of the author's professors at the University of Chicago. His book on the quantum theory of fields was first published in German. It was translated into English by Charlotte Houtermans – Fritz's wife.

Later on Feynman's techniques were to prove essential in calculating the effect of Compton scattering of γ rays by electrons in the plasma of a hydrogen bomb.[34]

11.7. Critical Assemblies

Arzamas-16 was a weapons lab, and weapons labs have one or more versions of 'critical assemblies,' that are like Flerov's idea for a uranium bomb – bring one subcritical mass close to another. In a laboratory set-up a critical mass can be approached in a controlled manner by varying the distance between two halves. An alternate scheme is to drop one slug through a donut hole in another, forming a critical mass for a short time during the descent. These setups are intense sources of fission neutrons that can be used for various experiments. They also were used to study the approach to criticality for various geometries, neutron reflectors, and amount of fissile materials. The Manhattan Project and its descendants after the war used similar rigs, a famous one called 'Lady Godiva.'[35] LA-13638 is co-authored by physicists from the Russian Federation, and contains accident reports from both the US and USSR weapons development programs. Criticality accidents have occurred not only at critical assemblies, where the hazards are obvious, but also at processing plants, where the assembly of enough fissile material in one place is considered unlikely (but sometimes happens). Processing plant accidents in the USSR are discussed in Chapter 12. A 6.2 kg bare sphere of nickel plated δ-phase ^{239}Pu in a critical assembly at Los Alamos was the cause of two fatal accidents in 1945-46, after the Hiroshima and Nagasaki bombs. This lump of plutonium was dubbed the 'Demon core.'[36]

Horace Smith, temporary charge d'affaires at the US embassy in Moscow in early 1946 delivered an invitation to Molotov, the Soviet

---

[34] B.L. Ioffe, HISAP96 Vol 2, *ibid.*, p 207ff.

[35] Thomas P. McLaughlin, et al., "A Review of Criticality Accidents" Los Alamos National Laboratory LA-13638, 2000 (available on line).

[36] Plutonium in the δ-phase has density 15.92 gm/cm,^{3} and is a face-centered cubic crystal lattice, like stacked grapefruit in the grocery store.

foreign minister, to send some observers to the planned nuclear tests in the summer of 1946 at Bikini Atoll in the Pacific Ocean.[37] The invitation mentioned the Soviet Union's membership in the United Nations Committee on Atomic Energy, and their obvious interest in nuclear weapons. Two delegates were invited to attend, with an additional member of the Soviet press corps. The tests were scheduled for July and August, 1946. A warship leaving San Francisco on June 12 would carry the Soviet delegation to Bikini, with a one day layover in Honolulu. The delegation would be on the warship for about three months, would have to share staterooms, and be prepared for hot weather. They should have all of their vaccinations against small pox, typhoid, etc., along with their passports. The US Navy regretted that it would not be possible to interrupt the voyage and return early. Per diem charges on the ship would be $1.50 per day per person. Stalin, Molotov, Beria and company mulled over the invitation. In a telegram from Foreign Minister Molotov in Moscow to Ambassador Andrei Gromyko in Washington, D.C. on February 2, 1946, the Politburo accepted, and awaited further instructions.[38]

Admiral Kuznetsov was enthusiastic about the tests. He wanted to know the effects of nuclear weapons on ships, and was anxious to send a navy officer. After some sifting and winnowing, the three delegates who went were a physicist from Laboratory #2 who was working on electromagnetic isotope separation described in Chapter 10, M. G. Mesheryakov, the naval officer requested by Admiral Kuznetsov who was already in the USA, A.M. Khokhlov, and S.P. Aleksandrov, an engineer in the Ministry of Internal Affairs (Beria's operation). Khokhlov doubled as a member of the press corps.

The Soviet Academy of Sciences together with the Navy mobilized to instrument the Pacific coast of the USSR with detectors for seismic,

---

[37] Atomnyy_proekt_SSSR._T.2.Kn.6.(2006).pdf, #38 Smith to Molotov.
[38] Atomnyy_proekt_SSSR._T.2.Kn.6.(2006).pdf, #20 Molotov to Gromyko.

atmospheric, and radiation signals originating from the US tests.[39] Their efforts proved unsuccessful, because they were too far away from the blast.

There were two tests in 1946, called 'Able' and 'Baker,' both about 20 kilotons of TNT equivalent. Able was dropped from an aircraft, and exploded in the air at an altitude of 100-200 m above the water. Able was made from the 'Demon core' plutonium of the Los Alamos critical assembly. Baker was an underwater explosion, resulting in a very interesting expanding sphere of water vapor.[40] There were several expendable US navy ships in the harbor, some of which were sunk by the blast. The rest were sunk later, being too radioactive for service. Some ships had live stock on board to assess biological effects. All of this information was duly reported back to Stalin by Mesheryakov, who witnessed the events, and had access to all of the data that were publicly released.[41] Mesheryakov became convinced of the need to study possible safety precautions in the event of nuclear war.

The Arzamas-16 critical assembly rig was called 'fast neutron boiler,' with the Russian acronym FIKOBYN. It consisted of two subcritical parts separated by a gap that could be varied. As the gap was decreased, the activity increased, giving an intense source of fission neutrons and fragment gamma rays. Some discussion of FIKOBYN and other critical assemblies is given in Appendix J.

The engineer in charge of the Arzamas-16 rig was one Shirshov, a skilled technician who also liked a drink now and then. Vannikov caught him at work with too much to drink, and fired him. He was immortalized by the unit 'Shirsha,' a measure of the gap between the shells – the spacers in Figure J.2 of Appendix J. The operators would say 'Set the gap at 5 Shirshas.' Critical assemblies were accident prone, and FIKOBYN was no exception.[42] An error in procedure allowed the active material to achieve

---

[39] Atomnyy_proekt_SSSR. T.2.Kn.6.(2006).pdf, #55, Semenov to Makhnev, 18 June 1946.

[40] en.wikipedia.org/wiki/Nuclear_testing_at_Bikini_Atoll.

[41] Atomnyy_proekt_SSSR. T.2.Kn.6.(2006).pdf, #101, Beria to Stalin, 7 March 1947.

[42] More than one accident, and two fatalities, occurred at Los Alamos.

low criticality, resulting in a very large neutron dose. Fortunately, no one was killed.

Another critical assembly incident was recounted by Vannikov in his oral interview.[43] He walked into the room where a critical assembly was being used, without the knowledge of the operators, who were presumably behind some suitable shielding. (Such carelessness would not be possible today.) As he approached the assembly, enough neutrons were reflected off of his body and back into the fissile material to exceed criticality, with an attendant large increase in radiation. Vannikov's white blood cell count dropped sharply in subsequent days, and he was indisposed for some time, but he recovered.

Sakharov was only a visitor at Arzamas-16, and he soon returned to Moscow to resume his fusion work with Igor Tamm. He and Klava were expecting their second child. Housing was grim, and the Sakharov family was growing. He and his first young daughter lived with his parents in Moscow while Klava was giving birth. On Zeldovich's recommendation, Sakharov appealed to Kurchatov for better housing. Like magic, the family obtained a three-room private apartment on the outskirts of Moscow. The entire Sakharov family was rapturous. Zeldovich astutely observed that it was the very first peaceful use of thermonuclear energy!

At the beginning of March, 1950, Sakharov received orders to return to permanent work at Arzamas-16. He was to remain there until 1968. His monthly salary was 2000 rubles, which was good pay. In Chapter 7 we noted that, in 1944, Kurchatov himself was paid 3000 rubles a month. By 1950 he was no doubt receiving much more, since the successful atomic test in 1949. Sakharov lived alone in the visitors' hotel. His family would join him in six months. Igor Tamm flew into Arzamas-16 in April, complete with skis and backpack. Soon Isaak Pomeranchuck and Nikolai Bogolyubov joined the theory group from the Moscow Institute (now

---

43 Raisa Kuznetsova, "Kurchatov v Zhizni," *ibid*, p 477.

ITEF). Some of Bogolyubov's students, including Dmitry Shirkov came along also, resulting in considerable theoretical firepower.[44]

11.8. Theoretical Work

What did the theorists do? The short answer is, they did anything required for the task at hand – the development of nuclear weapons. Nuclear reactors are complicated devices, and proper design and control require theoretical understanding. There are calculations to be done in nuclear physics. And there are shock wave calculations needed for explosive compression. Theoretical analysis of the thermodynamics of explosions, both conventional chemical and nuclear, was necessary, and very complex. A good example of a calculation that requires complete analysis of the explosion is the efficiency of the blast.[45] The energetics of chemical and nuclear explosions are discussed in Appendix F. The benchmark number for a chemical explosion is TNT, and 1 kg of TNT releases 4.18 MJ of energy.[46] Six kg of ^{239}Pu contains 1.5×10^{25} Pu nuclei. If every one of them fissioned, and released 200 MeV of energy, then six kg of plutonium would liberate 4.8×10^{14} J, or about 100,000 metric tons of TNT equivalent. The actual energy release is about 20% of this number, and it is obviously very useful for safety reasons to get a reliable estimate of the efficiency before testing the weapon.

Aleksandr A. Samarsky was an applied mathematician who joined the Soviet Atomic Project at age 29 in 1948, after being wounded in Red Army service during WWII. He was educated at MGU. He writes:

---

[44] Bogolyubov and Shirkov later wrote a popular text on quantum field theory.

[45] A.A. Samarsky," First calculation of the force of the explosion," HISAP96, Vol 1, *ibid.*, p 214.

[46] 4.18 MJ/kg looks precise, but is really arbitrary. 2.7 MJ/kg is the true free energy of the explosion that leaves two carbon atoms behind. Reaction with oxygen in the air gives an extra boost. 4.18 J/Calorie is a true constant, relating mechanical and heat energy, and that number was adopted as the standard.

"Stimulated by Igor Kurchatov, the Council of Ministers of the USSR setup a special laboratory under the direction of A.N. Tikhonov to perform the calculations necessary to understand the processes of a nuclear explosion. The laboratory was located in the Geophysics Institute of the Academy of Sciences in Moscow. At that time Tikhonov was head of the mathematics department at MGU and also the mathematics section of the Geophysics Institute. Finding the staff for the laboratory was not simple. The work required the union of physicists, mathematicians who specialized in differential equations, and numerical analysts. Qualified specialists in numerical methods at that time did not exist...Digital computers were still six years in the future, so the work had to be done with mechanical calculating machines."

There is a discussion in the chapter on Los Alamos, Chapter 6, on the Manhattan Project effort to handle the same problem. They did not have computers either until after WWII. The first computer at Los Alamos was called MANIAC, and did calculations for the design of the hydrogen bomb in 1951.

The first Soviet digital computer was called 'STRELA' (arrow), manufactured in Moscow from 1953-1957, capable of 2000 operations per second, with a 2048 word random access memory.[47] Seven units were built for use in computing centers around Moscow. STRELA came too late to work on early bomb design problems.

Several laboratories in the Moscow region got involved in the effort, including the Institute for Physical Problems – Kapitza's lab now run by Aleksandrov, Laboratory #2, KB-11, the Institute for Theoretical and Experimental Physics (ITEP), and the Institute for Applied Mathematics, (IPM) headed by M.B. Keldysh.[48] L.D. Landau, E.M. Lifshits, and I.M. Khalatnikov all worked at the Institute for Physical Problems, where they

---

[47] www.computer-museum.ru/strela
[48] Keldysh was also president of the Academy of Sciences. The research vessel that discovered the Titanic was named after him.

were able to combine classified research with work that could be published in the open literature on low temperature and solid state phenomena. Samarsky continues:

> "I was assigned the problem of developing a numerical method for the complete system of equations with partial derivatives describing a nuclear explosion. V.Ya Goldin and O.P. Kremer were assigned the task of calculating the explosion using the approximation of Landau, Lifshits, and Khalatnikov. They constructed a mathematical model of a nuclear explosion in the form of a system of ordinary differential equations for the relevant variables averaged in space."

The Landau group participated in the Soviet Atomic Project in response to specific requests from Kurchatov or his deputies. The calculation of the efficiency of a nuclear explosion was an early assignment in 1946. According to Khalatnikov, Landau set up a numerical computation center at the Institute, with 20 or 30 women operators of German electric calculating machines, under the direction of Naum Meiman, a mathematician who had met Landau at Kharkov in the 1930's.[49] L.D. Landau had a unique aura among Soviet physicists, and his expertise was eagerly sought to solve many complicated problems. However, Landau wanted to maintain his independence, and not be drawn in to the morass of the Project. Sakharov once asked Zeldovich why Igor Tamm was dedicated to the Project, while Landau preferred to remain detached. Zeldovich answered that "Igor Tamm has higher moral standards."[50] Zeldovich meant by 'moral standards' a willingness to devote all his strength to the Project. Sakharov did not know much about Landau's position on the subject, but he later had an opportunity in the mid 1950's to visit Landau's Institute in regard to some theoretical work ongoing on the hydrogen bomb. Many numerical calculations were required to understand the behavior of the hydrogen bomb during combustion. This was one of the challenges driving the development of digital computers. A fission bomb is a trigger for the fusion process;

---

[49] I.M. Khalatnikov, "Dau, Kentavr, I Drugie," *ibid.*, p 43.
[50] Andrei Sakharov, "Vospominaniya," *ibid.*, p 130.

fission supplies the neutrons and X-rays necessary to ignite the d-t reaction, so the explosive volume is a very busy place.

After completing their scientific discussion, Sakharov and Landau went walking outside in the garden, and Landau said: "I really do not like this activity. (Referring to nuclear weapons in general and his participation in particular.)" Sakharov asked why, and Landau replied: "Too much noise." Landau preferred peace and quiet, and would happily work on his own, untrammeled by external forces. On the other hand, Landau had served one year in prison in 1939, and knew that, as long as Stalin and Beria were in control, he had to produce or else. Not only was he obliged to respond to requests, but also he had to perform up to his high standards. When Stalin died in March, 1953, Landau said: "He has gone. I fear him no longer, and I will not do this work anymore."[51] Kurchatov, Khariton, and Sakharov soon offered the job of coordination of the theoretical work for the Project at the Institute for Physical Problems to Khalatnikov. By mid 1953 the job was well along, but there were always personnel problems, and Landau had not been particularly attentive to the daily lives of his workers. The mathematician N.N. Meiman later became active in the human rights movement together with Andrei Sakharov. In 1953, he was a crucial part of the numerical calculation effort at IPP, and Khalatnikov worried about his health and wellbeing. Meiman was sickly, always short of cash, and needed an extra teaching job to make ends meet, so he traveled to Ivanovo once a week to give math lectures. His housing was miserable (this seems fairly typical of anyone living in post war Moscow). Khalatnikov came to the rescue, and pulled strings to get him a better place to live, in the same building as Igor Tamm.

We return to 1946 and Samarsky's recount of his work in calculating the efficiency of a fission weapon.

"V. Ya Goldin was assigned the task of verifying the set of ordinary differential equations of Landau et al. He constructed the complete system of equations of the explosion in partial derivatives and an integral-differential equation for neutron transport. He was able to show that the ordinary differential equations could be

---

[51] I.M. Khalatnikov, "Dau, Kentavr, I Drugie," *ibid.*, p 50.

obtained from the complete system of partial differential equations by using an approximation supplied by Khalatnikov.

"Theoretical and physical aspects of the calculations, in particular stating the Cauchy boundary conditions for the system of ordinary differential equations and their singular points, was carried out by Goldin together with N.N Yanenko. All of these calculations were urgently needed, and their results were promptly put to use. A.A. Samarsky was occupied with working up the first calculation of a nuclear explosion.

"The calculation of a nuclear explosion required the simultaneous solution of the kinetic equation of neutron transport, and the equations of gas dynamics with heat conduction. In 1948, I set up a monotonic finite difference procedure to evaluate the transport equations in spherical coordinates, and in early 1949 we had the first calculations for a complete set of explosion equations, first for a plutonium sphere, and then with added uranium spherical shells (tampers). In addition, we developed averaged equations from the partial derivatives in space and angles, and used average values of the complete solutions to these equations, leading to the system of ordinary differential equations."

"Thus, in less than a year a group of three scientific workers and a group of operators working on hand powered adding machines, had managed to build methods, tune up the calculations, and obtain the first production results – all beginning from zero.

"More complicated calculations were done in 1949-1950. A differential-difference approximation was used; differential in time, and difference in space, that was solved by interation-finite difference methods. A method of parallel calculation was developed to speed things up, with 30 to 40 operators of calculating machines.

"Processing of the calculations obtained in 1949-1950 allowed me to formulate in 1950 general conservation laws on a discreet level for finite difference calculus.[52] Two years of experience showed that it was necessary to pay attention to the development of theoretical work. Tikhonov and Samarsky studied the conservation laws in the homogeneous finite difference scheme in detail. They found the conditions necessary to satisfy the conservation laws for the class of differential equations under study."

---

[52] The conservation laws are built into the partial differential equations, but it is possible to lose this important feature in making approximations.

The efficiency of a ^{235}U gun-type tampered bomb – the Hiroshima bomb – is discussed in Appendix F. The discussion is not very complicated, and the results are not very accurate, but it did not require several years to work it out. The problem Samarsky and company addressed is the ^{239}Pu implosion bomb, that is more complicated to describe because of the compression of the active material by the implosion lenses. Subsequently Samarsky's group worked on hydrogen bombs, where it is necessary to understand the products of the fission bomb used to ignite the fusion.

While the efficiency of a nuclear weapon is well defined – the number of nuclei undergoing fission divided by the total number available – it is not easy to calculate with precision, and it is certainly not easy to measure as the explosion develops.

Samarsky does not say what efficiency was predicted for the first test explosion, but inference from various documents would indicate a number like 10% was expected. RDS-1 was a surface test. The weapon was on a tower 37.5 m high. It was an implosion bomb, with 6 kg of ^{239}Pu.[53] It was a copy of the Trinity test bomb, that had an estimated energy of 20,000 tons of TNT, or an efficiency of about 20%. The energy of the blast is not all that easy to measure. The relevant parameters are the force of the shock wave, the radius and luminosity of the fireball, the crater dug at ground zero, the destructive forces as a function of distance, and clues to the number of nuclei that fissioned – the true measure of the efficiency – from fission fragments retrieved from the ground, as well as plutonium and uranium found as evidence of nuclei that did not fission. In other words, the theoretical calculation and the experimental result are disconnected. Theory attempts to calculate the number of fissile nuclei that undergo fission before the material expands enough to cease the chain reaction. What is measured for the most part is the destructive force of the blast – that is the parameter of military interest. Two eye witnesses to RDS-1 gave

---

[53] Six kg = 1.5×10^{25} nuclei. At 180 MeV per fission, 100% efficiency=4.3×10^{14} J. Assuming 1 metric ton of TNT = 4.2×10^9 J gives$\approx$100,000 tons of TNT.

estimates of the yield energy of 2000 tons[54], and 22,000 tons[55], differing by a factor of ten. The second estimate is closer to the accepted value, but the truth probably lies somewhere in the middle. Dokuchaev based his estimate on the small size of the crater, and the shorter radius of total destruction, compared to data from the Trinity test at Alamogordo. Mesheryakov, who witnessed the Bikini tests, was also at RDS-1, and assured everyone that the explosion was similar to Able.

The design of a fusion weapon, Sakharov's main contribution, requires elaborate numerical calculations, since a fission bomb is used to generate the heat and pressure to ignite the fusion component. The environment in which the detonation takes place is similar to that of the interior of a star, and not usually encountered on earth. The theory group at Arzamas-16 was occupied after 1950 with calculations pertaining to the hydrogen bomb.

The fusion reactions used to generate energy are $d + d \rightarrow {}^3He + n$, $d + d \rightarrow t + p$, and $d + t \rightarrow {}^4He + n$; ($d$ = deuterium ${}^2H$; t = tritium ${}^3H$). The $d + d$ reactions release 3.3 and 4.0 MeV respectively, and the $d + t$ reaction releases 17.6 MeV, roughly 1/10 of the energy released in fission, but only about 1/100 the molecular weight. Mixing tritium into the deuterium is therefore advantageous regarding the energy release, but tritium is expensive (it is made in reactors), and it decays with a 12-year half-life: ${}^3H \rightarrow {}^3He + \beta^- + \nu$. ${}^3He$ is stable.[56] Consider the $d + t$ reaction. Avogadro's number of d's and the same number of t's, which is 5 gm of material, gives 10^{25} Mev, or 380 metric tons of TNT equivalent. One kilogram of material equals 76 kT of energy, almost the same as 6 kg of ${}^{239}Pu$.[57] Fusion is not a chain reaction; the $d + t$ interact directly, without needing any secondary

---

[54] Y P. Dokuchaev, "The first Soviet Atomic Bomb," HISAP96, Vol 2, *ibid.* p 300.

[55] V.M. Gorbachev, "The first USSR Atomic Explosion – How the force was measured." HISAP96, Vol 2, *ibid.*, p 488.

[56] The energy released is only 18 keV, so this is a popular decay for attempts to measure the neutrino mass.

[57] 50 kg deuterium=1.5×10^{28} atoms. If all $d + d$ pairs fused liberating 4 MeV, 50 kg$\rightarrow 4.8 \times 10^{15}$ joules, or about 1000 kT of TNT.

particles, and there is no critical mass. Any amount of d + t that can be rapidly burned will contribute to the energy of the explosion. Expansion during the chain reaction can quench plutonium fission when the critical mass exceeds the sample mass, and this limits the explosive energy of a fission weapon. Deuterium-tritium fusion does not have this limitation. The problem is to get it to burn. While a large enough block of plutonium sitting in the laboratory will explode, a bottle of room temperature deuterium + tritium gas, or a cryostat of the liquids, no matter how large, is stable. The coulomb repulsion between two positively charged nuclei keeps them apart, and in order to overcome this, the gas has to be heated to a very high temperature – several hundred million degrees K. These temperatures are reached by a nuclear fission bomb, but are not needed to trigger fission in the first place. The analysis of the d + d reaction is similar, with an energy release for 4 gm of material of 75 T of TNT equivalent.

The high temperature requirement naturally leads to the idea of a two component weapon, in which the trigger is a fission bomb. Hydrogen bomb development started in both the USA and the USSR before a successful fission bomb was tested. Each group studied a 'tube' bomb, in which a cylinder of liquid deuterium (a cryogenic liquid, requiring refrigeration) was ignited by a plutonium compression trigger at one end of the cylinder. Theory groups labored over the energy balance in the environment of a fission explosion, and attempted to calculate whether the deuterium, if it ignited at all, would continue to burn down the cylinder fast enough to contribute significant energy to the explosion. There were several sources of energy loss, among them the γ rays mentioned in section 11.6 above. Accurate solutions to the differential equations were difficult to achieve without digital computers, but eventually on both sides of the Atlantic Ocean the conclusion was, without experimentally testing the idea, that the 'tube' design would not work.[58]

---

[58] B.L. Ioffe, HISAP96, Vol 2, *ibid.*, p 207. Ioffe discussed the calculations that led to the conclusion that the 'tube' would not work.

The transfer of information from the USA to the USSR ended in 1946, when Klaus Fuchs left Los Alamos for England. While he continued to give progress reports until his arrest in 1950, these were limited to work in England. Exchange of information between the USA and England dried up after 1946. So as a result, while Igor Kurchatov and some of the other scientists in the Soviet Project knew of the American efforts on the 'tube' design, they did not know of more recent developments, and a working hydrogen bomb was developed independently at Arzamas-16.

The next idea incorporated ^{6}Li deuteride, a solid form of deuterium rather than a gas or a cryogenic liquid- obviously more compact and easier to handle. The isotope of lithium is stable, and when bombarded by fast neutrons from the fission bomb, makes tritium: $n + {}^6Li \rightarrow t + {}^4He$, allowing the $d + t$ reaction to proceed as well as $d + d$, and avoiding the problem of the tritium half life. Sakharov's first design was a 'layer cake' of slices of lithium deuteride between slabs of uranium or plutonium, that was successfully tested in August, 1953.

The next refinement was to separate the lithium deuteride and the plutonium bomb, and use the fission radiation energy to initiate fusion by compressing the deuterium. In this design, which is still in use today, the fission bomb and the lithium deuteride share the same cylindrical casing that contains the fission energy long enough to ignite fusion, thus adding to the blast. In the West this is called the Ulam-Teller design, while in the Soviet Union it was independently developed by Andrei Sakharov, V.A. Davidenko, and Ya. B. Zeldovich. The rearrangement idea met with some resistance at Arzamas-16, and Yuli Khariton had to step in to support the change.[59] Figure 11.6 gives a sketch of a two component hydrogen bomb.

---

[59] Yu.B. Khariton, V.B. Adamsky, and Yu. N. Smirnov, "Making the Soviet Hydrogen Bomb," HISAP96, Vol 1, *ibid.*, p 200.

Two Component Hydrogen Bomb

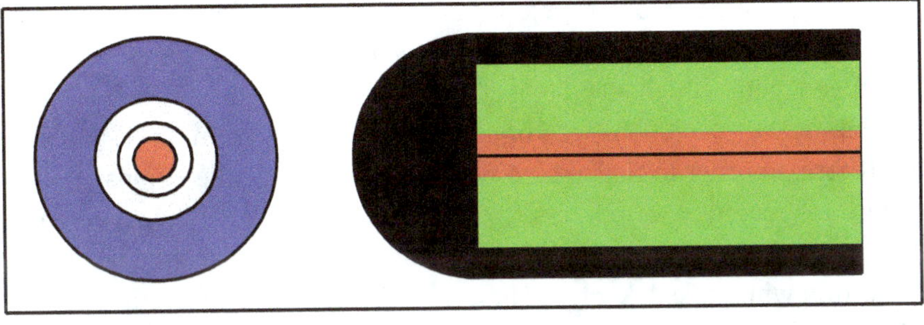

Figure 11.6. Sketch of a two component hydrogen bomb. The primary is an implosion plutonium bomb. The conventional explosive is blue, and the levitated plutonium core is red. Suspending the core improves the force of compression – an upgrade to the original design. The secondary is a uranium shell (black) surrounding lithium 6 deuteride (green) with a plutonium rod in the center. Fast neutrons from fusion cause the plutonium rod to fission.

11.9. Conventional Explosive Tests

Testing explosives is one of the many hazardous activities involved in nuclear weapons development. A Hiroshima type uranium bomb required a plugged gun barrel, with sub-critical hemispheres fired at each other to ignite the chain reaction. Gun barrel experiments had begun at Laboratory #2, but became too dangerous, and needed some isolated, open territory. In the Nagasaki bomb a chain reaction was triggered by compressing a sub-critical plutonium sphere. This was the weapon tested at Alamogordo, New Mexico, in July, 1945. The compression was achieved by surrounding the plutonium sphere with a shell of chemical explosives. Los Alamos is located on a mesa in mountainous country, with many canyons and caves that can serve as testing sites for explosives. Arzamas-16, like most of European Russia, is flat as a dime, so suitable bunkers had to be constructed for the purpose.

Working with explosives is inherently dangerous, and it is not so easy to learn what happened, since all you have after the blast is a pile of dust and perhaps some smoke. There is some discussion of TNT in Appendix F. One of the parameters of the material is its explosive velocity, which is the speed with which the detonation wave propagates. For TNT, this number is $v_e = 6900$ m/sec, or the wave travels 1 cm in 1.4 μ sec. Therefore, in order to understand what goes on during the explosion, one needs to record the process in microsecond time bins. Venaimin Tsukerman, who was legally blind from an illness, developed a technique to do this with pulsed X-rays. The principle is illustrated in Figure 11.7. The explosive device under study is in the middle of the array, and is surrounded by a thick iron shield that has collimator holes drilled in it. Pulsed X-ray tubes are arranged on one side of the setup, with X-ray film on the other side, connected through the collimators. An electric pulse triggers both the explosion and the first X-ray tube. Then subsequent tubes fire after 1, 2, and 3 μ sec, for example, and each takes its own picture spatially separated on the film. In this way, the development of an explosion could be recorded.[60] As time progressed, the X-rays sampled different angles perpendicular to the axis of the explosion, which was OK for a cylindrically symmetrical configuration. The shielding and collimators had to be robust enough to stand several shots without destroying the external apparatus or killing the experimenters.

---

[60] N.G. Makeev, "Birth of pulsed X-ray photography in the USSR," HISAP96, Vol 2, *ibid.*, p 463.

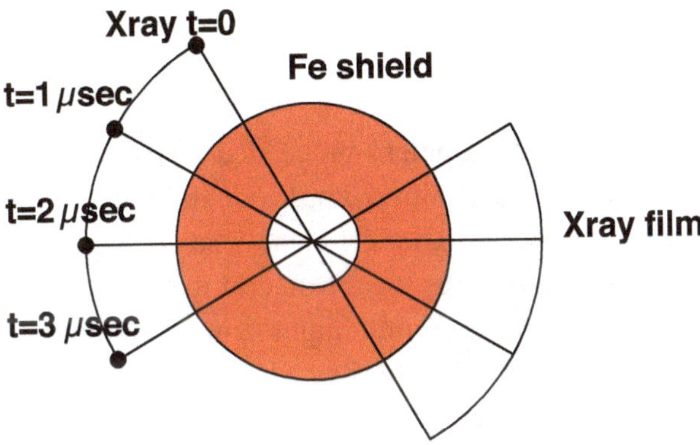

Figure 11.7. Schematic of setup used to record the development of an explosion using pulsed X-rays that sample the material at different angles at different times. Time is recorded as position on the film.

The critical mass depends inversely on the square of the density: $M_c/M_o = (\rho_o/\rho)^2$. Thus, if the uncompressed mass $M_o = \frac{1}{2}$ the critical mass, then the compression must increase the density to $\rho = \sqrt{2} \times \rho_o$ to make the same mass become critical. It is therefore of interest to know how much pressure is required to give an increase in density by $\sqrt{2}$, and to know whether or not the explosive configuration will supply that increase in pressure. The first quantity is the compressibility of plutonium. The second quantity, the pressure wave, depends on the explosive configuration. Figure F.1 in Appendix F shows data from Arzamas-16 on the density dependence of uranium under extreme pressures obtained from explosives. The uranium 'fluid' is twice as dense as the ordinary metal in a pressure of about 8 million atmospheres – a staggering pressure. They do not show a similar curve for plutonium – the substance of interest. Plutonium is probably similar to uranium; the metal has several different forms at atmospheric pressure, but at these extreme pressures that does not matter. Perhaps it is still classified.[61]

11.10. Later Developments

Work on weapons development proceeded in parallel with the main thrust to reproduce the plutonium implosion bomb. New designs for fission bombs incorporated higher efficiency, smaller linear dimensions, and less weight, and these ideas were tested in RDS-2 → RDS-5. RDS-6 was reserved for the hydrogen bomb. Engineering to produce a deliverable weapon that would fit in the bomb bay of an aircraft, or later could be carried in the nose of a rocket required coordination between Arzamas-16 and the Red Army. Part of the R&D involved automatic detonators for the conventional explosives, and an external neutron source trigger for the chain reaction to replace the Ra-Po source 'urchin' that was buried in the center of the plutonium sphere. This allowed more flexibility in the design of the shape of the nuclear charge.[62]

While one might wonder just how 'civilian' Lavrenty Beria was, he was not a marshal in the Red Army, and so it is possible to assert that the Soviet Atomic Project was under complete civilian control, with army officers only vaguely aware of what was going on, until the early 1950's. Then the requirements of practical military application of nuclear weapons brought in a strong military presence.

The Soviet rocket program has a history dating back to the 1930's, and had a significant impact on the cold war and the super-power status of the USSR in the 1950's and 1960's. Although it achieved international prominence under Nikita Khrushchev, it was humming along nicely under Joseph Stalin as well. Sergei Korolev is credited as leader of the program, together with the aircraft designers A.N. Tupolev, Artem Mikoyan

---

[61] L.V. Altshuler and K.K. Krupnikov, "Experimental Studies at Arzamas-16," HISAP96, Vol h`1, *ibid.*, p 184.

[62] A.A. Brish, "On Branch KB-11," HISAP96, Vol 1, *ibid.*, p 192.

(brother of Anastas Mikoyan) and P.O. Sukhoy.[63] Although the Soviet government supported the aircraft and rocket R&D, and successful rockets and nuclear warheads came together at the same time, the working engineers did not enjoy as much comfort as was afforded the bomb scientists. There is some discussion of their living conditions in Section 4.8 of Chapter 4. Sergei Korolev is pictured along with Kurchatov and Keldysh in Figure 11.8.

Figure 11.8. Rocket designer Sergei Korolev, Igor Kurchatov, and president of the Soviet Academy of Sciences, Mstislav Keldysh about 1950. Credit AIP Emilio Segre Visual Archives.

---

[63] The Soviet jet airliner Tu-104 was named after its designer Tupolev, the jet fighter MiG-15 after Mikoyan and Gurevich, and Sukhoy military aircraft (Su-27 for instance) are still produced today.

Chapter 12

Uranium and Plutonium

12.1. The Uranium Problem

Neither the Manhattan Project nor the Soviet Atomic Project got started with an adequate supply of uranium. Unlike iron or aluminum or carbon, uranium was not processed in industrial quantities before WWII. Its common ore is uraninite UO_2, also called pitchblende because of its black appearance. Pitchblende was refined to extract radium, a rare trace element. The refinement of the ore produces a uranium compound called yellowcake because of its appearance, that is mostly U_3O_8. Radium had several commercial and medical applications, and was very expensive because it occurred in such tiny amounts. The left over uranium was considered a material with little value, used as a component in ceramics.[1]

Uranium ore was discovered in a silver mine in Jachymov, in the NW corner of Czechoslovakia, near the German border. The relevant region is shown in Map #4. The element was first isolated by a German chemist named Martin Klaproth in 1789.[2] It was unknown that the element was radioactive until the discovery by Henri Becquerel, who exposed a sample of potassium uranyl sulphate from the Czech mine to a photographic plate in 1896. The Curies subsequently refined Czech pitchblende to discover radium.

---

[1] Nikolaus Riehl, "Stalin's Captive," ibid.
[2] education.jlab.org/itselemental/ele092.html

A gram of pitchblende is 13 times more radioactive than a gram of pure uranium, and the source of this extra radioactivity was the goal of the Curies' distillation.

All elements with atomic number greater than Z = 83 are unstable.[3] Lead has Z=82, and all three radioactive decay chains in uranium ore end up on the three naturally occurring stable isotopes of lead. Thus $^{238}U \rightarrow$ ^{206}Pb; $^{235}U \rightarrow$ ^{207}Pb; and $^{232}Th \rightarrow$ ^{208}Pb. The isotope ^{208}Pb is the most abundant. Since uranium has Z=92, and lead has Z=82, the transition U $\rightarrow$ Pb can be done in five steps, each one involving the emission of an α particle that has Z=2. This is what happens in the decay chain. There are some ancillary β decays that correct for the fact that uranium has too many neutrons, but the real chain is five α decays, changing Z by two units each time. Schematically U $\rightarrow$ Th (thorium)+ α; Th $\rightarrow$ Ra (radium)+ α; Ra $\rightarrow$ Rn (radon) + α; Rn $\rightarrow$ Po (polonium) + α; and finally Po $\rightarrow$ Pb + α. The atomic numbers of the daughters in the chain are all even – the odd atomic numbers are hopped over. All of this is taking place in the ore, creating the radioactive elements thorium, radium, radon, and polonium. Radon is a gas, and sometimes occurs in basements. The detection of radon in the atmosphere was used in prospecting for the ore. There was at least one instance of the discovery of a promising site in Eastern Siberia by prospecting from aircraft.[4] The α particles acquire electrons to become helium gas, some of which finds its way into natural gas wells.

Radium has atomic number Z = 88. The isotope $^{226}Ra \rightarrow$ $^{222}Rn + \alpha$ has a half-life of 1600 years, while its parent ^{238}U has a half-life of 4.5×10^9 years, a factor of 2.8×10^6 longer. Thus one gram of radium is 2,800,000 times more radioactive than one gram of uranium. Assuming that the factor of 13 enhancement in the activity of pitchblende is due to radium, then one gram of pitchblende contains 4.6×10^{-6} grams of radium. This demonstrates the feat of chemistry achieved by the Curies – to isolate a previously unknown element in concentration of a few parts per million.

---

[3] Bismuth is Z=83, and ^{209}Bi, the naturally occurring isotope, α decays to ^{205}Th, but the half-life is so long – 1.9×10^{19} years, that it can be considered 'stable.'

[4] Atomnyy_proekt_SSSR._T.2.Kn.4.(2003).pdf # 224.

Compared to radium, uranium is only mildly radioactive. However, if a sensitive detector is placed near a slab of uranium – either natural or depleted[5] - it will register plenty of activity. There are a number of energetic γ rays that are emitted by the various daughter elements in the decay chain.

Uranium ore is widely distributed in the earth's crust in several different mineral forms. The uranium content of the ore varies considerably, from as high as 20% in very rich ores to as little as 0.02%. "High grade" ore has a uranium content $\geq 0.1\%$. In prospecting, a survey instrument like a Geiger counter is used to detect the radioactivity that is the signature of the ore. Prospecting before WWII was limited, and as noted above the commercial element was radium, not uranium. Since the Czech mines are on the German border, there are also nearby mines in Southern Germany. The 'Erzgebirge' mountain chain runs along the Czech border. The best prewar mine was the Shinkolobwe in the Belgian Congo in Africa. Uranium was discovered in Canada's Northwest Territories in the 1930's. The known sources in the USSR were in Central Asia, and not developed.

Union Miniere du Haut Katanga was the Belgian mining company. One of its executives named Edgar Sengier deposited 1500 tons of high grade uranium ore in drums in a warehouse on Staten Island, New York in 1940.[6] Sengier wanted to keep the ore from falling into German hands. Neither Libby nor Groves comments on his motivation for doing this. He almost certainly knew nothing about possible military applications of the material. He probably figured that it was a valuable substance, and therefore it should be kept safe from the German army. In any event he did not manage to squirrel away all of the Belgian ore. A substantial amount remained in Brussels, and was captured for the Uranverein (the German atomic power project). Part was subsequently liberated for the Soviet Union by Kikoin and Khariton in Germany after the war, as recounted in

---

[5]   The isotope ^{235}U contributes about 5% to the radioactivity of natural uranium.
[6]   Leona Marshall Libby, "The Uranium People," ibid., p 83. Also mentioned by Groves in "Now It Can Be Told."

Chapter 7, and part was liberated for the USA by Sam Goudsmidt's Alsos mission. Hence the Union Miniere in the end furnished uranium for both sides of the cold war. But we are getting ahead of our story.

Fermi needed 50 tons of uranium for the CP-1 Chicago pile. Groves learned of the Staten Island supply, and sent Colonel Nichols to New York in September, 1942, to purchase all of the available ore, which was enough for CP-1, but not enough for the Manhattan Project as a whole – the gaseous diffusion plant at Oak Ridge, the plutonium production power reactors at Hanford, etc. Since Great Britain and Canada were partners in the Project, Canadian sources of uranium were vigorously exploited to meet the demand. Canada was the world leader in uranium production until 2009, when it was overtaken by Kazakhstan.[7]

12.2. Uranium and the Soviet Project

The Russian geochemist Vladimir I. Vernadsky prospected for uranium in central Asia before WWI and the Russian revolution. A copper-vanadium mine was found in Uzbekistan to have uranium ore. This was not the only known source in 1945, but none were large producers.[8] When Kurchatov's F-1 reactor required 50 metric tons of uranium, the USSR had no such amount on hand from the Belgian Congo or anywhere else. Kurchatov asked for some uranium metal from the USA via Lend-Lease, and this request was honored by Groves, as we have seen in Chapter 7. Kikoin and Khariton saved the day by capturing ore that the Germans took from Belgium, so that the F-1 reactor in Moscow and the CP-1 reactor in Chicago each enjoyed the product of mining in the Belgian Congo.

The Soviets knew that a lot more uranium was needed, and prospecting began in Eastern Europe as soon as the territories were occupied by the Red Army. Bulgaria was taken in the fall of 1944, and soon thereafter uranium was being extracted from a mine near the town of

---

[7] Canadian uranium mines are described in www.world-nuclear.org

[8] V.I. Ventrov and A.N. Eremeev, "The uranium ore base for the Atomic Project," HISAP96, Vol 1, p 101.

Buhovo. Mines in Poland were also exploited. However, neither the Bulgarian nor the Polish mines were as productive as those in Czechoslovakia and Eastern Germany.[9] The original Czech mines were shut down during WWII, but already in March, 1945, the Soviet government requested permission from the Czech government that had only recently returned from exile in London to bring the mines back into production of uranium ore. A secret agreement was signed in Prague in November of that year that set up a joint venture with the Soviet Union to reopen and work three Jachymov mines. The ore was to be split 90/10 between the USSR and Czechoslovakia. At the end of 1945 there were 237 miners at work, a number that increased to 3742 in 1947, and an annual yield of 49.1 metric tons of uranium.[10]

Working conditions in the uranium mines, both within the Soviet Union and outside it, in Eastern Europe, were severe. In the Soviet Union the miners were mostly prisoners. The majority of the workers in Europe were volunteers, but their living conditions were not that different from prison camps. In either case there was very little mechanization – the ore was extracted by hand with shovels and pick-axes, and hauled by wheelbarrows to conveyor belts that fed on site ore processing equipment. Typical ore from Jachymov was about 0.1% uranium, so 1000 tons of ore would give 1 ton of uranium. E.P. Slavsky, one of Boris Vannikov's deputies in the PGU, took some pictures of mining operations in 1945.[11] In one photograph donkeys carrying bags of ore are shown walking along. Slavsky asked, with a 0.1% purity, how much uranium one donkey could carry? Perhaps 30 or 40 g. Moving a metric ton of uranium would require many donkeys. As much elimination of residue rock, called 'tailings,' as possible was done at the mine to reduce the amount of material transported away by the railroad. Machinery began to appear in the mid 1950's, resulting in a decrease in the labor force of about a factor of two. Mining for uranium has the special hazard of radioactive material, but the miners

---

[9] A. Heinemann-Gruder, "Insufficient Uranium for the Soviet Atomic Project," HISAP96. Vol 2, p 331.

[10] A. Heinemann-Gruder, *ibid.*, p 333.

[11] E.P. Slavsky, in "Kurchatov v Zhizni," edited by Raisa Kuznetsova, *ibid.*, p 479.

did not wear special clothing or take other precautions to avoid ingesting radioactive dust. Their health suffered as a result. The Soviet Council of Ministers was aware of the hazards, and in December, 1949, it issued an order to establish treatment centers for miners suffering from silicosis and other lung diseases that resulted from occupational exposure.[12]

The Jachymov mines were located in the northwest corner of Czechoslovakia, next to the German border. The Erzgebirge mountain chain runs roughly along the border, and several locations in the German part of the chain were known to contain pitchblende, just like Jachymov. There were popular radium treatment baths at Oberschlema, and there was a town called Radiumbad just across the border from Czechoslovakia. So there was knowledge of radioactive ores in the Erzgebirge region before WWII, but no one would have guessed that it would soon be one of the major sources of uranium in the world. The region was contained in a triangle defined by the towns of Aue, Johanngeorgenstadt, and Annaberg, with a population of about 250,000.[13]

12.3. Vismut

A delegation of Soviet and German geologists visited the area in August, 1945, officially looking for tungsten ore. Soviet authorities occupied the area a short time later, and by December, 1945, extraction of uranium ore began under strict Soviet control. There were two outdated uranium factories in the region that were confiscated by the NKVD. In May, 1947, the Soviet Council of Ministers established a mining company called 'Vismut' (bismuth). Vismut was headed by NKVD Major General A.M. Maltsev in the town of Aue. They estimated that about 25000 miners were needed to do the work, and in Germany they were to be volunteers rather than prisoners, so a campaign was begun to attract the labor force. The surrounding towns could not absorb the working population with

---

[12] Atomnyy_proekt_SSSR._T.2.Kn.4.(2003).pdf, #130.

[13] Norman M. Naimark, "The Russians in Germany," Belknap Press of Harvard University Press, Cambridge, MA, 1995, p 238. See also A. Heinemann-Gruder, *ibid.*, p 335-336.

families, so housing had to be built, just like at a new laboratory for the Atomic Project. The housing quality was poor compared to German standards, and harsh living and working conditions soon gave Vismut a bad reputation in Germany. In the days before the Berlin Wall there was exchange of information between East and West, and West Germans also knew of the poor conditions in the Vismut mines. This was a source of friction between the USSR and East Germany that continued into the era of the DDR, when East Germany was a sovereign state. The Soviet Union in 1949 promised to do better in creating a better living environment, with schools, playgrounds, and movie theaters.[14] The voracious demand for workers placed a strain on an economy already short of labor.[15] Vismut became a 'zone,' like other closed towns in the Soviet Atomic Project, Arzamas-16 or Combine #817, except that Vismut was in Germany, not in the USSR. By 1949 Vismut employed 140,000 workers, all but 3386 of whom were German.[16]

Uranium ore destined for Kurchatov's reactors was first refined at the mine, and reduced in weight by about 1/1000, then purified further and turned into either UF_6 gas for the isotope separation plant or UO_2 or uranium metal slugs for reactor fuel. The second stage of purification reduced the amount of material by about a factor of two. Riehl's Elektrostal plant in Noginsk was the first processing plant to produce reactor grade uranium metal slugs, and it reached a production level of 1 ton per day. By 1949 there were three other factories in operation with similar capacity. A full power 1000 MW uranium-graphite plutonium production reactor burns about 50 tons of natural uranium annually.[17]

From 1946 through the end of 1953, 9500 metric tons of uranium were extracted from the Vismut mines, about 45% of the total of all the mines supplying the Soviet Atomic Project. Riehl's refinement cut this number in half, giving a total amount of reactor grade uranium metal from the

---

[14] Atomnyy_proekt_SSSR._T.2.Kn.4.(2003).pdf, #124.
[15] Norman M. Naimark, "The Russians in Germany," *ibid*, p 239.
[16] Atomnyy_proekt_SSSR._T.2.Kn.4.(2003).pdf, #276.
[17] One metric ton of $^{235}U = 8 \times 10^{16}$ joules, so $(50/143) \times 8 \times 10^{16} / 3.15 \times 10^7 = 900$ MW.

German mines in that seven year time span of about 4000 tons, or 570 tons a year – sufficient to power more than one plutonium production reactor.

Initially, because of a general feeling of anxiety, the uranium problem was attacked by a bewildering number of experimental mines spread over a vast continent. But things settled down eventually, with weak producers being closed, and stronger sources being exploited, so that by 1950 the USSR had adequate, reliable uranium supplies for its Atomic Project. 'Vismut' was turned over to the East German government, and decommissioned in the 1990's.

The original pitchblende mine in Czechoslovakia and the neighboring mines in Germany, together with Poland and other locations in Eastern Europe, being outside the Soviet Union, and in regions where the USSR wished to appear supportive, posed political and diplomatic problems that had to be handled with care. The Czechs wanted the uranium from their mines, and so did the Germans, and that led to some negotiations on the part of the respective governments, as the Soviet Union wanted ALL of the uranium, and had the Red Army to back it up. In the end, the local governments regained control.

In a detailed progress report on the Soviet Atomic Project submitted to Beria on August 15, 1946, Kurchatov summarized the uranium ore situation at that time.[18] Sites were known in Kazakhstan, Tadzhikistan, and the Baltic region of the USSR, with uranium content in the ore ranging from 0.02% to 0.1%. The total amount of uranium available from the mines at that time was a few hundred tons. Riehl's Elektrostal factory in Noginsk could refine a few tons of reactor grade uranium metal per week. More refineries, with higher capacity, were under construction. More portable radioactivity measuring equipment would be needed for further prospecting.

---

[18] work.atomlandonmars.com/APSSSR/Atomnyy_proekt_SSSR._T.2.Kn.2.(2000).pdf #233.

12.4. Combine #817; Reactor 'A'

The plutonium production reactor complex, called Combine #817, or Chelyabinsk-40, or Mayak (lighthouse), was built just east of the Ural Mountains at Ozersk north of Chelyabinsk and on the same north-south axis as Kikoin's gaseous diffusion plant #813 at Sverdlovsk. The locations were chosen for the same reasons. The area had been industrialized during WWII – Chelyabinsk was the site of a famous T-34 tank plant – and as a result already had a substantial amount of infrastructure like rail service, electric power, and water. It had remained outside of the war zone, so industries were not destroyed. It also had prison camps to supply the construction labor, and was sufficiently remote from European Russia that secrecy could be maintained. The Chelyabinsk region is shown on Map #3.

Isaak Kikoin headed factory #813, and Yuly Khariton headed KB-11 (Arzamas-16), but Igor Kurchatov retained direct control over Combine #817. He had built the first test reactor F-1 at Laboratory #2, and he wanted to supervise the construction of the first industrial power reactors 'A' to produce plutonium. His brother Boris had been working on the problem of chemical separation of plutonium initially working with uranium targets bombarded by neutrons from the Laboratory #2 cyclotron. The need for power reactors to produce plutonium was recognized before the F-1 reactor went critical in December, 1946. In early 1946 efforts were under way to design a uranium-graphite power reactor with numerous access ports to insert and remove uranium blocks. After it began operating the F-1 reactor was used to test many of the design ideas. Work on site construction began in 1947.[19]

Kurchatov and Vannikov spent several months together at Combine #817 under intense pressure from Beria, who remained in Moscow, to get the reactors operating at full power. Many of the extant accounts of Kurchatov's performance pertain to this period.

---

[19] N.S. Burdakov, et al., HISAP96 Vol 2., *ibid.*, p 398.

The 'A' plutonium production reactors had only a factor of two or so more graphite and natural uranium than the first F-1 test reactor, but were operated at more than 1000 times higher power levels, and so presented new engineering and safety problems. The reactors were water cooled, and all of the power generated was carried off by the water. Power station reactors heat the water to steam and drive electric generators, but the 'A' reactors simply allowed the water to cool down, and then restored it to the lake.

In August, 1948, Kurchatov gave a report on the status of Combine #817.[20]

"Combine #817 is intended for the production of the best known fissile material – plutonium. It is located about 150 km from Chelyabinsk in a forest at the side of a railroad, and consists or four factories and a town. All of this is by a lake where there earlier was a pioneer camp. The first factory, located on the lake shore, is for water treatment. The second factory, factory 'A', is the nuclear reactor, located about 1.5 km from the water plant. The third factory 'B' is the chemical plant, another 1.5 km from the reactor. The fourth factory 'V' is 10 km from the cluster of the first three, and is for refinement of plutonium metal. The town is on the shore of another lake – Irtyash – 15 to 20 km from the factory complex.[21] All of the factories and the town are connected by a railroad and paved all-weather roads.

"The working cycle of the combine is as follows: aluminum clad metallic uranium blocks are inserted into the reactor 'A.' Plutonium is produced as a result of nuclear transmutation of the uranium, the amount depending on the time the uranium stays in the reactor, and the reactor power level. Reactor 'A' is designed for 100,000 kW, and is expected to produce in three months 100 g of plutonium for every ton of uranium. This amount equals the daily output of reactor 'A', given its uranium load.[22][The amount of plutonium recovered when the reactor is unloaded depends on the exposure time.]

"Uranium blocks – highly radioactive with fission fragments after neutron bombardment in the reactor – are taken from reactor 'A' in special rail cars to the

---

20 Atomnyy_proekt_SSSR._T.2.Kn.4.(2003).pdf #191.
21 Boris Vannikov talked about the distance between town and factories in his oral history, quoted in Chapter 9.
22 It is shown in Appendix B that, assuming the same uranium generates the power and makes the plutonium, the plutonium production rate can be calculated given the power level, the energy release per fission, the capture cross section ratio, and the isotopic abundance ratio, independent of the amount of uranium present.

chemical factory 'B', where the plutonium is chemically separated as a salt together with lanthanum fluoride from the uranium and the fission products. The final separation is done in factory 'V', where plutonium salt is separated from all other compounds, the pure salt is reduced to plutonium metal, and it is cast into specially prepared hemispheres suitable for nuclear explosives. [The Russian alphabet goes 'A,B,V,G,D…']

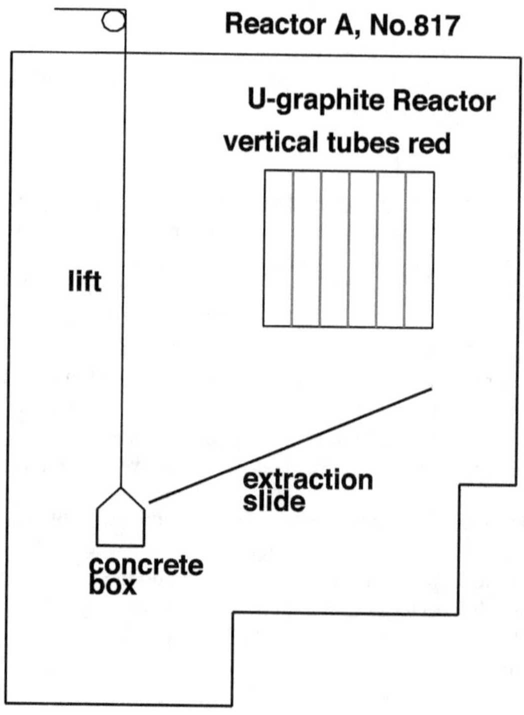

Figure 12.1. Schematic side view of Reactor 'A' at Combine #817 Chelyabinsk. The reactor is the square in the upper right corner, with its complicated suspension system not shown. Irradiated uranium slugs are dropped down into a water filled concrete box, and then hoisted to the top. The operation is described by Kurchatov in his report.

"The water treatment plant feeds cooling water to the reactor, where large amounts of heat generated by the nuclear reactions have to be carried away. Contaminants have to be removed from the water that could leave sediment on the cooling pipes in the reactor, or that could absorb neutrons and leave the water radioactive. Water is drawn from the bottom of lake Kyzyltash, goes through special quartz filters, then to a steel basin filled with activated charcoal, a substance that we are using in quantities not before seen in Soviet industry. The Ministry of Chemical Industries had to build an extra factory for insurance to meet our demand for activated charcoal.

"After chemical treatment, the water is pumped to a degassing basin, where dissolved gases are removed, and then pumped into holding tanks for cooling reactor 'A', from which the water remains in a closed system. Water extracted from the reactor is radioactive, and cannot be returned directly to the lake. It is stored in tubes in a hill separated from the lake by a dam. After 24 hours, the radioactivity has dropped enough to be restored to the lake. [Oxygen in the water captures neutrons, forming unstable isotopes of oxygen that β decay to fluorine with half-lives of several seconds, so waiting 24 hours should suffice.]

"Conditions for the water plant are the most stringent for any water purification plant in the Soviet Union. It has to handle 2500 m^3 of purified water per hour. At the present time the factory can handle 3200 m^3, and we plan in one or two months to increase this to 4000 m^3 of water per hour.

"I will give more details of the reactor factory 'A', where the nuclear reactor is located, built by Professor Dollezhal together with the construction bureau for building, electrification, and equipment, all under my scientific management.

"Two projections of the central part of the factory are shown in the sketches [see Figure 12. 1]. The central part is 80 m high, with 50 m located below ground, where the reactor itself is located, supported by a framework of steel girders resting on stanchions. The support structure had to be carefully designed to support the reactor which weighs more than 1600 metric tons. The reactor is surrounded by a 1.5 m thick layer of water in a steel tank, a 1.5 m thick layer of sand, and 2 m of reinforced concrete for protection from radiation. Twelve hundred thick walled aluminum tubes penetrate the graphite matrix vertically from top to bottom. The uranium blocks are suspended in these tubes, through which the cooling water also flows. The tubes have a diameter large enough to allow any single uranium block to be individually extracted out the bottom of the reactor if necessary. The blocks are held in place by a mechanical device called a cassette, and when released fall under their own weight into water that flows along an incline tube underneath the reactor to an underground container. The containers can be lifted by a remotely controlled crane to ground level of the building, where there are thick concrete walls and a roof. From there they are moved to the transport gallery, where the radioactive uranium blocks rest under water for two months before going to the chemical separation plant. The Optical Institute and factories of the Armaments Ministry have supplied the periscopes that are used to observe remote control operations of loading and unloading.

"Six thousand instruments are used for measurement of temperature, output water flow rate, and integrity of the 1200 aluminum tubes. The control room is located some distance from the reactor itself. Communications between the control room and the reactor building require more than 300,000 wires and 25000 electric circuits.

"Mounting of the reactor was roughly completed in May of this year. Using radium-beryllium sources it was confirmed that the graphite supplied by the Ministry of Ferrous Metals and the aluminum tubes from the Ministry of Aviation were of excellent purity. The purity of these materials has crucial importance for sustaining a chain reaction, as is the purity of the uranium metal blocks. As I have reported earlier, a chain reaction will follow if there is enough uranium and graphite.

"Loading uranium began in early June, and by June 8, 1948, 32.6 tons of blocks had been installed, and, in the absence of cooling water, this was enough to bring on criticality [water absorbs neutrons]. The chain reaction was quenched, and loading continued, and on June 10 there were 72.6 tons of the 120-150 tons planned, and the chain reaction resumed with cooling water in the tubes. On June 19, the power reached 100 Mw for a brief time, before an accident occurred in one of the cooling channels that required changes in our safety procedures. Afterwards work proceeded to increase the level and duration of the reactor power. We became convinced of the accuracy of the calculations for the amount of shielding, and that the handling of the radioactive uranium blocks was safe. Many of the relevant effects had been studied at the F-1 test reactor at Laboratory #2. We learned that the reactor could be maintained at any fixed power level, like 100 Mw to 0.5- 1.0% accuracy instead of the planned accuracy of 1-2%.

"Practice has uncovered a series of faults in the assembly of the project, for the most part consisting of insufficient quality control in cleanliness of the water handling plumbing, where foreign objects that impeded the water flow were found that caused two accidents. [Small pieces of construction paper were found to have stopped the water flow in one of the tubes, causing overheating and melting – a serious mess.] Accident detection and warning were also found to be deficient. But these defects have been successfully addressed.

"We have learned that in a high power uranium-graphite water cooled reactor the chain reaction process accelerates as the graphite heats up. [Caused by boiling of cooling water decreasing neutron absorption.] We also studied the production by uranium fission of radioactive xenon. Xenon makes its presence known by the time dependence of the neutron multiplication coefficient, and requires changes in techniques for handling the material. [The isotope is ^{135}Xe, discovered as a poison at Hanford, and described in Chapter 3.]

"We are now running reactor 'A' routinely, and have already accumulated about 1 kg of plutonium. We plan to increase production by increasing the power level of the reactor to 120-130 Mw. The successful construction and operation of an industrial power reactor is a significant step in the solution to the uranium problem. We now have the real possibility of an atomic bomb in 1949.

"We have gone down a long road in solutions to technical problems of the construction of nuclear reactors, and we have established a basis for exploitation of radioactive elements in technology, chemistry, biology, and medicine."

From the tone and content of his report, Kurchatov was addressing government ministers, and not technical personnel, because he has relatively little technical detail, and praise as well as criticism for Soviet industry. Nevertheless, it is a fair assessment of Combine #817 in August, 1948, with about one year to go before the plutonium bomb test. Partly due to problems in isotope separation, the pressure to bring reactor 'A' into plutonium production was immense. Vannikov and Kurchatov both were in residence in Ozersk during the commissioning. The accidents Kurchatov referred to were caused by interruptions in the distribution of cooling water. There were 1200 vertical aluminum tubes about 6 m long to handle the uranium slugs and the cooling water. Initial water flow rate from the lake was 2/3 m^3/sec, or 6.7×10^5 g/sec. If the water temperature increased by 50^0C, the water would carry off 140 Mw, which would keep a reactor of that power in stable operation.[23] The winter-summer variation in lake water temperature, from 5^0 to 20^0 C meant that the reactor power could be increased in the winter.[24] All of the heat was generated by the uranium slugs. If the reactor had 100 tons of uranium, and each slug weighed 1 kg, there were an average of 83 slugs spaced along each vertical tube. A 1 kg uranium sphere has a diameter of 5 cm – about the size of a lime. Kurchatov said that any slug can be dropped to the bottom, but we are not told exactly how this was done, except that the slugs were held in place by 'cassettes.' The obvious way to extract a slug in the middle would be to drop out all of the ones below it, but perhaps there was a mechanical rig that moved a slug to one side for it to drop to the bottom all by itself.

---

[23] Arithmetic: $(6.7 \times 10^5$ g/sec$) \times (50^0$C$) \times (4.18$ j/cal$) = 140$ Mw. One calorie heats one gram of water one degree centigrade.

[24] Atomnyy_proekt_SSSR._T.2.Kn.4.(2003).pdf, #228 Vannikov to Beria.

On the other hand, if a slug were to get stuck on the way down, there would be a very serious problem. And any mechanical device inside the reactor would have to operate in a very hostile environment.

There were two similar cooling water accidents in the startup of reactor in June and July, 1948.[25] In each case the temperature of the affected uranium blocks – the generators of the heat – rose quickly, melting the aluminum casing of the block and the aluminum cooling tube. Uranium metal surfaces came in contact and reacted with the graphite. Water also leaked into the graphite matrix. These were very serious problems that could lead to fires and destruction of the reactor. Affected parts of the reactor could not be accessed by personnel because of the high radiation. Kurchatov assumed a hands-on approach to the problem, and was scolded by I.M. Tkachenko, the same person that told Zhukov he could not carry out his military exercises in the Chelyabinsk region (Chapter 10). In a letter to Beria on 24 June, 1948, from Combine #817 Tkachenko wrote:

"At the present time after a preliminary start-up of reactor 'A' a series of malfunctions has resulted in very high radiation levels. Academician I.V. Kurchatov has ignored all safety rules and precautions (especially when nobody is looking) and personally goes into areas where the radioactivity is significantly higher than the specified norm. Comrade E.P. Slavsky has not been too careful either. On June 21 Comrade Kurchatov descended an elevator 21 m into the basement of the reactor room to read a humidity gauge and received more than 150 times the accepted maximum dose. His NKVD bodyguards were not with him, and the radiation safety officer did not prevent him from entering the reactor area because of Kurchatov's stature and authority."[26]

Tkachenko asked Beria for support in urging Kurchatov to abide by his own safety rules. Beria reprimanded Kurchatov and Slavsky. This letter was interesting for several reasons. First, the Soviet Atomic Project had a radiation safety system in place, with established dose limits for workers. Second, there were personal radiation monitors-film badges- that everyone was supposed to wear, but not everyone did. The overalls did not have

---

[25] Atomnyy_proekt_SSSR._T.2.Kn.4.(2003).pdf, #180, #181.
[26] Atomnyy_proekt_SSSR._T.2.Kn.3(2002).pdf, #396.

many pockets, and some workers simply left their film badges in a drawer at home. Once a week the safety people were supposed to estimate the amount of absorbed radiation by the darkness of the film. Third, Kurchatov was not crazy, but was determined to get the reactor repaired and in production mode, and prepared to take some risks. And fourth, but not least, it would be 100% impossible for this incident to occur today, either in the USA or the Russian Federation. Radiation safety is much more strict than it was 70 years ago, on either continent.

Kurchatov was not alone in receiving too much radiation exposure. According to an interview of A.K. Kruglov, a plant manager who survived into the era of 'glasnost,' a careful and orderly repair of reactor 'A' after the first accident would have required at least a year, and that was a totally unacceptable delay to the timeline of the Project.[27] So an army of workers was trained to extract radiated uranium blocks and disassemble the damaged part of the reactor by hand. In this way 39,000 highly radioactive blocks were extracted from the central hall of the reactor, the device was reassembled, and returned on line in two months, but at a cost. In the first year of operation of the reactor more than 30% of the personnel received doses between 100 and 400 rem, and some people received much more.[28]

The rem, and its basic unit the roentgen: $R = 2.58 \times 10^{-4}$ coulombs/kg for ionization created by x-rays in dry air, are doses, or integrated measures of radiation damage. The curie $= 3.7 \times 10^{10}$ disintegrations/sec is a decay rate, and rem/hour is a dose rate. The dose one gets is the decay rate times the ability for biological harm of the radiation emitted (ionization for most forms of radiation – neutrons require special treatment) times the time spent next to the source. Suppose a source of one curie emits 34 keV x-rays at the center of a sphere of radius 0.5 m – about

---

27 Vladislav Larin, "Unknown Accidents at Mayak," Energia, #1, 2000. An English translation was published earlier in Bulletin of the Atomic Scientists, 55 (5), Sept-Oct 1999, p 20-27.

28 'rem' is roentgen equivalent man. 1 Roentgen=2.58×10^{-4} coulombs/kg, ionization created in dry air. 'rem' makes adjustments for biological effects of different types of radiation. The dose from normal background is ~0.3 rem/year. A 450 rem dose in a short time is fatal in 50% of cases (DOE-HDBK-1130-2007).

1 kg of air. The average ionization energy for air is 34 eV, so an x-ray that deposits all its energy in the gas makes 1000 ion pairs, or 1.6×10^{-16} coulombs. If all of the x-rays are absorbed in the sphere the dose rate would be 6×10^{-6} coulombs/sec, and the received dose in the kg of air would be 1 R in about 45 sec. This exercise shows the relationship between decay rate and dose. Both terms are used to characterize radiation hazards.

The USSR Academy of Sciences setup several research centers to study the biological effects of radiation. Nikolaus Riehl worked at one in Sungul after he left the uranium refining factory in Noginsk. Sungul is located near Combine #817, between Chelyabinsk and Sverdlovsk. Igor Kurchatov was interested in the subject, and a biological division was established at Laboratory #2 in Moscow. People trained in genetics and radiobiology were among the German visitors, and groups were established at Sukhumi and Obninsk. One of the goals of this research was to establish safe levels of radiation for workers who were exposed to the hazard on a daily basis. This number is a moving target, generally revised downward as more is learned about radiation danger. Safe levels are higher for radiation workers, who are monitored for exposure, than for the general public, who are not. The present whole body dose limit per year for a radiation worker is 5,000 mrem, or 5 rem.[29]

Kruglov also recalled that the trolley cars used to transport the irradiated uranium from reactor 'A' to factory 'B' for chemical extraction of the plutonium, while designed to operate remotely without any intervention, often got stuck one way or another en route, and had to be attended by personnel, with accompanying radiation exposure. If the doses were carefully monitored with a pocket dosimeter that could be read in real time, and an individual worker was only in the danger zone for a short period, then by using many workers in series a task could be accomplished safely. But such procedure was rarely followed in the start-up years of the Project, when time was of the essence. The Project had radiation safety rules, radiation safety officers, and methods in place to measure exposure.

---

[29] www.nrc.gov/ information for radiation workers

However, radiation safety lacked absolute authority to veto any dangerous activity.

Instruments that could be remotely read out, remote manipulators with operators behind adequate shielding, and other provisions for safe handling in the hostile environment of a power reactor were slowly developed, and not all there on day one. Robotics, now so sophisticated, were very primitive if they existed at all. There was an overhead bridge crane along the storage area for the extracted uranium blocks, but the radiation levels in the operator's box that was suspended with the crane were too high, so Kurchatov had to arrange for a remote room next to the crane with good visibility and adequate shielding for the operator to spend his shift safely.[30]

Kurchatov proposed building a 'hot lab' at Laboratory #2 in Moscow. The hot lab began operation in 1949, after the first successful atomic bomb test in August of that year. The lab was designed to handle irradiated samples up to 10,000 curies from a research reactor. The design engineers had little guidance because there was at that time no similar industrial remote handling setup in the entire Soviet Union. Handling the most radioactive material was done with remote manipulators that could pick up bottles and pour liquids from behind concrete shielding. The first version of manipulator worked opposite to the operator's motion – move the handle up and the remote handle moved down. Although this is the simplest way to rig a remote arm over a wall, it turned out to be too confusing for everyone, and soon the engineers figured a way to make the motions go in the same direction. Operators could see what they were doing through a thick circular window 60 cm in diameter. The window was made of several layers of plastic coated lead glass to attenuate the electrons and γ rays. The field of view was increased by an arrangement of right angled prisms. The remote chambers were held at negative pressure to retard the escape of radioactive gases. All of the necessary resources – water, compressed air, vacuum lines and electrical outlets – were operated from the control room. The floor sloped to drain spills into

---

[30] B.V. Brokhovich in Raisa Kuznetsova, "Kurchatov v Zhizni," *ibid.*, p 499.

a safe receptacle. Less radioactive substances could be handled in glove boxes, where hands could be used to operate through rubber gloves. The box kept the operator from exposure to aerosols.[31] These working facilities at Laboratory #2 became models for other sites where radioactive materials were handled.

Recollections of people who were working with Kurchatov to bring reactor 'A' online were unanimous in their evaluation of his performance under stress. Kurchatov was always on call day or night from the 'A' control room. In the beginning, he shared a railroad car with Vannikov, and they would relax some in the evening by playing cards. Kurchatov did not smoke, and drank little alcohol. If he lost patience with his team, he did not show it, and he greeted every problem with equanimity - at least that is what is remembered.

During the 'A' reactor startup Beria exerted pressure from Moscow. He did visit Combine #817 later, when the second reactor was running, and entered its control room with Kurchatov during the evening shift. Beria's visit was planned for the day shift, and that crew had been outfitted with clean uniforms, but the evening shift was unfortunately in its regular clothes. B.V. Brokhovich was the crew chief, and he observed that Kurchatov seemed tired and nervous. Beria asked Brokhovich how things were going, and was told that everything was working smoothly. Beria then walked over to a panel of display lights for the temperature range of cooling water in the tubes, a few of which were blinking, and asked Brokhovich if he could make them all blink, which, by adjusting thresholds, he could and did, except that one light was burned out, and that upset Beria. That was a fairly typical visit of a dignitary to any control room. They try to ask intelligent and penetrating questions without any knowledge of what is going on.

Kurchatov enjoyed practical jokes. One winter day he and Vannikov walked into the cloak room of the cafeteria to leave their coats and eat lunch. Vannikov took off his boots, and walked away. Kurchatov produced

---

[31] I.A. Reformatsky and G.N. Yakovlev, HISAP96, Vol 2, *ibid.*, p 391.

two nails, and asked for a hammer. No hammer was found, but an ax was obtained instead. Kurchatov promptly nailed Vannikov's boots to the floor. After lunch Vannikov returned to try to put on his boots, to no avail. He immediately confronted Kurchatov, saying that no one else in the group would dare to nail his boots to the floor.[32]

Reactor 'A' survived, and the accidents led to careful monitoring of flow rates and temperatures throughout the complex of tubing. Operators learned what to look for as signs of trouble. In 1950 a second reactor 'AV-1' was operational nearby, with the same basic design and some improvements from operating experience with 'A,' particularly in the configuration of the cooling water headers and tubes, that had been a headache for the first reactor. As long as the uranium fuel remained in place, and the cooling water and controls behaved normally, the operation proceeded smoothly. New problems arose when the fuel blocks were extracted to remove the plutonium. The Combine eventually had several uranium-graphite reactors, all of roughly the same design, and became a training ground for reactor construction and operation throughout the Soviet Union. The Chernobyl reactor was also of this same type.

12.5. Combine #817; Plutonium Extraction

V.G. Khlopin's Radium Institute in Leningrad officially received the assignment of chemical separation of plutonium from spent uranium fuel elements in December, 1945.[33] Very small samples of plutonium – micrograms - and small quantities of fission products were produced by lengthy bombardment of uranium targets with neutrons from cyclotrons at the Radium Institute and Laboratory #2. The commissioning of Nemenov's cyclotron at Laboratory #2 for the explicit purpose of plutonium production before a nuclear reactor became operational was described in Chapter 7. Some insight into plutonium chemistry was gained from the known methods of extraction of uranium from its ore, and the assumption that plutonium would behave similarly.

32 B.V. Brokhovich in Raisa Kusnetsova, "Kurchatov v Zhizni," *ibid.*, p 502.
33 E.I. Ilyenko and N.A. Abramova, HISAP96 Vol 2, *ibid.*, p 351.

Fuel slugs removed from a reactor are ~98% uranium, with trace amounts of plutonium and fission products. Uranium and plutonium are mildly radioactive, but the fission products are copious emitters of β and γ rays, making the spent fuel elements too dangerous to handle. The unit of radioactivity is named after Pierre and Marie Curie, and defined as the activity of one gram of radium: 1 curie = 3.7×10^{10} disintegrations/sec. The activity of a 1 kg spent uranium fuel slug is the order of 100 curies. The fission products are about 30 elements in the middle of the periodic table, with atomic numbers ranging from 35 to 65, and widely varying chemical properties. Included are rare gases Kr and Xe, that are not chemically reactive. A few isotopes of fission products are useful for research or medical applications, ^{131}I for example, but most are considered disposable radioactive waste. Provisions were made at Factory B to capture iodine vapor from the fission products.[34] Safe storage of long lived fission products is one of the environmental problems facing today's nuclear energy programs.[35]

The slugs were clad in aluminum cans to prevent the cooling water from contacting the uranium. After extraction from the reactor, the cans had to be removed, either chemically or mechanically. The next step was to separate the heavy elements U and Pu from the lighter fission fragments. The idea was to use the high oxidation number states U(VI)$^+$ and Pu(IV)$^+$ or Pu(VI)$^+$ that were not accessible to the lighter fission product elements. First these states had to be created in the solution. If an organic molecule could be found that would reduce the U and Pu but not affect the fission fragments, then introducing this chemical into the solution would precipitate out the U and Pu, and the fission fragments would remain in solution to be disposed of later. This was called 'co-precipitation.' Then the Pu and U could be separated in the next stage. This was the basic strategy adopted by Khlopin at the Radium Institute, and is the same one used today for processing spent reactor fuel, although the modern

---

[34] Atomnyy_proekt_SSSR._T.2.Kn.4.(2003).pdf, #251.

[35] David Kramer, "Cleanup of Cold War nuclear waste drags on," *Physics Today*, July, 2017, p 28. Treatment of radioactive waste from Hanford, the Manhattan Project equivalent of Combine #817, is still very much a work in progress.

technique differs in detail.[36] Like many features of the Atomic Project, the chemical extraction of plutonium is a problem without a unique 'best' solution. There are many ways to do it, with advantages and disadvantages. Modern requirements place severe constraints on the amount of radioactive waste from the fission fragments that has to be handled, while in the 1940's in the USSR this was not a principal concern.

As Kurchatov said in his progress report, reactor 'A' was followed by factory 'B' and then factory 'V.' Factory 'B' had the task of removing the fission products, and producing a 40% concentration of plutonium salt in a mixture of salts of lanthanum fluoride. Then factory 'V' was to produce weapons grade plutonium metal hemispheres.

Aleksandra Kornienko, Kurchatov's secretary at Combine #817, arrived by train from Omsk to Kyshtym in April, 1949. She was greeted by a local woman who asked if she was going to work behind the barbed wire, where they were developing communism in secret! Kornienko had all of the correct papers, and was admitted inside the wire, where there seemed to be barracks everywhere, and prisoners at work on construction.

"They took us to the barracks where we were to live. I will never forget going to work wearing a brimmed hat just like it was springtime. They built a club for us named 'Leninsky Comsomol,' and I was walking past it, where prisoners sat on benches among mud puddles, it having just stopped raining. As I approached the puddle a rock skimmed over it, and I was doused with water from head to toe. I turned around and asked: 'What if it had been your own daughter?' A middle aged prisoner nodded. When I arrived at the office Vannikov took one look at me and said: 'Well, I am not the only one who has been drenched. You too, Aleksandra.' "[37]

---

[36] Separation methods are described by Lawrence Livermore National Laboratory: e-reports-ext.llnl.gov/pdf/799624.pdf. The modern procedure is called PUREX – plutonium uranium recovery by extraction. The organic molecule used is different from the one employed at Combine #817, but the idea is the same.

[37] Raisa Kuznetsova, "Kurchatov v Zhizni", *ibid.*,p 514. Interview with A.S. Kornienko.

The first uranium fuel slugs from reactor 'A' were delivered to the radiochemistry factory 'B' on December 22, 1948. After the aluminum cans were removed and the slugs dissolved in nitric acid, the solution according to Liya Sokhina contained about 100 mg/l of plutonium, one kg/l of uranium, and activity from fission products of 100 curies/l.[38]

To understand some of these numbers, assume that 100 Mw nuclear reactor runs for three months on 10^5 1kg natural uranium slugs (100 T of uranium, 700 kg of ^{235}U). Using 180 MeV per fission gives 5.3×10^{20} fission fragments/slug. Taking an average lifetime per fission fragment of 5 years gives an activity of 100 curies. The number of fissions is half the number of fission fragments, so 2.6×10^{20} ^{235}U atoms were burned, 0.1 g or about 1.4% of the fuel in the slug. Using the formula for plutonium production in Appendix B gives 0.07 g of plutonium per 1 kg uranium. So assuming 1 kg of uranium in 1 liter of solution checks with Liya Sokhina's numbers. It is useful to keep in mind that, although the solution contained 10,000 times as much uranium as plutonium, the α activity of ^{239}Pu dominated by a factor of 20, because the half-life of ^{238}U is 200,000 times as long as the half-life of ^{239}Pu.

There were a number of problems scaling up the chemistry from laboratory to industrial size. Laboratory experiments at Leningrad and Moscow with trace amounts of plutonium did not have to contend with the intense radiation field from fission products that heated the solution and created a lot of ionization. It was possible in the laboratory to reduce the presence of fission fragments by a factor of 10^{-6}, while the best that was at first achieved at Factory 'B' was 1/860, roughly a factor of 1000 more contamination. A.P. Ratner from Khlopin's Radium Institute came to Combine #817 and developed a procedure that exploited the ionization created by the fission products, thereby turning a problem into an advantage. Boris Kurchatov also worked to improve the performance of the separation chemistry in the industrial environment. The chemistry took place in a 'canyon,' a long concrete walled hall full of radioactive solution. It was frequently necessary for workers to enter the canyon, and

---

[38] Liya Sokhina, HISAP96, Vol 1, p 136.

consequently receive high doses of radiation. For the first two years of operation, the workers had about 100 rem/year exposure – a factor of 20 over the modern recommended maximum annual dose.[39] If spread uniformly over a year, 2 rem/week could probably be tolerated by a healthy young adult; the body recovers from gradual exposure. One hundred rem in one shot would have serious consequences.

On October 21, 1947, Boris Vannikov requested approval from Stalin for the requisition of gold, platinum, and silver for the manufacture of utensils resistant to corrosion. The amount of precious metal was rather modest, a few kg compared to the 12,000 metric tons of silver borrowed from the US Mint for calutron coils in the Manhattan Project. The request was probably to be used as a coating on containers and tools made of stainless steel or some other durable material. Stalin subsequently approved the request, and the equipment was fabricated.[40]

Liya Sokhina graduated from Voronezh State University in 1949, and started to work in the plutonium refinement plant 'V.' She observed that most of the young technicians were women, because of the war. Women could go to school and learn chemistry while men were in the Red Army. The 'V' plant had two parts, one to refine the plutonium from the 'B' plant, and the other to fabricate weapons grade metal. By February, 1949, plant 'B' had converted the first output from reactor 'A' into a solution containing 10 to 15 g/l of plutonium; 10 g/l of uranium-lanthanum; and 2-3 g/l of iron, chromium, and manganese. Residual radioactivity was 350 millicuries/l. This solution was to be delivered to the first stage of plant 'V.'

Unfortunately, the first stage of plant 'V' was not finished. Rather than accept a delay in the process until the plant was finished, the management of Combine #817 decided to convert an empty warehouse on site, formerly used by the Soviet Navy, into a chemical factory.

---

[39] Liya Sokhina, HISAP96, Vol 1, p 136.
[40] Atomnyy_proekt_SSSR._T.2.Kn.3.(2002).pdf #202, Vannikov to Stalin.

Sokhina described the refitted temporary first plant to be like an ordinary chemistry laboratory with sinks and cabinets, but with no special precautions for handling radioactive substances on an industrial scale. The plutonium concentrate solution described above was delivered from 'B' to 'V' in steel containers. As noted, the mix was still very radioactive.

The solution was filtered into glass jars by hand. Academician I.I. Chernyaev, plant chemist, called this phase the 'glass jar period.' Better facilities were under construction, and the glass jar period was replaced by equipment better designed to handle the materials in August, 1949. However, the first weapons grade plutonium for RDS-1 was prepared in the temporary converted warehouse.

There was an obsession with plant cleanliness. In the second part of the plant, where the final refinement was made for weapons grade material, the workers wore booties and smocks. Radioactive dust containing fission fragments was a health hazard, but plutonium dust was not considered dangerous, plutonium being an emitter of short range α particles. While modern radiation safety requirements are designed to avoid ingestion, no one at that time worried about breathing or swallowing plutonium. An alpha emitter inside the body can cause local damage and cancer growth.

The five senior scientists and academicians lived in a house about 200 m from Liya's barracks. She remembered going from her shift in the plant to that house any time of the day or night to ask for help. The techs called the house "The Academicians' Pickwick Club." She expressed some concern that these scientists, the best in the country, were stuck for 1.5 years in an isolated, radioactive place. The surrounding birch trees acquired radioactive dust. A leaf from one of the trees would cause a radiation monitor to chatter away.

Sokhina reported that in March, 1949, 8.5 g of high purity plutonium metal was delivered to Arzamas-16 from factory 'V,' Combine #817, and by the end of June there was enough metal for a bomb. It was all refined during the "glass jar period" that came to an end when a new factory came on line about the time of the first bomb test in August, 1949. The new factory looked good –polished stainless steel vessels and separated concrete chambers for chemical treatment. There was a learning curve to the actual operation, and intervention by hand with attendant radiation exposure was still necessary. Cans containing the plutonium in solution would occasionally explode. These were chemical, not nuclear explosions, but radioactivity was spread all over everywhere. An exhaust hood caught fire after one explosion, and the room was coated with a green powder (plutonium compound). A crew wearing gas masks came in to scrub the walls, floor, and ceiling, but the air remained radioactive. At that time, there were no personal dosimeters for individual workers. The workers in the factory had some experience with radioactive waste from storage that was brought to the factory for further processing, and sometimes exploded, which should have served as a warning of the 1957 disaster described below.[41]

There were spills, and plutonium was lost on the floor. The next year, 1950, the plant was rebuilt once again, and the loss of plutonium was minimized, but working conditions did not improve very much for the next several years. The principal health hazard was radioactive vapor. According to Liya Sokhina, who worked there for years, there were no sites in the Soviet Atomic Project more dangerous than Factories 'B' and 'V.'

In September, 1949, Combine #817 had 10,118 workers – 1954 technical staff, and the rest performing support work of various kinds. Construction had employed 45,000 prisoners. Factory #813 (gaseous diffusion) had 6335 workers, and Factory #814 (electromagnetic separation) had 1266. These numbers give the level of activity of the Soviet Atomic Project in the Urals region at that time. [42]

---

[41] Vladislav Larin, "Unknown accidents at Mayak," *ibid.*, p 7.
[42] Atomnyy_proekt_SSSR._T.2.Kn.4.(2003).pdf #276.

12.6. Aftermath

Eventually there were six plutonium production reactors at Combine #817. The present closed town of Ozersk contains what used to be the Combine. It has had various names, including 'Mayak,' Chelyabinsk-40, and Chelyabinsk-65. Sometimes it is referred to by the neighboring town of Kyshtym, because Ozersk was secret, and not on any Soviet map of that time, although it is clearly shown on map #3 in this book.

Despite Kurchatov's claims for careful handling of radioactive waste water from the power reactors, the surrounding small lakes and rivers, particularly the Techa River, became contaminated, affecting fish, plant and animal life. A Los Alamos study from 1982 suggested that small leaks in the aluminum cladding of the uranium slugs could have contaminated the cooling water.[43] The region as it appears today is shown in Figure 12.2, courtesy of NASA. Lake Kyzyltash was mentioned by Kurchatov as the source of the cooling water for the 'A' power reactor. Lake Karachai, a small lake within the Mayak site, also became polluted. While the operation of the power reactors to produce plutonium created some radioactive hazard, the main source of danger was the radioactive waste from the fission fragments that was removed from the uranium and plutonium by the chemical plant 'B.' These fragments were in a solution stored in underground concrete tanks with heavy concrete lids. The energy released by radioactive decay heated the material, and the tanks had to be externally cooled to control the temperature rise. On 29 September, 1957, the cooling to one of the tanks failed, and the temperature rise caused the tank to explode, blowing off the heavy concrete roof, and spraying radioactive solution over a very wide area. About 52,000 km^2, with about 270,000 inhabitants, received increased radiation doses, and some areas remain contaminated today. There were no immediate casualties. It was a

---

[43] Diane M. Soran and Danny B. Stillman, "An Analysis of the Alleged Kyshtym Disaster," LA—9217-MS. After a careful discussion of all possible sources of contamination, the authors reach the wrong conclusion – that it could have occurred without any explosion.

chemical, not a nuclear explosion;[44] it did not create a mushroom cloud in the atmosphere that could carry radioactivity from one continent to another. The accident was not detected in the West.

Nothing about the accident was publicly known in the West until Zhores Medvedev, a Soviet biologist who was at that time living in England, published an article in "The New Scientist" in 1976, almost 20 years after the fact. His claim of an explosion at Mayak that scattered radioactive material over a wide region was met with skepticism in the West, where advocates of nuclear power did not want to excite safety concerns regarding the storage of radioactive waste from reactors. He then wrote a book about the study of the environmental effects of the radiation, with some hypotheses regarding exactly what transpired, since the affair was still cloaked in secrecy in the USSR.[45]

Original documents pertaining to the response of the Soviet government to the 1957 accident at Mayak are now available on line, so there is no longer any mystery regarding what happened.[46] The evacuation of villages and collective farms exposed to the radiation was set in motion within a month of the accident, and a committee was established to determine the causes of the explosion, and to recommend actions to prevent a repeat occurrence. Subsequently the Mayak accident has been assigned third place in the list of nuclear reactor disasters, after Fukushima and Chernobyl. Two isotopes with half-lives of about 30 years, ^{137}Cs and

---

[44] Nitric acid used in dissolving the uranium slugs could form ammonium nitrate in the waste solution. $NH_4 - NO_3$ is explosive, decomposing into nitrous oxide and water vapor: $N_2O_3H_4 \rightarrow N_2O + 2(H_2O)$.

[45] Zhores A. Medvedev, "Nuclear Disaster in the Urals," W.W. Norton & Company, New York, 1979.

[46] www.nuclear.tatar.mtss.ru/"Dokumentov po Avarii 1957 g na Kombinate #817."

[90]Sr, have been among the most difficult contaminants. Despite cleanup efforts, parts of the region remain off limits for habitation and agriculture.

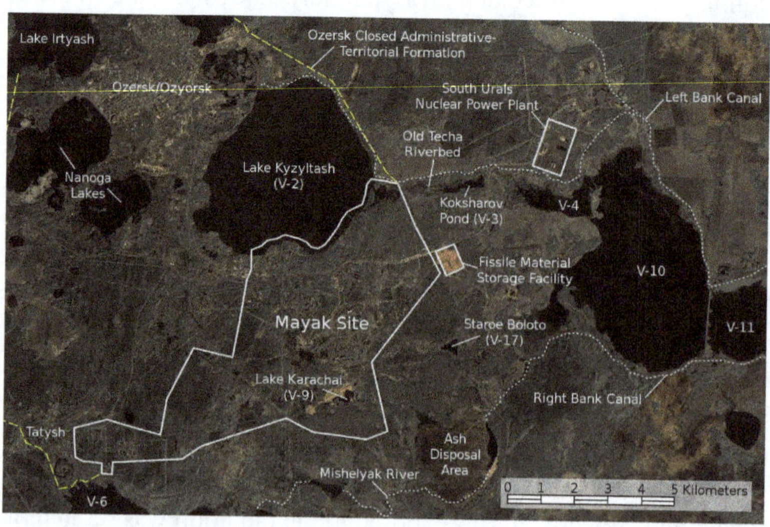

Figure 12.2. NASA world wind screenshot of the modern Mayak region, showing lake Kyzyltash, Lake Karachai (small lake inside the Mayak site boundary), the Techa River, and Ozersk. The fissile material storage building is a post-cold war facility.

Chapter 13

German Scientists and the
Soviet Atomic Project

13.1. Introduction

The story of the Soviet 'Alsos' mission to Germany, and enlistment of various German scientists to visit the Soviet Union for work on the Atomic Project has been told in Chapter 7. Early information regarding the German Uranverein and its participants was obtained from Austria by Igor Golovin, who visited that country shortly after its liberation by the Red Army, as this occurred before the invasion of Germany itself. Golovin was sent to learn about the Uranverein from physicists in Vienna who were participants in the project. His declassified report is translated in Chapter 7. Some of the principal players were identified by Golovin. In addition, many German scientists were well known before the war, and some had visited, or worked, in the Soviet Union. Not all of the Uranverein physicists were available to the Soviets. Some had surrendered to US forces and were collected by the Alsos mission of Sam Goudsmit.[1] In this way Werner Heisenberg, Carl Friedrich von Weizsäcker, Otto Hahn, Max von Laue, and six others were interned for six months at Farm Hall in Britain.[2] They were released to West Germany in December, 1945, and did not go to the USSR.

German industrial capability, machinery, and scientific instruments, in addition to personnel, were also exploited for the Soviet Atomic Project

---

[1] Samuel A. Goudsmit, "Alsos," AIP Press, New York, 1996.

[2] Jeremy Bernstein, "Hitler's Uranium Club, The Secret Recordings at Farm Hall," 2nd ed, Springer-Verlag, New York, 2001.

based on the principle of German war reparations. The Soviet industrial base was stressed by later phases of the Project - uranium isotope separation and plutonium production in power reactors – and East Germany was often called upon to supply necessary chemicals, alloys, pumps, etc. Scientific instruments such as voltmeters and oscilloscopes were in very short supply after WWII, and supplies from Germany were essential until Soviet manufacturing could catch up.

There are several memoirs written by German visitors. We have already extensively used the work by Nikolaus Riehl. Manfred von Ardenne, and Heinz and Elfi Barwich have written about their experiences in the USSR.[3] R.G. Pose was the son of Heinz Pose, an experimental nuclear physicist who had worked for the Uranverein, and lived and worked in the USSR from 1946 until 1952. He has written impressions of life from a child's perspective.[4] Pavel Oleinikov has written an excellent review article.[5] Norman Naimark has written a book about the Russians in Germany, the flip side of Germans in the USSR, but relevant for the description of conditions in the German uranium mines.[6] Often the most detailed information is available from declassified Soviet documents, rather than personal memoirs that tend to be reserved because of secrecy.

It is interesting that there were very few if any theorists in the German group. The Soviet Project's theory groups, around Zeldovich, Pomeranchuk, Sakharov, the Landau group, and others, did not have much support from German theorists. Maybe they didn't need it. In any case, the Germans were all experimenters, engineers, and technicians. One of the most important ones for the Project was Nikolaus Riehl.

---

[3] Manfred von Ardenne, "Erinnerungen," Droste Verlag, Dusseldorf, 1997; Heinz and Elfi Barwich, "Das rote Atom," Scherz Verlag, Munich, 1967.

[4] Rudolf G. Pose, "Vospominaniya ob Obninske," HISAP'96, Vol **2**, *ibid.*, p 280.

[5] Pavel V. Oleinikov, "German Scientists in the Soviet Atomic Project," The Nonproliferation Review, Summer 2000, p 1.

[6] Norman M. Naimark, "The Russians in Germany," The Belknap Press of Harvard University Press, Cambridge, Mass., 1995. See also A. Heinman-Gruder, HISAP'96, Vol **2**, *ibid.*, p 331.

13.2. Next Assignment for Nikolaus Riehl

By 1948 Riehl's uranium refinement operation at Electrostal was running smoothly, producing up to one metric ton per day of reactor grade natural uranium.[7] On April 26 of that year Makhnev wrote to Beria about re-assignment of Riehl to the plutonium production complex at Combine #817 located in the Urals near Chelyabinsk.[8] Riehl would become the deputy to Academician A.A. Bochvar, who was the director of the effort to produce weapons grade plutonium from the spent uranium fuel rods from the 'A' reactor. However, this transfer did not happen. Riehl did do some work on extraction and purification of plutonium, but it was done before 1950 in the Electrostal plant.[9]

In 1950 Nikolaus Riehl and his family moved from Noginsk to Sungul in the Urals. Sungul is in a lake region about 100 km south of Sverdlovsk (now Ekaterinburg), near Kyshtym where the 'A' reactor and Factories 'B' and 'V' for plutonium processing were located. Riehl became the director of a laboratory, but instead of plutonium refinement, he was to study biological effects of radiation, and methods to measure and quantify such effects on specific organs of the body, a complete change from heavy element metallurgy.[10]

13.3. Manfred von Ardenne in Germany

Manfred von Ardenne and his private research laboratory at Berlin-Lichterfelde was noted in Chapter 7 in connection with the Soviet 'Alsos' mission to Germany in May, 1945. Von Ardenne was an interesting man. It is fair to say that he was a member of the German aristocracy, and as an adult he was accustomed to dealing only with government at the highest levels – a characteristic that he retained during his stay in the USSR. He was also accustomed to being treated with the deference that was his due.

---

[7]   Nikolaus Riehl and Frederick Seitz, "Stalin's Captive," *ibid.*, p 102.

[8]   Atomnyy_proekt_SSSR._T.2.Kn.3.(2002).pdf #388, Makhnev to Beria.

[9]   Nikolaus Riehl and Frederick Seitz, "Stalin's Captive," *ibid.*, p 113.

[10] Nikolaus Riehl and Frederick Seitz, "Stalin's Captive," *ibid.*, Chapter 8, p 121ff.

However, he was an ingenious inventor, and continued to be very productive throughout his long life. He wrote his autobiography late in life, and died at age 90 in 1997 in Dresden, an East German city by then a part of re-united Germany.[11] Born in Hamburg in 1907 to upper class German parents in the society of Kaiser Wilhelm II, Manfred moved with his family to Berlin at age 5, and spent most of his time there until he went to the USSR in 1945. He was a precocious inventor; he built an elaborate radio transmission and reception setup in his parents' home in Berlin, and received his first patent for radio reception in 1923.[12] He attended school in Berlin, and Berlin University, but he was not patient with formal classwork, and was mostly self-educated. He founded his independent laboratory in his 20's, and kept it going with royalties from his patents, and some industrial and government contracts.

The German Post Office had jurisdiction over all forms of communication, and von Ardenne's interest in radio transmission led to R&D contracts with the Post Office Department. The Reich Postmaster General was Wilhelm Ohnesorge, who was a supporter of von Ardenne's research, and a highly placed minister in the Nazi government.[13] Von Ardenne developed equipment for radio and television transmission, and for radar. In 1940, he proposed a scanning centimeter wave length radar installation for detection of incoming enemy aircraft that was unsuccessfully presented to Reichsmarschall Göring.[14] He built a 60 ton magnet for a cyclotron that was not completed in Germany during the war, but was moved to Sukhumi with the rest of his laboratory in 1945. He invented the scanning electron microscope. He was more interested in building research equipment than in exploiting it for scientific measurements. In this sense, he had much in common with Ernest Lawrence. Indeed, von Ardenne was also very interested in isotope

---

[11] Manfred von Ardenne, "Erinnerungen," Droste Verlag, Dusseldorf, 1997. Obituary in the New York Times, May 29, 1997.

[12] Manfred von Ardenne, "Erinnerungen," *ibid.*, p 73.

[13] Albert Speer, "Inside the Third Reich," *ibid.*, p 226, stated that Ohnesorge was a close friend of Hitler's photographer, Heinrich Hoffman, and Hitler learned of Ardenne's work, and the Uranverein of Heisenberg, through this link.

[14] Manfred von Ardenne, "Erinnerungen," *ibid.*, p 194.

separation by the electromagnetic mass spectrometer method – Lawrence's calutrons. Von Ardenne invented various types of ion sources, which would later be used in EM mass separators of the Soviet Atomic Project. Von Ardenne was not trained in nuclear physics, and although he knew the members of the Uranverein, and participated in relevant discussions, he was only peripherally involved in it himself.

The story of Fritz Houtermans is told in Section 1.4 of Chapter 1, where we learned that Fritz was released from Gestapo prison on the request of Max von Laue. Fritz could not get a job in German industry or government in 1941 because of security concerns, but von Ardenne's laboratory was private, and Fritz was hired there. During the war years Houtermans worked on isotope separation, and problems of reactor design for the Uranverein.[15] Fritz's work was about the only direct contribution to the German atomic project made by von Ardenne's laboratory. Among the high level meetings that von Ardenne attended was the one organized by Albert Speer on June 4, 1942, at Harnack House in Berlin, to discuss matters of armament, including the possibility of atomic weapons.[16] During the war Fritz was asked to accompany a German delegation to Kharkov then under German occupation. He had been a nuclear physics researcher at the Ukranian Physico-Technical Institute (UFTI) in the 1930's. Because of this wartime visit Fritz was deemed disloyal by the Soviets, and not permitted to return to the post- war USSR, so he moved to Switzerland. While at Lichterfelde he continued design studies for nuclear reactors.

By 1944 it was clear that Germany would lose the war. The chief scientist at Siemens, Nobel Prize winner Gustav Hertz, visited Lichterfelde, and von Ardenne and Hertz decided to move to the Soviet Union after the war. Von Ardenne was offered the opportunity to leave Berlin with his family and laboratory for the West, but he declined, although at that time the trip could have been made safely. So they stayed

---

[15] Manfred von Ardenne, "Erinnerungen," *ibid.*, p 206.

[16] Albert Speer, "Inside the Third Reich," *ibid.*, p 225. At this meeting Heisenberg stated that atomic weapons, while feasible, would take too long to develop for WWII.

put. The Lichterfelde laboratory was in a bunker underground, and its entrance was camouflaged by a junk pile – so well that it was never discovered by the invading Red Army. In April, 1945, a German chemist and colleague of von Ardenne, Peter Thiessen, got out of a Soviet armored car in front of the Lichterfelde laboratory with an 'Off Limits' order, so that the laboratory would not be trashed.

13.4. von Ardenne Moves to the Soviet Union

On May 10, 1945, von Ardenne sent the following letter to Stalin:

"Manfred von Ardenne   Berlin-Lichterfelde-east

Research Laboratory   Jungfershtig, 19

Electron physics   10 May 1945

To the President of the Council of Peoples Commissars of the USSR [Stalin], Moscow, Kremlin

Referring to today's inspection of my research institute, the former institute for nuclear studies which I managed, supported by the Postmaster General of the Third Reich, I will welcome the opportunity for my institute, remaining independently capable, to work cooperatively with central scientific institutes of the USSR.

My institute is presently working on the following fundamental questions:

1. High resolution electron microscopy with two powerful electron microscopes.

2. Studies in the field of nuclear physics, in particular the use of radioactive and stable isotopes as tracers. The isotopes are made with a 1 MeV electrostatic accelerator, and studied with counting apparatus, magnetic isotope separation, and mass spectrometry.

3. Recording mass spectrometer for quantitative chemical analysis of gases, liquids, and solids.

4. Commissioning of the 60 ton cyclotron.

Principal tasks:

- Improve the resolving power of the electron microscope with the goal of resolving individual atoms.

- Separate isotopes in sufficient quantity to weigh them.

- Biochemical studies by the tracer method,

- Application of radiodetection over large distances [radar].

- Utilization of the already built large stereoscopic apparatus with polarized light for the training of young scientists. [not clear what this refers to].

Today I place my Institute and myself under the supervision of the Soviet government.

At your service, Manfred von Ardenne

Translated from the German by Al. Gumilev.[17]"

Although von Ardenne does not elaborate on many of the topics that he mentions, it is clear that no listed activity is related to nuclear weapons. There is no reference to his ion source development work, that was truly innovative, and that was to be used in the Soviet electromagnetic isotope separation machines. The use of isotopes as tracers probably refers to medical applications. He proposes to move his entire functioning laboratory, busy with highly technical but peaceful research, to the Soviet Union if that is desirable on the part of the Soviet government. In his view the move was part of German war reparations to the USSR. Needless to say, his proposal was enthusiastically received, and arrangements were made to transfer all of the equipment, including the cyclotron magnet, from Lichterfelde to a new facility in the USSR built especially for him.

On the same day, May 10, 1945, von Ardenne met with Makhnev, Artsimovich, Flerov, Kikoin, and Migulin to discuss moving his family, laboratory, and technicians. The letter to Stalin was probably given to

---

[17] Atomnyy_proekt_SSSR._T.1.Ch.2.(2002).pdf #346, von Ardenne to Stalin. Translated from Russian by the author. It would be interesting to translate this into German, and compare it to the original!

Makhnev for delivery. Later Makhnev brought General Avraami Zavenyagin to meet von Ardenne. Zavenyagin started out in metallurgy. In the early 1930's he was director of Magnitogorsk, and transformed the new center into a leading source of ferrous metals. He continued a successful career in metallurgy for defense during the war. In 1945 Zavenyagin was one of Beria's deputies involved with the Soviet Atomic Project, and working with Igor Kurchatov. Von Ardenne was favorably impressed with Zavenyagin's ability, and the two would work together during von Ardenne's sojourn in the USSR. Von Ardenne believed that Zavenyagin always had the best interests of the German visitors in mind.[18] A few days later von Ardenne and his wife flew from Berlin to Moscow on a Douglas DC-3. The children, and the Laboratory, remained in Berlin for the time being.

They landed at Vnukovo Airport, and were met by a woman who was to be their permanent companion.[19] They were housed outside the city, on the banks of the Moscow Canal in a small dacha with a garden. They had a full household staff to take care of them (and to keep an eye on them). Soon the family attended the Bolshoi Theater for Tchaikovsky's ballet 'Swan Lake' with other German guests. The war with Germany had ended only about two weeks before, but here were a group of Germans in the hospitality of the USSR, together with US and British military. Nikolaus Riehl also attended this event, and also commented on the mix of former enemies, but he recalled that the performance was Borodin's 'Prince Igor'.[20]

Zavenyagin visited von Ardenne with the news that Beria had decided that von Ardenne's original proposal as outlined in his letter to Stalin should be changed to emphasize work on the Soviet Atomic Project. Accordingly, von Ardenne attended a meeting in Beria's office with Kurchatov, Kikoin, and others. Beria said that von Ardenne must work on

---

[18] Manfred von Ardenne, "Erinnerungen," *ibid.*, p 227.

[19] Moscow has four airports: Sheremetevo, Domodedovo, Vnukovo, and Bykovo. The first two are now used for civilian passenger travel inside Russia and overseas. Vnukovo is used by government officials.

[20] Von Ardenne, "Erinnerungen,' *ibid.*, p 233. Riehl and Seitz, *ibid.*, p 82.

the atomic bomb, and gave him 10 seconds to think it over. Von Ardenne thought fast, and proposed to work on an ancillary project, isotope separation, instead of the bomb itself. Beria agreed to this assignment. One reason for his request was that von Ardenne did not wish to participate in the development of nuclear weapons, but another reason may have coincided with Riehl's position that too much entanglement with the Soviet Atomic Project would preclude their ever leaving the USSR to return to Germany. How deep into the Project Beria was willing to allow the Germans to penetrate is unknown.

The two children and other relatives arrived at the Moscow dacha after the von Ardenne parents had been there for three weeks. They brought news that the entire Lichterfelde laboratory, including the 60 ton cyclotron magnet and the 1 MeV electrostatic accelerator, had been loaded on freight cars for the trip from Berlin to Moscow. Red Army soldiers and the remaining lab staff did the packing. There were 750 crates. The packers ran out of wood, so they dismantled a neighborhood bowling alley for timber.

13.5. Gustav Hertz and Heinz Barwich

Gustav Hertz was born in Hamburg, Germany in 1887, and died in East Berlin in 1975.[21] His father was a lawyer and a brother of Heinrich Hertz. Heinrich Hertz performed several experiments to show that light was an electromagnetic phenomenon described by Maxwell's equations. The demonstration that longer wavelength radiation that behaved like visible light could be generated with ordinary electrical circuits changed the world. The unit of frequency – one hertz = one cycle per second – was named in his honor. He would have won the Nobel Prize in physics had he lived. Heinrich Hertz died in 1894 at age 37 years, and the first Nobel Prize was awarded in 1901. So Gustav Hertz had a very famous uncle. Gustav attended high school in Hamburg, and in 1906 enrolled in the University of Göttingen. He transferred to Berlin for his doctoral work, and completed

---

[21] www-history.mcs.st-andrews.ac.uk/Biographies/Hertz_Gustav.html

his degree in experimental physics there in 1911. He remained at Berlin, and began the collaboration with James Franck that would lead to their sharing the Nobel Prize in physics in 1925. The Franck-Hertz experiment was an early demonstration of quantized electron energy levels in atoms. They measured the low energy electron current passing through mercury vapor, and observed a sharp drop in transmission for a kinetic energy of 4.9 eV, an excitation energy level of the mercury atom. These data furnished important support for Bohr's quantum theory of atomic structure. Hertz was drafted into the German army in WWI, and wounded in action in 1915, but he survived the war. The young British physicist Harry Moseley, a brilliant student of Ernest Rutherford, was not so lucky. Moseley was killed in 1915 at Gallipoli.

Between the wars Gustav Hertz began work in industry at Phillips in Eindhoven, The Netherlands. He then joined the faculty at the University of Berlin, only to be forced out by the Nazis in 1933. He joined Siemens where he spent WWII. He became interested in the hyperfine structure of spectral lines, a phenomenon that is different for different isotopes of the same element, so he studied various techniques for isotope separation in order to obtain pure samples. He was not trying to separate isotopes of uranium. He retained enough connection with the University of Berlin-Charlottenberg to train some graduate students. Heinz Barwich was one of Hertz's students.

Heinz Barwich came from a middle class family. His father was a bureaucrat, a WWI veteran with pacifist and socialist sympathies. Heinz inherited the family's socialist orientation, although he was never a member of the communist party. In 1929 Barwich enrolled in an electrical engineering program at the technical high school of Berlin-Charlottenburg.[22] Barwich was attracted to physics by an impressive list of pioneers of modern physics at Berlin, including Einstein, Schrödinger, Heisenberg, and Born. He was especially interested in the lectures and demonstrations of Gustav Hertz, and signed on as his graduate student. Hertz's supervision continued after he left the University in 1933, Barwich

---

[22] Heinz and Elfi Barwich, "Das Rote Atom," *ibid.*, p 12.

finished his degree in 1936 with a dissertation on isotope separation. Hertz hired Barwich after graduation to work at Siemens. They did some development work on torpedo fuses for the German navy during WWII. Barwich moved his family to a farm outside Berlin to avoid the bombing raids, and he was there when Berlin fell to the Red Army in May, 1945. Siemens was no longer functional, and Barwich was out of work. He returned to Berlin looking for Gustav Hertz, only to learn that he had already left for the Soviet Union. He wrote a letter to Hertz addressed to the Soviet Academy of Sciences and gave it to the Soviet military commandant in Berlin. He was interviewed by a representative of the Soviet Alsos mission. When he said that he had worked in isotope separation, bells and whistles went off. On August 4, 1945, he and his family flew to Moscow. Barwich was 33 years old, married, and the father of three children with a fourth on the way. He was very pleased to be gainfully employed in his field of expertise.[23] Manfred von Ardenne and Gustav Hertz were already there.

13.6. von Ardenne and Hertz Move to Sukhumi on the Black Sea

The new von Ardenne laboratory was to be located in a converted sanatorium called Sinop less than one km south of the center of the resort town of Sukhumi on the Black Sea. The area is shown on the map of the Western Soviet Union, Map #1. Sukhumi is located at 43^0 N latitude, approximately the same as Rochester, New York, but it enjoys a mild, almost tropical climate – hence the sanatoria. Georgia's Black Sea coast was, and still is, a favorite vacation spot for Russians. The Germans sent there to work on the Soviet Atomic Project were fortunate. They received favored treatment, and good weather. They were required to work on assigned tasks, to report their progress regularly, and to interact sporadically with their Soviet colleagues. They were always under surveillance, and not allowed to wander about unescorted. Travel abroad was not possible, and even travel within the Soviet Union was restricted.

---

[23] Heinz and Elfi Barwich, "Das Rote Atom," *ibid.*, p 22.

There were the usual secrecy requirements – they could not discuss their work with outsiders, nor could they publish anything in the open literature. Relatives in Germany knew they were in the Soviet Union, but censorship was designed to remove any hint of what they were doing. The Germans were, however, a weak link in the strict Soviet security system, as is briefly described in Chapter 14. The von Ardenne family was allowed to go with an escort on vacation trips into the Caucasus mountains.[24] Probably other senior members of the team had similar privileges. The von Ardenne family lived there almost 10 years.

Von Ardenne in his memoirs said even less about his technical contributions to the Soviet Atomic Project, or about his interactions in meetings with the Soviet management than did Nikolaus Riehl, and Riehl did not say very much. Riehl had the advantage of fluency in Russian, making it easier for him to know what was going on. De-classified Soviet documents contain more information on the accomplishments of the German scientists than the memoirs of the participants, probably because of pledges of secrecy.

Gustav Hertz had his own laboratory in another sanatorium about 7 km further south of Sukhumi, named Agudzery. Heinz Barwich was one of Hertz' deputies. While Sinop received all of von Ardenne's equipment from his laboratory in Germany, Hertz was not so lucky. Siemens laboratory equipment that was usable had already been shipped to Kharkov, so Agudzery had to manage with what remained of the Kaiser Wilhelm Institute, and what could be obtained elsewhere.[25] When Barwich arrived the sanatorium was under construction for conversion into an institute by prisoners and Red Army soldiers. He was very impressed by the pleasant environment. One of his colleagues said that they, the German visitors, were in big trouble, because they could not possibly perform at a high enough level to justify their privileges. Barwich pointed out that before WWII there were many foreigners working to industrialize the

---

[24] Manfred von Ardenne, "Erinnerungen," *ibid.*, photo on p 282. The driver probably took
      the picture.
[25] Heinz and Elfi Barwich, "Das Rote Atom," *ibid.*, p 59.

Soviet Union, and that they were well received by the population, because industrialization raised the standard of living. Foreign participation ended with WWII, but the German specialists in Sukhumi benefitted from this favorable pre-war reputation. Even so, contact with the local citizens was very limited. All German personnel lived and worked at the laboratory, which was surrounded by a fence and guarded by soldiers. Leaving the site required a permit, and an escort. All of the necessities of life were supposed to be available inside the fence, typical of other laboratories and factories of the Soviet Atomic Project.

The two laboratories were to collaborate on the general problem of isotope separation. Sinop would work on electromagnetic mass separation, and in particular the development of an ion source. It also played a role in invention of a practical centrifuge for isotope separation. Agudzery and Sinop both contributed to the gaseous diffusion project.

13.7. von Ardenne's Ion Source Development

Von Ardenne had been developing a new type of ion source in Germany. He called it a 'duoplasmatron' ion source, and they are still in use today in accelerators, in plasma physics, and in many industrial applications of ion beam technology. A standard ion source design includes a hot filament cathode that emits electrons, a gas stream that the electrons ionize, and then a high negative voltage to accelerate the beam of positive ions. Von Ardenne used his knowledge of focusing charged particle beams with electric and magnetic fields gained through construction of high resolution electron microscopes, and his experience with ionized plasmas to improve the standard design by creating a plasma, configuring the accelerating electrodes to focus the beam, and adding an axial magnetic field to aid in collimation. In equilibrium, the ionized plasma continually fed positive ions through the electrostatic lenses. The result was a sophisticated ion source, that had a long operating life, and a very high efficiency – most of the gas was ionized, and most of the ions were accelerated into the vacuum chamber. A schematic diagram of the ion source is shown in Figure 13.1.[26] High efficiency was clearly of utmost importance in handling uranium gases in a mass spectrometer, where each atom of ^{235}U is precious. Von Ardenne the inventor made an important contribution to ion source technology.[27]

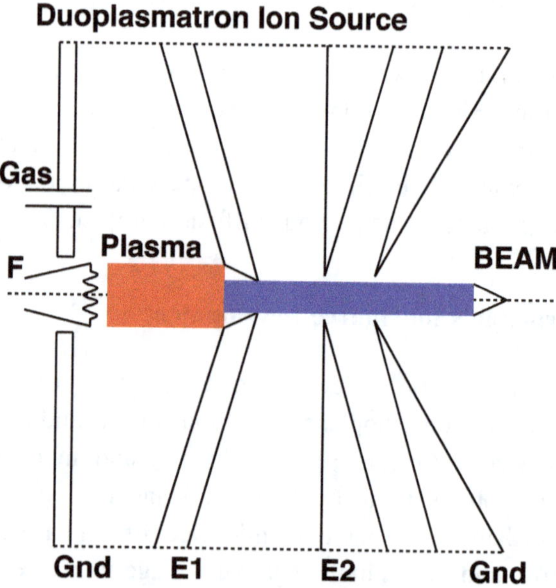

Figure 13.1. Schematic of a duoplasmatron ion source. Except for the gas inlet the device is cylindrically symmetrical about the beam axis. A coil surrounds the source to create an axial magnetic field. Electrodes E1 and E2 are suitably biased to contain the steady state plasma and extract the positive ion beam. F is the electron emitting filament.

---

[26] Adapted from Figure 3.2 in B.H. Wolf, "Handbook of Ion Sources," CRC Press, 1995 p 47.

[27] Manfred von Ardenne, "Erinnerungen," *ibid.*, p 292. For a modern treatment see B.H. Wolf, "Handbook of Ion Sources," CRC Press, 1995, p 47ff.

As time went on, Sinop was to develop the first practical ultra-centrifuges for isotope separation of UF_6 gas. Peter Thiessen designed the barriers used in the gaseous diffusion plants described in Chapter 10.

13.8. Ministry of Internal Affairs Reports on German Progress in 1947

The 9th Division of the MVD was organized in February, 1946, to manage the German guest scientists. The two sites, 'Sinop' and 'Agudzery,' were both former sanatoria in Sukhumi, on the Black Sea. There were 49 German specialists, 28 of whom were scientists, at work at the two sites in 1947.[28] Some German POW's with technical training were transferred from prison camps to the Sukhumi laboratories, but paid the price of a longer sojourn in the Soviet Union in return for more comfortable surroundings. Manfred von Ardenne headed the Sinop Laboratory, called Institute 'A,' for Ardenne. Gustav Hertz's laboratory at Agudzery was called Institute 'G'. All of the equipment in von Ardenne's private laboratory near Berlin had been moved to Sukhumi. The MVD reconstructed the sanatoria and built new work space for the physics institutes. The construction work was completed at the end of 1947, but research work had started already by the fall of that year.

Institute 'A' had three main tasks:

1. Electromagnetic isotope separation (von Ardenne). The main thrust here was the development of a suitable ion source for isotope separation. At first an arc source was made for uranium tetraflouride, but tests showed that the source was not an efficient user of material, and had a short life – the order of an hour. In addition, fluorine was liberated from the uranium tetraflouride by the source, to the detriment of the vacuum system. So Institute 'A' had to develop a new ion source, with better focusing and several other improvements, using metallic uranium. The resulting source was efficient, long lived,

---

[28] Atomnyy_proekt_SSSR._T.2.Kn.3.(2002).pdf #345.

and gave 100 milliamp of ion current. New similar sources are under development for cyclotrons.

2. Development of a suitable diaphragm for isotope separation (Thiessen). Two approaches were followed here as well. The first was a flat diaphragm made of copper and nickel. The diaphragms were of good quality, with appropriately sized holes and mechanical properties. The second approach was to fabricate nickel diaphragms in cylindrical shape, and this was done successfully. Laboratory #2 has tested parallel arrays of seamless cylindrical diaphragms 1.5 m long, with good results.

[Prusakov and Sazykin in HISAP96 Vol **1** p 165, and Plotkina and Voinov in HISAP96 Vol **2** p 199 say that the cylindrical filters had 10 nm diameter holes, were 15 mm in diameter, could withstand 100-200 mm of Hg pressure, and the area in one of the large compressors was 400 m^2. The area of a cylinder 15 mm in diameter and 1.5 m long = 0.07 m^2, so the array would have 5700 tubes.]

3. Development of a molecular method for isotope separation (Steenbeck). Again two approaches: condensation of droplets on a film of solvent. The film method has been tested with bromine isotopes with good results; and the ultracentrifuge method. Here a mechanical centrifuge has been built capable of 72,000 turns per minute.

In addition, Institute 'A' was to carry out work perfecting an electron microscope, to develop suitable methods for determining the isotopic content of uranium, and to study biological effects of radiation. Electron microscope development had achieved a resolution of 1.5 nanometers. Precision ionization chambers and mass spectrographs were developed to measure the isotopic content of uranium either by the range of emitted α

particles, or by direct measurement of the $^{235}U/^{238}U$ ratio.[29] In early 1947 Institute 'A' had 143 workers, including 64 scientists.

Gustav Hertz's Institute 'G' had about half as many workers. Both laboratories made use of a number of German POW's with technical training who had been transferred from the camps to Sukhumi.

13.9. Heinz Pose in Obninsk

German atomic scientists were placed in various other locations within the USSR. Heinz Pose was an experimental nuclear physicist who had also been a student of Gustav Hertz. Pose was unusual in that he had actually worked during the war in Germany for the Uranverein on reactor design – in particular on neutron moderation in materials, and neutron multiplication by uranium. After the war, he accepted an invitation to work in the Soviet Union for the reason given by many of his colleagues- that nuclear physics research was forbidden in Germany by the Allied Powers.[30] His family came from East Prussia, and he was familiar with Russian culture and customs. He had earlier corresponded with George Gamow when Gamow was in the USSR, and he knew other Soviet physicists, so he was comfortable with moving his family to the Soviet Union, which he did in February, 1946.

He returned to Germany to recruit workers for a new institute that was established in Obninsk, located about 100 km SW of Moscow, west of Serpukhov. It was called Laboratory 'V,' and its task was to develop a light water reactor fueled by enriched uranium. A large laboratory was planned, with 16 divisions. Eight were set up initially for various studies in neutron physics, including an electronics division, and a charged particle accelerators group (no computing division – remember, there were no computers). Progress was hampered by a lack of qualified personnel, and

---

[29] Werner Schutz's mass spectrograph was used for early measurements. Later the signature was a soft γ ray that accompanies the α decay of ^{235}U. See the section on gaseous diffusion in Chapter 10.

[30] Rudolf G. Pose, "Vospominaniya ob Obninske," HISAP'96, Vol **2**, *ibid.*, p 280

a shortage of research apparatus. Eventually, with recruits from Soviet graduate schools, the staff made progress in studying light water reactors. Their mission was only slightly related to weapons development, if at all.

Laboratory 'V' worked on the design of various types of nuclear reactors. They studied moderators of light water, heavy water, and beryllium, together with either natural or enriched uranium. By February, 1947, they had completed several theoretical studies, and facilities were still under construction for experimental testing.[31]

On March 21, 1949, Mikhail Pervukhin summarized in a report to Beria the status of all of the German laboratories. There were 515 scientists, engineers, and technicians, and 374 administrative staff in the entire German operation. Of the 515 technical workers 110 were German volunteers, 205 were released German POW's, and 184 were Soviet citizens, leaving 16 in a miscellaneous category. Of the Germans, 40 were PhD equivalent scientists. The Soviet government had decided to repatriate all German POW's by the end of 1949, and about 50 of the ex-POW's working for the Atomic Project had requested repatriation at that time. Pervukhin said that most of them probably could be sent home, and replaced by Soviet workers.[32] Some general housekeeping remarks included replacing Pose with Blokhintsev as director of Laboratory 'V,' and allowing German youth to attend high school with Soviet students (elementary schools were separate).

By 1949 work at Laboratory 'V' on a beryllium moderated reactor with slightly enriched uranium had progressed to the point of estimating the physical size of such a reactor (smaller than one moderated by graphite), and developing techniques to sufficiently purify the beryllium of contaminants like boron so that it would work in a reactor. The lack of adequate purification was holding up construction of a working model. Other efforts at Laboratory "V" included design work on a proton

---

[31] Atomnyy_proekt_SSSR._T.2.Kn.3.(2002).pdf, #345

[32] Atomnyy_proekt_SSSR._T.2.Kn.4.(2003).pdf #239 Pervukhin to Beria 21 March, 1949.

synchrotron for research in particle physics, led by Leipunsky, and development of instrumentation for work with radioactive materials.

Heinz Pose's son Rudolf was a boy when the family moved to Obninsk, and he has recorded some of his memories of those times. Obninsk consisted of two large stone buildings plus a few wooden houses. Rudolf did not say whether or not there had been any industry located there previously, as was often the case for site selection. The larger stone building became the laboratory, and the smaller structures furnished some shelter for the workers. However, housing was soon augmented by the ever popular Finnish houses, and even some three story wooden structures. In contrast to Laboratories 'A' and 'G,' located in the already existing town of Sukhumi, Laboratory 'V,' much like Arzamas-16, was outside Obninsk, and had to have a town created around it.

Like other German scientific groups, the Obninsk workers were treated relatively well in terms of living and working conditions. They signed yearly agreements, but were not given an end date to their term of service. They were under constant surveillance, and their movement was very restricted. The work was secret, and could not be discussed at home or anywhere outside the laboratory. Families could picnic in the neighboring woods, fish and swim along the Protva River,[33] and gather berries and mushrooms before the town was fenced, but afterwards they were confined to quarters. Shopping trips to Moscow were rare, and of course escorted. Tennis courts were built in town, and flooded for ice skating in the winter. The families settled into their routine.

The NKVD recruited some German workers as informants on the activities of others.[34] Russians also worked at the lab, and Russian children attended the same school as the German children. So the German children learned Russian. Rudolf demonstrated his Russian expertise by writing his article on Obninsk in that language. He commented that German and

---

[33] Protvino, the science village near Serpukhov that is a center for research in particle physics, is named after the Protva River.

[34] Von Ardenne also commented on this. "Erinnerungen," *ibid.*, p 294.

Russian children were not allowed to mix outside of school. While the Germans were valued and generally well treated, foreigners were never really trusted.

13.10. The Centrifuge

Figure 13.2 shows the array of gas centrifuges in Piketon, Ohio in 1984. These cylinders are 12 m tall. More modern models are shorter, but news photographs of the centrifuges in Iran look very similar. This is the technique that has completely replaced gaseous diffusion as the method of enrichment of ^{235}U from natural uranium. So why did it take so long?

The application of centrifuges to separate isotopes was proposed well before WWII. Jesse W. Beams at the University of Virginia succeeded in separating isotopes of chlorine, and published results on centrifuge development in 1936.[35] Beams' laboratory was an obvious choice for development of centrifuges on the industrial scale that would be required for the Manhattan Project. The problem was basically one of mechanical engineering. Centrifuges did not have the mechanical lifetime to serve in an industrial setting. The rotating machinery would become unstable and literally tear itself up. Groves soon pulled the plug on the project, and the effort switched to gaseous diffusion and electromagnetic separation.[36]

The leader in the Soviet Union was Fritz Lange. Lange was born in Berlin, and in 1924 he graduated from the University of Kiel, and attended graduate school in Berlin. Soviet physicists knew of his experimental work, and in 1933 when the Nazis came to power he left Germany, first going to London, then by invitation of Leipunsky to Kharkov to work at UFTI. The UFTI electrostatic accelerator was built under his direction, and in 1937 he became a Soviet citizen. He was transferred to the Ural Physico-Technical Institute in Sverdlovsk, where he met Isaak Kikoin, and became interested in centrifuges for isotope separation.[37] During the war

---

[35] J.W. Beams and F.B. Haynes, Phys Rev **50**, 491 (1936).
[36] H.D. Smyth, "Atomic Energy for Military Purposes," *ibid.*, p 169.
[37] O.D. Simonenko, HISAP96 Vol **2**, *ibid.*, p 438.

years Lange worked on a horizontal centrifuge with a steel rotor 250 mm in diameter and 600 mm long, with 5 mm thick wall, weighing about 21 kg. The rotation speed was 8000-10,000 rpm. (10000 rpm = 1.3×10^4 cm/sec speed at the periphery. See Equation C.23 in Appendix C.) The support axis was stationary, and three tubes were hooked up from outside to inside for the input gas, and the heavy and light output gases. A temperature gradient was established along the cylinder. This rig was used to study separation of ordinary gases, like nitrogen and carbon dioxide, effects of convection and turbulence, etc. It was a long way from a prototype uranium separator.

Figure 13.2. Array of centrifuges for separation of ^{235}U from natural UF_6 in Ohio. Courtesy of the US Department of Energy.

In December, 1945, the PGU established Laboratory #4 in Moscow, and Lange moved there to become director. He continued his centrifuge development work there, but he did not have much success.[38]

One of the physicists at Sinop was Max Steenbeck, mentioned above in Section 13.6. Like Gustav Hertz, he had worked for Siemens in Germany during the war, but unlike Hertz and von Ardenne he was arrested after the war by the NKVD and sent to a prison camp in Poland. He was released because of his technical background, and sent to von Ardenne's Laboratory A at the end of 1945. There he joined the efforts to separate the isotope ^{235}U from natural uranium. One of the members of his group was Gernot Zippe, an Austrian mechanical engineer who had also been in a POW camp after the war. Zippe was a highly skilled designer, and decided to take on the challenge of centrifuge development.[39]

The centrifuge developed at Sinop had little resemblance to the ones built by Fritz Lange. They were vertically mounted, with a small aluminum alloy cylinder resting on a needle bearing at the bottom, and constrained at the top by permanent magnets. The outer shell was evacuated; the cylinder was spun by an induction electric motor, and the bottom was heated to set up convection currents in the rotating gas. This complicated hydrodynamic system resulted in the lighter isotope, that was concentrated in the center, migrating to the top; and the heavier one, that was concentrated at the periphery, migrating to the bottom. Tubes running along the axis collected the enriched and depleted samples from the top and bottom respectively, and the enriched gas flowed to the next stage, like gaseous diffusion.

Appendix C has more technical detail regarding the operation of the centrifuge. The basic gas enrichment formula depends on the mass difference, in contrast with the other isotope separation methods like

---

[38] Atomnyy_proekt_SSSR._T.2.Kn.4.(2003).pdf, #222.

[39] H.G. Wood, Alexander Glaser, and R. Scott Kemp, *Physics Today* **61**, #9, p 40 (2008). See also William J. Broad, "Slender and Elegant, it Fuels the Bomb," New York Times, March 23, 2004.

gaseous diffusion or electromagnetic separation, that depend on the square root of the mass ratio. So the centrifuge is much more efficient, but the hydrodynamics of the rotating gas, and the necessity of avoiding liquefaction of the UF_6 limits the amount of gas in any given cylinder to a few grams. Thus many cylinders are connected in parallel for mass flow, but not so many stages are required in series to achieve high purity ^{235}U.

The new centrifuges began to appear after 1950 – too late to have an effect on the initial success of the Soviet Atomic Project. Centrifuge plants replaced gaseous diffusion plants in the USSR by the 1960's. Credit for the final design has been disputed. The Germans credited Steenbeck and Zippe, while the Soviets claimed that crucial modifications so that several centrifuges in series could handle UF_6 were made by Evgeny Kamenev and Isaak Kikoin.[40] One member of the Soviet group, N.M. Sinev, has written a book about it.[41] In any event, it was Zippe's memory that spread the word.

A drawing adopted from Prusakov and Sazykin of a needle base centrifuge of aluminum alloy supported at the neck by permanent magnets (not shown) is presented in Figure 13.3. The outer volume was evacuated and the bottom was heated to set up convection currents that carried the light isotope to the top, above the baffle.

In 1956 Gernot Zippe was released from the Soviet Union, and returned to Vienna. After attending a meeting in Amsterdam in 1957 Zippe flew to the United States. He had no documents related to his work in the Soviet Union, but he reconstructed the components of the centrifuge from memory. He went to the University of Virginia in Charlottesville, VA, the home of J.W. Beams, who was still there, and still interested in isotope separation by centrifuge. The design was truly revolutionary to the Americans. The Atomic Energy Commission got into the affair, and promptly classified what came from Zippe's head. Zippe then returned to

---

[40] V.N. Prusakov and A.A. Sazykin, HISAP96, Vol **1**, *ibid.*, p 156.

[41] N.M. Sinev, "Enriched Uranium for Atomic Weapons and Energy," Moscow, TsNIIAtominform, 1991.

Europe, to concentrate of peaceful uses of atomic energy. The Soviet → USA transfer of centrifuge design was a rare, but not unique example of reverse technology transfer in the field of nuclear science. The relative ease and efficiency of the centrifuges have contributed to nuclear proliferation.

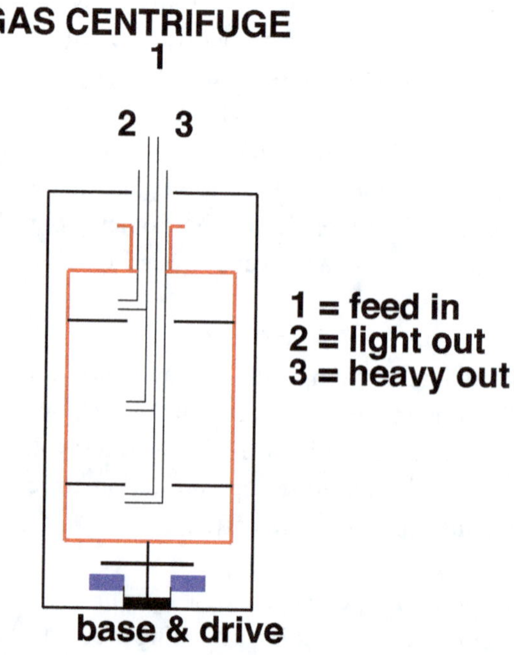

Figure 13.3. Schematic of a working gas centrifuge for enrichment of uranium hexaflouride.

13.11. Contributions of the German Scientists

The German groups at Sukhumi, Noginsk, Obninsk, and elsewhere fulfilled their assigned tasks for the Soviet Atomic Project. Their benign and even friendly treatment by the NKVD was clear evidence of satisfactory performance. Beria was too practical to tolerate them if they did not produce, and he certainly had no strong feelings of charity. His GULAG system always had room for more prisoners, and although the threat was always present, the German specialists moved out of the camps, not into them. Manfred von Ardenne was an aristocrat, accustomed to dealing with government ministers at the highest level, and also a talented inventor. His main contribution to the Project was a significant achievement for charged particle technology – the duo-plasmatron ion source. However, electromagnetic isotope separation did not play a key role in the Project. Von Ardenne was also a skilled manager, and under his guidance Peter Thiessen developed barriers for gaseous diffusion that were used in Soviet plants. Max Steenbeck and Gernot Zippe were also at von Ardenne's laboratory. Gustav Hertz derived considerable distinction as a Nobel Laureate, and was treated with deference. His contribution was in scientific management. Heinz Barwich was a good experimental physicist, and made significant contributions to understanding the operation of a gaseous diffusion plant. Barwich and Hertz spent some time at Factory #813, helping to recover the plant from its start-up pains.

The most important German contributor to the Atomic Project was undoubtedly Nikolaus Riehl. He was a native Russian speaker, and hard-driving supervisor for the refinement of tons of uranium needed to get the Project started in 1946. He had perhaps the best understanding of Russian customs and the Soviet system, and knew how to get things done. He was not shy.

Like all of the other researchers in the Project, the Germans published no scientific papers for a decade or so after the war. One measure of their achievement was Stalin Prizes given by the Soviet Government. The Stalin Prize was a substantial honor, involving a medal, a monetary award, sometimes furnished houses and cars, free education for the children, and free travel anywhere within the Soviet Union. Among the German Stalin Prize winners were Nikolaus Riehl and Gunter Wirths for uranium processing, Peter Thiessen, Heinz Barwich, and Gustav Hertz for isotope separation by gaseous diffusion, Werner Schutze for his precision mass spectrometer, and Manfred von Ardenne for isotope separation by the electromagnetic method. Von Ardenne used his prize money to buy land in East Germany for his new laboratory after release from the USSR. Riehl was hesitant to accept too many gifts, in order to minimize his commitment, and improve his chances of being repatriated to Germany, but he did wear the Stalin Prize medal, because it opened doors for him.

War reparations of German equipment, industrial capability, and instrumentation were perhaps more difficult to quantify, but also played important roles in the post-war Soviet Union. Many aspects of German technology were still superior, and according to Oleinikov it took several years for Soviet industry to catch up in the manufacture of scientific instrumentation that was critical to the Atomic Project, like oscilloscopes and multimeters.[42]

Did the Germans make important contributions to the Soviet Project? Yes. Could the Soviets have done it without them? Also, yes. Both statements are true. The Germans helped where asked to do so, but were never part of the main operation. Riehl and von Ardenne attended some meetings in Beria's office, but did not participate in the daily planning sessions of the PGU that determined the course of the Project. Their penetration of Arzamas-16, the actual bomb laboratory, was very small if it existed at all. As we have seen, the Germans remained at some distance from the big picture partly because of a reluctance on their part to become too involved, but it may also have been a mutual restraint, in that the Germans were a possible leak in the tight security of the Project.

13.12. Repatriation

A cool down period of a year or two away from secret work was required of all of the German scientists before repatriation. The purpose was to insure that their knowledge of the Project was somewhat stale. They were sworn to secrecy. All written records were kept at the laboratories

---

[42] Pavel V. Oleinikov, "German Scientists in the Soviet Atomic Project," *The Nonproliferation Review*, Summer 2000, p 24.

under lock and key. By the mid 1950's most of them had returned to Germany. At that time, before the Berlin Wall, it was possible if not easy to move from east to west, and the Riehl family settled in Munich, West Germany. By prior agreement with the Soviet government von Ardenne moved his entire laboratory to Dresden in East Germany, where he spent the rest of his life. Von Ardenne retained sympathy for socialist goals, despite his aristocratic origin. Perhaps it was his ready access to high levels of the East German government that made life tolerable. Gustav Hertz became a professor at the University of Leipzig in East Germany.

Heinz Barwich also returned to Dresden in East Germany in 1955. He became director of a nuclear physics research laboratory at Rossendorf. Klaus Fuchs became a member of Barwich's staff when he was released from British prison in 1959. In 1961 Barwich was invited to return to the USSR for a three-year term as deputy director of the Joint Institute for Nuclear Research – JINR- Dubna. He accepted, and Heinz and his second wife Elfi flew from Dresden to Moscow Sheremetevo airport in a TU104.[43] A laboratory automobile and driver took them to Dubna past Dimitrov, where there were (and still are) monuments commemorating the end of the German advance on Moscow in 1941. As deputy director Heinz was assigned one of the single family homes in village Dubna, and their next door neighbor was Bruno Pontecorvo, who had already been there for 11 years. Elfi was very impressed with Pontecorvo, who retained his considerable Latin charm. Dmitri Blokhintsev was director of JINR, and Georgy Flerov was head of the laboratory for nuclear reactions. Barwich enjoyed his return to the Soviet Union. He had met Elfi in Germany after divorcing his first wife, so a holiday trip to Sukhumi offered the opportunity to show her where he had spent 10 years. Sinop and Agudzery were still there, and still laboratories, but Barwich did not comment on their activities. There was an international conference on particle accelerators at Dubna in 1963, and Barwich participated in all of the preparations necessary to accommodate foreign delegates. In 1964, they returned to Dresden. Klaus Fuchs had acted as lab director in his absence.

---

[43] Heinz and Elfi Barwich, "Das Rote Atom," *ibid.*, p 193.

Heinz Barwich had always been sympathetic to socialist causes, and was supportive of the social experiment in the Soviet Union, where he felt welcome and reasonably secure. Communist East Germany however was a different environment for him, and a few years after their return the Barwich family decided to defect to the west. The Geneva conference on the peaceful uses of atomic energy in September, 1964, afforded the opportunity, and the Barwich's were soon in the United States. Heinz Barwich died of a heart attack on a return visit to Europe in 1966.

After his release from Laboratory "V" Heinz Pose continued his work in experimental nuclear physics for two more years at JINR, Dubna, before returning to research and teaching in Dresden, East Germany in 1959.

Chapter 14

Semipalatinsk Nuclear Test Range

14.1. Introduction

Arzamas-16 by the middle of 1948 had become a place of feverish activity. The various components of an atomic bomb were beginning to come together. Conventional explosive studies were ongoing both for compression and gun-type bombs, although in the end the latter was never made. Density of material under high pressure was measured to determine how much plutonium would be required for a compression bomb. They had to confirm that the choice, design configuration, and firing arrangement of the conventional explosive would in fact achieve the necessary compression. The design of a weapon that could be delivered by aircraft required the active participation of the Soviet Air Force. The sheet metal runway initially installed at the lab site, mentioned in Chapter 11, was not suitable for heavy bombers, and had to be upgraded. Dummy bombs of the estimated size and weight were made, and a four engine heavy bomber, the PE-8, was modified to accommodate the large size for use in practice bombing runs. Fissile materials, both ^{235}U and ^{239}Pu were starting to arrive from the factories in the Urals. A test site for a nuclear explosion was required, at some remote site away from European Russia.

14.2. Semipalatinsk

Yuli Khariton and his staff had all that they could handle, so fresh talent was recruited to work on the selection and preparation of a test site – a nuclear firing range. Academician Nikolai N. Semenov was a colleague of Yakov Zeldovich, and director of the Institute for Chemical Physics of the

USSR Academy of Sciences in Moscow. Semenov had a professional interest in chemical explosives. He would share the Nobel Prize in Chemistry in 1956. In 1946 Semenov became interested in the planned American tests at Bikini Atoll in the Marshall Islands, and proposed deploying various instruments on the Pacific coast of the Soviet Union and on ships of the Soviet Pacific Fleet to obtain relevant data.[1] Arzamas-16 capitalized on his interest in nuclear weapons detection, and recruited Semenov's Institute for Chemical Physics to work on plans for the nuclear firing range, including site selection, construction, and test instrumentation.

Site selection inevitably required a committee. The committee was composed of members of the PGU (Beria's management group), together with members of the Ministry of Defense. The Soviet Union was a very big place. After some initial screening, they narrowed the search to six sites. Four were in Kazakhstan, and two were farther east, near Lake Baikal. Whether any of the sites had previously served as a firing range was not reported, but the presence of military officers on the committee would make it seem likely. The site chosen was in the North East corner of Kazakhstan, 170 km west of the town of Semipalatinsk, near the Irtysh River. See Map #1 for the region, and Figure 14.1 for close up details.[2] The region was an uninhabited plain about 20 km in diameter, surrounded by low hills on all sides. There were natural dirt roads and fresh water. Light aircraft could land there without special runways, but heavy bombers would have to use the airport at Semipalatinsk. One disadvantage was that the nearest railroad was the north to south Novosibirsk-Semipalatinsk-Alma Ata line, 170 km to the east. Novosibirsk was served by the trans-Siberian railroad, with direct connection to the western Soviet Union. The committee suggested moving a Chinese consulate that was in Semipalatinsk to another, more remote, location.[3] In addition they

---

[1] Atomnyy_proekt_SSSR._T.2.Kn.6.(2006).pdf, #53 Semenov to Vannikov, 10 June 1946.

[2] Ya.P Dokuchaev, "First Soviet Atomic Bomb Test," HISAP96, Vol **2**, p 300. Figs 1 and 2 in this article are combined into Figure 14.1.

[3] Atomnyy_proekt_SSSR._T.2.Kn.6.(2006).pdf, #116 site selection committee to Beria, May 31, 1947.

estimated that about 1000 people would have to be moved for safety, and the total price tag for instrumenting the site would be about 100 million rubles.

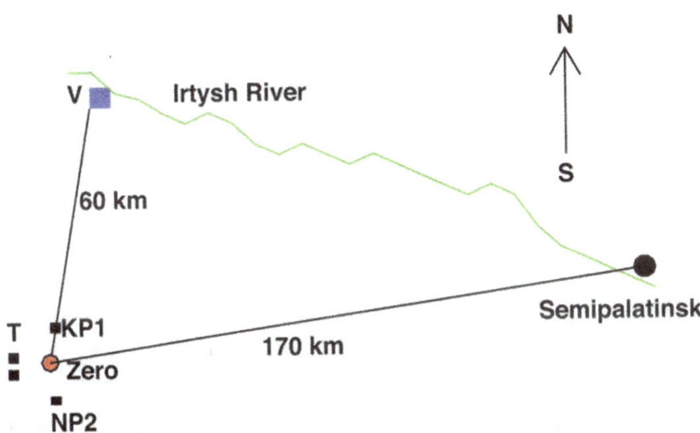

Figure 14.1. Sketch of the Semipalatinsk firing range. Zero was the location of the 30 m steel bomb tower. 'T' were two military heavy tanks 10 km from Zero equipped with radiation dosimeters. KP1 was the command bunker, 7 km from Zero, and NP2 was an observation point, 10 km from Zero. Two three story brick buildings not shown were 800 m from Zero in the direction of the tanks. 'V' was the village built to house the scientific staff. The course of the Irtysh River between V and Semipalatinsk is not accurately represented, but otherwise the figure is to scale. The Novosibirsk-Semipalatinsk-Alma Ata railroad line ran roughly north to south (not shown). The closest Chinese border is about 450 km south east of Zero.

As discussed in Chapter 11, the efficiency of the blast, defined as the ratio of the energy released divided by total energy content of the fissile material, was of prime importance in the layout of the test site. Some idea of the radius of destruction was necessary in order to place instruments so that they would be recoverable, and to locate observation bunkers so that the observers would survive. Too close would be disastrous, but too far away would not be good either. Figure F.4 in Appendix F shows circles of overpressure greater than one atmosphere, sufficient to cause extensive damage to structures, as a function of the blast yield in kilotons. The 20 kT circle radius is 1.9 km, and the circles scale as the cube root of the energy, so 40 kT $\rightarrow$ 2.4 km, and 100 kT $\rightarrow$ 3.2 km. The test weapon would have about 6 kg of ^{239}Pu which, if it all fissioned, would liberate 100 kT of TNT, so that value was the upper limit. Efforts to calculate the efficiency, or to infer it from critical assemblies, are discussed in Chapter 11, and in Appendix F. The compression bomb goes through three stages: compression and ignition; energy released by fission chain reaction; and expansion of the plutonium core to quench the process. An understanding of all three phases is necessary for an estimate of the efficiency. The energy appears in three forms listed in Appendix F:

1. The heat and light from the fireball   35%.
2. The destructive force of the shock wave   50%.
3. Neutrons and γ rays from the nuclear reactions   15%.

A chemical explosion does not achieve a high enough temperature to create a fireball, and of course it has no radioactivity[4], so item #2 is its only destructive effect. Nevertheless the 'TNT equivalent' refers to the total energy release.

Data existed from at least three air blast plutonium explosions done by the United States: the trinity test in New Mexico; the weapon dropped on Nagasaki; and the Able test at Bikini. The Able Bikini test was an air drop over water rather than land, and could have had different characteristics. All were consistent with an efficiency of about 20%, or a 20 kT bomb. In 1947 most of the details remained classified, although the Japanese cities were there for everyone to see. M.G. Mesheryakov was one of three Soviet delegates to witness the Bikini tests in 1946, and Beria forwarded his report, together with some photographs, to Stalin on March 7, 1947.[5] The report is qualitative rather than quantitative, perhaps because it was meant for Stalin's eyes, but Mesheryakov may not have had access to classified

---

[4] The Mayak accident in 1957 was a chemical explosion that involved radioactive materials.

[5] Atomnyy_proekt_SSSR._T.2.Kn.6.(2006).pdf #101 Beria to Stalin.

US data with more quantitative information. The Bikini air blast (a plutonium compression bomb) was at a height of 100-200 meters above the surface of the water, and had a more nearly spherical fireball than detonations at 30 m above dry land. The explosion sank a cruiser, a destroyer, and two transports that were anchored near ground zero. Other ships in a radius of 1 km were damaged by the shock wave and burned by the fireball to various degrees - decks buckled, wooden boats burned up, antennas and masts were broken, and smokestacks were bent over, but the ships did not sink. It is interesting that live ammunition stored on open decks did not explode.

The intense heat charred wood and burned paint on ships in a 1.5 km radius of ground zero. Screening provided by iron bulkheads or other structures prevented fire damage. (This effect was well documented in Hiroshima, where stone or brick building walls provided 'shade' from the blast.)[6] Mesheryakov concluded that, while on deck personnel would be badly burned, crew below decks, or inside gun turrets, would probably not suffer from the heat of the fireball.

A collection of 3519 livestock – goats, pigs, rats, and mice – were stored onboard the ships. 90% of the animals survived the blast. Lasting effects of the radiation on the health of the surviving animals were not known. Beria quoted a US communique regarding radiation damage to personnel:

"Measurements of the intensity of the radiation and examination of the condition of the shipboard animals have shown that the initial blast contained lethal doses of neutrons and γ rays, that would kill crews exposed on the decks of ships within a km of ground zero. Personnel protected by steel, water or other dense material would be relatively safe. Radiation would not disable a ship's crew instantaneously – even severely exposed personnel could remain at their posts for several hours. It is therefore possible that initial efforts to overcome the damage could keep the ship in action for some time, but it is clear that any ship within 1.6 km of an atomic bomb air blast would inevitably drop out because of harm to its crew."

---

[6] Most fires at Hiroshima and Nagasaki were started by damage to gas lines and other utilities, caused by the shock wave rather than the fireball.

This report is an example of the data on the effects of nuclear weapons available to the Soviet Atomic Project in 1947 when it was designing the test firing range. A resolution of the Council of Ministers, signed by Stalin on June 19, 1947, listed the preparations required for the firing range, and assigned responsibility to the Institute for Chemical Physics and the Ministry of Defense.[7] The plan was to determine as accurately as possible the efficiency of the blast by measuring the three channels of energy flow listed above. Physical measurements of the force of the shock wave, heat and light from the fireball, and penetrating radiation by neutrons and γ rays would be made by both ground based and airborne instruments, as well as equipment and structures assembled in the vicinity for that purpose. Livestock test animals would serve to evaluate the biological effects. Three methods of determination of the efficiency were outlined. The first would be measurements of the shock wave and the intensity of the light from the fireball, with emphasis on airborne instruments to avoid distortions caused by the earth's surface. An air detonation at an altitude of 1000 m or so, although involving other complications, would simplify the interpretation of both the shock wave and the fireball by removing the effect of the earth's surface. The fireball from a blast at 30 m is only a hemisphere. The second method would be measurement of the neutron flux as a function of distance which is directly related to the number of fissions. The third method would be collection of dust from the radioactive cloud with special filters by aircraft or projectiles, the goal being to measure the yield of fission products and the amount of remaining plutonium in the dust for an estimate of the efficiency.

Several types of pressure sensors with different sensitivity ranges and recording methods would be deployed to measure the shock wave. Optical recording instruments would be used to measure the temperature and size of the fireball as a function of time. High speed movie cameras capable of 10,000 frames per second or more, would be used to record the fireball time dependence. One of the ideas for measuring the neutron flux was to fire artillery shells containing elements that become radioactive when bombarded by neutrons through the fireball, and measure the induced

---

[7] Atomnyy_proekt_SSSR._T.2.Kn.6.(2006).pdf #123 Resolution of SM SSSR.

radioactivity afterwards! Balloons were also to be deployed with devices to measure radioactivity in the air. Like the Bikini tests, the Soviets proposed to use a collection of livestock distributed at various distances and in various structures for biological effect studies. There were few buildings in the area of ground Zero (one of the reasons for selecting it), so Beria's committee proposed an ambitious construction program to distribute structures around Zero to furnish cannon fodder for the shock wave, and to mimic the destruction to a populated area.

The analysis of the destruction at Hiroshima and Nagasaki by William Penney and coworkers, published in 1970 and discussed in Appendix F, was based solely on the shock wave damage. LA-8819 by John Malik, also mentioned in Appendix F, was based on all of the information he could find, and written in 1985. Obviously neither report was available to the Soviets in 1947, but they did have an impressive amount of detailed information both from espionage, and from the open literature. K. Way and E.P. Wigner published an article on the radiation from fission products in the Physical Review in 1948, based on an extensive review of the nuclear species created by fission written by 'The Plutonium Project' of Oak Ridge National Laboratory.[8]

The instrumentation for pressure measurements at Trinity serves as an example of the detail obtained by espionage. Short circuits caused piezoelectric pressure sensors to malfunction in the blast, but a carefully assembled passive box worked fine. The box had a row of holes in the lid covered by aluminum foils that would rupture at increasing overpressure from one hole to the next. After the blast, the highest pressure foil that did not break gave maximum pressure. If all of the foils were broken, one only got a lower bound on the overpressure. Such measurements confirmed that the maximum overpressure decreased with distance as $1/r^3$ as described in Appendix F.

---

[8] K. Way and E.P. Wigner, Phys Rev **73**, 1318 (1948). 'The Plutonium Project,' Rev Mod Phys **18**, 513 (1946). Both papers were available to Soviet scientists.

There was much more information available to Beria's committee regarding how to measure bomb damage, serving as an aid in planning the instrumentation for the firing range. But it also served as a cautionary tale, that the best laid plans to instrument a nuclear explosion often come up short, because things happen very fast, there are no second chances, and equipment either fails to work at the critical time, or is blown to pieces; and as detailed efforts have shown, it is not easy to obtain precise information after the fact.

14.3. Plutonium Metallurgy

Liya Sokhina worked in the plutonium refinement plant 'V' at Combine #817, the factory that was to supply weapons grade plutonium to the bomb assembly at Arzamas-16. As described in Chapter 12, Liya worked on plutonium chemistry in the temporary converted warehouse used in the 'glass jar period' of the operation, before the new plant was finished. After the chemistry came metallurgy. Laura Fermi said that the Metallurgical Laboratory at the University of Chicago had no metallurgists in 1942, but the metallurgists had the last laugh, because plutonium was a new artificial element, with no prior experience in handling it.

Plutonium metal is a complicated substance. It has six allotropic forms, labeled with Greek letters α, β, γ, δ, δ', and ε. Each has its own density and crystal structure, and phase transitions occur at atmospheric pressure with increasing temperature. The α phase is stable at room temperature. It has the highest density – 19.86 gm/cm^3, but is brittle and difficult to machine. Transition to the δ phase occurs above 300^0 C, but is stable at room temperature when alloyed with gallium or aluminum. The density is 16 gm/cm^3. This alloy can be machined and welded, and is the preferred form for nuclear weapons. During compression in an implosion bomb the δ phase changes to the α phase, and the increase in density aids in creating a fast chain reaction. Figure 14.2 shows the temperature dependence of the phases of plutonium metal. Much of these data on plutonium was supplied by NKVD to Beria's committee.[9]

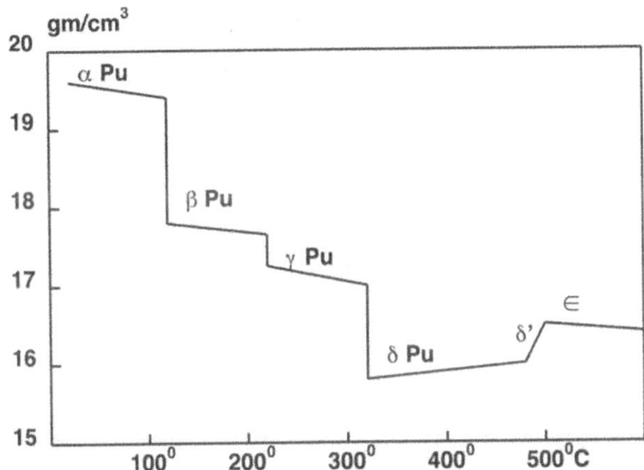

Figure 14.2. Phase transitions of the allotropic forms of plutonium metal. Density is shown as a function of temperature. The δ phase is alloyed with gallium for weapons. The melting point is 639 °C, off scale to the right. Derived from Wikipedia/plutonium.

Plutonium is radioactive: $^{239}\text{Pu} \rightarrow {}^{235}\text{U} + \alpha$, with a half-life of 24,100 years. Therefore, if one makes 2 gm of ^{239}Pu from ^{238}U by neutron capture in the usual way, and waits 24,000 years, one gets 1 gm ^{235}U! The half-life ^{238}U is 200,000 times longer than ^{239}Pu, so that even 0.1% plutonium by weight dominates the α decay activity. Table 14.1 summarizes some properties of the α emitters. The 5 MeV α particle ranges out in the material, and heats it up. For a 1 kg lump of plutonium the power is about 1.8 watts, enough according to Libby to feel like a live rabbit.[10] Plutonium is perhaps no more dangerous than any other α emitter – not good to ingest it, but not dangerous outside the body if it can be washed off or removed with protective clothing, because the α particles have a very short range,

---

9   Atomnyy_proekt_SSSR._T.2.Kn.6.(2006).pdf #337, 28 January, 1946, NKVD to PGU.
10 Leona Marshall Libby, "The Uranium People," *ibid.*, p 171.

less than 1 mm of water. However, enough of it can form a critical mass, and produce a very intense neutron source together with the heat generated by fission. Depending on the geometry and the surroundings, a critical mass of ^{239}Pu is not very large – a sphere the size of a grapefruit. Added to these safety and handling problems is the value of the substance. The amount of time and money that have gone into its creation means that losses incurred in handling it must be minimized.

Table 14.1. α Emitters

Isotope	half-life (years)	E_α MeV	α's/sec-gm
^{238}U	4.5×10^9	4.3	1.25×10^4
^{235}U	7.0×10^8	4.7	8.0×10^4
^{239}Pu	24,000	5.2	2.3×10^9

The first castings of plutonium metal were made in the same temporary warehouse where Sokhina did the chemistry, under very difficult conditions. N.I. Ivanov, who was deputy manager of the plant, has described the early days of plutonium metallurgy.[11] Solution containing plutonium was arriving from the 'A' reactor and 'B' chemistry operations in early 1949 in sufficient quantities to force operations in the temporary warehouse to ramp up before the necessary equipment had been installed. The installers and the technical workers got in each other's way. Vacuum furnaces and other heavy gear, together with support personnel, were shipped in from NII-9 in Moscow, and the first plutonium ingots were cast in NII-9's furnace that had originally been used for metallic coating of ceramics. A.A. Bochvar was in charge of the operation. They had no experience with macroscopic quantities of plutonium metal, and had about 100 days to solve all the problems and deliver enough product for the August test. Georgy Flerov and coworkers in Moscow had measured the relevant nuclear parameters of ^{239}Pu using milligram amounts of material.

---

[11] N.I. Ivanov, "First Nuclear Explosive," HISAP96, Vol **2**, *ibid.* p 431.

The critical mass was known – an important parameter for handling kilograms of it.[12]

The physical characteristics of the metal were determined in the shop. Phase transitions and melting point (Figure 14.2), thermal expansion coefficient (very large for a metal), and crystal structure were all determined in house. The melting point (639^0 C) is near that of aluminum (660^0 C), and much lower than iron (1538^0 C), which was perhaps the only advantage in working with plutonium over steel making. The α phase was found to be hard and brittle, and the δ phase could be alloyed with gallium to form a workable metal at room temperature. There was a tendency for the metal castings to crack, which was a problem because of the possibility of contamination. The casting shop began producing usable small ingots for tests, selecting those without cracks. Gamma ray sources were used to measure the uniformity of the casting. Workers wore rubber gloves and respirators, but had no other protection. The δ phase alloy could be welded and machined.

High level management descended on the plant as weapons grade product seemed imminent. Kurchatov, Khariton, Vannikov, Zavenyagin, Muzrukov, and Slavsky all came to town to see what was going on. They did not interfere with the work directly, but their presence was certainly felt as a sign of the utmost importance of the plant, and the necessity for its success in a timely fashion. Sometimes emotions ran too high, and management would pitch in as a sign of support, and to restore morale.

Plutonium casting equipment did not arrive until the fall of 1949, and technology for construction of crucibles with sufficient capacity came in even later, so the first weapon had to be assembled in a different way. Plutonium in powder form was too dangerous to handle. So it was decided instead to weld smaller pieces of the alloy that were cast under a hood in the plant. In early trials the plutonium adhered to the surface of the mold. E.P. Slavsky opened the first stuck mold with a hammer and chisel, and to

---

[12] Geometry is of prime importance. Six kg molded in a ring, with a large surface area for neutrons to escape, is safe, while the same mass in a sphere could become critical.

everyone's astonishment the plutonium casting was OK. Later molds were copper coated to prevent adhesion. The welding technique was first tested with aluminum, and after some improvements was used to assemble pieces of plutonium metal into kilogram masses, ready for machining.

Even the best machinist available was not permitted to work the plutonium without supervision after each and every cut. A.P. Zavenyagin stood over the milling machine and fussed about the dimensions, but in the end satisfactory product was obtained, and the sample passed mechanical tests.

A nickel coating on the plutonium was desirable to stop the surface α particles, to prevent oxidation in air, and to avoid plutonium contamination in handling. Rubbing plutonium metal off of the surface was bad for two reasons: it removed valuable material from the ingot, and it transferred radioactivity to the handler. The temporary converted warehouse initially did not have the equipment to do it, but A.I. Shalnikov arrived from Kapitza's Institute in Moscow (on whose orders unknown) to fix the problem. Shalnikov created an assembly to do nickel coating that looked more like a laboratory device than an industrial one, but it worked well for years. A long nickel whisker formed on the second sample that could not be cut off without affecting the continuity of the coating, but the entire coating was successfully removed and replaced. Nickel's ferromagnetic property was used to check the thickness for tolerance and uniformity.

Complete documentation accompanied each plutonium part that was to be delivered to Arzamas-16. A large motorcade with armed escort removed the samples from the shop. After the initial samples were completed procedures were setup at the plant for orderly production of weapons grade plutonium.

Ivanov remembered, as did others, that threats of punishment in case of failure by a cruel regime did not penetrate to those doing the work. He credited A.A. Bochvar, the plant manager, with maintaining a positive atmosphere, and stimulating cooperative effort.

14.4. "Look" Magazine

The issue of "Look" Magazine dated 16 March, 1948, contained an article by John Hogerton and Ellsworth Raymond entitled "When Will Russia have the Atomic Bomb?." Hogerton was chief engineer for Kellex at the Oak Ridge K-25 gaseous diffusion plant. His article describing K-25 in "Chemical and Metallurgical Engineering" in December, 1945, was discussed in Chapter 10. Raymond served in the US Embassy in Moscow during WWII, and was later in the War Department in Washington, D.C. He was an authority on Soviet industry and economy. The article was translated into Russian for Beria's PGU committee.[13]

The article was divided into two parts. The first part, written by Hogerton, described the complexities of the Oak Ridge gaseous diffusion plant for uranium isotope separation, and the scope of the industrial undertaking at Hanford for the production of plutonium. Hogerton correctly guessed that the first Soviet bomb would be made of plutonium, but he underestimated the effort that was applied to uranium isotope separation by gaseous diffusion. As we have seen, gaseous diffusion was a daunting assignment, and it lagged plutonium production by the power reactors at Combine #817 by several months, but it was definitely in the race. The second part, written by Raymond, was an assessment of the capabilities of Soviet industry in the aftermath of WWII destruction. Then the authors combined to reach the conclusion that the Soviet Union would have a plutonium atomic bomb by 1954 at the earliest.

Beria authorized commissioning the first stage of the full scale gaseous diffusion plant at the Sverdlovsk site (Factory #813) in May, 1948, and by the summer of 1948 the 'A' plutonium production reactor at Combine #817 was being commissioned, so both efforts were well under way in March, when the "Look" article appeared. While the article was in a popular magazine, and did not contain the latest classified intelligence, it probably was accurate in its portrayal of the lack of knowledge in the West

---

[13] Atomnyy_proekt_SSSR._T.2.Kn.6.(2006).pdf; #173 Translation of "Look" article with comments. The original article appeared in "Look Magazine," March 16,1948, p 27.

of the true status of the Soviet Atomic Project. This lack of any information on what the Soviet Union had already accomplished in three years of feverish activity is clear evidence of the effectiveness of Stalin's closed society. Winston Churchill called it an 'iron curtain.' The Soviet Union was a big place, with a generally harsh climate and a small population density, and travel by foreigners was tightly controlled, but Westerners have natural curiosity, work on the atomic bomb was known for sure to be taking place somewhere, and yet even the CIA had only vague knowledge. It seems that informants on the ground were non-existent. Some estimates were reported by the CIA In secret documents that have since been declassified, but no references to sources were given to back up the dates.[14] On December 15, 1947, the CIA stated that the first Soviet Atomic Test would probably not be before 1953, and almost certainly not before 1951. Then on July 1, 1948, the 'earliest possible' estimated date was moved to mid-1950, with the 'most probable' date mid-1953, again with no details regarding sources of information. Hence the CIA was more realistic than Look Magazine, but still about one year off in their earliest possible date. Then one year later, less than 2 months before the actual test at Semipalatinsk, the 'earliest possible' date was moved to mid-1951. In the US armed forces the Navy and War Departments favored a 1952 date, while the Army Air Force was less optimistic, forecasting a Soviet atomic bomb in 1949![15]

Hogerton granted that the Soviet Union had some obvious advantages; they knew that it was possible to build nuclear weapons, a fact learned by the Manhattan Project only in July, 1945, at the Alamogordo test, and they knew the scope of the project in general terms from the Smyth Report and other public descriptions. He discounted the value of espionage, as the full extent of penetration into the Manhattan Project was not known until 1950. The article also discounted the contribution of German scientists, stating that the German Uranverein made little progress during the war, and

---

[14] Doyle L. Northrup and Donald H. Rock, "Detection of Joe 1," Central Intelligence Agency Studies in Intelligence, September, 1966. Document available on cia.gov.

[15] Charles A. Ziegler and David Jacobson, "Spying Without Spies," Praeger Publishers, Westport, Connecticut, 1995, p 89. This book is devoted to the guesswork regarding the Soviet Project, and the ultimately successful detection of RDS-1 in September, 1949.

therefore the knowledge of the Germans was of little use. This was true, but they neglected the impact of German industry and German technical expertise. Through war reparations the Soviet Union gained access to German industrial capability, exemplified by Nikolaus Riehl and his uranium refinement factory, that helped the Soviet effort in many ways not directly related to nuclear physics, where the Soviets were already in good shape. Hogerton emphasized the advantage of the broad industrial base in the United States, an imbalance recognized by Beria, and overcome by extreme effort.

Raymond emphasized the limitations of Soviet industry, the lack of consumer goods and precision instruments, electric power requirements, and the shortage of uranium. He pointed out that the industrial capacity of the USSR was only 18% that of the United States, that building techniques were more primitive, and that a plant as complicated as K-25 would be impossible to build. Much of what he said was true, except for the 'impossibility.' Both authors recognized that the plutonium production avenue was more likely to bear early fruit, and that an implosion type plutonium bomb would be the first one to be tested. They recounted all of the hazards and difficulties in handling fission fragments and performing chemical reactions in large quantities of highly radioactive solutions.

General Groves had tried to seize all of the high grade uranium ore in the world, and he knew that the Soviet Union had few sources of supply. He concurred in the assertion that the United States should enjoy many years of monopoly of nuclear weapons based on this assumption alone. Groves' analysis was not available to Hogerton and Raymond, but because of his stature in the field of atomic energy it carried some weight in the US Government. The arguments presented in the "Look" article were therefore met with general approval in many quarters. Only the US Air Force thought 1949 to be a possible date for the first Soviet atomic test.[16]

While they overestimated the time required to do the job by five years, the authors were correct on many of the obstacles and difficulties that the

---

[16] Charles A. Ziegler and David Jacobson, "Spying Without Spies," *ibid.*, Chapter 2.

Soviet Project faced and overcame, and it did take a few more years for production to become routine in order to build a nuclear arsenal matching the one already possessed by the United States in 1948.

Donald P. Steury has written an article entitled; "How the CIA Missed Stalin's Bomb."[17] He confirmed the lack of clandestine intelligence information, and the optimistic and erroneous assumptions about the Soviet uranium supply, and capabilities of Soviet industry. He mentioned one interesting link, however - the Germans. Soviet secrecy was strict and successful. Foreign visitors were carefully watched and excluded from sensitive areas. The Germans were however foreign visitors, and they did have access to some sensitive areas, so they were potentially a security soft spot. While they did not know the scope of the Soviet Project, they did know a few components of it, and some information got out to East Germany through correspondence, or more direct means. In late 1946 a German physicist named Adolf Krebs walked over to West Berlin and reported that he had just interviewed for a job at Nikolaus Riehl's uranium refinery in Elektrostal. Krebs said that the plant was processing reactor grade uranium in large quantities. He also reported on the German groups at Sukhumi, saying that they were working on isotope separation by gaseous diffusion. This information was filed away by the CIA, without having an apparent impact on their estimate for the test date. The CIA had many problems in the early cold war period, and perhaps the Soviet Atomic Project was not its highest priority.

In 1960, more than ten years later, Francis Gary Powers' U-2 was shot down near Sverdlovsk while looking for ballistic missile launch sites. The high altitude U-2 reconnaissance aircraft with advanced cameras did not exist in 1948.

The "Look" article did not tell Beria's committee anything they did not already know, and its overestimate of the time required to surmount all of the difficulties must have given some satisfaction. On the other hand, in March, 1948, success was still 17 months away, and by no means assured.

---

[17] Donald P. Steury, Studies in Intelligence 49, no 1 (2005), p 19-26.

14.5. Arzamas-16, KB-11

Khariton's Sarov laboratory, KB-11 or Arzamas-16, was busy with the Soviet Air Force designing weapons deliverable by aircraft, working on conventional explosive lenses, developing new designs for atomic bombs after the initial test, and studying the possibility of fusion weapons. The PE-8 heavy bombers mentioned above were loaded with dummy nuclear weapons for bombing runs at a range near the Sarov lab, but any testing of real weapons had to be done at a more remote site.

Kurchatov recognized the possibility that the first test would fall short, and was anxious to have enough plutonium on hand for a second bomb should a repeat test be necessary. Incomplete detonation of the plutonium compression bomb was a subject of concern by the theory group as well, both with regard to the content of ^{240}Pu – a neutron source because of spontaneous fission, and other sources of neutron background, including the neutron source initiator at the center of the fissile sphere.

The explosive lens design was a critical component of the bomb. Lens components were chemical explosives. They were cast in liquid form into wedge shaped molds, to be assembled together like a soccer ball with 12 pentagons and 20 hexagons for 32 pieces total. Tolerances for the components were strict, so that the assembly fit together without any gaps. Conventional explosive propagation speeds were the order of 1 cm/μsec, and this number determined that the 32 lens components had to be triggered simultaneously within about 1 μ sec in order to coordinate the compression waves. Wartime development of radar and other technology had improved the timing capability of electronics, but the simultaneous trigger was still a challenge. Not only must 32 electrical pulses arrive at imbedded fuse wires simultaneously, but also the chemical explosives themselves had to ignite. A high voltage source charged a capacitor bank to 20 kV, and a trigger signal delivered this charge to the fuses via cables. After this trigger signal, the bomb would explode. Design and fabrication of the lenses and the trigger electronics were responsibilities of KB-11. As noted earlier, Lev Altshuler and Veniamin Tsukerman were responsible for compressibility studies, and recording the dynamics of the explosive lenses. Their plot of the compressibility of uranium shown in Figure 1F of Appendix F was a significant accomplishment. The theory group (Zeldovich) calculated an anticipated change in density for the plutonium sphere of a factor of two. Since the critical mass depends on $1/\rho^2$, an increase in ρ of a factor of two decreases the critical mass by a factor of four, so starting with ½ the critical mass gives twice the critical mass after compression. This is an advantage of compression bomb design – less of the expensive fissile material is needed. L.D. Landau's group at the Institute for Physical Problems in Moscow (Kapitza's Institute, now directed by Aleksandrov) was responsible for the efficiency calculations.

The neutron source initiator that was to be placed in the very center of the hollow ^{239}Pu sphere also had to be designed and tested at KB-11. A conventional neutron source was contemplated, for example ^{210}Po $\rightarrow$ ^{206}Pb + α; followed by ^{9}Be + α $\rightarrow$ ^{12}C + n. Other α emitters could be used, but polonium decay produced no γ rays close to the threshold for the unwanted photoneutron reaction ^{9}Be$(\gamma,n)^8$Be of 1.6 MeV. In a laboratory source the polonium and beryllium are mixed together as powders to improve the chance of an alpha particle from polonium finding a beryllium nucleus. In the center of the bomb this is not desirable, because the neutron background must be kept as low as possible to avoid predetonation (hence no spontaneous fission, and no (γ,n) reactions). So the two elements must be kept apart until the detonation pressure wave compresses the plutonium sphere to give a larger than critical mass. Obviously, the design of this device was a very delicate matter. Kurchatov figured that several tens of curies of polonium were needed. The half-life of ^{210}Po is 138 days, and ^{226}Ra is 1600 years (the definition of a curie), so ten curies of polonium is just 2.4 milligrams. However, the short half-life meant that the shelf life of any bomb neutron source was only a month or so – it had to be freshly made. This was a disadvantage for the use of polonium in stored weapons, and motivated the search for other suitable α emitters with longer half-lives.[18]

---

[18] Atomnyy_proekt_SSSR._T.2.Kn.6.(2006).pdf, #242, Kurchatov to Beria.

Kurchatov planned to make ^{210}Po by bombarding ^{209}Bi with neutrons in a reactor. Then ^{210}Bi $\rightarrow$ ^{210}Po + e$^-$ + ν with a half-life of 5 days. Kurchatov recognized that reactors would have to be running continuously during wartime conditions to produce the short-lived isotope for nuclear weapons. Among the other isotopes considered was the longer half-life ^{227}Ac. Actinium would also be made in a reactor, by bombarding ^{226}Ra with neutrons. However, the longer half-life and branching fraction of 1.4% to α emission meant that grams of actinium would be needed instead of milligrams of polonium. The Radium Institute in Leningrad was asked to look into the matter. The director of the Institute informed Beria that it would require more radium than the USSR possessed to use as a reactor target to produce enough actinium to make a useful neutron source.[19] The idea was dropped. Once installed in the bomb the neutron source initiator had to be monitored to insure that premature mixing had not occurred that would cause the weapon to predetonate. Vannikov and Kurchatov proposed to make three sources in the June-July time period for the August-September test. Monitoring the neutron source and energizing the high explosive firing circuitry were two of the many complications in handling a compression weapon in an airplane.

The test weapon would sit on a tower and be triggered remotely, but an airborne delivery weapon required a different trigger. There were two candidates, an altitude pressure sensor, and a radar measurement of the height. Both were under development at KB-11, and both were tested using aircraft to drop dummy bombs on which the two sensors operated independently. The radar device was preferred because it measured distance directly, independent of changes in atmospheric pressure, but the pressure sensor was more reliable, so more testing of radar prototypes was necessary. The bugs were eventually worked out of it, and it was successfully tested in dummy bombs in June, 1949.[20] An independent parallel timing circuit, synchronized to start when the bomb was released, allowed for a preset descent time before detonation. If released from an

---

[19] Atomnyy_proekt_SSSR._T.2.Kn.6.(2006).pdf, #264. Botsenyuk to Beria, 22 June 1949. A rare example of someone pointing out the impracticality of an idea.

[20] Atomnyy_proekt_SSSR._T.2.Kn.6.(2006).pdf, #234 and #266, Pervukhin to Beria.

altitude of 10,000 m, the descent time was measured in tests to be about 55 sec[21]. Safety on board the aircraft required activation of these firing circuits only when the bomb was dropped clear of the plane. Another concern for an airborne weapon was protection from the extreme cold encountered at high altitude. The bomb bay had to be kept warm.

Khariton estimated that it would take 24 hours for preliminary assembly, including the high explosives and the fissile material, and 4 hours to load it in the aircraft bomb bay.[22]

The gun type uranium bomb was still on the books as RDS-2, and consuming resources at KB-11, as late as March 3, 1949.[23] On that date the Council of Ministers (Stalin) postponed the testing of RDS-2, scheduled for December, 1949, until December,1950, because of delays in obtaining a sufficient amount of ^{235}U from the gaseous diffusion plant. RDS-3 was supposed to be an implosion bomb with a ^{235}U shell over a ^{239}Pu core, and was also delayed by the isotope separation, but since less uranium was needed, it could possibly be ready by December, 1949. (RDS-3 was actually tested in 1951.) At some time after this postponement the gun type RDS-2 was canceled altogether, and never brought to completion.

Critical assemblies are discussed in Appendix J. These experiments were carried out at KB-11 after enough ^{239}Pu was available to approximate a critical mass, i.e. in 1949, and were one more request competing for kg samples of the metal, joined by metallurgical studies, fabrication for the test bomb, and so on. Critical assemblies were inherently dangerous, and there were several accidents, including the one described by Vannikov when his body reflected enough neutrons back into the assembly to cause criticality. When digital computers were developed, more computer modeling was done, and fewer hazardous experiments. Khariton wrote to

---

[21] The Hiroshima bomb fell with an average speed of 200 m/sec, consistent with this number. John Coster-Mullen "Atomic Bombs" *ibid.,* p 36-37.

[22] Atomnyy_proekt_SSSR._T.2.Kn.6.(2006).pdf, #260, Protocol of 9 June 1949.

[23] Atomnyy_proekt_SSSR._T.2.Kn.6.(2006).pdf, #227, Order of the Council of Ministers.

Kurchatov on April 4, 1949, requesting 0.5 kg and 1.5 kg samples of pure ^{239}Pu to be used by KB-11. A 1.5 kg block would be sufficient for Georgy Flerov's group to make useful measurements of neutron reflection coefficients by various cladding materials of the bomb, including the uranium tamper. Comparing a 0.5 kg block to a 1.5 kg block would give useful data for extrapolation to the critical mass.[24] Flerov was still busy measuring the neutron energy distribution from plutonium fission. With the bomb explosion only about five months away there was still a shortage of fissile material for testing, and fabrication. The neutron background was too high in the first 1.5 kg plutonium sample delivered from Combine #817. The metal could be used in the bomb if mixed with plutonium of much lower neutron background. Presumably the high background came from too much ^{240}Pu.

While KB-11 worked on making a nuclear weapon that could be exploded in a test, Semenov at the Institute for Chemical Physics in Moscow was coordinating a group developing the necessary instrumentation to detect the effects of the explosion. The three energy channels listed above - shock wave, heat from fireball, and penetrating radiation (neutrons and γ rays) – were to be measured by pressure sensors, buildings and military equipment set up at varying distances from Zero, seismic detectors already in existence, livestock for radiation effects on the living, and conventional radiation measurement devices, some airborne either in aircraft or balloons. All of these data, together with what could be learned from the residual radiation by fission products, would contribute to an estimate of the efficiency of the blast. The amount of plutonium was to be 6 kg, which as we have seen corresponds to 100 kTons of TNT if it all exploded. The efficiency expected from Landau's calculations was about 10%, or 10 kTons of TNT equivalent.

On May 6, 1949, N.N. Semenov made a curious proposal to Yuli Khariton.[25] Semenov requested instrumentation to detect the prompt γ rays from the chain reaction as a function of time in μsec. He proposed to locate

---

[24] Atomnyy_proekt_SSSR._T.2.Kn.6.(2006).pdf, #237, Khariton to Kurchatov.
[25] Atomnyy_proekt_SSSR._T.2.Kn.6.(2006).pdf, #247, Semenov to Khariton.

an underground chamber in the vicinity of Zero with a buried cable running to a detector that recorded the current of electrons from conversion of the γ rays from the chain reaction. One or two prompt γ rays accompany each fission, so a 10% efficiency plutonium explosion would produce about 10^{21} γ rays, which is a whole lot – 100 coulombs of charge if converted into electrons. One problem with the idea was the practicality of doing anything close to Zero, but that was not the only problem. The main difficulty was that it would not work, because the prompt γ rays would all be absorbed by the uranium tamper that surrounded the plutonium core. For several μsec after compression the uranium tamper would remain intact; as it melted and expanded in the intense heat, the plutonium core would also expand, and the chain reaction would quench. Thus while the tamper held, and absorbed all of the heat from the chain reaction, no energetic γ ray would reach the outside world[26]. Only after the chain reaction ceased would radiation, now from fission products alone, be able to escape. Khariton of course knew all of this, but if he told Semenov, it is not recorded. Perhaps he called Semenov on the phone. In any event, this proposal gives us a rare look at what must have been a fairly frequent flow of ideas that had to be filed away and forgotten, sometimes quietly, but no doubt not always. One other example was described in Section 9.3.3 - a proposed method of isotope separation that Kurchatov knew would not work, but encouraged the inventor to pursue it further. As mentioned in Chapter 9, such instances have to be handled with care. Semenov was an academician, and a laboratory director, and had other responsibilities within the Project, so he could not simply be told to go away. L.D. Landau had a well-deserved reputation of scorching anyone who made a mistake. He probably did not know of Semenov's idea. Khariton approved some preparations, including the purchase of oscilloscopes from Germany, that could be of general use.

---

[26] Soft x-rays would be emitted from the temperature of several million degrees.

14.6. Test of RDS-1

KB-11 and Yuli Khariton had principal responsibility for the test. Igor Kurchatov spent most of his time at Combine #817, where the plutonium was made, and was not involved in the detailed arrangements for RDS-1, although as the time grew near, he found it difficult to think of anything else. The NKVD and the Red Army were responsible for security. Bomb components had to be transported from KB-11 at Sarov to the firing range 170 km west of Semipalatinsk in north-eastern Kazakhstan. See Map #1 and Figure 14.1. North-south rail service existed from Novosibirsk to Semipalatinsk. Sarov was not at that time on a main rail line, but was not too far from Gorky, where the trans-Siberian railroad headed east to Novosibirsk. The real bomb assembly and a dummy left KB-11 to go on two separate trains to Semipalatinsk for security.[27] Aircraft could also be used to transport parts to Semipalatinsk. The final trip to the firing range had to be made by truck for everything that was needed.

The test site had nothing to begin with. All of the instruments, supplies, utilities infrastructure, and laboratory personnel had to be brought in. The site was a desert, but the military wanted to know about destruction of buildings and equipment, so some structures had to be built specifically for the purpose of assessing damage. Reinforced concrete bunkers were erected close to Zero, and some 'ordinary' brick and wooden buildings appeared at varying distances. Some railroad track and a railroad bridge were installed. Military equipment including aircraft on the ground, and live ammunition were scattered around to measure destructive effects on the enemy. The military also erected a series of panels at varying distances for shock wave overpressure measurements. Several seismic stations in the region could detect tremors from the blast, and were on alert to supply useful data for the bomb's efficiency, without knowing specifically what they were looking at. As in the Bikini tests, a collection of sacrifice animals was used to measure the biological effects of the bomb. The victims were placed at different distances and with various shielding

---

[27] Raisa Kuznetsova, "Kurchatov v Zhizni," *ibid.*, Interview with E.P. Slavsky, p 491.

protection. Aircraft and balloons were to be used to sample radioactivity in the air and to make pressure measurements.

At Zero was a steel tower 37.5 m high, with a platform at 30 m for the bomb, and two elevators – one for freight and one for personnel. There were also ladders. The elevators were problematic. The freight elevator got the job done, but some people preferred to climb rather than use the personnel elevator. The laboratories for component assembly and post-mortem analysis were located near the housing village, 60 km away. Some supply storage bunkers were closer in, and there was an area for final assembly only 25 m from the tower. KP1 was the main control bunker to the north, and a second observation bunker was NP2 to the south, shown in Figure 14.1. All of the relevant control cables from the tower went to KP1.

A delegation headed by Pervukhin was sent to Semipalatinsk in late July to assess the readiness of the complex for the anticipated test. He reported to Beria on August 5, 1949.[28] The central tower and lifts, the workshops, and the laboratories were not finished. Utility hook-ups and roads were still in progress. Military sector was in reasonably good shape. The commission urged completion in a timely fashion. Instrumentation seemed generally to be in a better state of readiness than construction, except for the radiochemistry laboratory. There seemed to be no plan of operation either for the steps leading up to the test or for systematic evaluation of test results, a lack of preparation that extended to the biological sector as well. Training of the workers who were on loan from KB-11 and other laboratories and factories was proceeding according to plan, and Kurchatov was confident that by mid-August all would be good to go. Khariton gave Vannikov a list of 25 people from KB-11, mostly group leaders, who would have to go to Semipalatinsk, including Zeldovich, Frank-Kamenetsky, Flerov, Terletsky, and Tsukerman.

---

[28] Atomnyy_proekt_SSSR._T.2.Kn.6.(2006).pdf, #273 Pervukhin to Beria.

A resolution of the Council of Ministers of the USSR dated August 18, 1949, authorized the test firing of RDS-1, the first Soviet atomic bomb.[29] This resolution was prepared by Beria and presented to Stalin for his signature, but Stalin never signed it. According to Beria, Stalin's decision was based on the conviction that a resolution by the Council of Ministers was to be carried out successfully in full, and he believed that there was a good possibility that this test might fail, so it was better to proceed without a resolution.[30] In any event, the test had already been authorized by the Central Committee in 1947, and further documentation was unnecessary. Beria assumed responsibility for the test on his own. On August 26, just before he departed for Semipalatinsk, Beria prepared another resolution, similar to the one of the previous week, that contained specific dates of 29-30 August for the test. When he arrived in Kazakhstan Beria was assured that all was prepared for the test on August 29. He reported this to Stalin, who asked, "What if it doesn't work?" Beria answered that they would continue with further tests, but he knew that he, Beria, would bear the responsibility of failure, and that consequently punishment would flow down the ranks in case of a misfire. Fortunately, it was not necessary to take such remedial action.

Yakov Dokuchaev was assigned to the radiochemistry laboratory 'B' of Combine #817. His task was to study the various forms of radiation emitted by the solution of uranium, plutonium, and fission fragments delivered by 'A' reactor. The heavy elements radiated α particles, while the fission fragments were principally responsible for β and γ rays. Each isotope had unique signatures of type of radiation and lifetime, and a careful analysis of the radioactivity from the irradiated uranium slugs would be useful in attempting to measure the efficiency of the atomic bomb. On August 22, 1949, Igor Kurchatov told him that there would soon be a test of the atomic bomb, and that reliable measurements of the type that he had been making would be repeated on what radioactive remnants could be retrieved after the blast. He was to join a group of about 20 people from the Radium Institute of the Academy of Sciences that had flown into

---

[29] Atomnyy_proekt_SSSR._T.2.Kn.6.(2006).pdf, #275.
[30] Lavrenty Beria, "Secret Diaries, 1937-1953," *ibid.*, p 511.

Combine #817 from Leningrad. They all went in Douglas DC-3's on August 24 from Mayak to a temporary landing strip near the Semipalatinsk firing range, and were taken to the village on the banks of the Irtysh River.[31] Kurchatov and other dignitaries had two story houses on a hill overlooking the river. Down the hill were several one story houses and a cafeteria. Laboratories for analysis were in the neighborhood, about 1 km further west. Zero was 60 km south west of the village.

The weather was warm and sunny for the first few days, August 24-28, with cool nights. Two aircraft were on training flights, practicing retrieval of samples of radioactive dust that might accumulate on poles near ground zero. On August 28 Igor Kurchatov made sure that the measuring equipment was ready to analyze the radioactive remnants of the blast.

Dokuchaev was one of eight people assigned to the observation point bunker, NP2, 10 km south of Zero. A.P. Vinogradov, a chemist, B.A Nikitin, director of radiochemistry at Combine #817, N.L. Dukhov, deputy to Yuli Khariton at KB-11, an assistant to Dukhov, Ya. B. Zeldovich, head of the KB-11 theory group, and M.G. Mesheryakov, deputy director of Laboratory #2, were the others. Mesheryakov had been a witness to the Bikini tests in 1946, so his account would be particularly valuable. About 15 people, including Beria, Zavenyagin, Pervukhin, Flerov, Kurchatov and Khariton were all in the two room command bunker at KP1, 7 km north of Zero, and, one might think, uncomfortably close.[32]

Each occupant of NP2 was so informed by a colonel late in the evening of August 28. They were told that the planned time for detonation would be 6 am the following morning local time (3 am Moscow time). The weather was cloudy and cool, with occasional rain. The NP2 delegation arrived on site at 5:30 am. They were issued dark glasses, and told to put them on one minute before ignition. After ignition, the glasses could be removed, and they could stand up for 20 seconds to observe the fireball. After 20 seconds they were to lie down in a trench so as not to be blown

---

[31] Ya.P. Dokuchaev, HISAP96, Vol **2**, *ibid.,* p300.
[32] Where was Boris Vannikov?

over by the shock wave. A sound wave would take about 29 seconds to reach NP2; the shock would travel faster. After the wave had passed they were free to do as they pleased, but were not permitted to take notes. Dokuchaev tried on the glasses before the blast, and could not see anything. He looked at the sun poking through the clouds, and saw a faint red globe.

A.I. Burnazyan, deputy minister of health of the USSR, was in one of the tanks at 10 km west of Zero, shown in Figure 14.1. He was supposed to ride into Zero 30 minutes after the blast to make radiation measurements and collect samples. Major General S.G. Kolesnikov was also in this tank, together with the regular tank crew. They had the hatches closed waiting for something to happen.

At 6 am the occupants of bunker NP2 were informed over a loudspeaker from KP1 that the detonation had been delayed until 7 am, on the expectation that the weather would improve. The rain stopped, and the clouds parted somewhat, improving visibility and allowing some sunshine, but the reconnaissance planes that practiced a few days earlier did not fly, partly because of the weather. Bad weather seemed to accompany nuclear tests – it rained on the Trinity test at Alamogordo too, and although observation planes did take off in New Mexico, they were not as useful as planned.[33]

At 6:10 KP1 moved the time up to 6:30 to avoid predetonation. There was an indication from remote monitoring that the neutron background was increasing. So Dokuchaev and his compatriots had 20 minutes to wait for something to happen. They stood around, looking in the direction of the steel tower, 10 km away, and talking quietly among themselves. The signal for ignition was the Russian letter 'Ч', pronounced 'che', the first letter in the word час - 'Chas', which means hour, but in this sense 'time.' At one minute before ignition everyone put on their dark glasses, and the loudspeaker counted down in 10 second intervals until the letter 'Ч.'

---

[33] Leslie M. Groves, "Now It Can Be Told," *ibid.,* p 294.

There were several eyewitness descriptions of the light from the initial blast viewed from NP2 through their dark glasses. They were all looking at the same thing, yet their descriptions varied. Dokuchaev saw a bright white elliptic shaped flame like an electric discharge that lasted several seconds. Vinogradov saw a hemisphere four times as bright as the sun, golden in color, that lasted 4-5 sec. Zeldovich saw a bright cone shaped flash, with a tongue of fire coming out of the dust and smoke after one second.[34] All observers agreed that the bright flash lasted only a few seconds, after which a grey colored hemisphere of smoke and dust formed about 500 m in diameter, like a church cupola, and began to rise in a column. The column continued to rise, driven by the intense heat.

Everyone lay down in the safety trench awaiting the shock wave. After about 30 seconds they heard a loud noise, like a thunder clap, but they did not experience any large shock over pressure. They were too far away. After the fireball quenched, the column of hot gas formed and rose up, and a characteristic 'mushroom' cloud formed, to be carried by the prevailing winds. That information regarding the nuclear explosion was carried great distances in the atmosphere was demonstrated by the detection of 'Joe-1' described in Section 14.7. After an hour or so the column of hot gas had dissipated, and the sun came out, so Dokuchaev put on his dark glasses again to compare the brightness of the sun to his recollection of the initial flash. All he could say was that the two were about the same.

After the loud noise passed, everyone began congratulating each other. Dukhov said: "Long live Comrade Stalin. Hooray." Then he said: "Record the names of those present for history." Nikitin and Vinogradov shook hands, but Zeldovich and Mesheryakov were silent.

"The first impression was that we had let an evil jinni out of the bottle, and God knows when we will be able to return it, if ever. How much misfortune and death will this problem jinni bring to civilization?"[35]

---

[34] Atomnyy_proekt_SSSR._T.2.Kn.6.(2006).pdf, #292 and #293. These are reports to Beria on September 2, 1949.

[35] Ya. P. Dokuchaev, HISAP96-Vol **2**, *ibid.*, p 306.

Around 9 am Burnazyan and Kolesnikov lumbered in towards ground zero in their heavy tank. He recorded residual radiation levels at 10 am at Zero, and reported them to Beria the next day.[36] His numbers were as follows:

Table 14.2.

Location	Radiation μR/sec
At 10:00 am August 29	
Zero	500,000
500 m	8600
750 m	1000
1000 m	100
At 5:00 pm August 29	
Zero	150,000
200 m	100,000
500 m	2500
At 8:00 am August 30	
Zero	>50,000
250 m	25,000 to 60,000
500 m	400 to 1200

Burnazyan did not say how these measurements were made. Probably a dosimeter was held by hand at a fixed height above the ground (1 m?), and the counting rate was recorded. At Zero 0.5 R/sec would give a fatal dose in about 20 minutes. There is some discussion of radiation doses in Section 12.4 on the 'A' reactor.

Dokuchaev was a member of a four person reconnaissance party to Zero on August 30. They wore overalls, gas masks, military boots and underwear, and rubber gloves. They dressed for the task in a nearby tent

---

[36] Atomnyy_proekt_SSSR._T.2.Kn.6.(2006).pdf #286, Burnazyan to Beria, 30 August, 1949.

set up about 300 m from Zero for the purpose. Only N.A. Vlasov had a personal dosimeter. Near Zero his dosimeter read 50,000 μR/sec, consistent with Burnazyan's Table 14.2 above. They filled canvas bags with samples from the ground with small shovels for about five to seven minutes, and then retreated from the area, to change clothes in the tent. Total exposure time was about 15 minutes, and the dose was 50 R or greater. The samples went by truck to the analysis laboratory on the Irtysh River.

Gas masks that limited vision, and the mission to obtain samples from the dirt as quickly as possible in the presence of high radioactivity inhibited a careful study of the surroundings of Zero. No one wanted to hang around. Nevertheless, Dokuchaev made some observations:

1. The steel tower completely evaporated. It was about 30 hours after 'Che', and the sand around Zero for a distance of about 200 m had melted in the intense heat, and then had refrozen into a glassy state. Plant growth was sparse to begin with, but had vanished altogether in the glass.
2. Burnazyan's tank had made tracks in the sand at the edge of the glassy area, but did not appear to penetrate into Zero.
3. The earth's surface around Zero was compressed to form a saucer five to six meters in diameter, with a depth of about 50 cm.

Dokuchaev thought that the RDS-1 crater was small compared to the one at Trinity, and used this as one of the indicators that the TNT equivalent was less than 20 kT. The Trinity crater was described in Groves' book: "A crater from which all vegetation had vanished, with a diameter of 370 m and a slight slope toward the center, was formed. In the center was a shallow bowl 40 m in diameter and 1.8 m in depth. The material within the crater was deeply pulverized dirt. The material within the outer circle was greenish and could be distinctly seen from as much as 8 km away. The steel from the tower was evaporated."[37]

The two bombs were exploded at the same height above the ground, 30 m, and the soil was probably similar- desert sand in each case - so one might expect the craters to be similar, but if Dokuchaev's numbers are correct, the volume of the RDS-1 crater was only about 1% of the Trinity crater. On the other hand, the melted sand to glass areas that were created by the fireballs, were comparable in size. As they drove away he noticed that some of the structures built to test the force of the shock wave were damaged at 300 m from Zero, but the framework of the two red brick three story buildings at 800 m appeared from a distance to be OK.

Dokuchaev was tired and light headed after his shift in the 'zone.' He felt dizzy, and his vision was impaired. It seemed to him that it was already evening, when in fact it was midday. He ate lunch at the canteen upon his return to the village on the Irtysh, but still felt a bit sick. He went outside and sat on the river bank to get a breath of fresh air, and took a dip in the river before going to his dorm room and taking a 2-3 hour nap. After his nap, he regained his senses.

At the end of the day on August 30 the RDS-1 test had been declared a success based on preliminary evidence, with an estimated yield of not less than 10 kT of TNT. The next day a squad of about 20 Red Army soldiers was led by an officer into the radioactive zone to collect more dirt samples in canvas bags. They wore gas masks, and were dressed in the same type of clothes as Dokuchaev and his group the day before. Several soldiers complained of dizziness upon departure from the zone.

On August 31, they began the radio-assay of the samples that had been collected, in order to obtain more quantitative estimates of the efficiency of the blast. Dokuchaev came from Combine #817, so he had experience

---

[37] Leslie M. Groves, "Now It Can Be Told," *ibid.,* p 434, converted to meters. Groves' book was first published in 1962, and was available to Dokuchaev in Russian in 1996, when he wrote his report, but not in 1949.

with the radioactivity of uranium slugs extracted from reactor 'A', and dissolved in nitric acid. No one in the group had experience analyzing the remnants of a plutonium explosion. Samples collected from Zero were not the same as spent reactor fuel, but the two sources of radioactivity did share some similarities. The heavy elements ^{238}U, the main component of the reactor slugs, and also present in the bomb debris from the ~150 kg tamper, and ^{239}Pu, the product of the distillation of the reactor slugs, and the unexploded remnants of the bomb, were the main sources of α particles. Table 14.1 summarizes some of the properties of the α emitters (^{235}U was present only in trace amounts). The short half-life of plutonium made it the dominant source in both the reactor slug solution, where the plutonium to uranium weight ratio was 10^{-4}, and in the bomb remnants, where the ratio was higher.

Fission fragments were radioactive elements in the middle of the periodic table. They did not emit α particles, but were copious sources of β and γ rays. Fission fragments in spent reactor fuel came from burning ^{235}U for weeks or months, while those in bomb residue were from the instantaneous explosion of ^{239}Pu, so the mixes of short lived isotopes were different. The ratio of α activity to β activity was a measure of plutonium production per uranium fission in the slugs, while the same ratio was a measure of the plutonium non-fission to plutonium fission in the bomb, directly related to the efficiency of the explosion.

A method related to the analysis of residual radioactivity deposited on the ground was to measure directly from the fireball, within 10 sec of the explosion, the flux ~3 MeV γ rays. This flux is directly proportional to the number of ^{239}Pu that fission. The distribution of isotopes, half-lives, and transition energies for fission products was a complicated problem, but relevant formulas had been published in the Physical Review.[38] Measurements were made, and compared to the theoretical calculations

---

[38] K. Way and E.P. Wigner, Phys Rev **73**, 1318, (1948). This paper was a statistical analysis of the radioactivity of fission products, and gave average β and γ ray flux as a function of time after fission had occurred. Results of the calculations were compared to existing experimental data.

based on the known isotope content of the plutonium fission products. The results are shown in Table 14.3. The two measurements at 1000 m did not agree, so both numbers are quoted.

Table 14.3. 3 MeV integrated γ ray flux in Roentgens, 0 to 10 sec after 'Ч'[39]

Distance from Zero	Measured flux (R)	Calculated flux for 10% eff (R)	Inferred Efficiency
500 m	60,000	40,000	15%
1000 m	1300 & 2000	1800	10%

These numbers were consistent with a yield of between 10,000 and 15,000 T of TNT. Shock pressure measurements were compared to the formula of A. Sadovsky, quoted in Appendix F, and based on conventional explosives. The results were consistent with the two efficiencies quoted in Table 14.3.

On September 5, 1949, the Geophysics Institute of the USSR Academy of Sciences reported on seismic observations made after 4 am Moscow time on August 29, 1949.[40] They were asked to check their records, without any details as to why Comrade Beria was interested. Indeed, weak seismic signals were detected in Semipalatinsk, Frunze, Alma-Ata, Tashkent, Andizhan, and Stalinabad, all stations in Soviet Central Asia, well instrumented because of frequent earthquakes. Their findings were reported in Table 14.4.

---

[39] Atomnyy_proekt_SSSR._T.2.Kn.6.(2006).pdf, #289, Report of the Efficiency Committee 1 Sept 1949.

[40] Atomnyy_proekt_SSSR._T.2.Kn.6.(2006).pdf, #302

Table 14.4. Seismic signals of a small earthquake 0400 hours Moscow time, 29 August, 1949

Station	P	S	L	M	Δ
Semipalatinsk	4:00:33	4:00:56	–	–	180 km
Alma-Ata	4:02:18	4:03:52	–	4:04:28	800 km
Frunze	–	–	4:04:28	–	1000 km
Tashkent	–	–	–	4:06:30	1300 km
Andizhan	–	–	–	4:05:30	1100 km
Stalinabad	–	–	4:06:40	–	1600 km

P is the time of arrival of the longitudinal elastic wave; S is the time of arrival of the transverse elastic wave; L is the time of arrival of the surface wave, and M is the time of arrival of the maximum phase. Times are in hours:minutes:seconds Moscow time. Seismic speeds depend on elastic properties of the earth, and transverse waves are slower than longitudinal waves, but either speed is a few km/sec, so the Semipalatinsk arrival time of 4:00:33 is not consistent with a firing time of 3:30 as stated by Dokuchaev. Ч must have been very close to 4:00 Moscow time (three hours earlier than Semipalatinsk).

Drawing circles of radii Δ about the stations gave a point of origin west of Semipalatinsk 180-200 km. Quantitative evaluation of the strength of the 'earthquake' from these data was difficult. There was an explosion of 1000 T of TNT 20 m underground at Tula, 200 km from the Moscow seismic station, on March 26, 1949, that could be used as a comparison. If the present earthquake were underground, then its strength exceeded 10,000 T of TNT.

In the final report to Stalin the committee to evaluate the efficiency of RDS-1 concluded that it was 50% higher than the theoretical estimate – that is 15,000 T of TNT equivalent.

On September 3, 1949, the entire team of workers, scientists, and technicians received orders to leave Semipalatinsk and return to their normal duty stations. The firing range was subsequently used for nuclear weapons tests for 40 years, until 1989.

On October 29, 1949, the Council of Ministers of the Soviet Union awarded prizes for excellence in scientific work, and technical achievements in the utilization of atomic energy.[41] The list included:

1. As an expression of gratitude the Central Committee of the Communist Party of the Soviet Union confers on Comrade Lavrenty Pavlovich Beria the award of the Order of Lenin, and the title of Laureate of the Stalin Prize of the first rank.

2. Igor Vasilevich Kurchatov, academician, scientific manager of the project:
   - Title 'Hero of Socialist Labor.'
   - Prize money 500,000 rubles (in addition to earlier award of 500,000 rubles plus a ZIS-110 automobile).
   - Title 'Stalin Prize, first rank.'
   - Construct at State expense and furnish a private residence and a dacha.
   - Establish double salary for future work in the field of utilization of atomic energy.
   - Grant Kurchatov and his wife for life free transportation by railroad, water, or air anywhere within the boundaries of the USSR.

3. Yuli Borisovich Khariton, Corresponding member of the Academy of Sciences, principal builder of the atomic bomb
   - Title 'Hero of Socialist Labor'
   - Prize money 1,000,000 rubles, and a ZIS-110 automobile
   - Title 'Stalin Prize, first rank'
   - Construct at State expense and furnish a private residence and a dacha.
   - Establish double salary for future work in this field.
   - Grant the education of his children in any university at state expense.

---

[41] Atomnyy_proekt_SSSR._T.2.Kn.6.(2006).pdf, #314.

- Grant Khariton and his wife for life, and their children while dependent, free transportation by railroad, water, or air anywhere within the boundaries of the USSR.

Apart from Beria, but including Kurchatov and Khariton, 23 scientists and administrators in the program received the title of 'Hero of Socialist Labor.' Among them were:

- Andrey Anatolevich Bochvar
- Boris Lvovich Vannikov
- Nikolai Antonovich Dollezhal
- Nikolai Leonidovich Dukhov
- Avraamy Pavlovich Zavenyagin
- Yakov Borisovich Zeldovich
- Vasily Alekseevich Makhnev
- Boris Glebovich Muzrukov
- Mikhail Georgevich Pervukhin
- Nikolai Vasilevich Ril (Nikolaus Riehl in Russian form)
- Efim Pavlovich Slavsky
- Georgy Nikolaevich Flerov
- Vitaly Grigorevich Khlopin

Vannikov, Dukhov, and Muzrukov received the award for the second time.[42] Ya. B. Zeldovich, G.N. Flerov, L.V. Altshuler, and V.A. Tsukerman received substantial monetary bonuses as well. In all, 176 people received the Stalin Prize, 39 first rank, 112 second rank, and 25 third rank. In 1948 about 50,000 workers were employed in the factories and combines of the PGU, with more working in the uranium mines.

---

[42]   A.K. Kruglov, "Who participated in the Soviet Atomic Project?," HISAP96, Vol **1**, *ibid.,* p 62.

14.7. Detection of 'Joe-1'

In September of 1946 General Hoyt S. Vandenberg of the US Army Air Force, at that time director of Central Intelligence, wrote a letter to General Groves of the Manhattan Project proposing that US detection of possible nuclear explosions anywhere on earth be implemented, and that continuous monitoring of the atmosphere by weather aircraft be initiated. He asked if the technology existed for long distance detection of an atomic bomb.[43] Air sampling and acoustic detection had been deployed at the Bikini tests in July, without conclusive results. By that time the United States had acquired data on five nuclear explosions: the Alamogordo, New Mexico test in July, 1945, the two detonations over Japan, Hiroshima and Nagasaki, in August of that year, and Able and Baker Bikini tests. There were three known distant detection methods: acoustic from the shock wave, seismic from the impact on the ground, and airborne radioactivity carried away in the rising cloud. If the USSR is the most likely candidate for a nuclear test, then any detection scheme must necessarily be sensitive at a distance of a thousand kilometers or more. Airborne reconnaissance over Soviet territory was not feasible without retaliation. (The high altitude U-2 planes did not exist at that time.)

Zero was about 450 km from the closest part of the Chinese border, but diplomatic relations between China and the United States were not so friendly in 1947 as to allow a nearby listening post, and Central Intelligence was not aware of the location of the Soviet firing range in any case.

The signature of the blast depended on conditions. An air blast at an altitude of one km or so above the earth's surface would have only a shock wave, with no seismic effects, and no radioactive 'fallout.'[44] The shock wave would have a limited range, so that such a blast would be difficult to

---

[43] The airborne detection of RDS-1, named 'Joe-1' after Stalin by the US Atomic Energy Commission, is based on the CIA declassified report by Northrup and Rock, *ibid.*

[44] Improvements in detection technique offered the possibility of detection of airborne radioactivity from an air burst. Charles A. Ziegler and David Jacobson, "Spying without Spies," *ibid.*, p 138.

detect. An underground blast would have only seismic waves, and would be difficult to distinguish from other seismic signals – earth quakes, volcanic activity, or non-nuclear explosions. The under-water Baker test at Bikini was similar to an underground blast, in that the water retained most of the fallout. Fallout refers to airborne radioactive dust that is formed by interaction of the fireball with the ground surface. Melted debris from the earth mixes with the fission products and other bomb remnants, cools, and is carried to high altitude to be distributed over the earth's surface by winds. Detonation near the earth's surface can lead to increased atmospheric radioactivity detectable at great distances – several thousand kilometers. Alamogordo, Hiroshima, Nagasaki, and Bikini Able were all low altitude air blasts, at 30, 580, 500, and 160 meters respectively. (Joe-1 was at 30 m.) Nagasaki, because of local atmospheric conditions, yielded little fallout. An increase in atmospheric radioactivity is strong evidence of a nuclear explosion, although it could result from a reactor accident like Chernobyl, or a chemical explosion involving radioactive materials, like the accident at Mayak in 1957.[45]

If enough radioactive material can be retrieved from fallout, it could be representative of the remnants on the ground at the site itself, and therefore contain much valuable information. Suppose you were fortunate enough to grab a sample of 500 fission products and 2500 plutonium atoms from the air. If an accurate indicator, the efficiency of the bomb from these data would be 250/2750 = 9% (each fissioned plutonium atom gives two products). In addition, by measuring the fission product isotopic ratios, using the known ratios at production and correcting the shorter half-lives back to $t = 0$, one could determine the date of the explosion.

In January, 1947, Lewis Strauss, a member of the newly created Atomic Energy Commission, asked Secretary of Defense Forrestal if a program had been set up within the defense organization to monitor nuclear tests. The program got started as a result of Strauss' interest. The

---

[45] Charles A. Ziegler and David Jacobson, "Spying without Spies, Origin of America's Secret Nuclear Surveillance System," *ibid.*, The title of this book implies that there was no *in situ* western informant. It has detailed descriptions of the detection techniques.

US Air Force organized the Office of Atomic Energy, called AFOAT-1 for short, to set up acoustic stations, a seismographic network, and regular aircraft flights to monitor radioactivity in the atmosphere. The Office of Naval Research started two ground based detectors, one an array of Geiger counters to detect γ rays in the air, and a cistern arrangement to gather radioactive materials in rain water.[46] Two years were estimated to get all of this done, including the necessary communication between the various activities. In the meantime, backgrounds could be established to determine thresholds for a signal. A private company, Tracerlab, was contracted and trained by Los Alamos to do the analysis of radioactive elements collected. Detection methods were verified in the spring of 1948 at atomic tests on Eniwetoc Atoll in the Marshall Islands. Successful air sampling was done on filter paper mounted outside the aircraft. The US Navy set up a fallout water collector on Kodiak Island off the coast of Alaska. Several Geiger counter packages were deployed around the continental United States, including the roof of the Naval Research Laboratory in Washington, D.C. Fifty-five B-29's were equipped with air sampler filters, and flew daily missions from Guam and Bermuda to the North Pole. Prevailing high altitude westerly winds would carry a signal from Siberia across the Pacific.

Threshold was set at 50 counts per minute, and up to September 3, 1949, there had been 111 Alerts, all of which had been explained away as false alarms. Alert 112 came in on September 3, with a filter flown for 3 hours over the north Pacific that had radioactivity of 35 counts above threshold. A companion filter from the same plane had over 100 counts above threshold. These samples were sent to Tracerlab for analysis. While this was in progress, on September 5 a paper exposed on a weather flight east of Japan had over 1000 counts per minute. This sample was promptly flown to California for analysis.

The radioactive debris contained many clues. First one looked for fission products, and having found them, their relative abundances at the

---

46 The description of the Navy rainwater experiment comes from Herbert Friedman, Luther B. Lockhart, and Irving H. Blifford, "Physics Today," November, 1996, p 38.

time of observation. From that data, together with the known abundances at creation from the fission of plutonium and the half-lives, one could extrapolate back in time to give the approximate date of the explosion. The answer was that a nuclear explosion occurred between the 26th and 29th of August, 1949. The timing could also be used to rule out fission products from a reactor accident rather than a nuclear explosion. The radioactive sediment from the Kodiak Island water works – it had conveniently rained! – was chemically analyzed for plutonium, and the identification was verified by Maurice Shapiro of the Naval Research Laboratory, who measured the range of the α particles in photographic emulsion.[47] Given the estimated time of the blast, weak acoustic signals could be extracted from two stations, that pinpointed the location and time of the RDS-1 test as August 29, 1949, at Semipalatinsk in Kazakhstan, and an approximate energy release of 20 kT. That the explosion was known to be a plutonium bomb was classified Top Secret for many years.

These findings were reported to President Truman on 23 September, 1949, and he released a statement to the American public that same day:

"I believe the American people, to the fullest extent consistent with national security are entitled to be informed of all developments in the field of atomic energy.... We have evidence that within recent weeks an atomic explosion occurred in the USSR. Ever since atomic energy was first released by man, the eventual development of this new force by other nations was to be expected. This probability has always been taken into account by us.... This recent development emphasizes once again, if indeed such emphasis is needed, the necessity for that truly effective enforceable international control of atomic energy..."[48]

This announcement was immediately picked up by Soviet news correspondents, and translated into Russian.[49] The Soviet press had made no mention of the RDS-1 test in August, but this statement by the US

---

[47] The author knew Maurice Shapiro at the Naval Research Laboratory in 1958. For $^{239}Pu \rightarrow {}^{235}U + \alpha$, $E_\alpha = 5.24$ MeV.

[48] Truman Library, Public papers of Harry S. Truman.

[49] Atomnyy_proekt_SSSR._T.2.Kn.6.(2006).pdf #305 Truman's press release plus some interviews with members of the US Joint Committee on Atomic Energy in Russian.

President could not be ignored. TASS had to say something. This is what it said:

"On September 23 President Truman of the USA stated that, according to data obtained by the USA, an atomic explosion occurred a few weeks ago in the USSR. The same statement was simultaneously made by the British and Canadian governments. These simultaneous announcements by the Americans, British, and Canadians and also publication by several other countries have caused wide spread anxiety.

"TACC has been authorized to make the following clarifying statement.

"It is well known that the Soviet Union builds large scale works: hydroelectric plants, mines, canals, and highways that sometimes require new technology and big explosions. To the extent that these explosions often take place in various parts of the country, it is possible that they have attracted attention outside the borders of the Soviet Union.

"Regarding the subject of atomic energy, TASS believes it necessary to remind everyone that on November 6, 1947, Foreign Minister Molotov asserted that there is no longer any secret regarding the atomic bomb. Molotov meant that the Soviet Union knew the secret of atomic weapons, and it already had them in its arsenal. American scientists assumed that Molotov was bluffing, that Russia could not obtain nuclear weapons earlier than 1952. However, they were mistaken, because the Soviet Union already had the secret in 1947.

"There is no basis for anxiety among foreign countries. The Soviet government, despite its possession of nuclear weapons, will stand by its unconditional resolve not to use them.

"Regarding the control of nuclear weapons, it is necessary to say that control will be needed in order to verify that nuclear weapons are not produced."[50]

This cover-up was intended for consumption by the general public, and it may have assuaged concerns in some quarters. Knowledge of the

---

[50] TASS communication to Pravda, 25 September, 1949.

detection technique – fission products in radioactive fallout - would have spoiled an explanation based on conventional explosives, but that information was highly classified by the US. Groves had tried a similar press release after the Alamogordo test quoted in Section 7.15 – that the explosion was due to an accident at an ammunition dump – a ruse that did not fool everyone either.[51]

The British were kept informed of the work on Joe-1 detection, and were able to confirm some of the conclusions from air samples over the British Isles. Harold 'Kim' Philby, a member of the British Secret Intelligence Service and a Soviet spy, was in the British Embassy in Washington, D.C. and had full knowledge of the procedures used to detect Joe-1. This information was sent to Moscow, so that, although the detection methods were secret in the US, they were well known to Beria, although TASS kept silent.[52]

AFOAT-1 was a low budget operation, and because of a widespread sense of security from estimates for the 'earliest possible date' of a Soviet test, it was always threatened with extinction, but it was there and looking when something happened for it to see. On October 17, 1949, Vannevar Bush, head of the US Office of Scientific Research, wrote a letter of commendation to General Vandenberg, praising the work of AFOAT-1. The program became an established institution, continued for about another decade, through the International Geophysical Year (1957-58), and was an important contributor to Cold War intelligence.

14.8. Aftermath

Stalin had the bomb, but it would be several years before the USSR had enough weapons and a delivery system to join the class of 'Super Powers' in the cold war. There was support for Korolev's rockets, that were developed in parallel with the bomb project. Stalin favored jet aircraft for delivery until the rockets became practical. Rockets and the

---

[51] Leslie Groves, "Now it can be Told," *ibid.*, p 301.
[52] Ziegler and Jacobson, "Spying without Spies," *ibid.*, p 214-215.

space program made headlines in 1957 with the launch of 'sputnik.' More reactors were built at Combine #817 and elsewhere for the production of plutonium, and enriched uranium began to flow from the isotope separation plant, both for reactor fuel and hybrid uranium-plutonium weapons.

Work on the hydrogen bomb proceeded in parallel with the fission bomb project, principally at KB-11, with Andrei Sakharov as one of the inventors. A preliminary test of a thermonuclear device took place in 1953, after Stalin's death and Beria's arrest.

Stalin died March 5, 1953. Before his death he instituted yet another purge, this time labeled the 'doctors' plot' aimed at mostly Jewish Soviet doctors. This latest purge was stopped after his death, but it caused widespread anxiety in the medical community. Stalin's death created a power vacuum in the Central Committee that was initially filled by Lavrenty Beria and Georgy Malenkov. But the political situation remained fluid, and in June of 1953 Nikita Khrushchev outmaneuvered Beria and had him arrested by Marshal Zhukov. Beria was imprisoned for crimes against the state, and stripped of all awards and rank. He was executed in December, 1953. Beria was the last of the powerful Soviet ministers to be executed. He was too much of a threat to Khrushchev to be allowed to retire to his dacha and write his memoirs. But in 1965, when time came for Khrushchev to be moved aside, it was done bloodlessly, and subsequent general secretaries up to Gorbachev conveniently died of natural causes in office, as did Stalin. Gorbachev's position vanished out from under him.

Pavel Sudoplatov, one of Beria's go-to guys, was not executed, but did spend 15 years in prison. Sudoplatov well knew the system that had turned on him, and he tried repeatedly during his incarceration to get his sentence overturned. After his release in 1968 he became a translator of European languages into Russian. He spent several years trying to restore his reputation, finally succeeding in 1991, after the Soviet Union had collapsed. Sudoplatov aptly summarized the history of his time at the conclusion of his memoir. What he had to say applies to the period of the Atomic Project, and to those who worked on it:

> "The Soviet Union, to which I gave all my strength, and to which I would give my life, for the sake of which I tried not to notice the ever present cruelty, partly justified by the effort to industrialize a backward country. The Soviet Union for which I spent long months far from home, wife, and children – even 15 years in prison did not kill my devotion – has ceased to exist. And only when the USSR, once a super power, collapsed, was I at long last restored to my rights as a citizen, and my name returned to its rightful place in history.
>
> I hope that my story will help the present generation occupy a thoughtful, independent position in evaluation of our heroic and tragic past."[53]

After the arrest of Klaus Fuchs Aleksandr Feklisov retired honorably from NKVD agent operations. His connection to Fuchs was never identified. He remained in the intelligence service, becoming chief of North American operations because of his extensive experience in the United States. When Nikita Khrushchev planned a visit to the US in 1960, Feklisov was involved in plans for the security of the Soviet head of state. Then in 1962 Feklisov played a role in the Cuban missile crisis, discussed in Chapter 8. He visited Klaus Fuchs' widow in east Germany, and tried in vain to get Fuchs' contributions to the Project recognized by the Soviet government. There was tension between those who viewed the contribution of espionage as crucial, and those who considered it merely useful.

As was true in the West, nuclear weapons work after the initial inventions became the profession of specialists, and the physicists for the most part reverted to more peaceful research. Kurchatov had the vision to exploit the leverage of the Soviet Atomic Project into public support of basic science – a technique also successfully used in the West. Thus the synchrocyclotron, which was the anchor of JINR Dubna, was built in the mid 1940's as part of the Project, but was used to study nuclear and particle phenomena unrelated to weapons work. Subsequently the high energy

---

[53] Pavel Sudoplatov, "Razvedka I Kreml," *ibid.,* p 491.

physics laboratory at Protvino near Serpukhov was built for research in particle physics; Siberia was included in pure research by the creation of Academy Town and the Siberian Division of the Academy of Sciences after the XX Party Congress of 1956. Landau's group at the Institute for Physical Problems in Moscow flourished, winning Nobel Prizes for work in material sciences. Laboratory #2 has become the Kurchatov Institute, and continues to be a leader in research in controlled nuclear fusion – work instigated by Kurchatov himself and Lev Artsimovich. Yakov Zeldovich became even more famous, if that is possible, for his innovations in astrophysics. One of the few leaders who remained with the weapons project for the remainder of his career was Yuli Khariton, who stayed at KB-11 as director emeritus until his death at age 92. At the end of the 1980's Khariton wrote a letter to Mikhail Gorbachev:

> "I write you this letter because of a deep concern for the fate of our national nuclear weapons complex, that was built in very difficult times after World War II by millions of Soviet workers. The complex guaranteed strategic parity in the world. Soviet nuclear weapons were a powerful force for the absence of nuclear war for more than 40 years. The Soviet nuclear arsenal has enormous destructive power, and must be under strict, comprehensive government control – a central authority with clear lines of responsibility, and exclusive powers over its activities. I could not leave the project without appealing to you to meet with the scientists and managers of the nuclear weapons complex, in spite of your unbelievably busy schedule." [54]

With the notable exception of Andrei Sakharov, the scientists who worked on the Soviet Atomic Project were never punished by the system, even after Beria's fall from power. Igor Kurchatov continued his feverish pace at work on various aspects of nuclear energy, until his early death in 1960 at age 57. He had accomplished all that he set out to do. A.P. Aleksandrov in 1949 left the directorship of the Institute for Physical Problems to join Kurchatov at Laboratory #2, where he pursued nuclear propulsion for ships and submarines, working with the Soviet Navy with notable success. Peter Kapitza was restored to favor and his Institute, and

---

[54] Scientific American "In the World of Science" special edition on the 70th anniversary of the Kurchatov Institute, *ibid.,* p 31.

became a member of the presidium of the Soviet Academy of Sciences. He died in 1984 at the age of 89. In the early 1950's Isaak Kikoin became involved in the technology of detecting nuclear explosions remotely, and played a role in nuclear test ban negotiations.

Georgy Flerov joined the research staff at Dubna, and became a division head, leading an experimental team that used heavy ions to make new transuranic elements. The program has been very successful, and two of the discoveries have been named in recognition: Dubnium $Z=105$, and Flerovium $Z=114$. A street in Dubna bears his name. He died in 1990.

Was the Soviet Atomic Project a success for the communist system? The answer has to be yes, because Stalin's communism was the government, all of industry was state owned, and the Project succeeded. It placed a severe burden on the Soviet economy that had not recovered from the destruction of WWII, in effect prolonging wartime shortages of consumer goods. Igor Kurchatov was a member of the communist party, but many prominent scientists were not. Could it have been done without Beria? This question has to be dismissed as irrelevant, since Beria did orchestrate it, making countless technical decisions upon which the entire operation depended, without the required technical training himself. Beria checked with Stalin on a daily basis, but maintained tight control of the reins of the Project, and was confident that he knew what he was doing. He listened to Igor Kurchatov, but wrote the orders himself.

Could it have been done without the espionage? Here the answer has to be ambiguous, because the espionage supplied a whole lot of useful information. In addition, after 1945 there were several reports in the open literature that were also helpful – the Smyth report, and declassified articles in journals. The most important fact was that nuclear weapons were possible. The Soviet physicists understood nuclear physics pretty well, but the industrial scope of the undertaking, and which of the large operations would likely succeed were learned from the experience of the Manhattan Project. The task of producing enough fissile material for a bomb, either by isotope separation of natural uranium or by production of the artificial element plutonium in power reactors, strained the capabilities of Soviet industry, as the 'Look' magazine article claimed it would, but these difficulties were overcome in a timely fashion, with some compromises to favor speed over safety.

Did the German scientists play a crucial role? The answer is, I think, no – helpful, yes, but not crucial. Access to German industry and instrumentation through war reparations was also helpful. Beria was probably reluctant to couple the Germans too closely to Project operations – membership on the PGU for example – because of security concerns.

So that is the story. In 1945 the United States had nuclear weapons, and in 1949 the Soviet Union also had nuclear weapons. Was the world better off after the success of RDS-1? Certainly the Soviets thought so, because their lack of a nuclear arsenal was acutely felt. Fortunately for everyone in the course of the cold war no nuclear weapons were ever used. In 2018 the former superpower nuclear arsenals are mostly in storage, but other countries world-wide already have, or are developing a nuclear capability. The existence of nuclear weapons has not made us feel safer.

Two lines of a country music song of 1950, shown in Figure 14.3, form as fitting an epitaph as any to this book.

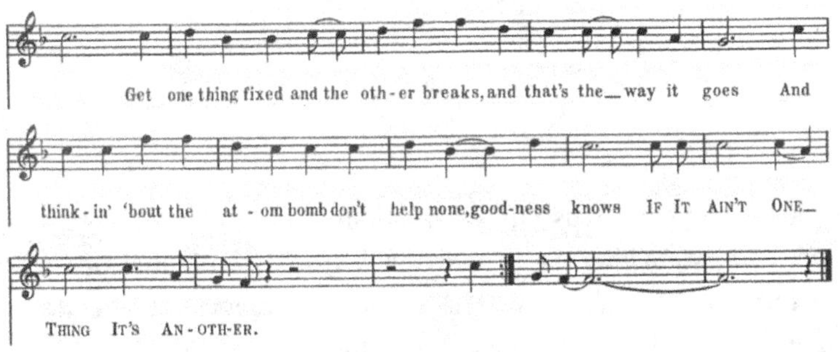

Figure 14.3. If it Ain't One Thing it's Another.[55]

---

[55] Copyright 1950 by Acuff-Rose Publications, 2510 Franklin Rd, Nashville, Tennessee, USA.

Appendix A

Nuclear Masses

A.1. Mass Formula

In 1935 C.F. von Weizsäcker proposed a formula for atomic masses based on the liquid drop model of the nucleus.[1] It is now called the semi-empirical mass formula, and for the ground state energies of all atomic isotopes, $M(Z,A)$ is written:

$$M(Z,A) = Z \times M(1,1) + (A - Z) \times M(0,1) - B(Z,A). \qquad (A.1)$$

Where Z is the atomic number (the number of protons), $M(1,1)$ is the mass of hydrogen, $(A - Z)$ is the number of neutrons, $M(0,1)$ is the neutron mass, and $B(Z,A)$ is the binding energy. The units are MeV. The formula for the binding energy is as follows:

$$B(Z,A) = a_1 \times A - a_2 \times A^{2/3} - a_3 \times Z^2 \times A^{-1/3} - a_4 \times (A - 2 \times Z)^2/A \pm a_5 \times A^{-1/2}. \quad (A.2)$$

There is physical reasoning behind the terms. The first term is the volume energy, which is proportional to A. The nuclear radius is proportional to $A^{1/3}$. The next term is the surface tension, accounting for a weaker binding for those nucleons on the surface. The third term is the coulomb repulsion, proportional to $Z^2/A^{1/3}$. The fourth and fifth terms are quantum mechanics, having to do with identical particles and pairing

---

[1] C.F. von Weizsäcker, Zetischrift fur Physik **96**, 431 (1935).

symmetry. The fifth term is positive for Z and (A-Z) even numbers, negative for Z and (A-Z) odd numbers (rarely occurring in stable nuclei), and zero for even-odd or odd-even. This term favors nuclei with even numbers of both protons and neutrons. B(Z,A) is treated here as a positive quantity, to be subtracted from the total mass in Eq. (A.1).

This formula is fitted to the data on the atomic masses of all of the isotopes. The fit does better for the higher masses than for the very low ones. In the higher mass region, the fit is generally good to $\pm$ 0.2 MeV. The fit parameters are:

$$a_1 = 15.67; \ a_2 = 17.23; \ a_3 = 0.75; \ a_4 = 23.2; \ a_5 = 12. \tag{A.3}$$

The formula is simple enough, but tedious to use for several different isotopes, so I wrote a FORTRAN program to handle it. That is a luxury that the people who figured all of this out in the first place did not have, since there were no digital computers. It makes their accomplishments even more impressive. Converting to atomic mass units, the results of plugging (Z,A) into the formula for isotopes of interest are given for uranium in Table A.1.

Table A.1.

Z	A	Atomic mass amu	+1n	BE/A MeV
92	235	235.0891	236.0977	7.41
92	236	236.0902	----	7.41
92	238	238.0946	239.1032	7.39
92	239	239.0978	-----	7.38

Note that the binding energy per nucleon – the last column - is almost constant. The interesting numbers are ^{235}U + n = 236.0977 and ^{236}U = 236.0902. This gives a difference in energy between ^{235}U + n and ^{236}U of 0.0075 amu = 7.0 MeV, which is the excitation energy of the compound nucleus ^{236}U formed when ^{235}U captures a thermal neutron with no kinetic energy. The isotope ^{236}U is even-even, which is favored by the a_5 term in Eq. (A.2), giving an extra boost to the binding energy. This makes the energy difference larger, which is the key to fission. The same arithmetic for ^{238}U + n compared to ^{239}U gives an excitation energy of 5.0 MeV. Seven MeV is enough to trigger fission, while five MeV is not.[2] The difference is small, but significant. The semi-empirical mass formula is all you need to get the excitation energies. However, you need the Bohr-Wheeler model to predict the energies needed for fission to proceed.

The same exercise with ^{239}Pu + n compared to ^{240}Pu gives an even larger excitation energy of 7.2 MeV, so it has enough energy to fission. As in the uranium case, the isotope ^{240}Pu is even-even, and as a consequence is more tightly bound. On the other hand, ^{232}Th +1 n compared to ^{233}Th gives an excitation energy of 5.1 MeV, too low to trigger fission.

From this exercise, we conclude that there are two fissionable isotopes with slow neutrons: ^{235}U and ^{239}Pu. Other isotopes of heavy elements do not quite make the grade.

The energy release in fission can be calculated from electrostatics. The Coulomb repulsion between the two fission fragments pushes them apart as they form into separate spheres. Coulomb's law in meter-kilogram units is:

$$V(r) = Z_1 \times Z_2 \times e^2/(4\pi\varepsilon_0 r). \tag{A.4}$$

Here Z_1 and Z_2 are the atomic numbers of the fission fragments, e is the elementary charge (the charge on an electron), and ε_0 is a constant of

---

[2] The Bohr Wheeler liquid drop model fission paper predicts 5.4 MeV needed for ^{236}U to fission, and 4.3 MeV for ^{239}Pu, so both are above threshold with thermal neutron capture. However, ^{239}U requires 6.05 MeV, so it doesn't make it. See Phys Rev. 56, 426 (1939). Of course, plutonium was not known at that time, but you can use their formulas to predict its fission energy threshold.

electromagnetic theory. The separation radius r is in meters. The potential energy is in joules. The elementary charge is:

$$e = 1.602 \times 10^{-19} \text{ coulombs.}[3] \qquad (A.5)$$

The numerical factor $(e^2/4\pi\varepsilon_0) = 2.3 \times 10^{-28}$ joules-meters. For the separation radius r it is reasonable to take the diameter of the parent uranium nucleus. Nuclear radii are proportional to $A^{1/3}$. Bohr and Wheeler use:

$$r = r_0 \times A^{1/3}, \ r_0 = 1.4 \times 10^{-15} \text{ meters} \qquad (A.6)$$

Using $A = 235$, we get $r = 8.6 \times 10^{-15}$ meters. Taking the atomic numbers for the fission products as $Z_1 = 50$, and $Z_2 = 40$ – typical numbers for uranium - we have $V(2r) = 2.6 \times 10^{-11}$ joules. To convert to MeV units, we need the relation 1 MeV $= 1.6 \times 10^{-13}$ joules.[4] Thus $V(2r) = 162$ MeV. This is the energy shared by the two fission fragments as they separate to a large distance. This number is ten times as large as the number for fusion of deuterium and tritium (we calculated 17 MeV for the $^2H + \ ^3H \rightarrow \ ^4He + n$ reaction in Chapter 3), but the fission energy release per kg is less, because of the small molecular weight of the hydrogen isotopes.

The total energy release for the fission of ^{236}U is about 200 MeV. The extra energy appears later, as the fission products β and γ decay after stopping in the material. The fragments have a very short range in uranium metal – less than 1 mm. So they stay close to the location of the

---

[3] We use exponential notation. $1000000 = 10^6$, and $0.000001 = 10^{-6}$. Exponents add when numbers are multiplied, and subtract when they are divided.

[4] This number comes from the definition of an electron volt and the elementary charge. In meter-kilogram units, e × 1 volt = 1 electron volt $= 1.6 \times 10^{-19}$ joules. Then 1 MeV $= 1,000,000 \times 1$ eV $= 1.6 \times 10^{-13}$ joules.

^{235}U which absorbed the neutron in the first place. About half of the energy released in β decay is lost to neutrinos which escape the reactor.

A.2. α Decay of Heavy Nuclei

All nuclei heavier than lead (Z=82) are radioactive, and the vast majority of isotopes emit α particles. A typical example is ^{228}Th (Z=90) → ^{224}Ra (Z=88) + α. The kinetic energy E_α = 5.0 MeV[5], and the half-life is 1.9 years. A glance at the heavy isotopes in Appendix A of Tipler and Llewellyn shows a variation in half-life of about 18 orders of magnitude (10^{18}) for a factor of three or so in α particle energy.[6] There are several puzzles relating to α decay. One is the wide range of half-lives. Generally, larger α kinetic energies correspond to shorter half-lives, but the energy spread is small. Then there is the problem of the coulomb barrier Refer to Figure A.1.

---

[5]   Actually E = 5.5 MeV; we have rounded off to make things simple.
[6]   Tipler and Llewellyn, "Modern Physics," 5th edition, Appendix A. *ibid.*

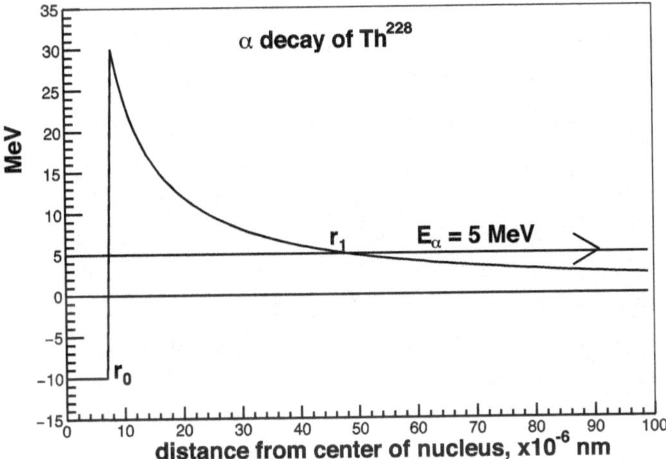

Figure A.1. Schematic energy diagram of α decay of ^{228}Th. The daughter nucleus is ^{224}Ra. The nuclear radius $r_0 = 8 \times 10^{-6}$ nm, and the barrier height is 30 MeV. The α particle emerges from the coulomb barrier at $r_1 = 45 \times 10^{-6}$ nm, and is repelled out to large distances, where it acquires the measured energy of 5 MeV.

This highly simplified picture of α emission by heavy nuclei was first proposed by George Gamow.[7] The daughter nucleus is shown in Figure A.1 as an attractive square well $V = -10$ MeV. The total energy is constant, represented by the horizontal line at +5 MeV. There are no forces on the α particle inside the well, but it gets kicked back by a restoring force at $r = r_0$. The kinetic energy inside the well is KE = 15 MeV. Classically the α particle is bound – it cannot climb over the coulomb barrier, which is 30 MeV high. But quantum mechanics allows the wave function to "tunnel" through the region $r_1 - r_0$, where the kinetic energy is negative, which makes no physical sense. This is a good example of quantum mechanical wave – particle duality. The wave can penetrate into the forbidden region, where the particle has negative kinetic energy, but the amplitude of the wave is exponentially attenuated, so that only a very small amount leaks through. Thus there is a small but

---

[7] G. Gamow, Zeitschrift fur Physik **51**, 204 (1928). Also, independently Gurney and Condon, Nature **122**, 439 (1928).

finite probability that the α particle appears outside the barrier at r_1 and is then accelerated by the repulsive coulomb field to infinity, where it has 5 MeV kinetic energy. The probability of tunneling depends exponentially on the area between the α particle line and the curve of the coulomb barrier. This exponential dependence explains the wide variation in half-life for the α emitters. The model does not address many questions. For instance, does the α particle bounce around in the nucleus waiting for a chance to escape, or is it assembled from protons and neutrons inside the nucleus shortly before emission? Despite this crude picture, it has been quite successful, and in its day it had an effect in convincing people of the applicability of wave mechanics to nuclear phenomena.

The inverse reaction can also occur. People who were constructing charged particle accelerators, like Cockcroft and Walton at Cambridge, thought that the height of the coulomb barrier – which for protons would be 15 MeV for ^{224}Ra – presented an insurmountable problem for the study of nuclear reactions caused by protons in the heavy elements, since high voltage generators could only reach a few MeV. Gamow visited Cambridge several times in the early 1930's, and convinced the experimenters that reactions could happen at lower energies by tunneling through the coulomb barrier from the outside.[8]

A.3. Spontaneous Fission

The following Table A.2 is taken from Wikipedia.[9] It shows that all nuclei of atomic number $Z = 92$ or higher decay by spontaneous fission. For many transuranic elements the decay by fission is a dominant branching fraction, which is interpreted as one reason why elements above $Z=92$ do not occur on the earth. When Petrzhak and Flerov discovered the spontaneous decay of uranium, they were not able to sort out the isotopic contributions, but it is now known that both isotopes ^{235}U and ^{238}U undergo spontaneous fission at comparable rates, so that their observations were due entirely to the dominant isotope ^{238}U.

---

[8] Oral History transcript of George Gamow, www.aip.org/history/ohilist/4325.html
[9] en.wikipedia.org/wiki/Spontaneous_fission

Table A.2.

Nuclide	Half-life	Fission prob. per decay	Neutrons per fission	Neutrons per gram-second
^{235}U	7.04×10^8 years	2.0×10^{-9}	1.86	3.0×10^{-4}
^{238}U	4.47×10^9 years	5.4×10^{-7}	2.07	0.0136
^{239}Pu	2.41×10^4 years	4.4×10^{-12}	2.16	0.022
^{240}Pu	6569 years	5.0×10^{-8}	2.21	920
^{250}Cm	6900 years	0.61	3.31	1.6×10^{10}
^{252}Cf	2.638 years	3.09×10^{-2}	3.73	2.3×10^{12}

The neutrons per gram-second, the right hand column in Table A.2, can be obtained with a little arithmetic. First convert the half-life of ^{238}U into a lifetime in seconds: ½ life of 4.5×10^9 years = lifetime of 2×10^{17} sec. Then the transition rate per nucleus for spontaneous fission is:

$$w = (5.4\times10^{-7})/(2\times10^{17}) = 2.7\times10^{-24} \text{ /sec.}$$

One gram of uranium contains 2.5×10^{21} atoms, and each fission gives 2 neutrons, so the neutrons per gram-second would be $2\times2.7\times2.5\times10^{-3} =$ 0.0135, in agreement with the Table within rounding errors. A kilogram of uranium therefore generates 13.5 fast neutrons per second by spontaneous fission without any external neutron sources. Petrzhak and Flerov measured fission products, not neutrons. And to detect the heavily ionizing fission products, the uranium atom had to be very close to the surface inside the ionization chamber, which limited the effective source volume.

Appendix B

Controlled Nuclear
Chain Reactions

B.1. Introduction

Suppose that each neutron absorbed by ^{235}U results in two more neutrons; that is:

$$^{235}\text{U} + \text{n} \rightarrow {}^{236}\text{U} \rightarrow 2\text{n} + 2 \text{ fission fragments.} \qquad (\text{B.1})$$

And suppose within 1/100 sec each neutron is also absorbed by ^{235}U, repeating Eq. (B.1) twice. If this were to repeat every 1/100 sec, there would be 2^{100} fissions in the first second. This is a staggering number: $2^{100} = 1.3 \times 10^{30}$.[1] With 200 MeV energy release per fission[2], the energy release in the first second would be 4×10^7 terajoules (4×10^{19} joules). This is a divergent chain reaction. Everything would melt.

The above example fortunately does not occur in nature. Although there are more than two extra neutrons emitted per fission, they are not always successful in creating two more fission reactions. For various reasons neutrons are lost before they have a chance to be absorbed by a uranium nucleus.

---

[1] Avogadro's number, the number of molecules in one gram molecular weight, is 6×10^{23}, so you would need 10^7 moles (2350 metric tons) of uranium not to run out of nuclei in the first second.

[2] See Appendix A.

594

Assume that the neutron density in the reactor is a function only of time. It is independent of location inside the reactor volume. This is not true, but it greatly simplifies the argument. Call the neutron density $n(t)$. Then the time dependence is governed by a simple differential equation:

$$dn(t)/dt = (\lambda_1 - \lambda_2) \times n(t). \tag{B.2}$$

Here λ_1 is the rate at which neutrons are produced, and λ_2 is the rate at which neutrons are lost. If $\lambda_1 = \lambda_2$, the neutron density is constant, and the reactor is 'critical.' Subcritical is λ_1 less than λ_2 – the neutron density damps out to zero as time goes on. Supercritical is λ_1 greater than λ_2 – the neutron density increases without limit as time progresses. The chain reaction is self-sustaining if the reactor is critical. Everything is in equilibrium. Just enough fission reactions occur to replenish the supply of neutrons. The power output of the reactor is constant. For a critical chain reaction exactly one neutron from the first reaction initiates the second reaction. The reactions keep going without an external energy supply in a linear fashion. In a practical reactor this 'multiplication factor' is very near unity.

The average energy of prompt neutrons accompanying fission of ^{235}U is 2 MeV. These are called 'fast neutrons.' The probability for capture on ^{235}U is much larger for slow neutrons, or thermal neutrons, which have energy $E_{therm} = 1/40$ eV. This energy is characteristic of the motion of molecules in a gas at room temperature. It also equals the vibrational energy of carbon atoms in graphite at room temperature.

B.2. Neutron Moderators

In order to increase the probability that a fission neutron subsequently triggers another fission, it is desirable to slow the fission neutrons down. This slowing down is accomplished by collisions with light weight nuclei in what is called a moderator. A simple design for a reactor is a three dimensional lattice of plugs of uranium metal, embedded in a volume filled with a moderator. The moderator is supposed to slow down neutrons from one plug before they reach neighboring plugs, thus improving the chances of subsequent fission. This was the original design of the world's first operating nuclear reactor CP-1, which went critical at Chicago in December, 1942.

Neutrons collide with nuclei in the moderator like billiard balls. If the moderator is hydrogen, which has the same mass as the neutron (to a fraction of 1%), it is possible to bring a neutron to a stop in one head-on collision. For a nucleus of mass number A, a neutron of energy E can lose in one collision $E \rightarrow E_{min}$, where:

$$E_{min} = ((A-1)/(A+1))^2 \times E. \tag{B.3}$$

The coefficient of E vanishes for $A = 1$. For carbon, $A = 12$, the coefficient is 0.716, so the maximum decrease per collision in carbon is about 30%. The average logarithmic decrement (natural logarithm, to the base e=2.71828) of the energy loss $E \rightarrow E'$ is:

$$\mu = \,<\ln (E/E')> \,= 2/(A + 2/3) = 0.1583. \tag{B.4}$$

To go from 1 MeV to 1 eV, $\ln (1,000,000) = 13.8$, and $13.8/0.158 = 87$ collisions.[4] To drop from 1 MeV to thermal energy 1/40 eV requires some 110 collisions. The mean free path for neutron collisions in carbon is $\lambda \approx$ 2.6 cm. Fermi wrote a formula for the mean distance traveled for N collisions as:

---

[3] The formula $\mu = 2/(A+2/3)$, where $A = 12$ for carbon, is an approximation. See J.R. Lamarsh, "Introduction to Nuclear Reactor Theory," Addison-Wesley, Reading, Mass, 1966. p 176.

[4] Production and absorption of slow neutrons by carbon, H. Anderson and E. Fermi "Collected papers of Enrico Fermi," Vol II, p 32. The treatment presented here is simplified.

$$\langle r^2 \rangle = 2 \times \lambda^2 \times N. \tag{B.5}$$

For a statistical random walk, r is proportional to $N^{1/2}$. For 100 collisions the mean distance is 36 cm. These numbers are relevant for the design of the carbon plus uranium lattice reactor. Hydrogen is the most efficient moderator, but is a gas at standard temperature and pressure, and not too convenient to handle. One could use water, H_2O, and indeed commercial reactors do use a light water moderator.[5] There is the disadvantage that hydrogen captures a neutron to form deuterium and emits a 2.2 MeV γ ray. In the early days, when no one was certain that anything was going to work, this disadvantage was considered fatal. In fact, a light water reactor requires enriched uranium to overcome the neutron loss to capture, and achieve criticality, while either heavy water or graphite can be used with natural uranium. Enrichment of uranium to get a reactor going would have been a significant delay. Deuterium is almost as good as hydrogen for a moderator – $E_{min} = 1/9 \times E$. And deuterium can capture a neutron to form tritium, but this has a lower probability than neutron capture on hydrogen. Knowing this, the German nuclear research project obtained heavy water from Norway. Some discussion of the industrial production of heavy water by electrolysis is presented in Chapter 7. The commando raids to disrupt the shipment of heavy water to Germany are the subjects of a Nova program.[6] There are modern power reactors that use heavy water as a moderator.

The production of heavy water D_2O was also eagerly pursued by the Soviet Atomic Project, because the use of deuterium could significantly decrease the physical size of a reactor. Evaluating Eq. (B.4) for A = 2 gives $\mu = \frac{3}{4}$, and to drop the neutron energy from 1 MeV to thermal takes about 25 collisions – roughly ¼ as many as with carbon. The mean free path in heavy water is not that different from carbon, $\lambda = 2$ cm, and the mean neutron path length to thermalize from Eq. (B.5) is $(\langle r^2 \rangle)^{1/2} = \lambda \times (2N)^{1/2} = $ 14 cm, about 40% of the corresponding number for carbon. The volume

---

[5] Hyperphysics.phy-astr.gsu.edu/hbase/nucene/ligwat.html. The water is typically pressurized so that it can be heated above 100^0 C without boiling.

[6] www.pbs.org/wgbh/nova/hydro/water.html

of the reactor scales roughly as the cube of this dimension; carbon moderated power reactors have about 1000 tons of carbon, while similar power levels can be achieved with a heavy water moderator of only 100 tons. Since either one would work with natural uranium, the more compact scale of heavy water was very appealing.

Graphite (carbon) was chosen as a moderator in the first reactors because it was reasonably effective; it did not capture neutrons; and it was cheap and plentiful, although it did have to be purified from commercial grade to be used in the reactor. Impurities in the graphite did capture neutrons, and therefore had to be removed.

We are now ready to look at Fermi's design for the world's first controlled chain reacting 'pile' CP-1[7] Carbon was chosen as the moderator, for the reasons given above. There were no digital computers, so every aspect of the design had to be tested in experimental mock-ups before the real pile was constructed. The same was true for the nuclear weapons, until the hydrogen bomb, which was one of the driving motivations for the development of digital computers.

B.3. Reactor Tests – Graphite Prisms

Neutron flux is not easy to measure. Thermal neutrons bounce around in all directions. Fast neutrons can be collimated, but not easily focused or deflected like charged particles. Neutron sources based on powder mixtures of α emitters like radium or polonium with beryllium (Ra-Be) or (Po-Be) were calibrated in the early days of neutron physics. The reaction produces fast neutrons: $^4\text{He} + {}^9\text{Be} \rightarrow {}^{12}\text{C} + \text{n} + 5.7\,\text{MeV}$. A moderator such as graphite, water, or polyethylene can be used to thermalize the neutrons.

The test setups used a radium-beryllium neutron source at the bottom of a rectangular stack of graphite blocks, with neutron absorbing foils inserted in various places to measure the neutron flux in the steady state.

---

[7] Experimental production of a divergent chain reaction, Am. Journal of Phys., **20**, 536 (1952). Reprinted in "Collected Papers of Enrico Fermi," Vol II p 272.

Then samples of uranium metal were inserted, to see what happened. A design for a graphite moderated natural uranium reactor was obtained from several experimental studies of this type. The reactor would be a large assembly, so the objective of smaller test setups was to confirm that the final design would work before attempting to build it. A parameter k_{inf} called the reproduction factor, is defined as the average number of new fast neutrons produced in an infinite lattice by one original fast neutron in the first generation.[8] The original neutron is thermalized by the moderator, and captured by uranium, to give more neutrons for the chain reaction. Starting with one neutron, one has a sequence of neutrons after n generations:

$$N(n) = 1 + k_{inf} + k_{inf}^2 + k_{inf}^3 + \ldots + k_{inf}^n \qquad (B.6)$$

As $n \rightarrow \infty$, this series diverges for $k_{inf} > 1$, corresponding to super criticality. If $k_{inf} < 1$, the series sums to $N = 1/(1 - k_{inf})$, which is sub critical. A real reactor, because of leakage and other effects, will have $k < k_{inf}$. In a test setup the multiplication factor will be even smaller. One has to rely on calculations to relate the neutron flux in a given finite sized test to k_{inf}. The reproduction factor can be expressed as a product of three quantities:

$$k_{inf} = \eta_T \times p \times f. \qquad (B.7)$$

Here η_T is the average number of fission neutrons emitted per thermal neutron absorbed, p is the fraction of neutrons not absorbed in the thermalization process, and f is the probability that a thermal neutron is absorbed by uranium.[9] The loss through the factor p can be due to capture on carbon, or resonance capture on uranium. The dominant isotope ^{238}U has several sharp resonances in the capture cross section around 50 eV neutron energy, that is, as the neutrons are slowing down to thermal

---

8 "Collected Papers of Enrico Fermi," Vol 2, p 120.
9 "Collected Papers of Enrico Fermi," Vol 2, p 121.

energies.[10] This is one factor in the lattice design of a uranium-graphite reactor. If the uranium lumps are kept small, they present a small target for neutrons that pass through the resonance energies enroute to thermalization.

Fermi, Anderson, and Szilard did graphite prism experiments at Columbia University, before moving to Chicago in 1942. Ivan Zhezherun describes similar experiments carried out near Moscow at Laboratory #2 in 1945.[11] Two such setups are shown in Figures B.1 and B.2. Figure B.1 is a horizontal prism to test graphite purity. The entire block, 310 cm long, is graphite. On the right side, in blue in the Figure, is the section under test. The rest is 'good' graphite. The BF_3 neutron counter (described below) is in the center. A Ra-Be neutron source is placed to the right for, say, one hour, to measure the absorption of neutrons by the test sample, and then moved to the left for one hour for comparison. If the test graphite absorbs more neutrons than the standard, it does not go into the reactor. In this way 600 metric tons of pure graphite blocks, 10cm×10cm×60cm were selected for the first reactor, F-1, at Laboratory #2 in the USSR. That equals about 60,000 blocks, hand stacked, in the test prism and then in the reactor itself.

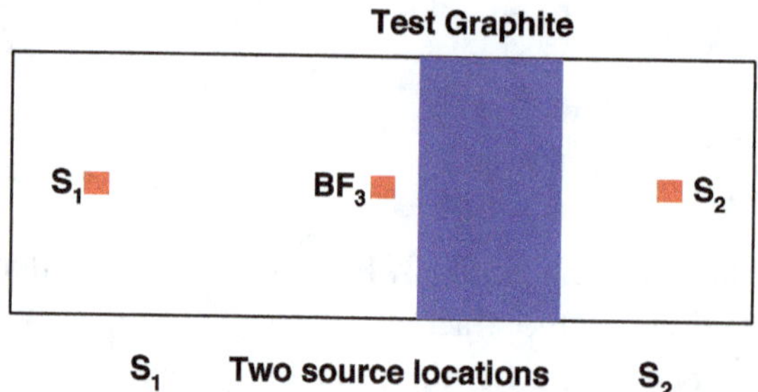

Figure B.1. Horizontal graphite prism to test graphite purity. The figure is to scale, 310 cm long, 120X120 cm cross section. The Ra-Be neutron source is moved from S_2 to S_1.

---

[10] John R. Lamarsh, "Introduction to Nuclear Reactor Theory," *ibid.*, p 38 Fig 2-12.
[11] I.F. Zhezherun, HISAP96 Vol 2, *ibid.*, p 68. Discussed in Chapter 7 of the main text.

Figure B.2 shows a vertical arrangement to test the purity of uranium. Thirty-six holes in the graphite, spaced in a 20cm×20cm square array, were filled with aluminum clad uranium slugs 10 cm long and 3.2 cm in diameter. Each hole held 12 slugs, giving 12×36 = 432 slugs altogether, about 650 kg. A Ra-Be neutron source was placed above the uranium array, and a BF_3 counter below. The uranium acted as an amplifier for the neutron flux. The ratio of counting rates with and without the uranium was a measure of the quality of the uranium sample. Quantitative measurements and theoretical calculations had confirmed that this ratio had to exceed 1.3 for the reactor to work.

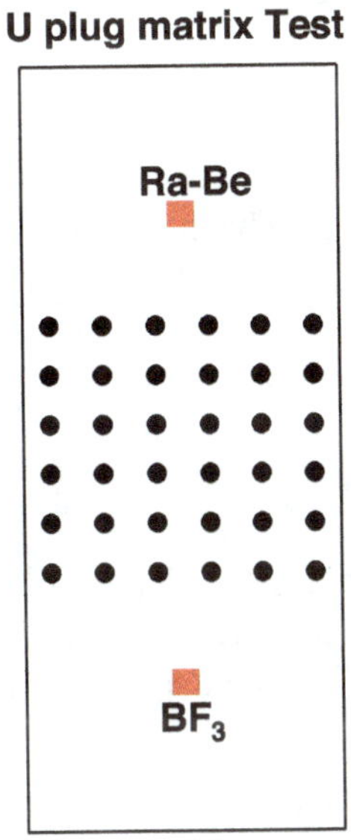

U plug matrix Test

Figure B.2. Vertical graphite prism from Laboratory #2 of the USSR Academy of Sciences to monitor the quality of uranium to go into the F-1 reactor. The prism was covered with a cadmium sheet to keep thermal neutrons from escaping.

Figure B.2 can be reconfigured to form a miniature reactor by arranging the uranium slugs in a three-dimensional matrix in space rather than in rods going all the way through, and removing the Ra-Be neutron source. Such a prism was called an 'exponential pile' by Fermi, because the neutron flux is exponentially damped as you move longitudinally away from the lattice, and the damping factor is related to k_{inf}, allowing one to determine if a given lattice design will become critical as the dimensions go to infinity.[12]

Interleaving descriptions of the Manhattan Project reactor design with the Soviet Atomic Project makes sense, since the two groups did much the same things. The prism test setups were very similar for reasons which were obvious to the experimenters. The two reactors, CP-1 in Chicago and F-1 at Laboratory #2, were virtually identical, partly because the Soviets had detailed knowledge of the Chicago work. Perhaps the most important fact known to Kurchatov and Zhezherun in 1945 was that the reactor would work. So we now return to the Chicago story in 1942. The Soviet work was three years later.

B.4. Fermi's CP-1 Reactor

It was decided to construct the Chicago pile in a sphere made out of graphite blocks, some with holes to contain lumps of uranium. The sphere wound up looking more like a rugby football, with polar radius 309 cm and equatorial radius 388 cm. The uranium slugs were placed at the corners of a square lattice, about 21 cm apart. This is about ½ of the distance needed to thermalize the neutrons. There was no effort made to enrich the amount of ^{235}U, so it was present in its natural abundance of 0.72%. The total amount of graphite in the pile was 350 metric tons, and the total amount of uranium metal plus uranium oxide was 42 metric

---

[12] "Collected Papers of Enrico Fermi," Vol 2, p 126. The analysis of an exponential pile is fairly complicated. It is a homework problem in Lamarsh (problem 9-23).

tons.[13] There were no blueprints. Every day during construction the plans were set for the next day. The graphite came in 11cm × 11 cm bars, 50 cm long, which had to be machined and drilled into smooth surfaced rectangular blocks 10×10×41.25 cm. The machining was done with woodworking equipment, and must have been a big mess. The assembly took three weeks of two a day shifts. A large rubber balloon shaped like a cube was purchased to contain the pile, which could be sealed and pumped out if it turned out to be necessary to get rid of the air, which it did not.[14] But the pile was assembled inside this balloon. There was no provision to cool the pile, so it could not be operated at high power for very long.

Control rods are an important part of reactor design, and the first reactor was no exception. Channels were placed horizontally – and one vertically – in the pile to accommodate the control rods. These were cadmium sheets tacked to wood strips. Cadmium absorbs thermal neutrons with high probability, so a sheet 1 mm thick would take the incident flux down to zero.[15] In this way the neutron activity in the pile could be quenched. Three of the horizontal rods and the one vertical rod were electrically controlled by the neutron intensity, so that they could be inserted in less than one second if the radioactivity became too high. These were called 'zip' rods. The other control rods were operated by hand.

Neutron intensities were monitored by several different techniques. Cylindrical proportional counters filled with BF_3 gas use the capture reaction: $^{10}B + n \rightarrow ^{11}B \rightarrow ^{4}He + ^{7}Li$. The α particles are counted as they ionize the gas. The capture reaction liberates about 2.3 MeV, and is sensitive to thermal neutrons. To count fast neutrons, the counter is inserted in a cylindrical shell moderator, like paraffin. In a graphite matrix the moderation occurs naturally. These counters are still in use for monitoring neutron intensities.

---

13 Only 273 kg of ^{235}U. The amount of uranium can be reduced by enrichment, which is done in modern power reactors. Note the similarity of the first reactors at Chicago and Laboratory #2 of the USSR Academy of Sciences. F-1 also matches the Hanford 305 test reactor.

14 Nitrogen, which is 80% of the atmosphere, was known to absorb neutrons.

15 To be precise, 1 mm thick cadmium reduces the thermal neutron flux by $e^{-10} = 0.000045$.

Indium foils were used to record the thermal neutron fluxes at various locations inside the pile. The dominant naturally occurring isotope is ^{115}In, which captures a neutron to form ^{116}In, which in turn β decays to tin: ^{116}In $\rightarrow$ ^{116}Sn $+ \beta + \nu + 3$ MeV. There are three states of ^{116}In which β decay with half-lives of 2 sec, 54 min, and 14 sec. The foils were inserted into slots in the pile, allowed to stay put for some period of time – like two hours – and then extracted and taken to a Geiger counter to record the induced radioactivity. The measured decay rates depend on the thermal neutron flux, and the time spent in the reactor.[16] A great deal of effort was spent in the absolute calibration of this technique. ^{115}In has a strong absorption resonance at 1.45 eV, above the thermal energy, and also strongly absorbs thermal neutrons (1/40 eV). Cadmium strongly absorbs thermal neutrons. Thus the difference in activities in a pure indium foil and a foil sandwiched between two cadmium sheets gives the thermal flux. The cadmium-indium sandwich is sensitive only to the 1.45 eV neutrons, which are a measure of the slowing down process. Thermal neutrons are everywhere, and you don't know their history, so in studying the slowing by a moderator, the 1.45 eV absorption peak is useful.[17]

Nuclear reactors are now well understood. The theory of operation is treated in introductory texts in nuclear engineering. There are several important components to the analysis:

1. The fissionable material, here ^{235}U, which emits fast neutrons and absorbs thermal neutrons. The absorption cross section $\sigma_a = 678$ barns.[18] The neutron multiplication factor $\eta = 2.06$, and the average fast neutron kinetic energy $E = 2$ MeV.[19]

---

[16] The formula for the number of counts recorded is $N(t) = R \times (\exp(-\lambda(t - t_0)) - \exp(-\lambda t))$, where R is proportional to the thermal neutron flux, $\lambda = 0.693/\tau$, τ is the half-life of ^{116}In. The foil is inserted at $t = 0$, and removed at $t = t_0$. Then counts are recorded for $t > t_0$. If $\lambda t_0 \gg 1$, the second term is zero, and the foil is 'saturated.'

[17] J.R. Lamarsh, *ibid.*, p 201.

[18] The cross section has units of area, and is defined as the transition rate divided by the flux. 1 barn $= 10^{-24}$ cm^2.

[19] These numbers come from Lamarsh, *ibid.*, Tables in Chap. 8 and Appendix A.

2. The moderator, here graphite, which has an absorption cross
    section $\sigma_a = 0.0034$ barns. Graphite requires about 110 collisions
    to thermalize a 1 MeV neutron.
3. An arrangement – the 'pile' – which is predominantly moderator,
    with a small volume of fissionable material distributed in a
    uniform pattern.
4. There is a parameter called the 'Fermi age' for thermalizing fast
    neutrons, which has units of area (not time), but is monotonically
    increasing with time. For graphite the age $\tau = 368$ cm^2.
5. After thermalizing, the neutrons have a thermal diffusion length
    in graphite, which is $L_T = 59$ cm.
6. There is a length parameter for the pile, which comes from
    boundary conditions on the differential equations. For a sphere
    of radius r, $B = \pi/r$.

Substituting B, η, τ, and L_T in Eq. 9-106 in Lamarsh gives the
parameter he calls $Z = 1.28$. Then Eq. 9-107 gives a critical mass of 180
kg of ^{235}U for the Chicago pile under ideal conditions. The actual pile had
273 kg of ^{235}U, so it should have worked, and it did.

B.5. Neutron Velocity Selectors

Mechanical selectors based on neutron time of flight were used in the
early days with Po-Be or Ra-Be radioactive sources of neutrons, and are
also used today at research reactors, which supply copious beams of
neutrons. The idea is simple enough. It takes a finite time for a neutron to
travel a distance L, given in terms of the neutron velocity as

$$t = L/v. \tag{B.8}$$

Now suppose the neutron travels in a straight line down a channel of length
L, which points off axis when the neutron enters, but continuously moves
so that the neutron remains in the center, and exits the channel at time t
given by Eq. (B.8). The channel walls are made of a material which
strongly absorbs the neutrons, like cadmium, so that neutrons which are

not synchronous with the sideways movement of the channel are lost. Figure B.3 shows a simplified drawing of the idea.

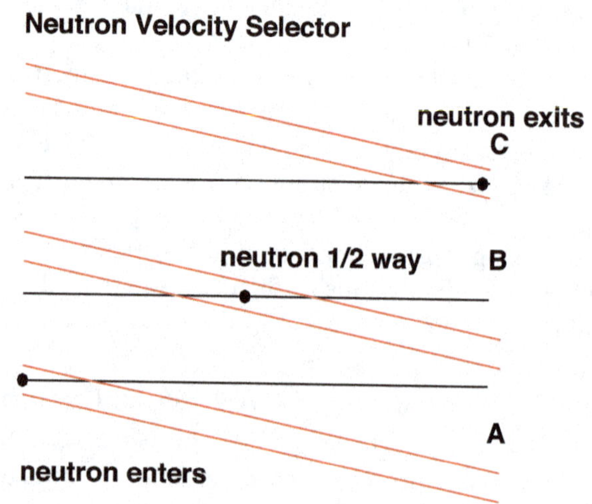

Figure B.3. A neutron synchronous with the lateral motion of the channel enters at A and exits at C. Time increases from bottom to top. The walls of the channel are made of a material which strongly absorbs slow neutrons.

The actual device can be made in several ways. Fermi and the Marshalls used a small (5 cm dia) cylinder that rotated about its axis, and contained a sandwich of cadmium and aluminum foils.[20] The neutrons were blocked by the cadmium, but passed by the aluminum. The stack of foils was parallel to the axis of rotation, so that once every 180^0 the stack was horizontal, and the neutron beam could enter. A sketch of the geometry is shown in Figure B.4. A light beam reflected off a mirror attached to the rotating cylinder, and detected by a photocell was used to count rotations, and to furnish a gate for the detector, so that it was only sensitive when the neutrons could pass through. The angular aperture was

---

[20] E. Fermi, J. Marshall, and L. Marshall, Phys Rev **72**, 193 (1947). See also T. Brill and H.V. Lichtenberger, Phys Rev **72**, 585 (1947).

about one degree vertically, in the direction of rotation. Only neutrons of the correct velocity would pass through the aluminum unscathed. The average thermal velocity is 2200 m/sec, which will pass through the diameter in t = 0.05/2200 = 23 μ sec. The angular velocity of rotation of the pipe required to move the sandwich one degree (1/57 of a radian) in 23 μ sec is: $\omega = 10^6/23 \times 57 = 763$ radians/sec. This translates into a rotation frequency of f = 7300 rpm. The maximum rotation frequency was 15000 rpm, so 2200 m/sec was well within the capability of the velocity selector. In fact, it could resolve a velocity of 4500 m/sec, or a kinetic energy of about 0.1 eV. This arrangement was used after the thermal column of the graphite reactor at Argonne National Laboratory. The neutron flux from a radioactive source would be too low to give reasonable intensities at the output of this rather small velocity selector.

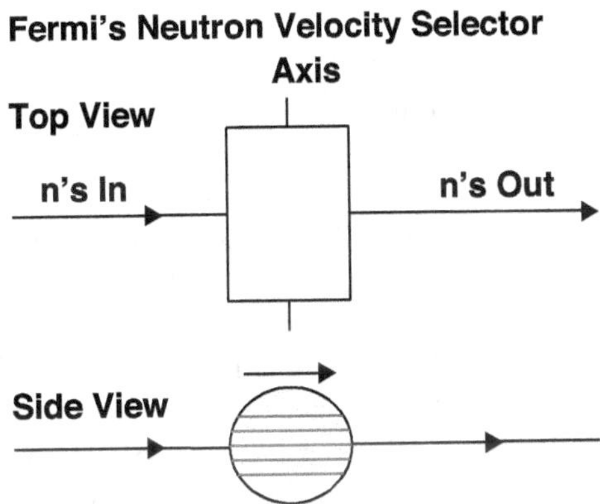

Figure B.4. Schematic of the velocity selector of Reference 14. A pulley not shown connects the cylinder to an electric motor. It rotates about its axis in the sense of the arrow. Slow neutrons from a reactor enter from the left, and only those with the synchronous velocity exit on the right. The cadmium sheets are shown in red (not to scale). Aluminum sheets fill the gaps.

There are other geometries – rotating turbine blades, or a system of toothed wheels, staggered to create a pitch, all rotating together. They can be made longer, one meter or so. But they all are designed for slow neutrons, with kinetic energies in the 0.1 eV range or less.

It is worth mentioning that an important application of neutrons with velocities in the 2200 m/sec range comes from the de Broglie wave length:

$$\lambda = h/p. \tag{B.9}$$

where h is Planck's constant, and p is the neutron momentum. In terms of kinetic energy $p = (2mE)^{1/2}$. It is convenient to work in eV units, so multiply and divide Eq. (B.9) by the speed of light c, to obtain:

$$\lambda = hc/(2mc^2E)^{1/2}. \tag{B.10}$$

Now $hc = 1240$ eV-nm, and $mc^2 = 940 \times 10^6$ eV, so for $E = 0.025$ eV (v = 2200 m/sec) a little arithmetic gives:

$$\lambda = 0.18 \text{ nm de Broglie wave length for thermal neutrons.} \tag{B.11}$$

This length is the same size as typical lattice distances in crystals, or atomic separations in molecules. For this reason, thermal neutrons are widely used in studies of the structure of matter. A 7 keV x-ray has the same wavelength, but it is sometimes advantageous to use neutrons rather than x-rays to study crystals. It may seem peculiar that slow neutrons behave just like x-rays, but they do. There are some differences in the diffraction because x-rays are scattered by the electrons which are bound to the nuclei in the material, while neutrons are scattered by the nuclei themselves.

Electronic time of flight from the pulsed beam of a cyclotron can handle higher energy neutrons if the timing resolution of the system is adequate. A 0.5 MeV neutron travels 10 meters in 1 μ sec (0.033 × speed of light). A flight path (in a vacuum pipe) L = 50 m gives a flight time of

5 μ sec. If the time measurement uncertainty is $\Delta t = 20$ nsec, one could measure the neutron energy near 0.5 MeV to an accuracy $\Delta E = 4$ keV. The ratio $\Delta E/E = (2c/L) \times (2E/mc^2)^{1/2} \times \Delta t$, so that it decreases (improves) as the neutron kinetic energy gets smaller. An overall timing uncertainty of 20 nsec is certainly achievable with modern electronics, but would have been difficult if not impossible before WWII. The proton or deuteron beam of a cyclotron can be deflected by pulsed electric plates into a metal target to create neutrons at $t = 0$. A cyclotron also produces a more intense neutron beam than is possible from a radioactive source. Time of flight systems have also been built at nuclear reactors, by inserting a 'chopper' in the beam pipe.

B.6. Plutonium Production Reactors

The fissile isotope ^{239}Pu has too short a half-life to occur in nature. It α decays to ^{235}U with a half-life of 24000 years. So plutonium has to be man-made. A high power nuclear reactor makes an intense bath of thermal neutrons. Neutron capture on the dominant naturally occurring isotope ^{238}U produces ^{239}Pu after two β decays:

$$^{238}\text{U} + \text{n} \rightarrow {}^{239}\text{U} \rightarrow {}^{239}\text{Np} + \beta^- + \nu; \; {}^{239}\text{Np} \rightarrow {}^{239}\text{Pu} + \beta^- + \nu. \tag{B.12}$$

The uranium β decay half-life is 23 minutes, and the neptunium β decay half-life is 2.35 days, so after the reactor starts up the amount of ^{239}Pu increases linearly with time.

Igor Kurchatov is quoted in Chapter 12 as having said that 100 Mw natural uranium reactor would make 100 g of plutonium per day. To check this number, we need three quantities in addition to the reactor power. First is the energy release per fission, $E_{fiss} = 180$ MeV $= 2.9 \times 10^{-11}$ joules; second is the cross section ratio $\sigma_{cap}(\text{n}+^{238}\text{U} \rightarrow {}^{239}\text{U})/\sigma_{fission}(\text{n}+^{235}\text{U} \rightarrow \text{fission}) = 2.68/582 = 0.0046;$[21] and finally there is the natural isotopic abundance

---

[21] www-nds.iaea.org gives a table of best values for thermal neutron cross sections.

ratio $^{238}U/^{235}U = 143$. The number of plutonium atoms produced per second is then:

$$dn/dt = ((\text{reactor power}/ E_{fiss}) \times \sigma_c/\sigma_f) \times (^{238}U/^{235}U) = 2.2 \times 10^{18} \text{ Pu atoms/sec.} \tag{B.13}$$

Since each atom weighs 4×10^{-22} g, this amounts to a production rate of about 1 mg/sec, or 86 g/day. This calculation assumes that the same spectrum of thermal neutrons is responsible for both reactions, so that the neutron flux cancels out. The amount of uranium does not appear either, nor does the volume of the reactor – only its power level. The reactor burns about 120 gm of ^{235}U per day running at 100 Mw, so if its charge is 100 tons of natural uranium, or 700 kg of ^{235}U, it would run out of fuel in 16 years. Every atom of ^{235}U burned produces 2/3 of an atom of ^{239}Pu. As pointed out in Chapter 12, the power level in a smoothly running reactor can be calculated from the temperature rise in a known water flow rate. Kurchatov's number of 100 g/day could come from different average cross sections, or a more elaborate calculation by folding the neutron energy distribution at the uranium target into energy dependent cross sections. On the other hand, perhaps Kurchatov rounded the number up to make it easy to remember (100 MW = 100 g/day). In any event, Eq. (B.13) is a simple way to calculate plutonium production from a uranium reactor. If enriched uranium is used, the factor 143 should be adjusted downward accordingly.

Appendix C

Isotope Separation

C.1. Isotope Separation

Isotopes by definition have the same chemical properties, but different masses, because the nuclei have different numbers of neutrons. Isotopes must be separated by mass, which is a physics problem rather than a chemistry problem. There are many practical applications. For example, specific isotopes are used for medical purposes, often extracted from fission products.

It is almost true that chemistry depends on the number of valence electrons, and not on the mass of the nucleus. However, there are some small isotopic mass effects. This is particularly true for deuterium, which weighs twice as much as ordinary hydrogen. The ionization potential for deuterium is 0.03% greater than for hydrogen. This results in a shift in the visible spectrum (the Balmer series) in hydrogen, which Harold Urey and co-workers used to discover deuterium.[1] Liquid deuterium has a slightly higher boiling point than liquid hydrogen. Heavy water is not biologically active – it won't satisfy your thirst.[2] But such effects are amplified for ^{1}H vs ^{2}H, because the factor of two mass difference holds for no other pair of stable isotopes.

There are spectral line shifts even for uranium. They are very tiny, but lasers can be sharply tuned, so that it is possible with laser light to ionize ^{235}U and not ^{238}U, thus achieving separation with an electric field. But

---

[1]   H. Urey, F. Brickwedde, and G. Murphy, Phys Rev **39**, 164 (1932).

[2]   en.wikipedia.org/wiki/Heavy_water

611

lasers did not exist in 1942, and even today laser isotope separation, although proven effective, is not used for production of large amounts of ^{235}U.[3]

So we return to the problem facing the pioneer uranium people, who had to obtain kilogram amounts of ^{235}U. For each kilogram, you have to process 143 kg of natural uranium, if your extraction procedure is 100% efficient.

Uranium metal is obtained from uranium ore, which may contain 0.2% or more of uraninite, UO_2, also called pitchblende.[4] Several chemical reactions produce yellowcake, U_3O_8, which can be converted with hydrofluoric acid to UF_6, uranium hexafluoride. This substance exists in three phases, solid, liquid, and gas, at temperatures and pressures that are not extreme. Isotope separation by gaseous diffusion uses the gas phase of UF_6. It is fortunate that ^{19}F is the only stable isotope of fluorine.

C.2. Thermodynamic Considerations

Thermodynamics is a very general subject, treating macroscopic bodies with parameters like temperature, pressure, volume, and heat flow, with little or no attention to the details of what is really going on. It therefore has the power of producing general theorems based on conservation laws. No known process violates the conservation of energy. On the other hand, the devil is often in the details, which happens to be the case for isotope separation. Rudolf Peierls and Klaus Fuchs applied thermodynamics to isotope separation in a paper that was declassified in the 1990's.[5]

---

[3] llnl.gov/str/Hargrove.html (Lawrence Livermore National Laboratory)

[4] web.ead.anl.gov/uranium/guide

[5] "Selected Scientific Papers of Sir Rudolf Peierls," R.H. Dalitz and Sir Rudolf Peierls, editors, World Scientific, Singapore, 1997, p 303.

Mixing of two non-reactive gases is a common problem in thermodynamics. Suppose you start out with two gases – neon and argon for example – in separate volumes V_1 and V_2 shown in Figure C.1. The gases are each assumed to obey the perfect gas law:

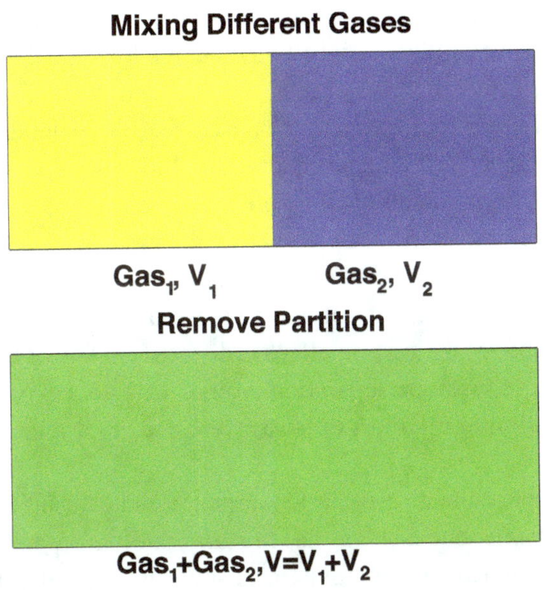

Figure C.1. Box containing two different gases at the same temperature and pressure. The partition is removed, and the gases mix, increasing the entropy of the system.

$$pV = nRT. \qquad (C.1)$$

where p is the pressure in pascals, V the volume in m^3, T the absolute temperature in Kelvin, R is the gas constant (R = 8.31 joules/K-mole), and n is the number of moles of the gas. One mole of a gas contains Avogadro's number $N = 6 \times 10^{23}$ atoms, and at 'standard temperature and pressure' (300 K and 1 atmosphere of pressure = 101.3 kilopascals) occupies a volume of 22.4 liters. If the two different gases are at the same temperature and pressure, the ratio of the number of moles must equal the volume ratio:

$$n_1/n_2 = V_1/V_2, \tag{C.2}$$

If the volumes are equal, so are the numbers of moles, or the argon, with molecular weight 20, weighs twice as much as the neon, molecular weight 10. Imagine that the neon gas is yellow, and the argon is blue. Then when the partition is removed and the gases mix, the volume $V = V_1 + V_2$ is filled with green gas. Isotope separation amounts to parting the green gas into its two constituents, putting all of the yellow gas in one box, and the blue in the other, as was the case before the gases were allowed to mix. For argon and neon this is not too hard, because the gases have very different boiling points, but for the two isotopes of UF_6 it is quite a challenge because of the very small mass difference. Any of the techniques discussed here for isotope separation would work to separate argon and neon – in fact similar mixtures of gases were used to test the procedures.

The mixing occurs at constant temperature and pressure, and the total volume is the same, but each gas has the larger volume V to move around in. What has changed? Thermodynamics says that the order has changed. The initial state was ordered, and the mixed state is disordered, and the mixing is, in the thermodynamic sense, irreversible. The variable associated with order is called the entropy – S. The entropy of the system is increased by the mixing. The change in entropy at constant temperature is defined as:

$$dS = p \times dV/T = nR \times dV/V. \tag{C.3}$$

where the perfect gas law was used to obtain the second formula. Integration gives a logarithm of the volume ratio:

$$\Delta S_1 = n_1 R \times \ln(V/V_1). \tag{C.4}$$

Since V is larger than V_1, the entropy increases. The formula for ΔS_2 is the same, with 2 replacing 1. So the total increase in entropy is the sum:

$$\Delta S = R(n_1 \times \ln(V/V_1) + n_2 \times \ln(V/V_2)). \tag{C.5}$$

If the initial volumes are equal, and there is one mole of each gas, then we get the simple formula $\Delta S = 2R \times \ln(2)$. Peierls and Fuchs use formulas like C.5 to estimate the change in entropy and the associated work that has to be done to separate the mixed isotopes of uranium. For the real problem, the initial volume of ^{238}U would be 143 times as large as that for ^{235}U. If $^{238}UF_6$ were yellow, and $^{235}UF_6$ blue, the mixture would be pure yellow! The ease of separation depends on the mass difference, $\Delta M/M$, a factor not present in Eq. (C.5). For UF_6, $\Delta M/M = 0.0086$, a small number, complicating the problem further. Peierls and Fuchs were able to estimate some relevant parameters, like the total area of the barrier needed in gaseous diffusion to separate 1 mole (235 gms) of ^{235}U at 90% purity in one day – about 4000 m^2.

C.3. Gaseous Diffusion

In thermal equilibrium at absolute temperature T (in Kelvin) the average kinetic energy of molecules in a gas is given by the formula:

$$E = \tfrac{1}{2} Mv^2 = 3/2 \, k \, T. \tag{C.6}$$

where M is the mass of the molecule in kg, v its velocity in m/sec, and k is Boltzmann's constant $k = 1.38 \times 10^{-23}$ joules/K. For two molecular masses in thermal equilibrium the velocity ratio therefore depends on the square root of the mass ratio, thus:

$$v_1/v_2 = (M_2/M_1)^{1/2}. \tag{C.7}$$

For the gas of interest, uranium hexafluoride this ratio is:

$$v_1/v_2 = ((238+6\times19)/(235+6\times19))^{1/2} = 1.0043. \qquad (C.8)$$

The lighter isotope is 0.43% faster moving through a surface in space.

Now suppose there is a barrier with small holes which separates the gas from a vacuum, as shown schematically in Figure C.2. One half of the gas on the high pressure side is allowed to flow to the low pressure side.[6] Then the enriched gas at low pressure is re-pressurized and sent to the second stage. The depleted gas from the second stage is also re-pressurized and fed back into the first stage, and the depleted gas from the third stage is fed back into the second stage. Each compression heats the gas, which has to be cooled to the operating temperature. Since the enrichment per stage is small, this technique mixes gases with the same fraction of the rare isotope. The first stage depleted gas is removed to storage for future use. In this toy model, the amount of gas reaching the output of the third stage is described by an infinite series. This is what is meant by the requirement that the system has to 'settle in.' Let M_{in} be the mass of gas input to the system, and M_{out} be the output of stage 3. The series then is:

$$M_{out} = M_{in} \times (1/8 + 1/16 + 1/32 + 1/64 + \dots) = M_{in}/4. \qquad (C.9)$$

---

[6] If it all went through, there would be no enrichment.

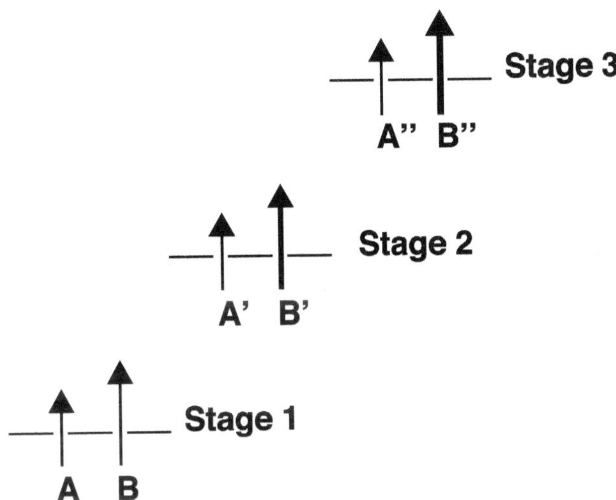

Figure C.2. Toy model of 3 stages of gaseous diffusion. A is the slower (heavier) isotope, and B is the faster (lighter) one. The lengths of the arrows are exaggerated relative speeds. The thickness of the B arrow increases with the stage, indicating enrichment of that isotope.

The series is $1/8 \times (1/(1 - x))$, where $x = \frac{1}{2}$. It converges rapidly. This is the total mass, so apart from the enrichment factor the output rare component is only $\frac{1}{4}$ of the input amount. In a real system with several thousand stages this fraction is much smaller. Now suppose each stage enriches the rare isotope by a factor $1 + \varepsilon$, where we take $\varepsilon = 0.004$, a small number. Then if m_{in} is the mass of the rare isotope at the input, and m_{out} is the mass of the rare isotope at the output, the two are again related by an infinite series:

$$m_{out} = m_{in} \times (1 + \varepsilon)^3 \times (1/8 + (1 + \varepsilon)/16 + (1 + \varepsilon)^2/32 + (1 + \varepsilon)^3/64 + ...). \quad (C.10)$$

Every time the gas goes through a barrier, it picks up a factor of $(1 + \varepsilon)$. This series is the same as Eq. (C.4), with $x = (1 + \varepsilon)/2$, which can be summed for small ε as:

$$m_{out} = m_{in} \times (1 + \varepsilon)^4/4. \qquad (C.11)$$

So by feeding ½ the gas through again and again, you gain an overall extra factor of $(1 + \varepsilon)$ in enrichment, and a factor of two in total mass. For $\varepsilon = 0.004$,

$$(1 + \varepsilon)^4 = 1.016. \qquad (C.12)$$

The fraction of the rare isotope is increased by 1.6%! An increase in the fraction from the naturally occurring value of ^{235}U of 0.7% to 50% would require $(1 + \varepsilon)^N$, where $N = 1069$. Hence, we anticipate that a very large number of stages will be required to achieve substantial enrichment of ^{235}U. In a real device, because of backward leakage and other degrading effects, ε is even smaller.[7]

The holes in the barrier should be smaller than the mean free path for a collision between two molecules of the gas, because collisions tend to erase differences in molecular velocity. The mass flow rate through a hole which is smaller than the mean free path λ for scattering of a molecule in the gas - $\lambda \approx 100$ nanometers for UF_6 – is given in texts on the kinetic theory of gases as[8]:

$$G = A \times (p \times \rho/2\pi)^{1/2} \text{ kg/sec.} \qquad (C.13)$$

---

[7] H.D. Smyth, "Atomic Energy for Military Purposes," *ibid*, p 174, quotes $\varepsilon = 0.003$, based partly on the fact that as more input gas flows through the barrier its rare gas fraction decreases.

[8] Leonard Loeb, "Kinetic Theory of Gases," McGraw Hill NY, 1934 p 302.

where A is the area in m^2, p is the pressure in pascals, and ρ is the density of the gas in kg/m^3. The gas is assumed to flow from pressure p into a vacuum.

UF$_6$ is a gas at atmospheric pressure (100 kPa), and 340 K, about 40 K above room temperature. Its density at these values of P and T is:

$$\rho = 12.5 \text{ kg/m}^3 = 2.1 \times 10^{25} \text{ molecules/m}^3. \tag{C.14}$$

This is ten times the density of air at STP. UF$_6$ is a very heavy gas. For a hole 10^{-8} m across, Eq. (C.13) predicts a mass flow of G = 4.5×10^{-14} kg/sec. This is very small, but it is only one hole! Suppose the barrier area is 0.3 m^2, and it is 10% holes. Then the number of holes is 3×10^{14}, and G = 13.5 kg/sec mass flow rate through the barrier, which is about 1 m^3 of the gas. UF$_6$ is 2/3 uranium by weight, so the flow rate is 9 kg of uranium per second. The flow rate itself should therefore not be a problem. It is the pumping, circulating, cooling, and the huge number of stages that complicate the process.

The material of the barrier, the fraction and size of the holes, are still classified after 75 years.[9] The holes cannot be very large, or else the gas would simply leak through, and there would be no pressure differential across the barrier. The holes cannot be too small either, or the UF$_6$ would be blocked. The right size is about 10 nanometers. The UF$_6$ molecule itself is less than 1 nanometer in diameter.

A gaseous diffusion plant for uranium isotope separation has several thousand stages, each one with a compressor to pressurize the uranium hexafluoride, and a heat exchanger to extract the heat caused by the increased pressure. Since the mass of gas decreases as it goes through the chain, the separator volume and the compressor capacity could also decrease, but a practical design has only a few specific sizes. Because UF$_6$

---

[9] They are redacted from the declassified document "Manhattan District History Book II – Gaseous Diffusion (K-25) Project" Vol **3** – Design, 1947, available on osti.gov

chemically reacts with water, a different liquid refrigerant was used.[10] Both functions require electricity. The K-25 plant was sited at Oak Ridge, Tennessee, because of the electric power sources supplied by the Tennessee Valley Authority. The total cost of the K-25 plant was about $250 million in 1945 dollars, or about 10% of the Manhattan Project cost. It was fully operational in early 1945. The analogous Soviet plant 70 km north of the city of Sverdlovsk on the east side of the Ural Mountains was in full production of enriched uranium in 1950.

In this simplified discussion, we have ignored problems of 'corrosion,' that is removal of uranium from the system due to chemical reactions on surfaces, in packings and bearings, in seals, and everywhere else. Corrosion is a serious problem because of the chemical reactivity of UF_6, and new techniques had to be developed to avoid it.

C.4. Thermal Diffusion

Isotope separation by thermal diffusion has no simple explanation; "neither its existence nor its sign can be derived from elementary considerations".[11] Nevertheless, it does work. It was suggested in the prescient paper by Frisch and Peierls[12] as a means of separating uranium isotopes, based on the work of a German physical chemist Klaus Clusius.[13]

A temperature gradient in a mixture of two gases can result in a gradient in the relative concentrations of the two constituents. In a vessel composed of two concentric cylinders, the inner one is heated to a high temperature, and the outer one is cooled. UF_6 gas is bled in between the cylinders. The lighter isotope ^{235}U moves towards the hot inner wall, and the heavier isotope ^{238}U moves towards the cold outer wall. Convection then moves the light isotope to the top, and the heavy one to the bottom, where they can be collected. The direction of motion relative to the

---

[10] P.C. Keith, "Chemical and Metallurgical Engineering," February, 1946, p 112.

[11] W.H.Furry, R.C. Jones, and L. Onsager, Phys Rev **55**, 1083 (1939).

[12] www.stanford.edu/class/history5n/FPmemo.pdf (the Frisch-Peierls memorandum of March, 1940.)

[13] K. Clusius and G. Dickel, Naturwiss. **26**, 546 (1938).

temperature gradient depends on the force of interaction between the molecules, and is hence not simple to derive.

The technique was used in the later stages of the Manhattan Project. Thermal diffusion towers were constructed near the gaseous diffusion plant, which served as a source of steam. They were used to enrich the gas fed into the gaseous diffusion plant. The electromagnetic separators also profited from output of thermal diffusion.

C.5. Electromagnetic Separation

Electromagnetic isotope separation is the easiest to understand conceptually, since it is based on the mass spectrograph, which has been widely used to study the isotopic composition of various substances – the amount of ^{14}C for dating of an archeological sample, for instance. It is based on the mass separation of ions which results from combined electric and magnetic fields. The substance is placed in an ion source, ionized, and accelerated by a voltage of several kV. The source is placed in a constant magnetic field. Figure 3.C. shows the setup. The accelerating voltage V gives an ion with charge q a kinetic energy E:

$$E = \tfrac{1}{2} M v^2 = q\, V. \tag{C.15}$$

The radius of curvature of the orbit perpendicular to a constant magnetic field is:

$$R = Mv\,/\,qB. \tag{C.16}$$

If Mv is in units kg × m/sec, q in coulombs, and B is Tesla, then R is in meters. Momentum in kg × m/sec is not convenient for calculations involving ions, because they have a very small mass. However, momentum can be measured in MeV/c. For q = e, the elementary charge

($e = 1.6 \times 10^{-19}$ coulombs), and $Mv = p$, the momentum, in MeV/c, a useful formula is:

$$R(\text{meters}) = p\ (\text{MeV/c})/(300 \times B\ (\text{tesla})). \qquad (C.17)$$

and the momentum in MeV/c can be calculated directly from the mass of the ion in MeV and the accelerating voltage in MV, thus:

$$p\ (\text{MeV/c}) = (2 \times \text{accelerating voltage in MV} \times \text{ion mass in MeV})^{1/2} \qquad (C.18)$$

The elementary charge and the speed of light are absorbed in the MeV units.[14] For example, suppose singly charged uranium ions are accelerated by 50 kV = 0.05 MV. Using Eq. (C.18), the momentum of the ions is 300 MeV/c. Eq. (C.17) then gives the radius of curvature in a 1.5 T magnet: R = 0.66 m. These units greatly simplify calculations involving ions.

If two isotopes of ionized uranium are accelerated by the same voltage, the ratio of their momenta is given by:

$$p_1 / p_2 = (M_1 / M_2)^{1/2} \qquad (C.19)$$

As in gaseous diffusion, the effect depends on the square root of the mass ratio.

For a 180^0 bend, the displacement of the two ion species is given by:

$$\Delta x = 2 \times \Delta R = 2 \times R \times \Delta p/p = 8\ \text{mm}. \qquad (C.20)$$

---

[14] For kinetic energy E in MeV, $cp = (2 \times Mc^2 \times E)^{1/2}$ in MeV, and p in MeV/c is the same number.

for the two isotopes of singly ionized uranium accelerated through 50 kV and deflected by a 1.5 T magnetic field. As mentioned in the text, this looks like a big effect, and should work great. The square root of the mass ratio is 1.006, but the combined electric (in the source) and magnetic (between the poles of the magnet) fields result in a macroscopic separation of the two isotopes – almost one centimeter. Imagine two cans placed 1 cm apart, the outer one for the heavier ^{238}U and the inner one for the lighter ^{235}U, after 180^0 bend in the magnet. Run the device for a while, and then turn the problem of extraction of the isotopes from their respective cans to the chemists, who know how to do that sort of thing.

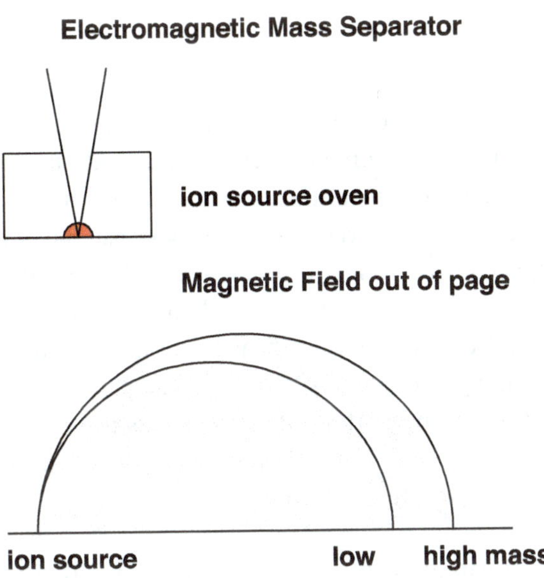

Figure C.3. Calutron Electromagnetic Mass Separator.

The problems arise when it is necessary to work with kilograms of material. Ion sources must be designed to handle large ion currents, and space charge effects inside the magnet volume must be controlled to prevent the separation and focusing of the beams from being destroyed.[15] Ref. (15) shows that the space charge limited current of ^{235}U is 4×10^{-5} mA/cm, or 10^{-13} kg/sec –cm of ^{235}U, which for a 60 cm long source would require 5000 years to acquire 1 kg! So this limit has to be overcome.

Considerable effort went into the design of the ion source. UF_6 would be a convenient material, since it was used in gaseous diffusion, but UCl_4 was used instead because it seemed not to degrade the source as rapidly. Manfred von Ardenne preferred to use metallic uranium that required a high temperature oven. Uranium chemistry played a big part in isotope separation. The conversion between uranium compounds is one example. Extraction of the uranium deposited in the containers inside the calutron is another.

Space charge effects are due to the interactions between the ions. If the ion density is high enough, the coulomb repulsion between the positive charges in the beam can overcome the effects of the external fields which are there to define the orbit, resulting in loss of spatial separation of the two isotopes at the focus. For uranium, the heavier isotope is much more abundant. The space charge effect can degrade its larger radius, pushing some ^{238}U into the ^{235}U cup – obviously very bad. The authors of Ref. (15) discovered that electrons from the residual gas in the vacuum chamber of the magnet travel along with the positive ion beam, partially screening the space charge effect, and permitting higher currents to be separated than previously thought possible. The calutrons were operated at a specific vacuum pressure to exploit this screening.

There were two variations of calutron, called α and β. The α version came first. The magnetic field was 0.35 T, the accelerating voltage was 35 kV, and the bending radius was 1.2 m. These numbers all check for singly ionized uranium ions. The beam height (parallel to the magnetic field) was 0.4 m. From Eq. (C.20) above, the separation was:

---

[15] L.P. Smith, W.E. Parkins, and A.T. Forrester, Phys Rev **72**, 989 (1947). This paper was submitted in 1942, but published after WWII. A popular article on the Oak Ridge mass separation project is: W.E. Parkins, Physics Today **58**, 45 (2005).

$$\Delta x = 2 \times 1.2 \times 0.006 = 1.4 \text{ cm alpha calutron.} \qquad (\text{C.21})$$

The constant magnetic field required a coil of about 1.4×10^5 ampere turns, which is a lot of current $-$ 3500 amperes and 40 turns, for example. The current was carried by silver buss bars from the US mint. [16]The Tennessee Valley Authority supplied the power.

The β calutrons were half the size, with twice the magnetic field $-$ same number of ampere turns. The accelerating voltage was the same, 35 kV. The β units came later, and were tuned to handle uranium samples already enriched in ^{235}U. Much of the enriched material came from the thermal diffusion plant.

The performance of a calutron is greatly enhanced by the enrichment of the feed material. For example, suppose you get 1 gm per day of ^{235}U, contaminated by 1.5 gm of ^{238}U, for a total of 2.5 gms in the inner cup (40% pure ^{235}U) from naturally occurring uranium, which is 0.7% ^{235}U. For 1 gm of ^{235}U, you have processed 143 gms of uranium total, so the contamination, caused by scattering, beam self interactions, focusing problems, etc. of outer cup material in the inner cup is 1%. Now suppose the feed is increased in purity by a factor of 2, so that the ^{235}U is now 1.4%. In a day the same 143 gms of uranium go through the system but now there are 2 gms of ^{235}U. What's more, the contamination is the same, namely 1.5 gms, so now the output purity is $2/3.5 = 57\%$ instead of 40%. Increasing the feed purity is a win-win situation. The 1.5% spill over from heavy to light also occurs backwards, from light to heavy. The percentages may not be equal. This effect can be studied by varying the magnetic field to sweep the heavy ions across the inner detector. Let us suppose that the spill over is symmetrical, that is 1.5%, or 15 milligrams (for 1gm of product) of ^{235}U remains in with the 143 gms of ^{238}U. This is small, but not negligible $-$ it would be recovered and reprocessed. It demonstrates, however, that as the feed material becomes more enriched, the spill over is significant, and the entire vacuum chamber has to be scrubbed to recover all of the product.

---

[16] Buss bars are thick pieces of metal. The coils have a small number of turns.

The Manhattan Project's calutrons at Oak Ridge – called the Y-12 plant – processed all of the ^{235}U used in the Hiroshima bomb. The Soviet Atomic Project had a similar R&D program in electromagnetic mass separation led by Lev Artsimovich, with a factory located at Sverdlovsk -44 near the gaseous diffusion plant, a place Heinz Barwich called 'Oak Ridge in Siberia.'[17] The Soviet calutrons benefitted from von Ardenne's duo-plasmatron ion sources, and were used for enrichment to 90% from early product of gaseous diffusion at a lower concentration. However, when the gaseous diffusion plant started operating at full efficiency, the electromagnetic separators were no longer used for production of weapons grade ^{235}U.

C.6. Centrifuges

Centrifuges were not used for isotope separation in either the Manhattan Project or the initial stages of the Soviet Atomic Project, although the principle was tested by both projects. It was Gernot Zippe, a German engineer working in the Soviet Union after WWII, who developed a practical centrifuge device for uranium enrichment.[18]

The principle is simple enough. The force on a mass M rotating with velocity v in a circle of radius r is:

$$F = M\, v^2/r. \tag{C.22}$$

Other forces, like viscosity, which tend to cancel the centrifugal force, do not depend on the molecular weight, so the different isotopes separate in space. The effect is proportional to the mass, rather than the square root of the mass, which improves the sensitivity. A simple formula for the enhancement in the ratio of light/heavy isotopes of a perfect gas at the

---

[17] Heinz and Elfi Barwich, "Das Rote Atom," Scherz Verlag, Munich, 1967, p 103.
[18] en.wikipedia.org/wiki/Zippe-type_centrifuge

center of the centrifuge compared to the periphery was given in a paper by J.W. Beams:[19]

$$K_0/K = \exp(v^2/2RT) \times (M_2 - M_1).$$ (C.23)

K_0 is the light to heavy ratio at the center, K is the same ratio at the periphery, v is the linear velocity at the periphery of the spinning device, M_2 is the molecular weight of the heavier isotope, and M_1 is the molecular weight of the lighter one. If $K_0/K > 1$, the light to heavy ratio is larger at the center. T is the absolute temperature in Kelvin, and $R = 8.3 \times 10^7$ ergs/mol-K is the gas constant in cgs units. Beams used a very high velocity $v = 8 \times 10^4$ cm/sec; then at room temperature $T = 300$ and ($M_2 - M_1$)=3 (the difference in uranium isotopes) the formula gives $K_0/K = 1.44$, which is a very promising number, if a technique can be found to exploit it.

The device is a spinning cylinder filled with UF_6 gas. Just like the cream separator, the heavier ^{238}U tends to migrate towards the outside, leaving the ^{235}U on the inside. The bottom is heated, setting up a thermal circulation, which brings the light isotope to the top and the heavy one to the bottom, similar to the action in thermal diffusion, where they are scooped up. The gas pressure is higher at the outer rim of the cylinder, and if the UF_6 were to liquefy the result would be disastrous. This limits the amount of gas a single cylinder can handle to only a few grams. On the other hand the enrichment factor given by Eq. (C.23) above is much higher than for gaseous diffusion. A few thousand centrifuges may be needed, but they are hooked up in parallel to increase the mass flow, and only 30 or 40 stages in series are required to obtain high purity ^{235}U.

Centrifuges are the modern isotope separators of choice.[20] They are relatively inexpensive to operate, and have a smaller footprint than a

---

[19] J.W. Beams and F.B. Haynes, Phys Rev **50**, 491(1936).
[20] H.G. Wood, A. Glaser, and R.S. Kemp, Physics Today **61**, 9, p 40 (Sept 2008).

gaseous diffusion plant. As such, they represent a challenge to world efforts at non-proliferation of nuclear weapons.

International attention to the nuclear program of Iran has brought uranium enrichment back into the news.[21] A measure of the effort required to enrich a sample of F kg of natural uranium with assay $x_f = 0.007$ (the 'feed'), into P kg of product with assay $x_p > x_f$, and T kg of tail with assay $x_t < x_f$ is given by the separative work units formula:

$$W = P \times V(x_p) + T \times V(x_t) - F \times V(x_f). \tag{C.24}$$

$V(x)$ is called the value function, and is defined as:

$$V(x) = (1 - 2x)\ln((1-x)/x) \quad 0<x<1. \tag{C.25}$$

$V(x)$ is symmetrical about $x = 0.5$, and $V(0.5) = 0$. $V(0.1)=V(0.9) = 1.75$. This function is used in economics. Why it disfavors a 50-50 mix is not obvious, but the separative work units formula as shown in Eq. (C.24) is widely used in comparing different uranium enrichment plants. One separative work unit = 1 SWU. Note that SWU's do not have units of energy.[22]

One thousand kg of natural uranium (one metric ton) contains 7 kg of ^{235}U. Suppose a plant produces 100 kg of product enriched to 4.5%, or 4.5 kg of ^{235}U, leaving 2.5 kg in the tail. Since the tail is 900 kg, the tails assay is 0.28%. So for our formula for SWU we have F = 1000, $x_f = 0.007$; P = 100, $x_p = 0.045$; T = 900, $x_t = 0.0028$.

Substituting in the numbers and turning the crank gives:

$$W = 278+5256-4880 = 650 \text{ SWU effort}. \tag{C.26}$$

---

[21]   Science, Vol **348**, #6241 (19 June 2015) p 1320.
[22]   The formulas come from en.wikipedia.org/wiki.Separative_work_units

Figure C.4, taken from world-nuclear.org,[23] gives the functional form of the effort required to purify 1000 kg of natural uranium to yield the content in the product given on the x axis. The curve rises steeply at first, but at around 20% it flattens out. Eq. (C.23) gives about 1250 SWU for the 'research reactor 20%' point, assuming that the value for the feed – the subtracted amount – is always the same.

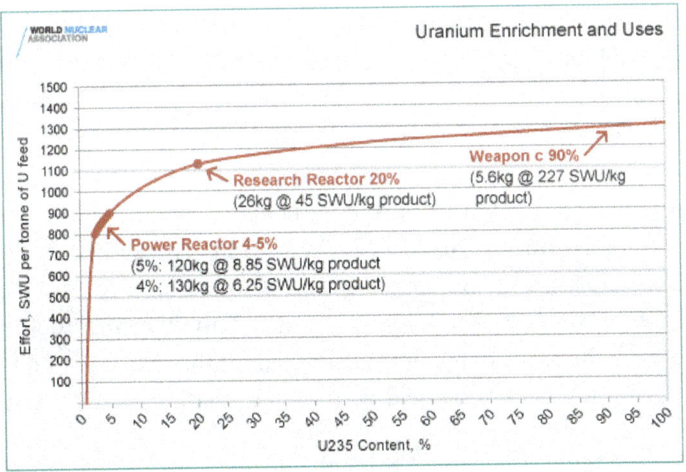

Figure C.4. Plot of the SWU required to yield the % of ^{235}U beginning with 1000 kg of natural uranium.

While the effort is constant for all separation methods, the required electrical energy is not. A gaseous diffusion plant requires about 9×10^9 joules per SWU, while a gas centrifuge plant requires only about 2×10^8 joules per SWU, or about 2% as much energy. A centrifuge typically has a much smaller gas volume than a gaseous diffusion stage, so several centrifuges are connected in parallel to form a single stage. However, the centrifuge process is much more efficient. The total number of stages may only be 10-20, compared to 1000 for gaseous diffusion.

---

[23] www.world-nuclear.org/info/Nuclear-Fuel-Cycle/Conversion-Enrichment-and-Fabrication/Uranium-Enrichment. A similar plot is in the Science article quoted.

C.7. Plutonium Extraction

A uranium fueled reactor makes plutonium through the processes:

$$^{238}U + n \rightarrow\, ^{239}U \rightarrow\, ^{239}Np + \beta + \nu;\ ^{239}Np \rightarrow\, ^{239}Pu + \beta + \nu. \qquad (C.27)$$

^{239}U half-life is 23 minutes, and ^{239}Np half-life is 2.3 days, so these reactions are reasonably fast. $^{239}Pu + n \rightarrow$ fission most of the time (~70%), but neutron capture also can make ^{240}Pu. ^{240}Pu has a half-life of 6561 years, decaying predominantly by α emission to ^{236}U. However, ^{240}Pu also undergoes spontaneous fission (a rare decay mode, less than 10^{-6} branching fraction).[24] The longer the fuel remains in the reactor, the larger the amount of ^{240}Pu there is.[25] The spontaneous fission leads to neutrons, which can cause an unwanted chain reaction in a nuclear weapon made from ^{239}Pu. Spontaneous fission is discussed further in Appendix E, and its implications for pre-detonation of active material are discussed in Appendix F. Therefore, for weapons grade plutonium, the fuel must be removed from the reactor in a time optimized for the ratio of ^{240}Pu to ^{239}Pu. Isotope separation is not practical with such a small mass difference.

The following table was copied from chemcases.com[26]

Table C.1.

Material	Fresh Fuel %	Spent Fuel %	Type of Waste
Transuranic elements	0.00	0.065	Transuranic
U-236	0.00	0.46	Depleted uranium
Pu isotopes	0.00	0.89	Transuranic
Fission products	0.00	0.35	High level
U-235	3.3	0.08	Depleted uranium
U-238	96.7	94.3	Depleted uranium

It shows how a fuel rod, which starts out with ^{235}U enriched to 3.3%, evolves in three years inside the reactor. Note that only about 1/40 of the

---

[24] periodictable.com/isotopes/094.240/index.full.dm.html
[25] The amount of ^{239}Pu grows linearly with time, while ^{240}Pu grows quadratically.
[26] chemcases.com/nuclear/

^{235}U remains, and almost 1% of the material has become plutonium. Assuming 20 kg of ^{235}U has been burned in three years, a little arithmetic gives an average reactor power of 16 Megawatts.[27]

In the example considered in Ref. (26), the fuel rod assembly consisted of 264 rods weighing 660 kg (2.5 kg per rod). Almost 6 kg total of plutonium can be extracted from the rods by chemical means. For weapons grade plutonium the rods must be removed sooner, giving a much smaller total plutonium fraction, typically around 0.025%. Smaller plutonium fractions are more difficult to handle chemically.

Equation B.13 in Appendix B gives a simple formula for plutonium atoms produced per second given the reactor power and the isotope ratio. Substituting a 16 Megawatt power level and an isotope ratio of 30 (3.3% enriched ^{235}U) in Equation B.13 gives 3 kg of plutonium in 3 years — a factor of two smaller, not too bad considering the number of variables in reactor performance.

Ref. (26) also describes the chemical process, called the bismuth phosphate process, used for by the Manhattan Project for plutonium extraction. The Soviet Atomic Project at Combine #817 used an organic molecule to precipitate the heavy elements.[28] The modern procedure still in use is called 'PUREX.' We will not reproduce the reactions here. There are several steps, resulting in a claimed plutonium recovery of 95%.

---

[27] One ^{235}U atom mass = 4×10^{-25} kg, and each fission is 200 MeV.
[28] E.I. Ilenko and N.A. Abramova, HISAP96, Vol 2, *ibid.*, p 351.

Appendix D

Charged Particle Accelerators[1]

D.1. Electrostatic Accelerators

Electrostatic accelerators maintain a DC high voltage on a metal terminal which is supported above ground potential by an insulating column. The column is usually an evacuated tube made of ceramic insulators cemented to metal field shaping electrodes that create a uniform voltage drop from the terminal down to ground. While the design of high voltage equipment began with X-ray machines, and power lines for distribution of electrical energy, high voltage electrostatic accelerators required substantial R&D to learn how to handle the conductors and insulators safely. For the acceleration of protons, the terminal polarity is positive, and for electrons it is negative. Voltages are typically 5 million volts (MV), leading to kinetic energy of 5 MeV for particles with one unit of the elementary charge, like protons. Note that the 'e' has been inserted between the M and the V. MV is a voltage, and MeV is an energy.

To understand how electrostatic generators work, we first must review simple circuits. Figure D.1 shows a battery charging a capacitor through a resistor. No current flows through the capacitor – it is an open circuit. But it does store charge. Current flows through the resistor R

---

[1] A good description of particle accelerators of the WWII period is given by M. Stanley Livingston and John P. Blewett, "Particle Accelerators," McGraw-Hill Books, New York, 1962. Livingston was one of Lawrence's first students to help build the cyclotron.

until the capacitor is charged to the full voltage of the battery. The current I(t) decreases exponentially as the capacitor charges up.

$$I(t) = V \times e^{-t/RC}/R \tag{D.1}$$

The final charge on the capacitor is $q = C \times V$, where C is called the capacitance, measured in farads. One farad = 1 coulomb/1 volt. Volts are ordinary household units, but a coulomb is a very large stored charge. The elementary charge $e = 1.6 \times 10^{-19}$ coulombs. Microfarads, abbreviated μf, are more common units of capacitance. The larger the capacitance C, the more charge can be stored for a given voltage drop. The unit of current, the ampere, is also an ordinary unit, and is defined as 1 ampere = 1 coulomb/sec. Power is easy; 1 watt = 1 volt×1 ampere. The battery supplies the energy lost in the resistor as the capacitor is charged:

$$\text{Energy loss in R} = \text{energy stored in C} = \tfrac{1}{2}\,CV^2. \tag{D.2}$$

The energy is independent of the resistance. Resistance is measured in ohms, and 1 ohm = 1 volt/1 ampere. Linear circuits obey Ohm's Law, $V = I \times R$.

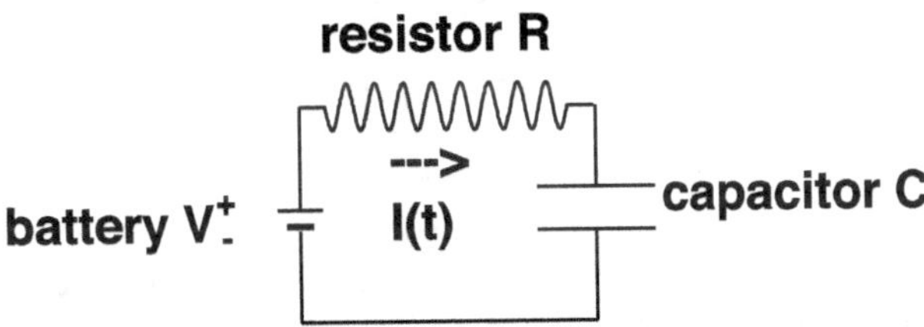

Figure D.1. Simple battery circuit to charge a capacitor through a resistor.

Figure D.2A shows two capacitors in a parallel hookup, and Figure D.2B shows them in series. In parallel the two capacitors charge to the same final voltage: $V = q_1/C_1 = q_2/C_2$. The total charge is $q = q_1 + q_2 =$

C×V, where $C = C_1 + C_2$ is the capacitance of the combination. Capacitors in parallel add, like resistors in series. In the series combination, the two interior plates, one of C_1 and the other of C_2 connected by a wire, have no net charge. The charges on the two plates of a capacitor are equal in magnitude and opposite in sign. The charge in the middle must separate, $+q$ going to C_2 and $-q$ going to C_1. Each capacitor carries the same charge in the series hookup. After the battery has charged the capacitors up, the total voltage appears across the two in series: $V = q/C_1 + q/C_2$, or $1/C = 1/C_1 + 1/C_2$. Capacitors in series add as reciprocals, as do resistors in parallel. Hence the combination rules for resistors and capacitors are reversed. In parallel the capacitor voltages are equal, and in series the capacitor charges are equal.

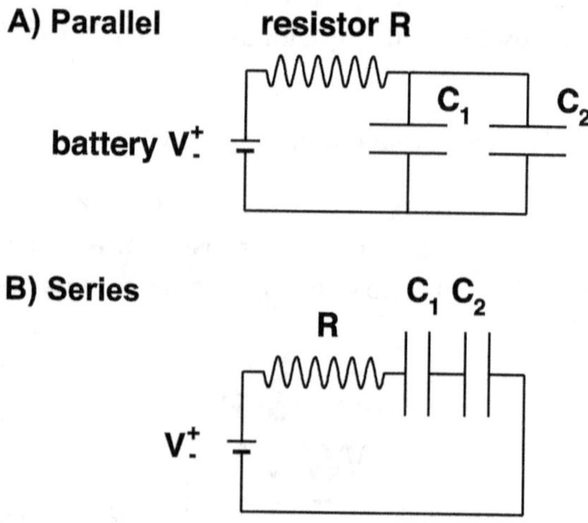

Figure D.2A/B. Top, charging two capacitors in parallel, bottom in series. Parallel voltages are equal, and series charges are equal.

Figure D.3 shows the principle of the Cockcroft-Walton circuit. Charge two capacitors C_1 and C_2 in parallel to the same voltage V. (In practice, the capacitors usually have the same capacitance value.) Now carefully unhook the capacitors from the battery (use very heavy rubber gloves, because the stored charge can be lethal), and hook the + plate of

C_1 (a) on to the − plate of C_2 (d). Now from b to c you have $V + V = 2V$, and you have doubled the voltage at no cost.

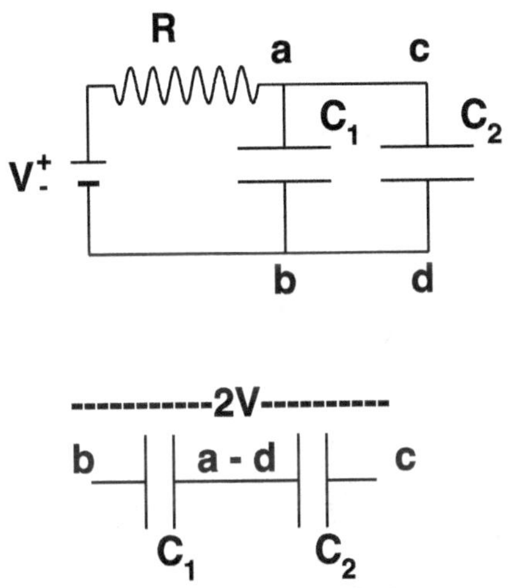

Figure D.3. Charge in parallel, then hook up in series. Top of C_1 goes to bottom of C_2.

A Marx generator[2] is a commonly used circuit to generate high voltage pulses using the parallel-series trick. Such a circuit is shown in Figure D.4. A DC voltage supply charges up the capacitors in parallel. The components in a diagonal pattern are spark gaps, which connect the bottom of C_1 to the top of C_2, and so on up the chain. The voltage is not sufficient to trigger the gaps. Now the first gap is fired, by pulsing a trigger needle, or by some other means. That doubles the voltage on the second gap, and it fires rapidly, as do all the others, resulting in a voltage at the top of N×V, where N is the number of stages. The conducting spark gaps can be thought of as wires, so that all of the capacitors are now hooked up in series. As charge is drawn off the top, the gaps quench from top to bottom, and the parallel charging begins again, so what you get is a high voltage pulse. This is a cheap way to create a pulsed X-ray source, or to test high voltage breakdown in materials. Generators in the

---

2 Invented by Erwin O. Marx in 1924.

MV range have been built and operated.[3] It is not so convenient for a charged particle accelerator, because the useable high voltage is only on for a short time – a fraction of a second, before the stack has to be charged up again.

Marx Voltage Multiplier

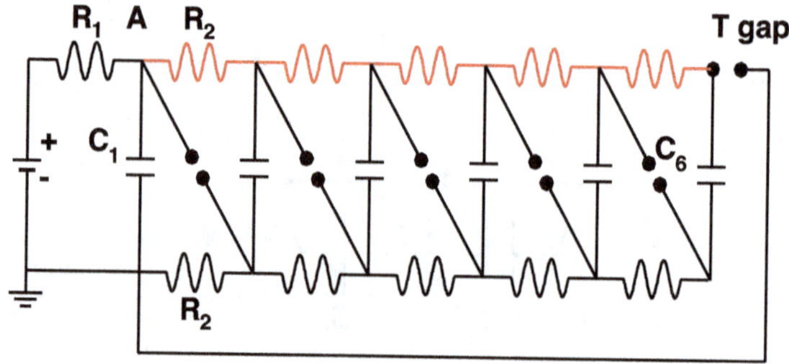

Figure D.4. The first spark gap (T gap) is triggered, and then all of the others fire, giving n×V at the output point A, where n is the number of stages, and V is the input DC voltage. R_1 = 10 Megohms, and all others R_2 = 1 Megohm. If V = 10 kV, fairly typical, then the voltage at point A = 60 kV. Drawing charge from A will quench the spark gaps, and start the parallel charging of the capacitors again. For C_1 through C_6 = 0.01 μfarad high voltage capacitors the circuit would be fully charged in about five seconds.

The Cockcroft-Walton circuit was first proposed by Heinrich Greinacher in 1921.[4] It was Cockcroft and Walton however who turned the idea into a practical machine for studying nuclear physics.[5] Figure D.5 shows a 4× multiplier circuit, which can be replicated indefinitely. Unlike our previous examples, this circuit uses alternating current as a source, and diodes to perform the switching. A diode is a non-linear element, which conducts in the direction of the arrow, but not in the

---

[3]   I.C. Sumerville, "A Simple Compact 1 MV 4kJ Marx," Proceedings of the Pulsed Power Conference, Monterey, CA, June 11, 1989, p 744-746, XP000138799.

[4]   H. Greinacher, Z. Phys **4**, 195 (1921).

[5]   J.D. Cockroft and E.T. Walton, Proc Roy Soc (London) **A136**, 229 (1932) **A144**, 704, (1934).

reverse direction. In the forward direction it is a wire, and in the backward direction it is an open circuit. The operation of this circuit is not all that obvious. We begin with no voltage anywhere. Let the first oscillation of the voltage source be negative with respect to ground. Then the capacitor C_1 charges through the diode D_1 so that the point a is at $+V$. The diode D_2 is turned off, because both sides are at ground. Now the voltage source changes to $+V$ with respect to ground. The charge on C_1 has to reverse, meaning that $+q \rightarrow -q$, so charge $2q$ flows through D_2 onto C_2, which is now at $2V$. The next negative swing on the voltage source moves the action to D_3, which charges C_3, and on the next ½ cycle C_4 is charged to $4V$. Subsequent cycles do not change the output voltage, unless there is current drawn from point d by leakage or a charged particle beam. In the case of a current load, the AC voltage source keeps topping up the output. The normal configuration for a Cockcroft-Walton accelerator is two high voltage towers. One consists of the multiplier circuit, with diodes in a zig-zag pattern between two stacks of capacitors, called the pushing column (top row in Figure D.5) and the smoothing column (bottom row). The other tower is the accelerating tube, connected to the multiplier tower at discreet points, containing the ion source inside the HV spherical terminal, and with an evacuated ceramic tube for the beam. Figure D.6 shows an example of a two tower Cockcroft-Walton, which until 2012 was the source for the accelerator complex at Fermi National Accelerator Laboratory. The voltage generating tower with the zig-zag rectifiers is clearly shown on the left, and the accelerating terminal is on the right. The beam tube is horizontal in this case. This machine operated at 750 kV. Such machines are usually set up in (dry) atmospheric pressure air, which limits the voltage on the dome to about 1 MV. Electrons, which are unavoidable around machines of this type, are accelerated upwards towards the HV terminal, where they create energetic X-rays. The radiation requires shielding for protection of experimenters. Cockcroft and Walton built the multiplier tower and the acceleration tower on the floor above the experimental area, so that the beam came through the ceiling, and there was room for shielding for radiation safety. The experimenters must have done a lot of running up and down stairs to keep everything working. These machines are still in use today.

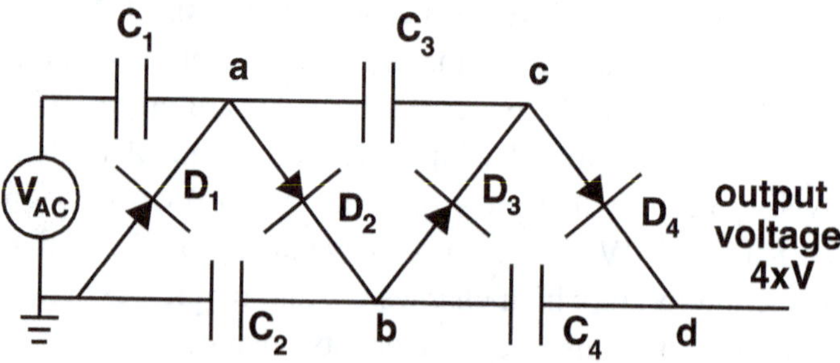

Figure D.5.   A four stage Cockcroft-Walton voltage multiplier driven by an alternating voltage source. The first two diodes push twice the peak voltage onto the capacitor C_2, at the point b. The pattern is replicated to increase the output voltage: $2V \rightarrow 4V$ at the output point d.

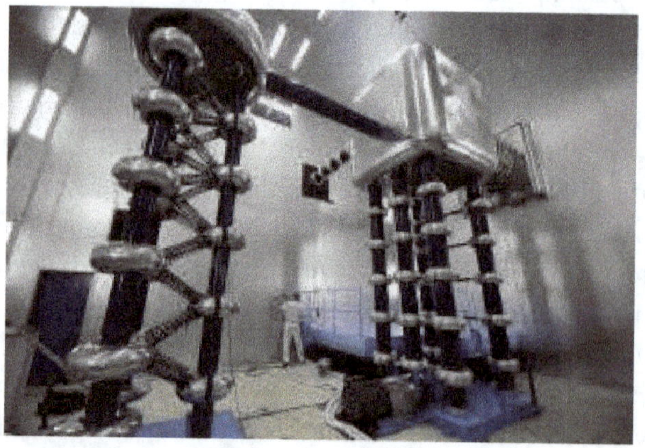

Figure D.6. Cockcroft-Walton accelerator at Fermilab. Note man in corner for scale. Courtesy Fermi National Accelerator Laboratory.

In the electrostatic accelerator invented by R.J. Van de Graaff,[6] a single spherical dome carries the high voltage and contains the ion

---

[6]   R.J. Van de Graaff, Phys Rev **38**, 1919A (1931).

source for the charged particle beam to be accelerated. The charging of the dome is based on a fundamental principle of electrostatics. All of the charge on a hollow conducting sphere resides on its outer surface. There is no charge inside. Suppose there is a hole in the bottom of the spherical shell, so that charge can be inserted into the interior by a smaller conducting ball attached to an insulating stick. Touching the ball to the metal surface inside immediately transfers all of the charge from the ball to the outer surface of the large sphere. This process can be repeated indefinitely, regardless of the amount of accumulated charge on the outer surface. The resulting voltage on the sphere is given by:

$$V = q/C, \text{ where } C = (1.11 \times 10^{-10} \text{ farads/m}) \times r. \tag{D.3}$$

where r is the radius of the sphere in meters.[7] Suppose you have a 1 m radius sphere, and a charging current of 0.001 amperes. Then in 1 minute the sphere charges up to 540 MV! Of course, you wouldn't get there; something would cause a breakdown at a much lower voltage – a few MV. But the numbers look promising, and lead to Van de Graaff's idea. He proposed to use a charged insulating belt to serve as a continuous source of metal balls on a stick.[8] Figure D.7 shows the essential components.[9] A power supply at ground potential uses a metallic comb to spray positive charge onto the moving belt, which is driven by an electric motor and pulley system at the base of the machine. A second comb at a slave pulley at the top inside the sphere extracts the charge, which promptly flows onto the outer surface, as described above. The insulating support column for the HV dome is an evacuated tube of glass or ceramic cylindrical spacers separated by metal electrodes. The electric field inside the tube accelerates and guides the charged beam from the

---

[7] F. Hinterberger, Electrostatic Accelerators, cds.cern.ch/record/1005042/files/p95.pdf. This is a very nice article summarizing the design and operation of electrostatic accelerators.

[8] The author's former colleague, the late Ray Herb, invented the pelletron, where the charging belt is replaced by metal beads on a nylon rope – very close to our metal ball on a stick. Pelletrons are still made by National Electrostatics Corp in Middleton, Wisconsin. See www.pelletron.com

[9] Copyright Encyclopedia Britannica 1995.

dome to the experimental target area below. The design of the beam tube represents one of the critical aspects of machine performance. The metal electrodes are typically connected together by precision high voltage resistors, which establish a uniform voltage drop along the tube. The voltage on the dome can be monitored by measuring the current that flows down the resistor chain. The voltage can be regulated by equating the charging current to the total load current on the dome – beam plus leakage currents. Voltage regulation to $\pm 0.1\%$ is typical. A single HV dome gives a compact footprint that can be enclosed in a pressure vessel to control the environment and improve the HV performance. The pressure vessel also provides some shielding from the x-rays mentioned earlier.

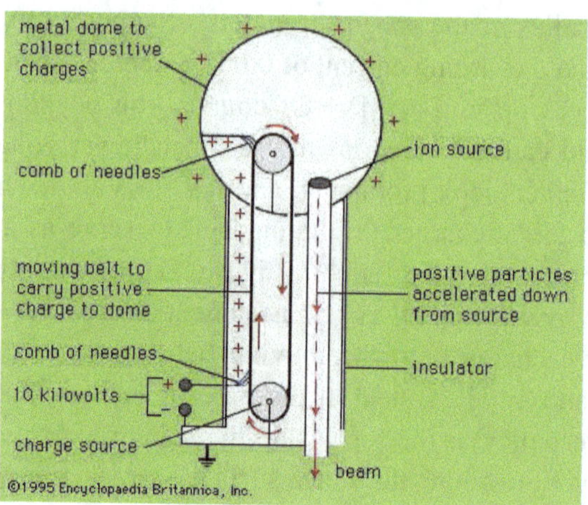

Figure D.7. Schematic of the components of a Van de Graaff electrostatic accelerator. A moving dielectric belt carries positive charge up inside the dome, where it is all transferred to the outside. The voltage on the dome accelerates the protons from the source to the experiments below.

Figure D.8 shows a publicity photo of an early Van de Graaff machine assembled in a dirigible hangar in Dartmouth, MA, near MIT. It is not clear whether this machine was configured with a beam pipe for experiments with accelerated protons or deuterons. Such apparatus would be below the base of the picture.

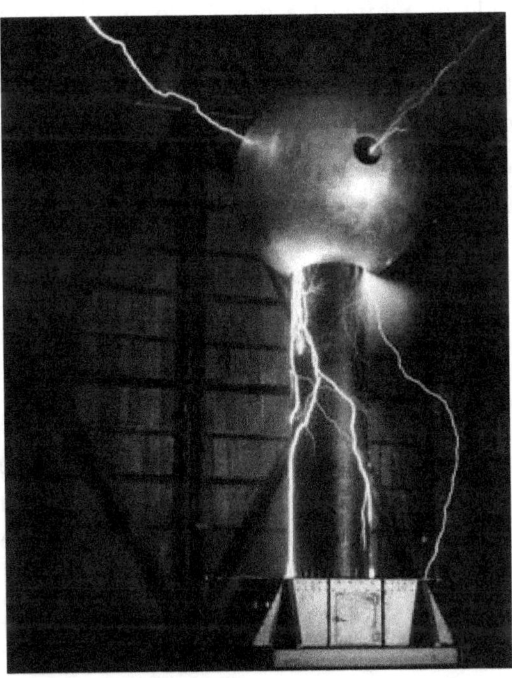

Figure D.8. Early MIT Van de Graaff generator. The dome voltage is 7 MV. Credit MIT.

The maximum voltage for an electrostatic accelerator, and hence the maximum energy gain for a charged particle beam, is determined by breakdown and sparking, which can occur from the dome or along the beam/support tube. In most cases this limit is a few million volts. In order to reach higher charged particle energies, it is necessary to gain the energy incrementally, rather than in one big step.

D.2. Cyclotrons

E.O. Lawrence proposed to achieve high energy protons and deuterons by incremental acceleration. If you have a pair of capacitor plates at 20 kV, say, and you send protons through a hole perpendicular to the plates 100 times, you should get 2 MeV kinetic energy. You might use a magnetic field to confine the protons to ever expanding circular orbits. Use conducting screens instead of solid capacitor plates, and stretch them half way across the magnet, to accelerate the beam as it crosses in one direction. There is only one thing wrong with this idea – it won't work. The reason is that a static electric field is a conservative force for charged particles, like the gravitational field for masses. If you go around a circle through plates with a DC voltage as described, you gain exactly zero energy. What you gain between the plates is lost in the rest of the circle.

The fix is to replace the DC voltage with an alternating voltage, which has one sign to accelerate the beam in one direction. As the beam goes 180^0 at constant energy in the magnetic field, the voltage changes sign, and accelerates the beam again as it crosses the gap going the other way. The particles gain energy with each turn in the time dependent electric field. The frequency of the alternating voltage is the cyclotron frequency defined in Chapter 2, and given in Eq. (D.4). It is independent of the particle energy, and hence constant during acceleration.

$$f = \omega/2\pi = eB/(2\pi m).\tag{D.4}$$

Numbers for the elementary charge, and the proton mass as quoted in footnote 15 of Chapter 2 give f = 22.8 Mhz for a 1.5 Tesla magnetic field. These are radio waves, with a wave length of 13.1 meters. This means that the oscillators required to power the AC voltage come from radio engineering. This is one example of technological developments leading to advances in scientific instrumentation. No cyclotron would work without radio frequency oscillators.

Figure D.9. RF system for the MIT cyclotron, showing the D's, the ion source half way in the middle, and the two $\lambda/4$ co-axial lines which connect to the D's. The RF oscillator drives the co-axial lines. Linear dimension of the vacuum chamber is about 80 cm. Credit MIT.

There are three main components to a cyclotron: a constant magnetic field over a circular area; an RF accelerating field applied to a gap between two "D's," as shown in Figure D.9; and an ion source at the center of the magnet to produce ionized hydrogen, deuterium, or helium for acceleration in the machine. A drawing from Lawrence's 1934 patent application is shown in Figure D.10.[10] The vertical pole tips are a feature of the drawing. Lawrence's cyclotrons all had horizontal pole tips, and a vertical magnetic field. Eq. (C.12) in Appendix C gives the momentum in MeV/c for a particle with the elementary charge as p = 300 B×R,

---

[10] en.wikipedia.org/wiki/Ernest_Lawrence

where B is in Tesla and R is in meters. The non-relativistic kinetic energy in MeV is then given by:

$$E = (300B{\times}R)^2/(2{\times}mc^2). \tag{D.5}$$

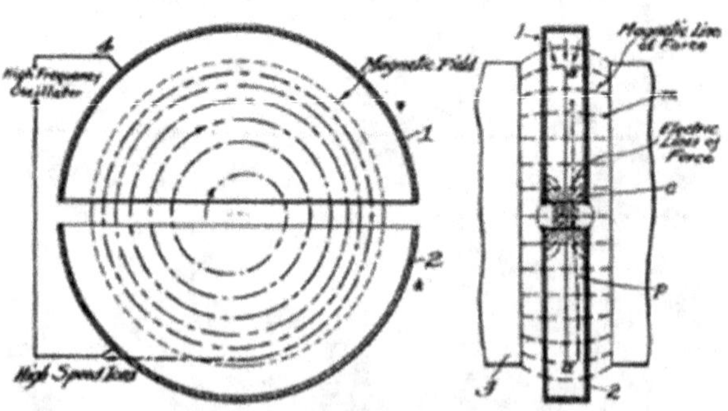

Figure D.10. Cyclotron patent application, showing the spiral orbits of the accelerating particles, the hollow D's with the alternating electric accelerating field in the gap, and the pole tips of the magnet.

where mc^2 is the particle mass in MeV. Given the magnetic field strength B and the radius R of the magnet, this formula yields the maximum beam energy achievable. Substituting some typical numbers R = 0.5 m, B = 1.5 Tesla, mc^2 = 938 MeV (protons) gives a maximum kinetic energy of E = 27 MeV. This would be a 40" cyclotron in the language of Livingston and Blewett, where all of the magnet dimensions are in inches. An important feature of the magnetic field, not obvious in the drawing, is a decrease in field strength with increasing radius from the center, where the ion source resides. The fringe field near the edge of the pole tip bows outward, but so does the field inside the gap, ever so slightly. This decrease in field is achieved by inserting iron shims near the center, to strengthen the field at small radii. A radial decrease in field strength focuses the beam, keeping wayward protons moving towards the median plane. Without this feature most of the beam would spiral into the D surface and be lost, resulting in a very low beam current reaching the outer radius of the machine. This effect was discovered experimentally by Livingston when he was a graduate student at Berkeley, working in

Lawrence's group. He found that by hand tuning the magnetic field he could substantially increase the beam current accelerated to full energy.

A cyclotron magnet is a large piece of iron, with copper coils which carry current to generate the field. The magnetic field to form the charged particle orbits is in a gap between two pole pieces. Soft iron is ferromagnetic, which means that the iron adds to the magnetic field of the current, increasing substantially the field in the gap. The iron becomes completely polarized at a field strength called 'saturation,' usually around 1.2-1.4 Tesla for good iron. Above the saturation field strength, the iron no longer contributes extra field, and any gain in field strength must come from the coils. Hysteresis is the retention of a remnant field by the iron after the current in the coils has been set to zero, giving a permanent magnet. This effect is present in good quality soft iron, but is minimized. Figure D.11 shows a typical 'window frame' cyclotron magnet. A simplified design is based on a few simple principles. The product of the magnetic field and the cross-sectional area is called the flux, and is conserved – that is, it is a constant.

$$\Phi = B \times A = \text{constant}. \tag{D.6}$$

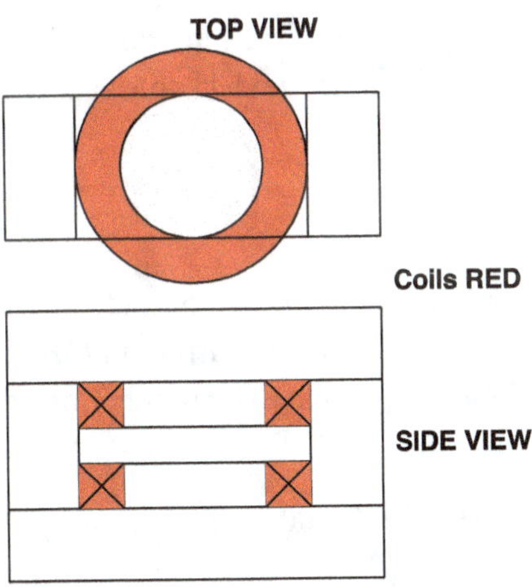

Figure D.11. Window frame magnet for a cyclotron with 40 cm radius pole tips. The frame is 210 cm long, and 150 cm high. Top and bottom are 210X80X40 cm^3, and the side pieces are 70X80X40 cm^3. The drawing is to scale.

The thickness of the magnet iron is chosen to handle the flux. In this design ½ of the flux goes to the right, and ½ to the left, so the area of the iron yoke is ½ of the area of the pole tip. The pole tip here is a circle, so the cross section of the iron is a bit over size. The coils drive the magnetic field. The B field is continuous. In a ferromagnetic medium like iron it is convenient to introduce a secondary field for calculations, called H. The line integral of the H field is equal to the number of enclosed ampere turns:

$$\int H dl = NI. \tag{D.7}$$

The loop of integration goes around the coils, through the iron yoke, and across the gap. The magnetic field is proportional to H:

$$B = \mu H = \mu_0 (1 + \chi)H. \tag{D.8}$$

Here μ_0 is a universal constant of electromagnetism, $\mu_0 = 4\pi \times 10^{-7}$, and χ is called the magnetic susceptibility. χ is a very large dimensionless number for iron – several thousand. Since $H = B/\mu$, and B is continuous, the contribution to the integral in Eq. (D.7) inside the iron is negligible, and the magnetic field in the gap is given by the simple formula:

$$B \times d = \mu_0 N I. \tag{D.9}$$

Where d is the height of the air gap in meters, B is the field in Tesla, N is the number of turns, and I is the current in the coil in amperes. For B = 1.4 T and d = 0.2 meters, we have the number of ampere turns:

$$NI = (1.4 \times 0.2/4\pi) \times 10^7 = 223{,}000 \text{ ampere turns} = 2230 \text{ amps and 100} \tag{D.10}$$
turns.

That designs the magnet and coils for 1.4 Tesla in an air gap 20 cm high, over an area of 5000 cm.^{2} Total weight of the iron is 16 metric tons, and of the copper coils is 1.0 metric tons. Current flows through an extruded copper conductor, 2 cm × 2 cm in cross section, with a 1 cm diameter hole in the center to carry cooling water. The coils are made in 'pancakes' 10 turns each, five top and five bottom. Both ends of the pancake coil come to the outside. Current flows in series through all pancakes, top and bottom. Cooling water flows in parallel, for greater efficiency. The complicated plumbing for all of this is on the outside.

The resistance of the coils can be calculated from the formula:

$$R \text{ (ohms)} = (\rho \times l)/A. \tag{D.11}$$

Where l is the length of the wire in m, A is the cross-sectional area in m^2, and ρ is the resistivity in ohm-m, which is a property of the conductor. For copper $\rho = 1.7 \times 10^{-8}$, giving $R = 0.02$ ohms. The power $P = I^2 R \approx 100$ kWatts.[11]

The maximum momentum of a charged particle beam is proportional to the radius of the cyclotron and the strength of its constant magnetic field. The weight of the magnet scales roughly as the cube of the radius, so to double the momentum of the beam, you need a magnet eight times as big! So cyclotrons have practical upper limits to achievable energy because of the mass of magnetized iron required to cover the orbits. The cyclotron built at the University of Chicago after WWII accelerated protons to a momentum $p = 1000$ MeV/c, and the magnet weighed 2000 tons, close to a practical limit for such a device.[12]

---

[11] Any text on electricity has these formulas. I used David J. Griffiths, "Introduction to Electrodynamics, 3rd edition," Prentice Hall, New Jersey, 1999.

[12] The cyclotron frequency has to decrease as the energy increases, a relativistic effect that does not change the scale of the magnet. The formula $p \times f = e \times v \times B/2\pi$ is relativistically correct, but $p = \gamma \times m \times v$, where $\gamma = 1/(1-(v/c)^2)^{1/2}$ is called the Lorentz factor, and c is the speed of light. If v/c is small, $\gamma \approx 1$. For Chicago 28 Mhz$<f<$18Mhz, or a decrease in frequency of 2/3 as the beam reaches full energy. This change was achieved by a variable shunt capacitor across the transmission line. Phase stability keeps the particles synchronized as the frequency changes. It is self correcting: particles travelling too fast gain less energy, and those travelling too slow gain more.

The D's are hollow metal cans, usually made of copper. When the beam is inside the can, there is no electric field. The accelerating field is only in the gap. As the proton bunch travels 180^0 inside the D, the RF voltage changes sign, so that the beam is accelerated again as it crosses the gap in the opposite direction. The orbit is kicked out by the incremental energy gain, remains circular for 180^0, and then is kicked out again. The copper tubes attached to the D's in Figure D.9 are parts of a transmission line, in which the D gap forms the open end, and there is a node at the rear of the tube, which is $\lambda/4$ in length. For our example RF frequency of 22.8 Mhz this length is $\lambda/4 = 3.28$ meters. The RF oscillator which drives the lines is not shown. Typical accelerating voltages are in the range 100-200 kV. One hundred kV accelerating voltage gain every half turn would require 135 turns to reach 27 MeV at the outer edge of the magnet. These are not small RF voltages, and can lead to voltage breakdown problems inside the cyclotron vacuum.

The ion source has the task of injecting ions with a few keV kinetic energy into the center of the machine. Some of the ions will be captured by the RF voltage, and spiral out. There are several different source designs. The most popular is a low voltage hot cathode arc discharge. Electrons from the cathode move across a stream of hydrogen gas. Some of the gas is ionized, and accelerated out of the source and into the vacuum chamber of the cyclotron. Note that the vacuum system of the cyclotron has to handle the unionized gas from the source.

At the outer radius an electrostatic deflector removes the beam from the machine, giving an extracted beam in the experimental area that can be directed onto a target of interest. For neutron time of flight, deuterons from the cyclotron can be pulsed by the deflector onto a stripping target which defines t=0. Then the detectors can be placed tens of meters downstream, and the time of arrival of the neutron gives its energy.

In summary, electrostatic accelerators and cyclotrons are each very sophisticated research machines. The task of constructing and operating the machine, together with successfully performing the nuclear physics experiments, was a daunting one, not to be attempted by the faint of heart.[13] It is a fact of modern life at large accelerator laboratories such as Fermilab or CERN that the scientists are split into two groups. One group builds, maintains, and operates the particle accelerators, and the other performs the experiments that use the accelerated particle beams. The operations are too complex for one group to do both jobs, although there is some crossover.

D.3. Experiments

No discussion of the accelerators would be complete without some remarks about the experiments. A scan of the Physical Review for 1938-39 shows that the techniques were primitive. Experimental apparatus was much the same whether an accelerator or a radioactive source furnished the beam. We have already learned that there are radioactive sources of α particles, γ rays, and neutrons, but not for protons or deuterons. So the accelerators gave some new particles for study, and also gave a much more intense and controlled beam. Particles were detected with cloud chambers ionization chambers, Geiger counters, and photographic emulsions. Cloud chambers detect tracks of individual ionizing charged particles passing through the atmosphere of the chamber. The ionization causes droplet formation in the saturated gas. Placing the chamber in a magnetic field gives curved tracks. The radius of curvature is proportional to the momentum, giving a measurement of particle energy. The tracks are photographed parallel to the magnetic field, to display the curvature. A pulsed light source illuminates the chamber perpendicular to the magnetic field. A diffusion cloud chamber can be made at home.[14] The ionization chamber is a direct descendent of the electroscopes used by the Curies. Charged particles ionize the gas in the chamber, and the ions are collected on a metal plate inside, producing a measureable

---

[13] You can build a cyclotron in your basement, and many people have. For a report on an amateur cyclotron conference, see Symmetry, a joint Fermilab/SLAC publication, Vol 7, p 24 (2010). For a successful project see "Building a cyclotron on a shoestring" by Toni Feder, Physics Today **57**,11,30 (2004).

[14] www.ins.cornell.edu/~adf4/cloud.html.

current. The current from a single particle is usually too small to detect, so the chamber is used either to measure a particle flux, giving a larger current, or external electronic circuits are used to amplify the signal from a single particle. Materials of interest can be inserted into the gas volume of the ion chamber. This technique was particularly useful in the early days of fission studies, when the uranium target was placed directly inside the ion chamber gas, allowing the heavily ionizing fission fragments, which gave large ionization signals, to be detected directly. Oscilloscope traces were used to register the voltage pulses from the chamber.

Direct gas amplification can boost the signal from an individual particle. The anode collector plate is replaced by a wire that is at high positive voltage – a few kV. Wire diameters are in the 20 μm range. The electric field within a few diameters of the wire is strong enough to accelerate electrons in one electron mean free path in the gas to an energy high enough to ionize the gas, creating more electrons. Electrons produced by the incident ionizing particle are thereby multiplied creating an avalanche of negative charge, which results in a measureable signal on the wire.[15] The properties of the amplification cascade can be varied, and it is possible to obtain a signal that is proportional to the energy deposited in the gas by the incident particle. These counters are called proportional counters. Lowering the voltage below gas amplification gives ion chambers, and raising the voltage above the proportional range gives a Geiger counter. These gas counters are all still in use as radiation monitors, and as charged particle detectors in experiments.

Ionization chambers, photographic emulsions, and thin foils were used to measure neutron fluxes. For neutrons BF_3 was a typical ion chamber gas. The isotope ^{10}B captured a neutron, and emitted an α particle that ionized the gas and was detected. Neutron flux and energy were detected in photographic emulsions by measuring with a microscope the number and length of recoil proton tracks. The foils were

---

[15] The electric field inside a cylindrical gas counter with outer radius r_1 and wire radius r_0 is $E(r) = V/(r \times \ln(r_1/r_0))$. The electron mean free path in an atmospheric pressure gas is $\lambda \cong 0.5\mu m$. If $E(r) \times \lambda \geq 15$ eV, then each λ doubles the number of e's.

activated by the neutron flux. Some processes were sensitive to fast neutrons, and some to thermal neutrons. Foil activation is described in Appendix B. Electronics was in a very early stage. Bruno Rossi had invented the electronic coincidence circuit in 1930, and applied it to the study of cosmic ray shower development.[16] Coincidence detection of different phenomena from a nuclear reaction became common when scintillation counters were introduced with faster time resolution.

Beam currents from an accelerator were measured by a "Faraday cup," which was an insulated electrode in a vacuum at the end of the beam pipe that was struck by the entire beam. A micro ammeter between the cup and ground gave a direct reading of the beam current. A thin target of some element of interest – carbon or beryllium or lithium – was inserted into the beam. Absorption cross sections could be measured by varying the thickness of the target, and using the formula $\lambda = 1/n\sigma$, where n is the number of target nuclei per cm^3 and σ is the cross section in cm^2. The beam is attenuated by an exponential:

$$N(x) = N(0) \times \exp(-x/\lambda). \tag{D.12}$$

A logarithmic plot of $\ln(N(x))$ versus x has slope $-1/\lambda$. If the number of target atoms per cm^3 is known, one can solve for the absorption cross section, which has the dimensions of an area.[17]

Let's describe briefly a specific experiment using a Cockcroft-Walton electrostatic accelerator at the University of Chicago to measure α particles from the proton bombardment of ^{9}Be, in the notation of nuclear reactions ^{9}Be(p,α)^{6}Li.[18] The proton beam was accelerated up to 400 keV by the electrostatic accelerator, and struck a beryllium target. α particles emitted at 90^0 to the beam were deflected through another 90^0 by an

---

[16] Bruno Rossi, Nature **125**, 636 (1930).

[17] Given the density of the material, ρ in gm/cm^3, the number density $n = \rho \times N_{AV}/AW$, where $N_{AV}=6\times10^{23}$ atoms/gram atomic weight, and $AW=$ atomic weight.

[18] S.K. Allison, Lester Skaggs, and Nicholas Smith, Jr., Phys Rev **54**, 171 (1938).

electrostatic analyzer. An electrostatic analyzer measures the kinetic energy directly. For this geometry – cylindrical plates, mean radius 25.4 cm, gap between plates = 0.635 cm, the formula is

$$KE(\alpha \text{ particle}) = 40 \times V. \tag{D.13}$$

where V is the voltage across the plates.[19] The α particles entered an ion chamber at the output of the electrostatic analyzer. The signals were amplified by external electronics, and displayed on an oscilloscope, where they were counted by experimenters. The number of counts per second was plotted as a function of the plate voltage on the electrostatic analyzer, giving a peak in the count rate at 36 kV, or $KE(\alpha) = 1.44$ MeV. From this number, the proton beam energy of 320 keV, and simple kinematics the experimenters obtained the mass difference between ^{9}Be and ^{6}Li.

Conservation of energy and momentum for ^{9}Be$(p,\alpha)^6$Li, where the α particle is emitted perpendicular to the proton beam (a particularly simple geometry), gives the following formula for the kinetic energy of the α particle:

$$E_3 = ((E_1 \times (1 - m_1/m_4) + \Delta_{13} + \Delta_{24})/(1 + m_3/m_4). \tag{D.14}$$

The subscripts label the particles: one = proton, two = ^{9}Be, three = α, and four = ^{6}Li. The Δ's are mass differences in MeV, and $m_1/m_4 = 0.167$, $m_3/m_4 = 0.665$ are mass ratios. The masses are of course now all known, so one may solve for E_3, using $E_1 = 0.32$ MeV, $\Delta_{13} = -2789.11$ MeV and $\Delta_{24} = 2791.24$ MeV, to obtain

$$E_3 = \alpha \text{ Kinetic Energy at } 90^0 = 1.44 \text{ MeV.} \tag{D.15}$$

---

[19] $\ln(r_2/r_1) = 0.0252$ is in the denominator. The elementary charge is absorbed in converting volts to electron volts. See footnote 15.

in agreement with the number obtained in 1938. Note that one must keep at least six significant figures in the mass differences, because they nearly cancel.[20] Equation D.14 can be solved for m_4, the mass of ^{6}Li nucleus, using the experimental measurement for $E_3 = 1.44$ MeV. Multiply both sides by m_4 and rearrange terms to obtain a quadratic equation. This gives two roots for m_4. Only the positive root makes sense, and gives the book value for the mass of the ^{6}Li nucleus (5601.55 MeV).

Equation D.14 can be applied to the reaction ^{7}Li(d,n)^{8}Be discussed in Chapter 7 for neutron production at 90^0 to the deuteron beam of the 70 cm cyclotron at Laboratory #2 in Moscow. In this case 1 = deuteron, 2 = Li; 3 = n; and 4 = Be. The sum $\Delta_{13} + \Delta_{24} = 15$ MeV, so the reaction is exothermic. The neutron energy at 90^0 for 4 MeV deuterons is:

$$E_3 = 8\times(3E_1/4 +15)/9 = 16 \text{ MeV neutron KE at } 90^0. \tag{D.16}$$

---

[20] The masses come from Appendix A of Tipler and Llewellyn, "Modern Physics" 5th edition, ibid. The numbers are nuclear masses. Take the atomic masses from the Table in amu, multiply by 931.5 to convert to MeV, and subtract $n\times0.511$ MeV, where n is the number of electrons in the atom.

Appendix E

Spontaneous Fission of Uranium, K.A. Petrzhak and G.N. Flerov, JETP 10, 1013, (1940)

A sensitive method for study of the phenomenon of uranium fission by detection of fission fragments is presented. The spontaneous fission of uranium, without external neutron sources, is established. The observed effect corresponds to spontaneous fission with a half-life of $10^{16} - 10^{17}$ years for the isotope ^{238}U.

E.1. Introduction

Bohr and Wheeler [1] calculated the probability of spontaneous fission of uranium, and got a half-life of the order of 10^{22} years. The calculation was for the isotope ^{238}U, and used the formula for a particle penetrating the coulomb barrier. The thickness of the barrier was taken as the radius of a fission product nucleus, an assumption which can only be verified by experiment.

The question of the possibility of spontaneous fission of uranium was experimentally studied by Libby [2], who looked for secondary neutrons from spontaneous fission, in analogy with those from fission induced by neutron bombardment. He used a BF_3 counter, sensitive to slow neutrons. Based on a null result, Libby quotes a limit for spontaneous uranium fission of $T > 10^{14}$ years.

E.2. Experimental Method

We use the method first proposed by Frisch [3] for observing the fission of nuclei. An ionization chamber with plates covered by a layer of uranium oxide, coupled to a linear amplifier, adjusted so that the more numerous α particles from uranium decay do not register. The pulses from fission fragments, somewhat larger than from α particles, fire a thyratron and trigger a mechanical relay counter. The thickness of the working layer of uranium oxide and the area of the plates of a typical ionization chamber (2 plates 30 mm in diameter), gave enough sensitivity to confirm the limit set by Libby. In order to increase the sensitivity of this method, it was necessary to increase the area of uranium oxide. A special ionization chamber was constructed, in the form of a multi-layer chamber with total area 1000 cm^2 and 15 plates. The spacing between plates was 3 mm; each plate was covered by a layer of uranium oxide 10-20 mg/cm^2 thick. The collector potential was 350 V. The ion chamber and the first stage of the amplifier were mounted on rubber cushions to minimize sensitivity to vibration.

Insert into the text by LGP

Typical α particle energy from decay of a heavy element like uranium is 5 MeV. Typical fission product energy from $^{239}U \rightarrow {}^{140}I + {}^{99}Y$ (Z=53 and Z=39 respectively) is 100 MeV, or 20 times the α energy. Fission products are back to back, with the same momentum, so the total energy released is $E = E_1 + E_2$, where $E_1/E_2 = m_2/m_1$, the lighter nucleus has the higher energy. There are cloud chamber pictures of back to back fission products, ranging out in 5 or 6 cm of air (Boggild, Bronstrom, and Lauritsen, Phys Rev **59**, 275 (1941)). This is the same range as a 5 MeV α particle! Hence the energy loss per cm in a gas, and the resulting number of ion pairs and total charge collected, is 20 times larger for the fission products as for the α's. This difference in pulse height allows spontaneous fission to be observed in the presence of 10^6 times as many α particles. Getting two heavy ions at once is a bonus, improving the chances of detecting at least one of them. The slowing down process for fission fragments, called dE/dx, differs from α's or protons in two ways: a.) The heavy ions pick up electrons as they go along, while the p's and α's have constant charge; and b.) The heavy ions lose energy near the end of their range by collisions with the atomic nuclei of the gas, while the p's and α's continue to fritter away their kinetic energy by collisions with electrons.

The P&F text resumes.

The amplifier had to be carefully designed in order to record the fission fragments in the presence of the large background of pulses from α particles. Pile up of α pulses in an ordinary amplifier can give signals equal to those from fission fragments, an effect that can be suppressed by good time resolution. In order to increase the time resolution of the system the grid of the input amplifier tube was grounded through 10^5 ohm resistor. Together with the capacitance of the ionization chamber of 150 pf, this gave a time constant of 15 μ sec. A pulse on the anode of the first tube had the form shown in Figure E.2. The usual form of a pulse without the leakage resistor is shown by the red line in the figure. The greatest increase in resolution of the amplifier was achieved by decreasing the coupling capacitor between the first and second stages to 10 pf (usually 100-1000 pf). This set up shifts the band width to much higher frequencies. This greatly sharpens the pulse, because the band pass of the amplifier favors the higher frequencies in the Fourier decomposition of the signal. The intended frequency band pass of the amplifier gave a number of other advantages, in particular it allowed the suppression of the low frequency flicker effect and microphonic noise from the ion chamber. The optimum value of the coupling capacitance between the first and second stages approaches experimentally the relation between the capacitance of store bought tubes and the choice of capacitance that maximized the fission fragment signal in the presence of amplifier noise.

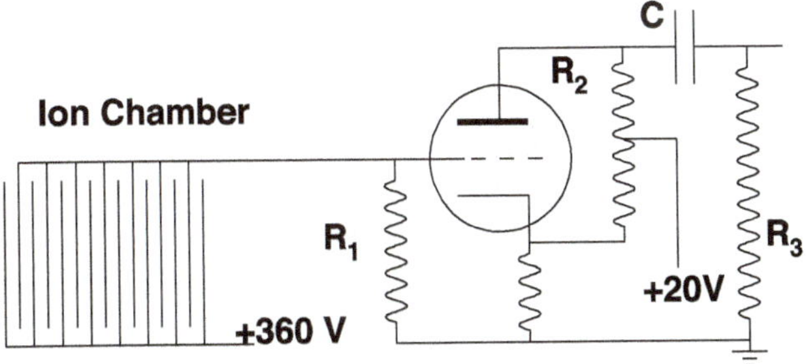

Figure E.1. R1 = 0.1 MΩ, R2 = 50 kΩ, C = 10 pf, R3 = 0.3 MΩ.

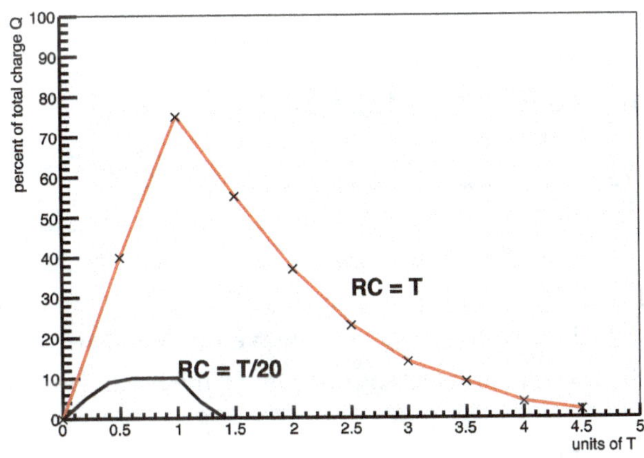

Figure E.2. Unclipped and clipped pulses at output of the first stage of the amplifier.

The increase in the ion chamber capacitance and the process of sharpening the pulse coupled with the drain resistance strongly decreased the amplitude of the voltage signal from particles, such that the α particle pulses were of the same order as the Johnson and shot noise in the first stage. Fission product pulses exceeded the background of the amplifier. In order to make the signal larger, the final tube of the amplifier, a 6-F-5, operated in a non linear region, after which the fission product pulses were 4-5 times larger than background from the superposed α particles and voltage fluctuations from Johnson and shot noise. The average gain was $\sim 10^7$.

Pulses from the amplifier output went to an oscilloscope for visual observation, and to a register built of an array of glow tubes for counting the number of pulses.

Operating the linear amplifier in a non-linear region required special care to maintain a constant gain. In our experiment, the gain was held constant to 1-2% over 5-10 hours of operation. Constant gain was maintained by comparing the anode and cathode supply voltages to a standard cell.

E.3. Experimental Results and Control Experiments

The experimental setup made it easy to observe the process of uranium fission under neutron bombardment. The sensitivity of the chamber, as one might expect, was 30-40 times greater than the sensitivity of ordinary chambers. The amplifier and ion chamber, tuned for the observation of fission fragments, gave one pulse per minute from a 1 mC (Rn+Be) source of fast neutrons placed next to the chamber.

Insert into the text by LGP

One curie is named in honor of Pierre and Marie Curie, the discoverers of radium, as the decay rate of 1 gm of $^{226}\text{Ra} \rightarrow {}^{222}\text{Rn} + \alpha$. The half-life of ^{226}Ra is 1600 y, and a little arithmetic gives 1C = 3.7×10^{10} decays/sec. The neutron flux is from $^9\text{Be} + \alpha \rightarrow {}^{12}\text{C} + \text{n}$. The α capture probability varies with the preparation of the source, and is not given in the paper. Placing the source 'next to' the chamber is not too precise either, but based on information given later in the article, the neutron flux was 50 n's/cm^2 –sec, incident on 15 g of uranium oxide. The fast neutron fission cross section for ^{238}U is $\sigma = 0.3 \times 10^{-24}$ cm^2, so if their detector was 100% efficient, they would see:

Number of fissions/sec = $(3.3\times10^{22}\,U)\times(0.3 \times 10^{-24}\,cm^2)\times50/cm^2\text{-sec}$
Fissions/sec = $0.5 \rightarrow 30$ counts/minute.

P&F observe 1 count/minute, which gives an overall efficiency of about 3% with these numbers. This efficiency is important in converting the observed spontaneous fission rate into a transition rate per nucleus.

Return to the main text.

In the first experiments with the chamber-amplifier tuned for fission fragments, but without the neutron source, the output was connected both to the oscilloscope and to the register. The count rate was low – 6 counts per hour – but this was expected, since a normal ion chamber setup would not see anything at all.[1] However, the observed pulses looked very much like the pulses observed when the neutron source was present. It was then necessary to perform a series of control experiments to determine whether or not the effect was real. The origin of the observed pulses could be caused by any of the following:

1. Reception of external perturbations of the amplifier.
2. Pile up of pulses from α decay.
3. Presence of gas amplification in separate regions of the ion chamber.
4. Accidental discharges on the surface of the uranium oxide.

Special experiments showed that not one of the above causes could have explained the effect. The chamber with plates but without uranium oxide did not give a single pulse in five hours. This showed that the observed spontaneous pulses depended on the uranium oxide, and not on external perturbations.

For the verification of the second possibility for the source of the pulses a special chamber was built in which uranium oxide was replaced with thorium oxide with additional polonium so that the α particle flux

---

[1] The neutron induced fission rate was 60 counts/hour, a factor of ten higher. [LGP]

was twice that of the uranium, but the spontaneous fission was zero. In 10 hours only three counts registered. However, the surfaces of uranium oxide and thorium oxide may not have been identical. Therefore, another control experiment was performed in which the uranium oxide chamber was filled with thorium emanation [^{220}Rn] in sufficient quantity to double the α particle flux. In this experiment, there was no increase in the observed effect, from which it follows that α particle pile up contamination could not explain the signal, thanks to the discriminating power of the amplifier.

From these special experiments, it follows that the chamber does not have a region of gas amplification, because increasing the amount of ionization would have increased the number of false pulses. Increasing the voltage on the chamber from 360 to 600 V did not noticeably increase the effect, which also excludes the possibility of gas amplification. The uranium oxide surface was covered with a 1 μm bronze foil, which decreased the amplitude and number of spontaneous pulses, but the same effects were observed from the pulses caused by neutron bombardment from a Rn+Be source. Thus all four possible sources of false signals are ruled out. It follows to note in addition that by carrying out the various test experiments in three different chambers with the same results, we have excluded the possibility that the observed effect was due to the chamber assembly.

The distribution of pulse amplitudes was recorded to confirm that the spontaneous pulses came from the fission of uranium. This was done by observing the dependence of the counting rate on the grid bias of the input tube. [Increasing the negative bias on the grid raised the threshold for input pulses.] The curves are shown in Figure E.3. The y axis is the number of counts/hour, and the x axis is the negative bias voltage on the grid. Curve one was obtained without an external neutron source, and represents pulses from α particles and Johnson noise background. For curve two an extra α source, thorium emanation [^{220}Rn], was added to the chamber. For curve 3 an external neutron source was used to stimulate uranium fission. The curve shows the distribution of amplitudes of pulses from fission products registered by the non-linear amplifier. The shape of curve 3 coincides with the distribution of pulses from spontaneous decay, in the absence of the neutron source. In the figure the number of spontaneous pulses per hour [multiplied by an appropriate normalization factor so as to lie on the curve] is shown for three different grid biases. All of these control experiments were done with the 1000 cm^2 area chamber.

Insert into the text by LGP

Figure E.3 shows pulse height distributions. Curve 1, with no neutron source, actually has a long tail like curve 3 – only at a much lower count rate. (the spontaneous decays are there.) Curve 3's long tail with approximately constant count rate as the threshold is raised demonstrates that the pulses from fission fragments are much larger than the α particle background. The three spontaneous decay points shown with the error flags are scaled up a factor of 10 or so to account for the low rate of spontaneous decays. The pulse height distribution, however, is the same as it is for induced fissions, which is the point of this plot.

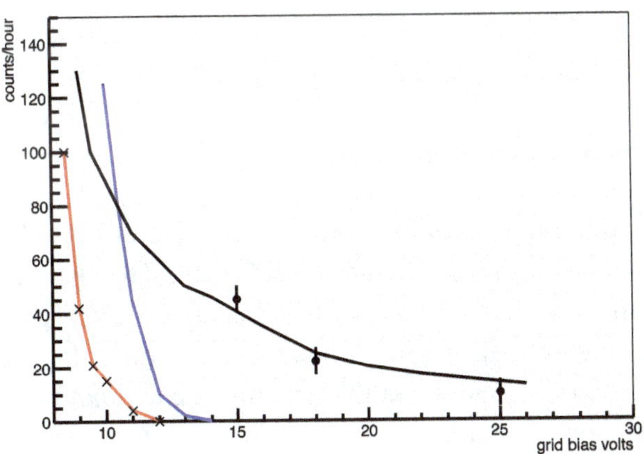

Figure E.3. Counting rate as a function of grid bias on the input vacuum tube. The data points are from spontaneous fission. The curves are explained in the text.

Recently a larger chamber, again with 15 plates, but with a total area of 6000 cm^2 was constructed. The chamber was filled with dry argon. The number of spontaneous decays observed with this new chamber increased to 25-30 counts per hour, compared to 6 counts per hour before. The counts versus grid bias curve obtained with the new chamber is shown in Figure E.4. On the basis of all of these experiments we conclude that we have observed spontaneous fission of uranium.

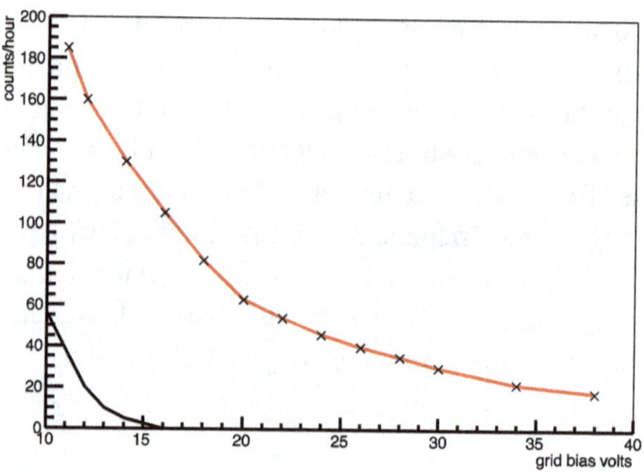

Figure E.4. New chamber with more uranium. Black curve no source. Red curve induced fissions.

E.4. Discussion of the Results

Could cosmic ray neutrons cause the observed effect? [That is, could there be no such thing as spontaneous fission. The observations were simply neutron induced fission, with the neutrons coming from cosmic rays.] Experimental data on the flux and energy of cosmic ray neutrons on the earth's surface show that these neutrons could not be the cause. In order to explain the effect with cosmic rays, we need a flux of 5 neutrons/cm^2 –sec.[2] This number is much larger than the upper limit for the flux of cosmic or earth origin neutrons. In addition, the absence of an effect in the chamber coated with thorium oxide (3 pulses in 10 hours) convinced us that it was not possible to ascribe the uranium effect to cosmic rays. It follows from the work of Petrzhak and Flerov [4], and Nikitinskaya and Flerov [5], that neutrons below about 1 MeV do not cause fission in uranium or thorium.[3] The effective cross section for thorium fission is about 5 times smaller than for uranium.[4] Consequently cosmic neutrons, unable to cause fission in thorium, cannot explain the effect observed in the uranium chamber.[5]

Insert into the text by LGP

The interview with Petrzhak translated in Chapter 5 talks about repeating the experiment in the Moscow subway, where the cosmic ray flux was attenuated by a factor of 20. It is strange that no mention of the underground measurements is made in this paper. We do learn the relevant numbers from this paper – 6 counts/hour from spontaneous fission. This very low rate would make measurements in the subway tunnel difficult if the apparatus could only be operated during quiet time, when no trains were running. In four hours you would get 24 ± 5 counts, with statistical fluctuations. It would be advantageous to take data for 24 or 48 hours straight.

Return to the main text

Can electrons or mesons in the cosmic radiation cause the fission of uranium? There is no evidence for such an interaction between electrons

---

[2]   The flux 50 n's/ cm^2 –sec used in the induced fission calculation comes from this number, and the factor of ten difference in fission rates source/no source. [LGP]

[3]   The text refers to ^{238}U. Thermal neutrons can cause fission in the isotope ^{235}U.

[4]   Pretty close. Modern numbers are σ =0.3 barns for ^{238}U fission and σ = 0.07 barns for ^{232}Th. [LGP]

[5]   They said earlier that the ^{232}Th α activity in the special test chamber was twice as much as for the ^{238}U chamber. The lifetime ratio Th/U = 3, so 2× the α activity requires 6× as much Th, just about compensating for the smaller fission cross section. But they saw only 3 events from Th in 10 hours. [LGP]

and heavy nuclei. But even if you suppose that such an interaction is possible, the effective cross section would have to be $\sigma \sim 10^{-23}$ cm^2, which is implausibly large.

Ion discharges of the Hoffman type could have given similar large pulses. The number of such discharges in the large chamber, filled with high pressure gas, was smaller than our observed effect. Indeed, in the uranium chamber the breakdown pulses would not be as large as the fission fragments, because the number of ion pairs would be much smaller. Experiments without uranium in the chamber confirmed this.[6]

Is it possible to ascribe the observed effect to neutrons which came from nitrogen gas, bombarded by uranium α particles?[7] The observed rate of 6 fissions per hour corresponds to 1/20 mC Rn+Be source.[8]In order to obtain such a number of neutrons, one would need 100 times the flux of α particles. It is possible to use the data of Libby to refute this hypothesis. Measurements with a BF$_3$ counter showed, that 1.5 kg uranyl nitrate did not emit as many neutrons as 1/10 mC (Rn+Be) source. In our experiments we had 15 grams of uranium oxide, so that only 1/50 of the observed effect could be explained by neutrons emitted by the uranium.[9]

We are inclined to think that the effect that we have observed is the result of spontaneous fission of uranium.

The question arises: is it possible that the spontaneous fission observed comes from daughter nuclei in excited states after uranium decay? We carried out a special experiment with a layer of U$_3$O$_8$ on the chamber plates enriched in UX$_1$ [10] enriched 12 times in comparison with its equilibrium concentration in natural uranium, and we did not see any increase in the effect. Therefore, we conclude that the fission comes from

---

[6]   This seems repetitive, since the effect has already been dismissed. [LGP]

[7]   The authors may be referring to the reaction ^{14}N(α,n)^{17} F, which requires kinetic energy from the α particle to go, and would give low energy neutrons. [LGP]

[8]   Should be 1/10. [LGP]

[9]   Libby's number was a limit. It is not clear what the source of decay neutrons from uranium is, other than spontaneous fission. [LGP]

[10]   Thorium 234. [LGP]

one of the natural isotopes of uranium in its ground state, with the following estimated half-lives:

$$^{238}\text{U} - 10^{16} \sim 10^{17} \text{ years}$$
$$^{235}\text{U} - 10^{14} \sim 10^{15} \text{ years}$$
$$^{234}\text{U} - 10^{12} \sim 10^{13} \text{ years}$$

The decay of ^{235}U is more probable because the height of its potential barrier lies about 1 MeV lower than for ^{238}U.

We wish to thank Professor I.V. Kurchatov for his close attention to all aspects of this experimental work.

Literature cited:

1. N Bohr and J. Wheeler, Phys Rev **56**, 426 (1939).
2. W.F. Libby, Phys Rev **55**, 1269, (1939).
3. O. Frisch, Nature, **143**, 276 (1939).
4. K.A. Petrzhak and G.N. Flerov (JETP in press).
5. T.I. Nikitinskaya and G.N. Flerov (JETP in press).

Final remarks by LGP

Regarding the three isotope 'half-lives,' each range is shorter by a factor of 100 going from 238-235-234. This simply reflects the approximate composition of natural uranium. The authors did not know which isotope was giving the signal. If the spontaneous fission signal was from a rarer isotope, like ^{235}U, then its decay rate must be higher, or its 'half-life' shorter. The signal in fact comes from the dominant isotope, ^{238}U, which has a half-life of 4.5×10^9 years.

This is a good place to discuss more fully lifetimes and decay rates, which were briefly mentioned in Chapter 2. A radioactive substance obeys an exponential decay law. The decay rate is proportional to the number present:

$$dN(t)/dt = -\lambda \, N(t), \text{ integrated gives } N(t) = N(0) \times \exp(-\lambda t). \qquad (E.1)$$

The parameter λ has units 1/time, and is called the decay rate. Its reciprocal $1/\lambda = \tau$, the lifetime. The half-life is shorter than the lifetime: $T_{1/2} = 0.693 \times \tau$, where $\log_e (2) = 0.693$. Many radioactive particles or nuclei decay into several different final states. The example before us is ^{238}U, which has at least two final states, α decay and fission. Each final state has its own transition rate λ_i and the total transition rate is the sum: $\lambda = \sum_i \lambda_i$. The lifetime is $1/\lambda$. The parent decays at a constant rate, regardless of which final state is being observed. There is no such thing as a partial lifetime. Quoting a rare decay mode in years, as Petrzhak and Flerov did – they were in good company, Bohr and Wheeler did the same thing – implies that the rare mode lives longer, which is not true. Suppose an unstable nucleus decays 90% of the time A$\rightarrow$ a+b, and 10% into A$\rightarrow$ c+d, and the lifetime of A is two days. Then if you plot the number of final states a+b for a few days, and compare it to the number c+d, the first rate is 9 times the second, but the time dependence is the same.

So let us attempt to convert P&F 's 6 counts per hour into a decay rate for ^{238}U $\rightarrow$ two fission fragments. We have 15 gm of uranium oxide, or 3.8×10^{22} nuclei, virtually all 238. The efficiency for observing fission fragments was estimated above to be about 3% (P&F do not give this number, which is of critical importance in evaluating the branching fraction.) Correcting the observed rate we have true rate/sec = $6/(3600 \times 0.03) = 0.055$ events/sec. So the true decay rate is:

$$\lambda_{\text{fission}} = (0.055 \text{ events/sec})/(3.8 \times 10^{22}) = 1.4 \times 10^{-24} /\text{sec}. \qquad (E.2)$$

The lifetime of ^{238}U is $\tau = 6.5 \times 10^{9}$ years, or $\lambda = 5 \times 10^{-18}$ /sec. The ratio of these two numbers gives the branching fraction (^{238}U $\rightarrow$ fission)/ (^{238}U $\rightarrow$ everything) $= 0.3 \times 10^{-6}$. This result is close to the book number of 0.5×10^{-6}.[11] The uranium half-life divided by the smaller 'partial half-life for spontaneous fission' quoted by P&F above gives: $(4.5 \times 10^9)/(10^{16}) = 0.45 \times 10^{-6}$, which is very close to the correct number. I do not know why P&F did not present these calculations in the paper. Their result agrees well with modern measurements, and they could have been more precise. In any event, this was a significant experiment, carefully performed in the USSR in 1940.

---

[11] Serber in 1943 quotes a number for ^{238}U spontaneous fission of 15 neutrons/kg-sec, which is consistent with modern numbers (2 neutrons per fission), indicating that the Manhattan Project had studied the problem. Robert Serber, "Los Alamos Primer," University of California Press, Berkeley, 1992. p 52.

Appendix F

Nuclear Weapons

We have discussed nuclear fission, chain reactions, reactors, and isotope separation. Now we come to the main event: nuclear weapons. What constitutes an explosion?

F.1. TNT

Let us begin with chemical explosives, TNT in particular, since its explosive yield is taken to be the standard of comparison with others, including nuclear weapons. A chemical explosive has oxygen built in to the molecule. The explosive reaction is an oxidation, and having local oxygen atoms allows the reaction to proceed fast enough to explode. Slow oxidation allows the heat generated to be carried away in an orderly fashion, with no explosion. The molecular configuration of the explosive is stable- it has a binding energy. However, the molecules formed from the parent in the final state are more stable. Such reactions are called exothermic. The difference is the explosive energy. The transition from initial to final state is like a rapid phase transition – it involves the bulk substance in a collective fashion. It is important that the final state contain gases – CO, CO_2, H_2O (vapor), or diatomic gases like N_2. The gases in the final state convert the energy of the reaction into high temperature and pressure, which serve as the medium to carry the force of the blast, and to do mechanical work.[1]

---

[1] Much of the information on TNT was obtained from: www.fas.org/man/dod-101/navy/docs/es310/chemstry/chemstry.htm

TNT has the chemical composition: $C_7H_5N_3O_6$. Six carbon atoms form a hexagonal benzene ring, with three NO_2 radicals, one CH_3 radical, and two hydrogens attached to the six corners of the hexagon. The chemical formula in a form more suggestive of its structure is written: $C_6H_2(NO_2)_3CH_3$. The molecular weight of TNT is 227 g/mol, and the molecular binding energy (also called the heat of formation) is -54.39 kJ/mol.

TNT breaks down into its final state upon detonation as follows:

$$2(C_7H_5N_3O_6) \rightarrow 3N_2 + 5H_2 + 12CO + 2C.$$

The diatomic nitrogen and hydrogen molecules are gases, with a negligible molecular binding energy. The carbon monoxide in the final state, also a gas, is the source of the exothermic energy released. The molecular weight of carbon monoxide is 28 gm/mol, and the binding energy is -111.8 kJ/mol. Taking the nitrogen, hydrogen, and carbon along for the ride, the energy release is:

Energy release/mol = -54.39 + 6×111.8 = 616.41 kJ/mol.

This converts to 2.7 MJ/kg, which is the energy in the primary explosion. In practice some extra energy is obtained by oxidizing the remaining carbon using atmospheric oxygen, and the 'standard' used by NIST for TNT is 4.18 MJ/kg.[2] This number is used for energy equivalent calculations. One mole of TNT produces 10 moles of gas in the final state (everything except the left over carbon atom, which if it does not react with atmospheric oxygen, just forms some soot at the blast site). This amount of gas is effective in converting the energy into shock waves and heat to do the work on the target substance. It is interesting to convert the explosive energy into eV at the molecular level. 1 kg of TNT has $(1000/227)×6×10^{23} = 2.6 \times 10^{24}$ molecules. So 2.7 Mj/kg $= 2.7×10^6 / (2.6 \times 10^{24} \times 1.6 \times 10^{-19}) = 6.5$ eV per molecule! This is a whopping amount of chemical energy. The final state has 10 gas molecules, so each one gets

---

[2] en.wikipedia.org/wiki/Trinitrotoluene

0.65 eV, which from KE = 3/2kT gives a temperature of 5000 K, about the temperature of the surface of the sun. For diatomic nitrogen molecule, the velocity would be v = 1.3 km/sec. The pressure of 10 moles of gas in a volume of 10^{-3} m^3 at 5000K from the perfect gas law PV = nRT would be 4.1×10^8 pascals, or 4100 atmospheres.[3]

TNT is a stable substance, which has to be detonated by a small explosion, which can be a dynamite cap or some similar fuse. One gram of TNT will completely explode in about 1 μ sec.[4] The explosive velocity, which is the speed of propagation of the explosive wave, is 6900 m/sec.[5] The blast signal would move 1 cm in about 1.4 μ sec, comparable to the time required for a 1 cm^3 block to explode. Since the speed of sound is 343 m/sec at atmospheric pressure and 20^0C, the explosive gas moves at Mach 20! This creates shock waves in the surrounding air. Explosions are characterized by the generation of a lot of energy in a very short time.

F.2. Chain Reaction Nuclear Weapons

A reactor works on slow neutrons, and only ^{235}U fissions when bombarded by slow neutrons. Each fission process produces about 2.6 neutrons, but they are not slow – fission neutron kinetic energy average ~ 2 MeV. To thermalize them in a moderator takes about 1 μ sec, which stretches the chain reaction out too long to cause an explosion. So thermal neutrons are good for reactors, but not for bombs. Both ^{235}U and ^{238}U fission when bombarded by fast neutrons. The ^{238}U cross section was discussed in Appendix E in the context of the spontaneous fission. The ^{235}U fission cross section is $\sigma = 1.2 \times 10^{-24}$ cm^2, about 4 times the ^{238}U cross section. The uranium bomb is based on fast neutron fission of ^{235}U. The mean free path for neutron capture $\rightarrow$ fission is: λ = 17 cm. The average neutron velocity v = 2×10^9 cm/sec, and the time between collisions is $\tau = 8.5 \times 10^{-9}$ sec. One kilogram of ^{235}U is 2.5×10^{24} atoms. Each fission gives 2.6 neutrons, so to take care of 1 kg of uranium would require $(2.6)^n = 2.5 \times 10^{24}$, or n $\approx$ 60 collision. Each collision takes time

---

[3]   R = 8.3 J/mol-K

[4]   geophysics.nmsu.edu/chapter01.html#3

[5]   Wikipedia, *ibid.*

τ, so it all takes place in time $t = 60 \times 8.5 \times 10^{-9}$ sec $= 0.5\ \mu$ sec, comparable to the times for explosion in TNT.

This discussion follows that of Robert Serber, who gave a series of introductory lectures to scientists arriving at Los Alamos in 1943.[6] Let $n(r,t)$ be the number of neutrons/cm^3 in a sphere of uranium metal. In a large enough sphere, there are sources and sinks of neutrons, and there is no preferred direction. Across a given surface there may be a net mass flow, because of the proximity of sources. In contrast with charge flow, where the current density $J = \rho v$, where ρ is the charge density and v the charge velocity, here the current density is proportional to the gradient of the neutron density. This is called diffusion. The current density is directed opposite to the concentration gradient – the flow tends to equalize the concentration. Heat flow, which is the transfer of kinetic energy through a medium, behaves in a similar fashion. Both processes are irreversible – the flow in time cannot run backwards. If $n(r,t)$ is the neutron number density in m^{-3}, then the current density $J = -D \times \nabla n$, where D is called the diffusion coefficient, and is assumed constant. Because of the presence of sources, the 'conservation of neutron flux' is written:

$$dn/dt + \nabla \cdot J = (\nu - 1) \times n/\tau. \tag{F.1}$$

Here ν is the average number of neutrons per fission, and τ is the average time between fissions. Using the relation between J and n then gives the diffusion equation for the neutrons:

$$dn/dt - D\nabla^2 n = (\nu - 1) \times n/\tau. \tag{F.2}$$

---

6   Robert Serber, "Los Alamos Primer," University of California Press, Berkeley, 1992. p 25 ff. The cross sections and other experimental numbers used here are modern, and in general differ from those of 1943. (taken from www.kayelaby.npl.co.uk/atomic_and_nuclear_physics/4-7/4_7_2.html)

The coefficient on the right hand side $(v-1) \times n/\tau$ is the time dependent growth of the neutron flux. The first derivative with respect to time indicates irreversibility. Now assume a time dependence $n(r,t) = n(r) \exp(v' t/\tau)$, where v' is called the 'effective neutron number.' Then Eq. (F.2) becomes time independent:

$$D\nabla^2 n = (v' - v + 1)n/\tau. \tag{F.3}$$

Defining $k^2 = (v - v' - 1)/(D\tau)$. $\tag{F.4}$

Then we have the differential equation in classic form:

$$(\nabla^2 + k^2)n = 0. \tag{F.5}$$

In the case of spherical symmetry, the Laplace operator becomes:

$$\nabla^2 = d^2/dr^2 + (2/r)\, d/dr. \tag{F.6}$$

And substituting $n(r) = f(r)/r$, gives the simple harmonic equation for $f(r)$:

$$d^2 f/dr^2 + k^2 f = 0. \tag{F.7}$$

Solutions to this equation are of the form $f(r) = A \times \exp(ikr) + B \times \exp(-ikr)$, where A and B are arbitrary constants, to be determined from the boundary conditions. If $n(r)$ is finite at $r = 0$, then $f(0) = 0$, or $A = -B$. If the sphere has a finite outer radius R_o, then there is a boundary condition at the surface. $f(R_o) = 0$ means that no neutrons leave the sphere. If that is true, then $f(R_o) = A \times \sin(kR_o) = 0$, with the first solution being $k = \pi/R_o$. A little algebra then gives the equation for v':

$$v' = v - 1 - D\tau \, (\pi/R_o)^2 \qquad\qquad \text{(F.8)}$$

If $v' > 0$, the neutron density grows exponentially with time. If $v' < 0$, the neutron density is exponentially damped, and if $v' = 0$, the neutron density is constant. For ^{235}U the number of extra neutrons is $v = 2.6$, so $v - 1 = 1.6$, and the critical value $v' = 0$ occurs for $D\tau(\pi/R_o)^2 = 1.6$. In order to solve for R_o, the critical radius, we need to evaluate the expression $R_o = (D\tau\pi^2/1.6)^{1/2}$. The diffusion coefficient D depends on the total mean free path for the neutrons:

$$D = \lambda \, v/3. \qquad\qquad \text{(F.9)}$$

where λ is the transport mean free path, v is the average neutron velocity, and 1/3 comes from the number of space dimensions.[7] The transport mean free path is inversely proportional to the total neutron cross section:

$$\lambda = 1/N\sigma_t \qquad\qquad \text{(F.10)}$$

$N = 0.48 \times 10^{23}$ is the number of uranium nuclei per cm^3, and σ_t is the sum:

$$\sigma_t = \sigma_{elastic} + \sigma_{n\,gamma} + \sigma_{fission} + \sigma_{inelastic}. \qquad\qquad \text{(F.11)}$$

These cross sections for fission neutrons on ^{235}U are listed in a table in www.kayelaby.npl.co.uk, and are reproduced here for convenience. The isotope ^{233}U is included in the table, although its application as a weapon

---

[7] There is a not entirely satisfactory derivation of this in Serber, *ibid*, p 67ff. In the context of neutron diffusion in reactors, see J.R. Lamarsh, "Introduction to Nuclear Reactor Theory," *ibid.*, Chapter 5.

is not discussed. It is an artificial isotope like plutonium, made in a nuclear reactor.

Table F.1.

Nucleus	σ in barns averaged over fission spectrum				# fission n's
	σ_{nn}	σ_{ngamma}	σ_{nf}	$\sigma_{nn'}$	ν
^{233}U	4.23	0.063	1.84	1.13	2.6
^{235}U	4.4	0.095	1.22	1.92	2.6
^{239}Pu	4.6	0.065	1.80	1.37	3.1

Note that the ^{239}Pu fission cross section is about 50% larger than for ^{235}U, and $(\nu - 1)$ is about 30% larger. Both numbers contribute to a smaller critical mass for ^{239}Pu.

The elastic cross section (second column in Table F.1) should be decreased by the factor $<1-\cos\theta>$, which accounts for the forward scattering of the neutrons. Because the angular distributions of MeV neutrons elastically scattered by uranium and plutonium are peaked forward, the elastic cross section is not fully effective in removing neutrons from the flow. If the scattering is isotropic, $<\cos\theta> = 0$, and the elastic cross section is not modified. Taking $<\cos\theta>=0.7$ for fission neutrons gives:

$$\sigma_t = 4.6 \times 10^{-24} \text{ cm}^2, \text{ and } \lambda = 4.6 \text{ cm} \tag{F.12}$$

The time between fissions $\tau = 1/(N\sigma_{fission} v)$, so in the formula for the critical radius the neutron velocity cancels out:

$$R_o = \pi \times (\lambda/(4.74 N\sigma_{fission}))^{1/2} = 12.7 \text{ cm} \tag{F.13}$$

The critical volume $V_o = 4\pi R_o^3/3 = 8600 \text{ cm}^3$, and, using the density $\rho = 19 \text{ gm/cm}^3$, gives a critical mass $M_o = 163 \text{ kg}$. This number is too large because of the boundary condition $n(R_o) = 0$. A more careful treatment extends the zero to $n(R_o + 0.7\lambda) = 0.$[8] This reduces $R_o = 9.5$ cm, and the critical mass to $M_o = 68$ kg. Repeating this exercise for ^{239}Pu gives a reduced radius $R_o = 5.8$ cm, and $M_o = 15$ kg. Modern numbers for the critical mass of an isolated sphere of material are:[9]

Table F.2

Isotope	critical mass	with U tamper	density	λfission
^{233}U	16 kg	--------	18.9 g/cm^3	11 cm
^{235}U	52 kg	15 kg	18.9 g/cm^3	17 cm
^{239}Pu	10 kg	5 kg	19.4 g/cm^3	13 cm

F.3. The Tamper

The 'tamper,' a spherical shell of heavy metal outside the critical mass, can serve two purposes in the explosive device. The efficiency of the explosion, discussed below, is defined as the number of active nuclei that undergo fission divided by the total number available. As the sphere of active material heats up, it expands, and eventually the density drops below the critical value, and the reaction stops. The inertia of the tamper, in contact with the expanding sphere of active material, can retard the expansion, and therefore increase the increase the efficiency. The tamper can also reflect neutrons back into the exploding material, and effectively decrease the mass required.[10] Critical assemblies, discussed in Appendix J, are not designed to explode, so only the second feature of the tamper, the reflection of neutrons back into the active material, comes into play. In the bomb case, the inertial effect may be the more important, because the times involved are so short that some of the neutrons reflected back arrive after it is all over.

---

[8] J.R. Lamarsh, "Introduction to Nuclear Reactor Theory," *ibid.*, Eq. 5-51, p 134.

[9] Nuclearweaponarchive.org/Nwfaq/Nfaq2.html paragraph 2.1.2

[10] The U tamper values in Table F.2 come from R. Serber, "Los Alamos Primer," *ibid.*, p 33.

A typical spherical shell tamper made of depleted uranium with inner radius $r_1 = 15$ cm, and outer radius $r_2 = 30$ cm would weigh 2000 kg, or 2 metric tons. The diameter of a 50 kg uranium sphere is about 16 cm, roughly one fission mean free path. A uranium tamper drops the critical mass to 15 kg, so twice the critical mass, 30 kg, would be a sphere 14 cm in diameter.

The derivation of the critical radius with Equation F.8. was based on the boundary condition that the neutron density vanish at the surface of the sphere. Obviously if this is the case no neutrons escape for the tamper to reflect back. The presence of the tamper thus changes the boundary conditions on the neutron density, effectively increasing the density at larger radii inside the active sphere, and increasing the number of fissions. The analysis follows that presented by Serber. Diffusion theory is used in both the active material and the tamper. The active material of mass M and radius R is surrounded by an inert tamper (usually a heavy metal, but for critical assemblies beryllium is sometimes used) spherical shell of outer radius R.' Let n' (r,t) be the neutron density in the tamper. The neutron diffusion equation in the tamper is:

$$\partial n' (r,t)/\partial t = D' \nabla^2 n' (r,t) - (\alpha/\tau) n '(r,t). \qquad \text{(F.14)}$$

D' is the diffusion coefficient in the tamper, and the (α/τ) term allows for neutron absorption in the tamper. Assume that the time dependence in the tamper is the same as in the active material $\sim \exp(v't/\tau)$, where the effective neutron number v' vanishes for criticality. Then the differential equation in space reads:

$$\nabla^2 n' (r) + k'^2 n' (r) = 0, \quad k'^2 = -(v' + \alpha)/(D'\tau). \qquad \text{(F.15)}$$

The diffusion current density $j = D \nabla n$ is conserved at $r = R$, the interface between the active material and the tamper. Continuity equations at the interface are:

$$n(R) = n'\ (R);\ D\ (dn/dr)_{r=R} = D'\ (dn'/dr)_{r=R} \tag{F.16}$$

To simplify the problem, assume that the reaction is critical ($v' = 0$), and there is no absorption in the tamper ($\alpha = 0$). Then $k' = 0$, and the diffusion equation in the tamper reduces to Laplace's equation:

$$\nabla^2\ n'\ (r) = 0. \tag{F.17}$$

The origin $r = 0$ is excluded from the tamper, so a suitable solution to equation F17 in spherical coordinates is:

$$n'\ (r) = A/r + B,\ r > R. \tag{F.18}$$

where A and B are constants to be determined. If at the outer radius R' the neutron density vanishes, $B = -\ A/R.'$ Now further assume that $D = D'$, that is the neutrons have the same mean free path in the active material and the tamper. Then the matching conditions at the interface $r = R$ become:

$$\sin(kR)/R = A(1/R - 1/R'),\ \text{and}\ k\cos(kR)/R - \sin(kR)/R^2 = -A/R^2. \tag{F.19}$$

Eliminating A gives a transcendental equation for k:

$$(R' - R)k\cos(kR) + \sin(kR) = 0. \tag{F.20}$$

For a very large outer radius $R' \gg R$ and $kR' \gg 1$ Equation F.20 reduces approximately to $\tan(kR) \approx -kR'$, or $kR \approx \pi/2$. The simplest solution with no tamper was $\sin(kR) = 0$, or $kR = \pi$, a factor of two larger. Since k^2 defined in Eq. (F.4) is fixed, the tamper in this simple model has halved the critical radius, and decreased the critical mass by a factor 1/8. This has all happened because the tamper has moved the zero boundary condition on the neutron density out to a much larger radius R.' Serber says that a more realistic treatment gives a factor of ¼ rather than 1/8.[11]

The critical radius depends on the density ρ, the average number of fission neutrons ν, and the total and fission cross sections σ_t and σ_f as follows:

$$R_o \sim (1/\rho) \times (1/\sigma_t \sigma_f (\nu - 1))^{1/2} \qquad \text{(F.21)}$$

Since $M \sim \rho R^3$, we have for the mass dependence:

$$M_o \sim (1/\rho^2) \times (1/\sigma_t \sigma_f (\nu - 1))^{3/2} \qquad \text{(F.22)}$$

The cross sections and ν are constants, but ρ can be varied. An implosion bomb is based on the fact that if the density can be increased a factor of two by compression, the critical mass will decrease a factor of four, thereby requiring less of the very expensive critical material. This technique is almost universally used in nuclear weapons design. A sub critical mass before compression is surrounded by a spherical shell of conventional explosive, which is detonated in such a way as to create a spherical compression shock wave surrounding the mass on all sides simultaneously. The shock pressure can reach 4×10^5 atmospheres, and by focusing inward can exert pressures of several million atmospheres, squeezing a spherical device by a factor of two to three, in one to four μ sec. The maximum compression lasts about one μ sec, and results in a shorter λ, a faster reaction time, and a higher yield.[12]

Figure F.1 shows a plot of the density of uranium as a function of external pressure. This curve was obtained at the Soviet weapons lab, Arzamas-16.[13] Note that uranium density increases a factor of two under

---

[11] R. Serber, "The Los Alamos Primer," *ibid.*, p 31.

[12] nuclearweaponarchive.org *ibid.*, paragraph 2.1.4.

[13] L.V. Altshuler and K.K. Krupnikov, "Experimental Studies at Arzamas-16," in HISAP96, Vol **1**, *ibid.*, p 184. We reproduce Figure 6.

about 8 million atmospheres. This is an unbelievable, unworldly pressure, and indicates some of the conceptual difficulty in understanding the mechanics of an implosion weapon. 'Adiabatic' means that the curve is for constant entropy. Such extreme pressures are well above the yield point of the material, so the substance becomes some sort of fluid. The thermodynamic state of the material in the presence of the shock wave involves the pressure, temperature, density, volume, shock velocity, and fluid velocity. The curve in Figure F.1 was derived from velocity measurements on a uranium sample subjected to an implosion shock wave from chemical explosives. The curve was obtained by measuring various velocities, and using the formulas of shock wave theory (the Rankine-Hugoniot equations).

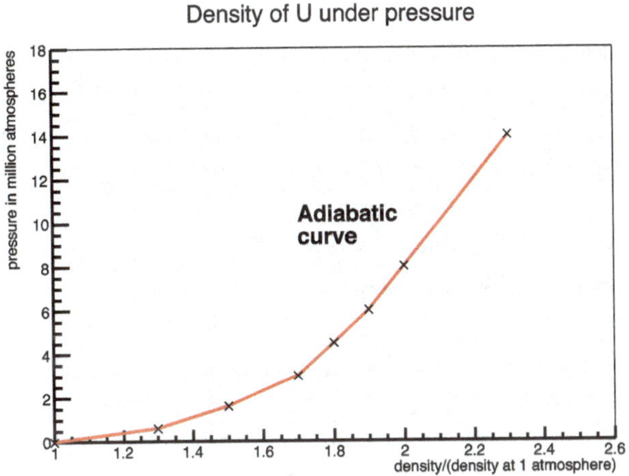

Figure F.1. Compressibility of uranium under extreme pressures, measured at Arzamas-16.

Equation F.8 gives the formula for v', which was set to zero for the criticality calculations. The time dependence of the neutron density was assumed to be $n(r,t) = n(r) \exp(v' t /\tau)$, and $v' > 0$ means exponential growth of the neutron density. If $v'(R_o) = 0$, then:

$$v'(R)/\tau = ((v - 1)/\tau) \times (1 - R_o^2/R^2). \tag{F.23}$$

This is the time constant in the exponential. If the radius is $R = R_o \times (2)^{1/3}$ – twice the critical mass – then $v'(R)/\tau = 0.6 \times 10^8$ sec^{-1}, using $\tau = 10^{-8}$ sec. Hence the time required for the neutron density to increase by a factor of $e = 2.718$ is 2×10^{-8} sec. The reaction grows very fast for a super critical mass. This is $0.02 \, \mu$ sec. Recall that it takes about one μ sec to explode one gm of TNT.

This is a book about the Soviet Atomic Project, with the Manhattan Project woven in as a framework. Both projects were huge intellectual and engineering challenges, with interesting physics problems, interpersonal relations, government sponsorship and interference, large investments in human and industrial resources, national pride, etc., continuing for several years. The subject is one of endless fascination. But at the end there were terrible explosions: three in the Manhattan Project, two of which were directly applied to human populations, and many more developmental tests during the 1950's and 1960's. The Soviet Union only carried out tests, but the Soviet citizens nevertheless paid a big price for the nuclear arsenal. The mutual destruction weapons arsenals assembled by the adversaries formed the basis for the stalemate in the cold war. We are still living today with its consequences.

We have come to the point where we must briefly discuss the detonation and destructive effects of nuclear weapons. This is not a pleasant topic. In the section above on TNT we learned that the energy liberated in the chemical explosion is carried by expanding gases in the final state. A nuclear explosion has more channels for the energy to flow. There are conventional pressure and shock waves, but also a high flux of neutrons, and radiation from the intense heat of the fireball. These components are more or less instantaneous with the blast, lasting a few seconds. Then there are the fission products, which have longer half-lives – sometimes years – and which can hang around as radioactive contamination long after the explosion. A detonation in the air can distribute the radioactive contamination over a wide area, depending on the winds. This airborne radioactivity has served as evidence of a remotely detonated nuclear blast. It is interesting to note that a meteorite, entering the earth's atmosphere and burning up, exhibits both the fireball and the shock waves of a nuclear explosion.

The source of energy is the fission process, which has the same energy channels in the final state in a reactor or a bomb. The energy feeds into three main effects:

Thermal radiation and light from the fireball   35%
Shock waves in the atmosphere from the blast   50%
Energy from nuclear processes – both prompt and delayed   15%

About 85% of the total energy released becomes kinetic energy of the fission fragments.[14] This energy is converted to heat as the fragments range out in the metal, and feeds the first two entries above. The remaining 15% is divided up among the kinetic energy of the prompt neutrons, the 'delayed' neutrons and gamma rays from de-excitation of the fission fragments, and finally the β decays of the fission fragments into stable nuclei. Fission fragments are neutron rich, so the β decay is: $n \rightarrow p + e^- + \bar{\nu}_e$. Half of the energy goes into the undetected electron anti-neutrino, which is a few % of the total amount.

The middle entry is the only one for TNT, but the 'TNT equivalent' refers to the total energy of the explosion. Therefore, the shock wave destruction of a 20 kTon nuclear explosion would match that of 10 kTons of TNT. Most of the destruction is caused by the overpressure of the shock wave, but if one estimates the energy of the explosion based solely on this effect, one must remember to multiply by a factor of two. The energy per unit volume is important for the extreme pressure and temperature generated by a nuclear explosion. A 50 kg ^{235}U weapon in the shape of a sphere would have a radius of about 8.5 cm, and perhaps an energy of 20 kTon of TNT. The same amount of TNT – density 1.65 gm/cm^3 - would fill a sphere of radius 14 meters! It is the increase in energy density by a factor of 10^6 or so that gives the high temperature needed to create the

---

[14] Detection of fission fragment kinetic energy is discussed in Appendix E.

fireball. Typical temperatures for TNT are around 5000 K, comparable to the surface of the sun, while in a nuclear explosion the temperature is comparable to the interior of the sun (more than 10^7 K – see Figure F.2). Chemical explosions do not create fireballs. The intense heat of the fireball, lasting a few seconds after the blast, can cause fatal burns, but compared to the shock wave it does relatively little damage to structures. The shock wave is generated at the outer edge of the fireball.

Assuming a total energy release of 200 MeV per fission, 85%, or 170 MeV, goes into heat inside the bomb material. Complete burning of one kilogram of ^{235}U would then liberate 6.4×10^{13} joules. Using the TNT standard of 4×10^6 joules/kg, and one metric ton = 1000 kg, one kg of ^{235}U is converted into 16,000 tons of TNT. The destructive force of nuclear weapons is almost impossible to grasp. The evolution of the explosion is described in Figure 3, p 41 in Serber. That figure is reproduced as Figure F.2.

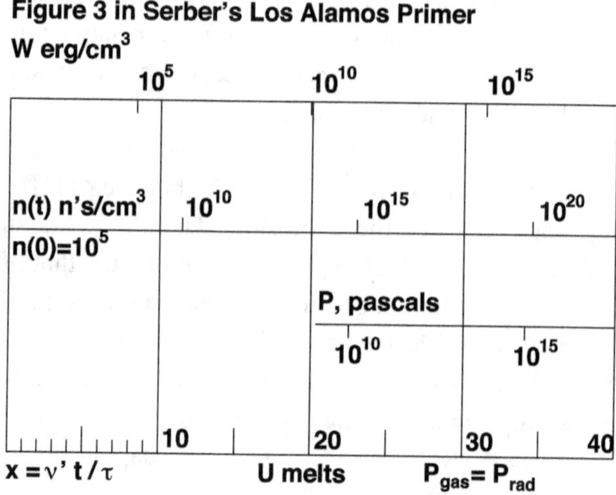

Figure F.2. Abscissa is the exponential in the neutron density growth equation. W is proportional to n. The numbers are discussed in the text. Uranium melts around 1400 K, and the radiation pressure equals the gas pressure at $T = 1.5 \times 10^7$ K.

There are a few formulas needed to understand Figure F.2. The first is mentioned in the text, but not listed as a numbered formula. It is the formula for the time dependence of the neutron density, which is the driver for the entire process:

$$n(t) = n(0) \exp(v' \, t/\tau). \tag{F.24}$$

The reduced multiplication factor v' is defined in Eq. (F.23). Equation (F.24) is the driving formula for the explosion. The energy created by each fission is $\varepsilon = 170$ MeV $= 4.3 \times 10^{-4}$ ergs, where we use ergs instead of joules to match Serber's units. Remember that 1 joule $= 10^7$ ergs. The energy density W is simply:

$$W(t) = \varepsilon \times n(t). \tag{F.25}$$

Equations (F.17) and (F.18) give the n(t) and W(t) horizontal lines on Figure F.2, with the initial condition $n(0) = 10^5$, which was used by Serber. At $x = v't/\tau = 20$ the absolute temperature, derived from the kinetic energy of the atomic constituents, reaches the melting point of uranium. If N is the number of atoms per unit volume, then the kinetic energy of each atom is given by:

$$KE(t) = W(t)/n = kT. \tag{F.26}$$

where k is Boltzmann's constant $k = 1.38 \times 10^{-16}$ erg/K, and T is the absolute temperature. Therefore, given n, the top horizontal axis can be converted to absolute temperature. $x = 20$ corresponds to $W = 10^{10}$ erg/cm^3. Assume 50 kg of ^{235}U, which, because of a tamper, is 2 $\times$ the critical mass M_o.[15] The uranium is normal density $\rho = 19$ gm/cm^3, so the

---

[15] For this exercise I assume a critical mass of 25 kg, a bit larger than in the table.

volume is V = 2630 cm^3. The molecular weight is 235 gm. So the number of moles = 213.[16] The total number of uranium atoms is:

$$N_{tot} = 213 \times 6 \times 10^{23} = 1.2 \times 10^{26} \text{ atoms.} \qquad \text{(F.27)}$$

The total number per unit volume is $N_{tot} / V = 4.8 \times 10^{22}$ atoms/cm^3.

The total reaction time is very short. The end of the graph is x = 40 = v't/τ. Equation F.16 gives the formula for v.' For $R_o/R = (1/2)^{1/3}$ i.e. two critical masses, and (v - 1) = 1.6, we have v' = 0.59. Taking $\tau = 10^{-8}$ sec as the time between fissions, the time t = 1.7 × 10^{-8} x, or the end of the plot is t = 0.7 μ sec. In this short time it is reasonable to assume that the volume of the uranium, and the number of atoms per cm^3 remains constant. The assumption of constant volume has to be relaxed to estimate the time that the expansion caused by the generated heat quenches the chain reaction. This effect will be discussed below.

The top axis can then be converted into temperature:

$$T = W \, (V/N_{tot}) \times (1/k). \qquad \text{(F.28)}$$

For x = 20, this formula gives T = 1500 K, which is close enough to the melting point of uranium – 1400 K. (It is perhaps worth emphasizing that this discussion is only approximate, and makes no claim to high precision. That is the reason for Serber's logarithmic plot.) At x = 20, the middle of the plot at about .35 μ sec, W = 10^{10} erg/cm^3, corresponding to $N \approx 10^{14}$ neutrons/cm^3.

After x = 20, we have a gas, and we can apply the perfect gas law to calculate the pressure:

––––––––––––––––––––––––––––

[16] Serber does not give this information, but we shall see that the numbers work out.

$$PV = nRT. \tag{F.29}$$

Here n is the number of moles (213), R is the gas constant $R = 8.3$, T is in K, V in m^3, and P is in pascals. One pascal $= 1$ newton/m^2. One standard atmosphere is 10^5 pascals. After the uranium melts, the pressure can be figured from the temperature and the perfect gas law, and these numbers are shown on the pressure line in the plot. The temperature comes from the top line – W, and Eq. (F.28). Then the pressure comes from Eq. (F.29).

The energy source is the fission of the uranium, but as the system gets hotter, black body radiation becomes important. The pressure from black body radiation is given by the formula:

$$P = \sigma \, T^4 / c. \tag{F.30}$$

where $\sigma = 5.7 \times 10^{-8}$ joules/sec $-m^2 - K^4$ is Stefan's constant, and $c = 3 \times 10^8$ m/sec is the velocity of light. As the temperature increases, the radiation pressure overtakes the gas pressure, and radiation dominates, although the source of energy remains the uranium nuclei. The pressure cross over occurs when:

$$T = (nRc/\sigma \, V)^{1/3}. \tag{F.31}$$

For these numbers $T = 1.5 \times 10^7$ K, around $x = 30$.[17] For longer times, the radiation dominates. The end point $x = 40$ corresponds to $W = 10^{19}$ erg/cm^3, which is complete combustion of 50 kg of ^{235}U, that is an explosion with 100% efficiency.[18]

---

[17] $T = 1.5 \times 10^7$ K is comparable to the core of the sun, where particle kinetic energies are in the keV range, necessary to ignite nuclear fusion.

[18] Total energy density for 50 kg is 1.45×10^{19} erg/cm^3, or endpoint eff at $x=40$ is 70%, rather than 100%, but close enough.

Nuclear weapons do not achieve 100% efficiency. Even with a tamper, which helps contain the original volume, the assumption of constant volume of the fissile material does not hold. The material expands, and the chain reaction stops. The Hiroshima bomb (Little Boy in Los Alamos jargon), which was 90% ^{235}U, had an efficiency of only about 2%. Plutonium weapons are more efficient, partly because the critical mass is much smaller. The Nagasaki bomb (Fat Man), an implosion ^{239}Pu weapon, was 20% efficient.

Efficiency = (energy released)/(fission energy of all of the material), which is the same as (number of nuclei that fission)/(total number of fissionable nuclei) The critical mass varies as $1/\rho^2$ Eq. (F.22). Suppose we start with twice the critical mass. Then the density has to drop by $1/\sqrt{2}$ to reach one critical mass, whereupon the reaction ceases. The expanding material has a constant mass, so the density is inversely proportional to the cube of the radius:

$$\rho_1/\rho_2 = (R_2/R_1)^3. \tag{F.32}$$

Let R_1 be the initial radius of the $2 \times M_c$ material – $R_1 = (2)^{1/3}\,R_o$, the critical radius. Then at radius $R_2 = (2)^{1/6} \times R_1$ the density has dropped enough to give $1 \times M_c$, setting $v' = 0$, and switching off the explosion.[19] $(2)^{1/6} = 1.12$, so the radius of the sphere has increased 12%. The radius is several centimeters, so this change is not all that small – a centimeter or so. The end point of the time scale in Figure F.1 is only 0.7 μ sec, so to cut off the reaction prematurely by expansion, the radial velocity has to exceed $\sim 3 \times 10^6$ cm/sec, or about 0.0001 $\times$ velocity of light. That is pretty fast for a lump of uranium metal, but past $x = 20$ in Figure F.1 we are dealing with a gas. So how fast does the sphere expand in the range $20 < x = v'\,t/\tau < 40$?

Take $x = 30$, where the radiation pressure equals the gas pressure, at $T \approx 1.5 \times 10^7$ K. At this temperature the velocity of a uranium ion is about v

---

[19] If $x = (2)^{1/6}$, then $x^6 = 2$.

= 10 cm/μsec, which is fast enough to cause problems. The time between x = 30 and x = 40 is 0.2 μ sec, so at x = 35 the radius of the sphere could increase by as much as a centimeter, stopping the explosion. At this point in time the energy released would be about 1% of the total. A close fitting tamper would decrease the rate of expansion, and hence increase the efficiency.

Serber derives a formula for the efficiency. Let R_o be the critical radius at normal density, and R_1 be the initial radius of the sphere of material, also at normal density ($R_1 > R_o$). Define Δ:

$$\Delta = (R_1 - R_o)/R_o, \text{ or } R_o/R_1 = 1/(1 + \Delta). \tag{F.33}$$

Then substituting (R_o/R_1) in Equation F.23 gives a formula for v' to first order in Δ after a little algebra:

$$v' = 2\Delta \times (v - 1) \tag{F.34}$$

If Δ vanishes so does v', and the reaction ceases.

Assuming we start with $2 \times M_o$ as above,

$$R_2 - R_1 = ((2^{1/6} - 1) \times (2)^{1/3} R_o \approx (\Delta/2) \times R_o^{20} \tag{F.35}$$

R_1 is the initial radius for twice the critical mass $R_1 = (2)^{1/3} R_o$, and R_2 is the final radius at which the chain reaction stops $R_2 = (2)^{1/6} R_1$. A little arithmetic gives $R_2/R_1 = (R_1/R_o)^{1/2} \approx 1 + \Delta/2$ to first order; so $R_2 - R_1 = (\Delta/2) \times R_1$ - Eq. (F.35) to first order.

Serber then uses the velocity of the gas molecules, which is comparable to the velocity of sound, to estimate the time required for the gas to expand

---

[20] R. Serber, "Los Alamos Primer," *ibid.* second formula at the top of p 42.

sufficiently to reach the critical radius at the lower density, when $v' = 0$. If the expanding gas reaches the critical radius in time τ/v', then

$$v = (v'/\tau) \times (R_2 - R_1) = (v'/\tau) \times (\Delta/2)R_o \qquad \text{(F.36)}$$

Then the energy/cm^3 = $\frac{1}{2} \rho v^2 = (1/8) \rho (v'/\tau)^2 \Delta^2 R_o^2$, and using Eq. (F.34) for v' gives

$$\text{energy/cm}^3 = \frac{1}{2} \rho (v - 1)^2 \Delta^4 (R_o/\tau)^2 \qquad \text{(F.37)}$$

The total energy/cm^3 = $\varepsilon \rho N_{av}/MW$, where ε is the energy release per fission (170 MeV), N_{av} = Avogadro's number = 6×10^{23} atoms/mole, and MW = 235 gm. The density cancels out in the ratio, which is the efficiency:

$$f = (\frac{1}{2} (v - 1)^2 \Delta^4 (R_o/\tau)^2 \times MW)/(\varepsilon N_{av})^{[21]} \qquad \text{(F.38)}$$

Substituting in the numbers for ^{235}U and $2 \times M_o$: $v - 1 = 1.6$, $\Delta = 0.26$, $R_o = 8$ cm, $\varepsilon = 2.9 \times 10^{-4}$ ergs, gives $f = 0.5\%$. This low efficiency for twice the critical mass without a tamper agrees with Serber's estimates. As already mentioned, the tamper will increase the efficiency, both by returning neutrons that escape back to the reactant, and by inhibiting the expansion of the active material.

In order to improve on the estimated efficiency, I wrote a one page program in C++ to model the time dependence of the expansion. I was encouraged to do this by a similar calculation presented by Bruce Reed in

---

[21] Note that this formula is dimensionally correct. Mass $\times v^2$ in the numerator is canceled by ε in the denominator. A similar, but not identical, formula is in the Frisch-Peierls memorandum of 1940. See www.atomicarchive.com

his book, although this result is entirely my own invention.[22] The input constants (not always in C++ format) are as follows:

umass = 50000. gm.

critmass = 25000. gm.

$n(0)$ = 100000. Serber's initial neutron density.

ε = 0.00029; fission energy release in ergs.

wmax = 1.45×10^{19} ergs/cm^3.

$\rho(t_0)$ = 19 gm/cm^3; initial density of uranium.

$V(t_0)$ = umass/$\rho(t_0)$; initial uranium volume, vzero.

$r(t_0)$ = $(3*\text{vzero}/4*\pi)^{1/3}$; initial radius.

$n(t_0)$ = $n(0)*\exp(v't_0/\tau)$; initial neutron density number/cm^3.

w = $\varepsilon*n(t_0)$; initial energy density.

v = 2.6; avg number of fission neutrons.

v' = $(v\text{-}1)*(1 - (\text{critmass}/\text{umass})^{2/3})$; effective neutron number. As the uranium expands, critmass increases, and v' vanishes when critmass = umass, stopping the chain reaction.

τ = 10 nsec; avg time between fissions.

x = $v't_0/\tau$ = 20; start time for the calculation. t_0 = 340 nsec.

Δt = 10 nsec; time increment in the loop.

The loop has 100 steps, incrementing the time by 10 nsec each step, starting from 340 nsec. First it calculates the speed of the uranium atoms from the energy density:

$$u(t) = 0.006*(w(t)*V(t_0))^{1/2}.$$

Then the change in volume in time Δt:

$$\Delta V = \pi*r(t)^2 *u*\Delta t,$$ where I divide by 4 to account for the outward radial direction. Then increment the volume:

$$V(t+\Delta t) = V(t) + \Delta V.$$

---

[22] Bruce C. Reed, "The Physics of the Manhattan Project," 3rd ed. Springer, Heidelberg, 2015, Chapter 2. A very nice book at undergraduate physics level.

Calculate the new density:

$\rho(t)$ = umass/V(t), and the new critical mass: newcrit = $\rho(t_0)^2$ * critmass/$\rho(t)^2$; and finally ν' = (ν-1)*(1-(newcrit/umass)$^{2/3}$). The chain reaction is driven by $\nu.'$ The neutron density and energy density are then incremented:

$$\Delta n = n(0)*(\nu'/\tau)*\Delta t*\exp(\nu't/\tau); \text{ and}$$
$$n(t+\Delta t) = n(t) + \Delta n.$$

w = ε*n, ends the loop. Energy flow into radiation, which becomes significant around x=30, is ignored. The results of this exercise are shown in Figure F.3. There are two curves. The lower one uses ΔV as above, and has an efficiency of 0.01%. The volume change was suppressed by an arbitrary factor of 1/25 for the second curve, which has an efficiency of 4%. This illustrates the dramatic effect that a tamper has for the efficiency of the explosion by delaying the volume expansion.[23] Of course this exercise is only a model under very limited assumptions, and not a true calculation. It is based on Serber's primer, as illustrated in Figure F.2. It only adds a time dependent volume term.

This step calculation could be done by hand. It would probably take about one hour. The 'computer' would write down a series of numbers at each step – n, w, ν', etc., and then plot the curves. Changing any parameter would require another hour of calculation. Hand written numbers are subject to human error.

---

[23] For comparison both Serber and Reed estimate the uranium bomb efficiency at around 1%.

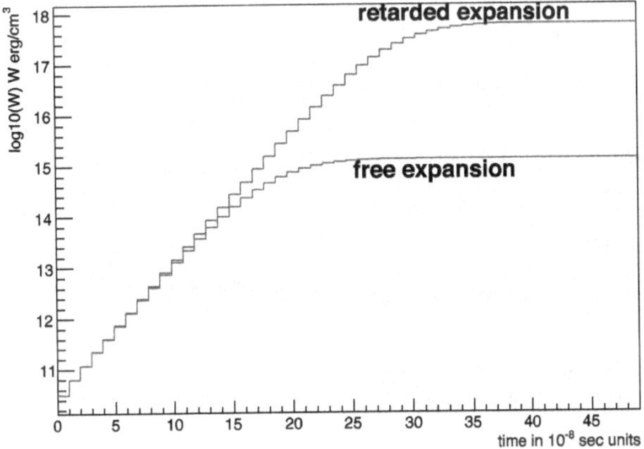

Figure F.3. Time dependence of the log of the energy density for the parameters used in Figure F.2. The time axis is offset by 340 nsec, i.e. 30 = 640 nsec after t=0. The efficiency of the top curve, 4%, is 40 times larger than for the lower curve, showing the dramatic effect of a tamper in retarding the expansion of the uranium melt.

The next topic in the 'Primer' is pre-detonation and detonation. The former is to be avoided, and the latter is to be achieved in a timely fashion. Detonation requires neutrons to seed the chain reaction, and simultaneously $v'>0$, or the assembly of more than one critical mass. Neutrons are always around at some level. Cosmic rays produce neutrons by collision with the surrounding material. Petrzhak and Flerov worried about this as a source of background to their measurement of spontaneous fission of ^{238}U as described in Appendix E. Spontaneous fission is an obvious source of background neutrons. If the tamper is made of ^{238}U, there is a nice neutron source built in to the weapon, ready to cause an explosion at any time. A chain reaction is avoided by the absence of a critical mass until the system is triggered. Table A.2 in Appendix A lists the number of neutrons per gram – second for the uranium and plutonium isotopes of interest. ^{240}Pu stands out with 920 neutrons/gm – sec, about 70,000 × the flux from ^{238}U! We have seen that (Ra,Be) or (Po,Be) powder mixtures are neutron sources, where the heavy element emits an α particle, which is absorbed by beryllium to produce a neutron: $^{9}\text{Be} + \alpha \rightarrow {}^{12}\text{C} + \text{n}$. The normal decays of uranium and plutonium are also by α emission, so light elements like beryllium present as a contamination could create background neutrons. The tolerable number of background neutrons per second is a statistical question.

Not every neutron initiates a chain reaction, even if $v' > 0$. Some are captured without fission, and some diffuse outside and are lost (which amounts to the same thing as capture for our purposes). Let p be the probability of success – fission; and q be the probability of failure – loss. Then $p + q = 1$. In this game fission gives you one extra neutron, or $v = 2$. Success gives 2 neutrons, failure gives zero. If you start out with n neutrons, the probability of losing them all is $(q/p)^n$. Now Serber shows that $q/p = 1 - v'$.[24] So the probability of starting a chain reaction with n neutrons is:

$$\Pr\{success\} = 1 - (q/p)^n = 1 - (1 - v')^n = nv', \text{ for } v' \ll 1. \qquad (F.39)$$

The background neutron rate is a constant dn/dt. Then n in Eq. (F.39) in time Δt is $n = (dn/dt)\,\Delta t$, and the probability for the background to create a chain reaction is $\Pr\{success\} = (dn/dt)\Delta t\, v'$. Δt is a measure of the assembly time to exceed the critical mass. For a gun type uranium bomb $\Delta t \sim 10^{-4}$ sec, which would cancel a background neutron source $dn/dt = 10^4$ n/sec, leaving the probability of pre-detonation $\Pr\{\} = v'$, which is the order of 15 – 20% as the active material is being assembled. Obviously it is desirable to hold the background neutron rate considerably below 10^4 n/sec. Spontaneous fission of ^{240}Pu gives about 1000 neutrons/gm sec, so 10^4 n/sec is not a ridiculous number, making the gun type plutonium bomb impractical. This fact led to the development of implosion, which compresses the material much faster, $\Delta t \sim 10^{-6}$ sec, and allows ^{240}Pu as high as 7% in 'weapons grade' plutonium.

A picture can be reconstructed of the Trinity Test plutonium weapon from a combination of several sources of information: Serber[25]; Libby[26];

---

[24] R. Serber, "Los Alamos Primer," *ibid*, pp 47 and 48.

[25] R. Serber, "Los Alamos Primer," *ibid*., p 52.

[26] Leona Marshall Libby, "The Uranium People," *ibid*, p 222.

and Rhodes.[27] Take a cast sphere of ^{239}Pu metal the size of an orange, with less than 7% ^{240}Pu. Cut it in half, and scoop out a small amount in the center to accommodate a pea sized lump of polonium, encapsulated by a spherical shell of beryllium. Now surround the 'orange' with a shell of natural uranium tamper. The whole assembly, about the size of a volleyball, is enclosed within the conventional explosive lens array necessary to achieve implosion. The final bomb is perhaps 1m in diameter. The Po-Be initiator is mixed as the plutonium sphere compresses.

For a discussion of the effects of a nuclear explosion, suppose we take 2% for the efficiency of a 50 kg ^{235}U bomb, which means that the reaction ceases around x = 38, W = 3×10^{17} erg/cm^3, and n = 1×10^{21} neutrons/cm^3. In a volume of 2630 cm^3 there would be a total energy of 8×10^{20} ergs, or 20,000 metric tons of TNT equivalent. There would also be 2.6×10^{24} neutrons with an average energy of about 2 MeV, and a velocity of 2×10^7 cm/sec. This is an intense pulse of neutrons radiating out from the center of the blast. The integrated exposure in neutrons/cm^2 falls off as $1/r^2$, and at 1 km distance – with no absorption by the atmosphere – would be 2.0×10^{13} neutrons/cm^2. This is a substantial radiation dose. In addition, a comparable number of 3 MeV γ rays are emitted within one minute of the explosion by the fission fragments.[28] Neutrons and gammas combined for about 10% of the Hiroshima casualties according to Serber. The residual radioactivity from fallout was negligible a few days after the blast. The incidence of leukemia, which occurred in less than 6 months, was 3700/million for survivors within 1 km of the blast center at Hiroshima, about 10 times that for an unexposed control population.[29]

The gas in the atmosphere plays a key role in the destructive effects of the bomb. Most of the structural damage is caused by the shock waves, extreme pressures, and winds from the blast, like a tornado. The atmosphere is the medium that carries the destructive forces. The

---

27 Richard Rhodes, "The Making of the Atomic Bomb," *ibid.*, p 655.

28 R. Serber, "Los Alamos Primer," *ibid.*, p 34. See also K. Way and E.P. Wigner, Phys Rev **73**, 1318(1948).

29 www.world-nuclear.org/info/Safety-and-Security/Radiation-and Health/Radiation-and-Health-Effects/

mathematical description of shock wave propagation in the atmosphere is very complicated. Yakov Zeldovich has written a book on the subject, which has been translated into English by the US Air Force.[30] He gives an empirical formula for the maximum pressure as a function of distance r and total explosive mass M as follows:

$$p_{max} = p_0 + A \times M/r^3 + B \times ((M)^{2/3}/r^2 + C \times (M)^{1/3}/r. \qquad \text{(F.40)}$$

Here A, B, and C are suitable constants, and p_0 is atmospheric pressure. Since the total energy is proportional to M, which is in turn proportional to the volume of the bomb $\sim R_0^3$, the three terms can be written as $(R_0/r)^3$, $(R_0/r)^2$, and (R_0/r) respectively, with appropriate adjustment to the constants. The first term after p_0 dominates close to the blast, and the third term is all that remains at large distances. A. Sadovsky used a variation of Eq. (F.40) for the analysis of the shock wave from RDS-1, in which the first term was dropped, and coefficients for the next two terms were specified:

$$p_{max} - p_0 = 3.4 \times (M)^{1/3}/r + (3.4 \times (M)^{1/3}/r)^2 \qquad \text{(F.41)}$$

The overpressure is in atmospheres, r in meters, and M in kg.[31] This handy formula gives a quadratic equation for $(M)^{1/3}$. If the overpressure is one atmosphere at 1500 m ($p_{max} = 2p_0$), Eq. (F.41) gives $(M)^{1/3} = 271$ kg$^{1/3}$, or M = 20000 T of TNT.

Alexander Glaser, in a talk on effects of nuclear weapons, shows circles which contain overpressures of 15 psi or greater (one atmosphere of overpressure) for air blasts of varying strength.[32] The circles are reproduced in Figure F.4. The radii scale as the cube root of the yield, and

[30] Ya.B. Zeldovich, "Theory of Shock Waves and Introduction to Gas Dynamics," Foreign Technology Division, Air Force Systems Command FTD-HT-66-258, 1967.

[31] Atomnyy_proekt_SSSR._T.2.Kn.6.(2006).pdf, #289 Efficiency of RDS-1, 1 September, 1949.

[32] Alexander Glaser, www.princeton.edu/~aglaser/lecture2007_weaponseffects.pdf

this scale is greater than Eq. (F.41), ie 20 kT is at 1900 m rather than 1500 m. This difference may reflect an uncertainty in associating overpressure with explosive energy.

An overpressure of 15 psi is sufficient to destroy most commercial buildings, so these distances are destructive radii. Serber says that one ton of TNT has a destructive radius of about 70 m, so 20 kilotons of TNT equivalent would destroy everything within a radius of ~2 km - $(20,000)^{1/3}$ = 27. This number is consistent with the Hiroshima and Nagasaki explosions,[33] and with the circles in Figure F.4.

The intense heat of the blast creates a fireball in the atmosphere. The energy released ionizes the air, and light is emitted upon recombination. Serber says that a 20 kTon explosion creates a fireball which after 0.3 sec has a radius 125 meters, and a temperature of 7300 K. Applying the perfect gas law Eq. (F.29) to this sphere, assuming a pressure $P = 10^5$ pascals (one atmosphere), gives $n = 1.6 \times 10^7$ moles. Taking the ionization energy of air to be 1.4 MJ/mole, the air in the volume would be ionized by 2×10^7 MJ, about 25% of the energy of a 20 kTon explosion. The fireball remains at high temperature for about ½ second, and then fades out three seconds after the explosion.[34] Incendiary effects from the fireball radiation do not contribute substantially to the destruction caused by the bomb. Most of the fires are started as a result of the blast damage – broken gas lines, etc. But unprotected human skin exposed to the radiation can be severely burned. Serber says that 20-30% of the fatalities at Hiroshima and Nagasaki were due to flash burns from the fireball.

---

[33] R. Serber, "Los Alamos Primer," *ibid.*, p 37.
[34] R. Serber, "Los Alamos Primer," *ibid.*, p 38.

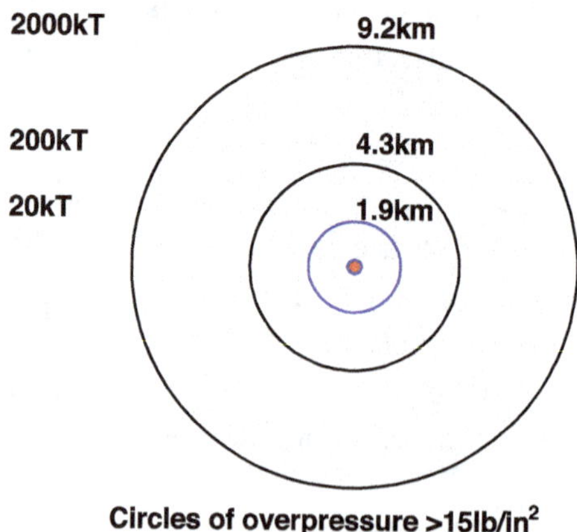

Figure F.4. Circles of overpressure > 15lb/in² (1 atmosphere) for explosive yields of 20, 200, and 2000 kTons. The radii scale as the cube root of the yield. Sadovsky's formula Eq. (F.41) gives 20 kT at 1.5 km.

William Penney, a British physicist, accompanied Robert Serber and US Navy Ensign George Reynolds in September, 1945, to Japan to study the effects of the bombs. The energy of the blast, or the efficiency of the fission process, is neither easy to predict, nor to measure after the fact. Penney and co-workers continued to pursue the question of just how much energy was released in the Hiroshima and Nagasaki bombs, and they published a study several years after the war.[35] They based their estimates solely on the blast damage, which depends on the amplitude and duration of the overpressure.

On location in 1945 the team first thought of drawing circles that characterized a particular type of destruction – snapped telephone poles, for example. This turned out not to work, because specifying a specific destructive act did not give uniform distances. There were too many local variables affecting the pressure waves. Instead Penney took back to Britain

---

[35] Lord William Penney, D.E.J. Samuels, and G.C. Scorgie, "Nuclear Explosive Yields at Hiroshima and Nagasaki," Philosophical Transactions of the Royal Society of London, A266(1177) 357-424 (1970). This journal first published in 1666.

some carefully selected objects, with detailed descriptions of the surroundings, orientation relative to ground zero, distance, etc. The height of the blast was important for the propagation of the shock waves. This was measured at Hiroshima to be 1890 feet (567 meters) by observing the angle of shadows of window frames inside a building of known distance from ground zero.[36] Snapped telephone poles were taken to Britain, along with squashed metal drums and cans, bent flag poles from the top of buildings that were not completely destroyed, memorial funeral stones that were toppled over, and other miscellaneous objects. Hydrostatic pressure in the laboratory squashed similar cans and drums. Wind tunnel tests determined elastic limits on flag poles, and tipping points for funerary stones. Finally, two test setups were built, one to 1/54 scale, and another to 1/92 scale, to measure the blast force on scale model buildings and poles. The 1/54 scale had 29 kg of Composition B explosive suspended on wires above two aluminum model buildings with scaled flag poles on top. The 1/92 scale used 6 kg of the same chemical. The explosive yield should scale as the cube of the dimensions, so roughly 29 kg @ 1/54 → 4600 tons of TNT, and 6 kg @ 1/92 → 4700 tons of TNT. To compare with nuclear weapon equivalent, multiply by two for the fraction of energy that a nuclear explosion channels into the blast, resulting in each case being approximately 10 kTon of TNT. (These calculations are not precise. Composition B by weight may not be exactly the same as TNT, but this is a small effect. In any event, these arguments are not better that 10-20% accuracy.) From all of these studies Penney quotes Hiroshima = 12±1 kTon, and Nagasaki = 22±2 kTon, where the uncertainties can be viewed with some skepticism.

In Los Alamos Report LA-8819 John Malik in 1985 studied the Hiroshima and Nagasaki yields based on all of the data he could find.[37] Canisters were deployed from aircraft to sample the debris in the atmosphere to add to the other sources of information regarding the energy

---

[36] All of the studies related to the Manhattan Project work in feet, pounds per square inch, etc. 1 atmosphere = 14.7 lb/in^2 = 10^5 pascals = 10.7 metric tons/m^2 (units popular with the Soviets).

[37] John Malik, "Yields of Hiroshima and Nagasaki Nuclear Explosions," LA-8819, Los Alamos National Laboratory, 1985.

in the blasts. Malik quotes various numbers for Nagasaki ranging from 19-24 kTon, with an 'official' yield of 23 kTon; and for Hiroshima the numbers range from 6-23 kTon, with an official yield of 13 kTon. The uncertainty assigned to these numbers is between 10% and 20%. The results are consistent with Penney's findings.

To summarize, the neutron density drives the explosion. The parameters used in this discussion, 50 kg of ^{235}U, twice the critical mass, normal density, various cross sections and mean free paths, and the energy released per fission are intended to be typical numbers, used to lead the reader through the nuclear explosion process. The volume of critical material remains roughly constant – its expansion ultimately causes the reaction to cease. Typical efficiencies are 20% or less. There are several destructive forces created by the blast: 1.) conventional shock waves propagating in the air; 2.) an intense burst of neutrons and gamma rays from the nuclear reactions; 3.) heat radiation from the atmospheric fireball created by the extreme temperature. Residual radioactivity, or 'fallout', existed for the Hiroshima and Nagasaki bombs, but was widely dispersed by the atmosphere, and was hardly detectable in the target areas. Precise values of the yields are difficult to predict a priori, and also difficult to measure a posteriori.

This presentation has been at the level of an undergraduate physics course. If the reader works through this Appendix, and follows the arguments, it does not mean that he/she can build a nuclear weapon in the basement. Weapons design drawings require much more careful analysis than is presented here.

Appendix G

Encryption and Decryption

There are many ways to scramble a message. There are look-up tables in code books for the correspondence between groups of letters, or numbers, or both, and the language of interest. There are devices like Enigma that scramble letters in a seemingly random way. And there are schemes that change letters into numbers, alter the numbers in a way known to the receiver, and transmit the message as a series of scrambled numbers. In each case the procedure has to be reversed for anything to make sense to the receiver. A message can be intercepted if transmitted wirelessly, or by eavesdropping on cable communications.

A famous example of an encrypted message of significance intercepted in wartime is the Zimmerman telegram, described in a book by Barbara Tuchman.[1] It was sent over a transatlantic cable from the German Foreign Secretary, Arthur Zimmerman, in Berlin, to the German Ambassador in Washington, Count Johann von Bernstorff, on January 17, 1917. At that time the United States was a neutral country, as was Sweden. The British had cut the German transatlantic cable at the start of the war in 1914, but both Sweden and the United States allowed Germany to use their cables for diplomatic traffic. The routing of the transatlantic cables from the continent of Europe to the United States went through the English Channel, where there was a station with booster amplifiers to send the signals across the ocean. British Naval intelligence, referred to as 'Room 40,' had tapped into this booster station, and was intercepting all of the German diplomatic traffic to the

---

[1] Barbara Tuchman, "The Zimmerman Telegram," Ballentine Books, New York, 1966. The actual telegram and its transcription are shown in the illustrations after p 148.

United States. The boss of Room 40 was the chief of naval intelligence, Admiral William Hall.

The Zimmerman telegram had two forms. The transatlantic form was a two part code, more difficult to decipher. However, the second version, a relay between the German Embassy in Washington and the German Embassy in Mexico City, was a simpler form. The Washington-Mexico City message was a Western Union telegram, an array of about 150 groups of four and five digit numbers. Each number was a German word.[2] British Naval Intelligence had experience with code of this type, had acquired some code books for interpretation, and was able to decrypt the message in a few days – almost in real time. It authorized the German ambassador in Washington to contact Mexican authorities through the German Embassy in Mexico City, and offer Mexico the states of Texas, New Mexico, and Arizona in return for joining Germany in victorious conflict against the United States! And it was just a Western Union telegram! The code of the Zimmerman telegram is a mapping between numbers and words – about 10,000 possibilities in this case.

Other mappings can exchange individual letters, or groups of letters, and the rules for the mapping can change from letter to letter (the Enigma machine). Letters can also be converted into numbers, then scrambled (the one-time pad). Both examples are described below.

The German Enigma machine, which was widely used by the German army and navy in WWII, was an electro-mechanical typewriter device that exchanged individual letters. It had a keyboard with 26 letters, and in its simplest form three wheels, each with 26 positions. At the far end was a 'reflector' wheel that caused the signal to turn around and retrace its steps along a different path. Each day the three wheels were set to that day's initial conditions, information that was supplied by a book, or transmitted freely. Assume the initial conditions are ABC. The sender sets the wheels on his/her machine to ABC, and types in twice the settings to be used for the message. JKL JKL for instance (repeat to be

---

2   David Kahn, "The Codebreakers," Scribners, New York, 1996. p 286.

sure of successful transmission). The first J→C; K→M; L→N. After each letter is keyed in, the wheels rotate, so the next J is not C, but H, say. Perhaps JKL JKL becomes CMN HYF. The receiver gets this string, CMN HYF, sets up the machine initial conditions ABC, types in the string, and gets out JKL JKL. The Enigma diagram in Figure G.1 shows how this works.

Figure G.1 is a simplified circuit diagram, showing the switches, keys, and lights for six letters JKLMNC. The three wheels plus reflector, and the plug board (an added feature on some enigmas) can be replaced by a black box, which exchanges the letters. The device is really just an automated plug board. After each key stroke the black box is rearranged, so that the mix of letters is different - there would be different branches within the box. The Figure does not show how this is done. It is an important feature of the real device, but doesn't matter for our discussion. The keys are on single pole double throw switches. Depressing the J key connects the battery power line to the J output to the black box, while each key at rest is in contact with that letter's light. The black box patches the J line to the C line, and the circuit is completed through the C light bulb, which lights up. The other lights along the top string are off.

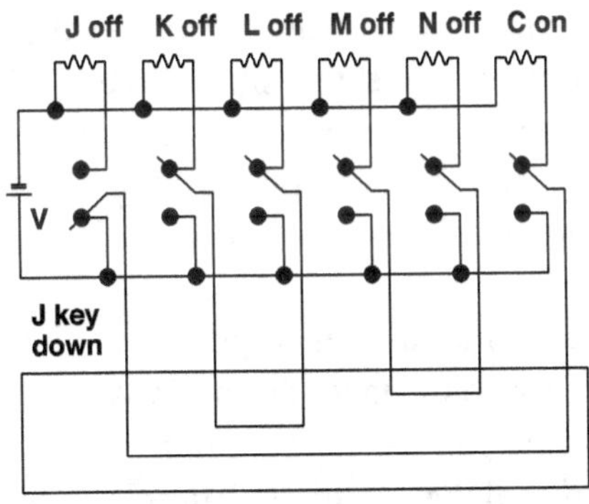

Figure. G.1. Six letters of a toy Enigma machine, showing the current path from the depressed J key switch through the mixer to the energized C light. All other keys are up, and all but the C light are off. V is the battery.

This drawing makes the reverse operation clear. With the same black box, depressing the C key instead of the J key will light up the J bulb, meaning that the encoding and decoding are done with the same keyboard. The circuit also clearly shows that it is impossible for a letter to remain unchanged - J→J - since the single pole double throw switches have to be in one of two positions, either up for the light or down when the key is depressed, but not both at the same time.

What goes on in the electro-mechanical black box mixer is determined by the orientation settings of the toothed wheels on the keyboard, but the end effect is that 26 letters are interchanged in pairs, meaning 13 wires are hooked up, connecting the switches as shown in the example in Figure G.1 (which has three wires for six letters). The hookup is changed by pressing a key, a feature not shown in Figure G.1, but you don't have to know what goes on in the black box, as long as sender and receiver start out in the same configuration.

After transmitting the setup pattern, the sender sets his/her wheels to JKL, and types in the rest of the message. With the receiver's wheels set to JKL, the decoded message, one letter at a time, can be recovered. The fact that a letter was never mapped into itself was useful for the code breakers. If the coded message was transmitted by wireless Morse code, a common form in combat, it could be intercepted by anyone. The message has to be written down from the decoded letters, one by one.[3] Bletchley Park in England was the site of the ULTRA group, where Enigma coded messages were decoded with success throughout the war.[4] One of the brains behind the operation was Alan Turing, famous for his early conceptualization of digital computers.

The 'one-time pad' is a basic encryption technique used extensively by the USSR and other combatants during WWII, and indeed is still in use today. The one-time pad has many forms. We will describe in detail a form that takes the alphabet, turns letters into numbers, shifts the

---

[3] It is not obvious how Enigma works. I learned from Wikipedia, as I often do. En.wikipedia.org/wiki/Enigma_machine#Basic_operation.

[4] Ronald Lewin, "Ultra goes to War," McGraw Hill, New York, 1978.

numbers around with a key (the one-time pad), and transmits the string of shifted numbers. The example presented here is adopted from a paper written by Dirk Rijmenants.[5] The enigma scrambles letters. This procedure converts letters to numbers, then scrambles the numbers. The one-time pad is the key to scrambling and unscrambling. The number of digits in each random 'word' is arbitrary, but is selected to be five in this case.

The key is a set of five digit random numbers. By random numbers, we mean that any single digit from 0 to 9 (ten digits) is equally likely. In 1000 random digits, there should be 100 of each of 0 to 9. In real life there are statistical fluctuations, and even a 'perfect' random number generator will give different occupations for a finite number of digits. There are computer codes that generate random numbers, and in any irrational number, like $e = 2.71828...$ or $\pi = 3.14159...$ the infinite series of digits should be random. Specific techniques for generating random numbers electronically during WWII – when there were no digital computers - are not known to the author.[6]

The first table of 50 5 digit random numbers was generated by a computer algorithm.

One Time Key Pad from computer code

```
72906  41702  43599  55079  96702
22199  89286  76308  87342  10374
77132  18026  15416  77770  51394
84881  22329  29466  65037  97533
58813  48724  20846  51729  96001
```

---

[5] Dirk Rijmenants, "Cipher Machines & Cryptology," http://users.telenet.be/d.rijmenants. The author is grateful to Mr. Rijmenants for his aid in understanding this subject. An example is also in Lamphere and Shachtman, "The FBI-KGB War," *ibid.*, p 83.

[6] David Kahn, "The Codebreakers," *ibid.*, states that shipping manifests with columns of numbers were used. The subject is very much alive. See T. Lunghi, et al., "Self-testing Quantum Random Number Generator," Phys Rev Lett **114**, 150501, (2015).

```
87012  30793  42572  72901  86375
64414  28605  85888  24851  38561
45805  72850  94449  38477  54688
40768  25092  37454  11505  83484
98901  78383  11348  17490  30096
```

The sums for the 250 digits are as follows:

```
n0 = 25
n1 = 23
n2 = 25
n3 = 22
n4 = 29
n5 = 23
n6 = 18
n7 = 29
n8 = 33
n9 = 23
nsum = 250
```

The average should be 25 for each integer. Taking $\sigma = 25^{1/2} = 5$ as the standard deviation, most values should be in the range from 20 to 30, and they are. Only two values, $n6$ and $n8$, are outside this range. This deviation is to be expected for ten numbers. So this set looks OK.

The digits in the irrational number - π form the second set. This is not secure enough for a real pad, because although random, the digits are well known, but they will serve for our discussion. Here is a pad of 50 5 digit random numbers, copied from the 1000 places of π:[7]

---

[7] www-history.mcs.st-and.ac.uk/HistTopics/1000_places.html

One time key pad from π

```
14159  26535  89793  23846  26433
83279  50288  41971  69399  37510
58209  74944  59230  78164  06286
20899  86280  34825  34211  70679
82148  08651  32823  06647  09384
46095  50582  23172  53594  08128
48111  74502  84102  70193  85211
05559  64462  29489  54930  38196
44288  10975  66593  34461  28475
64823  37867  83165  27120  19091
n0  =  22
n1  =  26
n2  =  29
n3  =  24
n4  =  28
n5  =  24
n6  =  22
n7  =  17
n8  =  31
n9  =  27
nsum  =  250
```

Here $n7$ and $n8$ are out of the 25 ± 5 range, but the variations do not look different from the computer-generated numbers, so we can assume that they are just as good. The length of the pad determines the length of the message, because the message when converted to digits cannot be longer than the pad. This puts a premium on brevity. For this reason, among others, technical information on the Manhattan Project was rarely sent by code – it was simply too voluminous.

To use a numerical pad, one needs to convert the alphabet to numbers. Here we present Rijmenants' algorithm for the English language. A different table would be used in Russian.

Character to digits checkerboard for English

Code	A	E	I	N	O	T			
0	1	2	3	4	5	6			
B	C	D	F	G	H	J	K	L	M
70	71	72	73	74	75	76	77	78	79
P	Q	R	S	U	V	W	X	Y	Z
80	81	82	83	84	85	86	87	88	89
FiG	(.)	(:)	(')	( )	(+)	(-)	(=)	REQ	SPC
90	91	92	93	94	95	96	97	98	99

The first six letters A,E,I,N,O,T are most frequently used in English, and hence merit single digits. Double digit letters begin with 70, so there is no confusion between single digit and double digit letters in a word. FiG = 90 means that the following digit is actually a number, which is written three times to insure validity. Thus 904449122290 means 4.2. 93 serves both as an apostrophe and a comma. 94 opens and closes parentheses. REQ, the request code, is a question mark. SPC = 99 is a space between letters or words. Code 0 at the beginning is used as a prefix for a lookup code book, a feature that we are not covering. A lookup code book gives three digit numbers for frequently used words, thus considerably shortens a message that contains these words.

We are now ready to tackle a message:

HELLO WORLD, IT IS MAY 1

Write down the numbers using the checkerboard:

75278785998658278729399369938399791889990111190

Whew! Now split them up into 5 digit groups, because we have arbitrarily chosen 5 digit groups for our key pad:

75278 78599 86582 78729 39936 99383 99791 88999 01119 09191
14159 26535 89793 23846 26433 83279 50288 41971 69399 37510

The chain of 46 numbers is not divisible by 5, so at the end are two periods 9191. Written below the 5 digit groups are the first 10 5 digit random numbers from the one-time pad. The reader may be pretty confused already, but so far there has been no encryption. Now comes the encryption part. If the sender subtracts the bottom numbers from the top, then the receiver, with the same key pad, adds the numbers received to the pad. The rules for addition and subtraction guarantee no negative numbers, and there is no carry over from one column to the next. If the sender adds, the receiver subtracts. The no carry over rule may be unfamiliar, but it is actually easy to use. It guarantees that five digits remain five digits after the operation. For the first group under addition 75278+14159 = 89327, and under subtraction 75278-14159 = 61129. An advantage of this rule is that you can start anywhere – each column is independent of its neighbors. Begin with the most significant digits if you like, or even in the middle! Let's say the sender adds them:

```
MSG 75278 78599 86582 78729 39936 99383 99791 88999 01119 09191

KEY 14159 26535 89793 23846 26433 83279 50288 41971 69399 37510

SND 89327 94024 65275 91565 55369 72552 49979 29860 60408 36601
```

The bottom row of numbers can be sent by short wave radio, using Morse code. The receiver who has the one time key pad, and who knows that to obtain the original message the key pad numbers must be subtracted from the message received, gets the following:

```
REC 89327 94024 65275 91565 55369 72552 49979 29860 60408 36601

KEY 14159 26535 89793 23846 26433 83279 50288 41971 69399 37510

MSG 75278 78599 86582 78729 39936 99383 99791 88999 01119 09191
```

Re-assemble the words, looking for the 99's which are the spaces, and leaving them out:

```
75278785 86582787293 36 383 79188 90111909191
```

Now use the checkerboard. 75=H, 2=E, etc

```
75278785 = HELLO
86582787293 = WORLD,
36 383 79188 90111909191 = IT IS MAY 1..
```

Messages must be short, and both sender and receiver must have the one-time pad, and the algorithm for conversion. Each digit received C = A + B, where B is the one-time pad, and A is the original number. If you do not know A or B, but you have C, you are faced with one equation in two unknowns, which cannot be solved by the world's greatest mathematician. If B is a random number, then so is C. Hence the procedure is impossible to break without the one-time pad.

Appendix H

Soviet Intelligence

In this Appendix we discuss some of the secret information obtained by Soviet espionage, and described by Sudoplatov in "Special Tasks," compared with ancillary information in Russian obtained from other sources. At the end we talk about some of the discrepancies between the English and Russian versions of Sudoplatov's memoirs.

Voprosy Istorii Estestvoznanya I Tekhniki, 1992 #3 p 103ff A.A Yatskov "Atom I Razvedka" There are 14 documents in Russian regarding comments and questions from Intelligence information on the Manhattan Project and Tube Alloys. Most are written by Kurchatov. Two of them are translated by the author in full below.

Sudoplatov in the English version of his book, "Special Tasks"[1] translates in Appendix 2 Yatskov's Documents #1 through #11, and #14. He leaves out #12 and #13, which are available in VIET, and will be translated below. Documents #7 through #11 are also available in Russian in /APSSSR/.[2]

The first six Documents are too early to be in our part of APSSSR, which starts in 1944. English and Russian versions are available in "Special Tasks" and VIET respectively. The English translations are satisfactory.

---

[1] Pavel and Anatoli Sudoplatov, with Jerrold and Leona Schecter, "Special Tasks," Little, Brown and Co, Boston, 1994.

[2] Work.atomlandonmars.com/APSSSR/Atomnyy_proekt_SSSR._T.1.Ch.2.(2002).pdf

#1 Sept 25, 1941 Gorsky from London about Tube Alloys from Lord Hankey's committee, reported by 'Leaf' identified as Donald Maclean. This report in Russian is 'Document #1' in Yatskov's article in VIET.

#2 Oct 3, 1941 Gorsky from London, also reported by Leaf, about a uranium committee document submitted to the war cabinet about the atomic project. This report in Russian is 'Document #2' in Yatskov's article in VIET.

#3 March 1942 Beria to Stalin about overseas interest in the uranium problem as learned by intelligence. This report in Russian is 'Document #3' in Yatskov's article in VIET.

#4 March 7, 1943 Kurchatov to Pervukhin forwarded to Merkulov. Kurchatov's evaluation of intelligence material on isotope separation, nuclear explosions, and plutonium. This report in Russian is 'Document #4' in Yatskov's article in VIET.

#5 March 22, 1943, Kurchatov to Pervukhin questions needing to be clarified by espionage on plutonium cross sections, spontaneous fission, lifetime. The Russian version is 'Document #5' in Yatskov's article in VIET.

#6 July 3, 1943 Kurchatov to Pervukhin. Kurchatov has reviewed 286 titles of American works on the uranium problem, and he submits questions to be forwarded to the NKVD. The Russian version is Document #6 in Yatskov's article in VIET.

Items 7 through 11 in "Special Tasks" are also available in Russian in the online source /APSSSR/.

Table H.1.

'Special Tasks' English	/APSSSR/ Russian	VIET Russian
#7 p 456-57 Merkulov to Beria	#316 p 234 28 Feb 1945	#7
#8 p 458-59 signed Kurchatov	#321 p 245 16 Mar 1945	#8
#9 p 459-60 signed Kurchatov	#320 p 244 16 Mar 1945	#9
#10 p 460-61 signed Kurchatov	#329 p 261 7 April 1945	#10
#11 p 461-62 signed Kurchatov	#333 p 268 11 April 1945	#11

Sudoplatov omits #12 and #13, implying that they are too hot to handle. He notes that VIET 1992 #3, p 107-134 was published and distributed in Russia, but then withdrawn from circulation because of the secrecy of the technical data they contain about building an atomic bomb. I obtained an intact version of VIET 1992 #3 from interlibrary loan via the University of Michigan. The University of Wisconsin has VIET 1992, but not issue #3! So we can translate into English the missing Documents. As we shall see, Document #12 gives a broad description of the Trinity test plutonium bomb, while #13 gives all of the details.

H.1. Document #12 High Explosive Bomb

The first explosion of an atomic bomb is expected in July of this year (1945). Bomb construction: The active material is ^{239}Pu. ^{235}U is not used. In the center of a sphere of plutonium weighing 5 kg is a neutron initiator – a Po-Be source which is mixed at the time of explosion. Surrounding the plutonium is a uranium tamper, then an aluminum shell 11 cm thick. The aluminum is surrounded by a layer of conventional explosives, 46 cm thick. The explosives are pentalit or composition B. The diameter of the bomb assembly is 140 cm, and weighs about 3 tons. The expected force of the explosion is 5000 tons of TNT – corresponding to an efficiency of 5-6%, and 75×10^{24} fissions.[3]

Supplies of active materials

- Uranium 235 by April of this year equaled 25 kg, with a production rate of 7.5 kg per month.
- Plutonium from Camp 2 (Hanford) 6.5 kg in hand. The production rate has exceeded the projections.

Test explosion is anticipated by 10 July, 1945.

Document #13
October 18, 1945
From: V.N. Merkulov
To: Comrade Beria

General description of an atomic bomb:

An atomic bomb is a pear shaped object, with maximum diameter 127 cm, and total length including stabilizers of 325 cm. Total weight is about 4500 kg. A bomb consists of the following parts, going out from the center:

1. Initiator
2. Active material
3. Tamper
4. Aluminum shell
5. Conventional explosives
6. 32 explosive lenses
7. Detonation device

---

[3]   5 kg of Pu is 1.2×10^{25} atoms (not 7.5×10^{25}). Fission at 5% efficiency would give $0.05\times1.2\times10^{25}\times200\times1.6\times10^{-13} = 1.9\times10^{13}$ joules, taking 200 MeV as the average energy release, and converting MeV into joules. From Appendix F 1 metric ton of TNT $= 4.18\times10^{9}$ joules, giving about 5000 T of TNT equivalent. The number of fissions is incorrect in the text. The actual efficiency of a plutonium implosion bomb is greater than 5%.

8. Duralumin outer shell[4]
9. Armor plate steel cover
10. Stabilizer.

The first six items are pairs of hollow hemispheres, centered on the initiator. The duralumin shell protects the detonation assembly. Outside the shell is the armor plate steel bomb cover.

H.2. Description of the Separate Components of the Bomb

1. The initiator is a hollow beryllium sphere 1 cm in radius, which is indented with parallel wedge shaped grooves. The grooves are covered with 0.1 mm layer of gold, and a layer of polonium. A second smaller beryllium sphere with the same grooves, gold, and polonium sits inside. It is called the 'urchin.' Dimensions of the urchin:

Outer radius of the hollow beryllium sphere 1.0 cm
Radius of the wedges at the base 0.4 cm
Radius of the wedges at the top 0.609 cm
Radius of the inner beryllium sphere 0.4 cm
Number of wedges 15
Polonium on the hollow sphere 30 curies
Polonium on the inner sphere 20 curies.

The hollow sphere is prepared in two halves in a nickel-carbonyl atmosphere, which coats the beryllium with nickel to absorb the α particles from polonium decay. There is no source of neutrons until the urchin is compressed by the implosion. The shock wave from the outer layer of explosives is focused on the center of the bomb, passing through the aluminum layer[5], the tamper, and the active material to reach the initiator. The resulting pressure breaks the sphere along planes, proceeding through the apex of the wedge shaped grooves, opening the hollow beryllium sphere to the α particles from the polonium surface on

---

4　Duralumin is an aluminum alloy that can be hardened – widely used in aircraft.
5　Called the 'pusher.'

the central sphere of the initiator. This creates a flow of neutrons [via $^9Be(\alpha,n)^{12}C$]. Neighboring grooved surfaces collide, forming 'Munroe jets'[6] which penetrate the central sphere through the layer of polonium and gold, bringing the polonium in contact with the inner surface of the hollow beryllium sphere and the continuous beryllium inner sphere, also creating a neutron flux. These neutrons in turn trigger the chain reaction in the active material.

2. Active Material

The active material is the element plutonium in the delta phase with a specific gravity of 15.8. It is prepared in the form of two halves of a hollow sphere in a nickel-carbonyl atmosphere like the initiator. The external diameter of the sphere is 80-90 mm. All of the active material, including the initiator weighs from 7.3 to 10 kg. A corrugated gold foil coating 0.1 mm thick prevents high velocity jets[7] from penetrating between the halves of the plutonium into the initiator, which could cause the initiator to emit neutrons prematurely. There is a hole in one of the hemispheres about 25 mm in diameter for insertion of the initiator into its bracket from the outside. After insertion the hole is plugged with plutonium.

3. Tamper

The tamper is a hollow sphere 230 mm in outside diameter, made of metallic uranium. There is an opening in the sphere for the insertion of the active material which is plugged with a uranium cap. The tamper reflects neutrons back into the active material, and has the effect of decreasing the critical mass. The outer surface is covered with a layer of boron to absorb thermal neutrons that could cause predetonation.

---

[6] A 'Munroe Jet' is formed by a conical shaped explosive charge with a metallic coating on the inside of the cone. The metal vaporizes and forms a penetrating stream to cut through armor plate.

[7] These jets presumably come from the implosion.

4. Aluminum layer [the pusher]

The layer of aluminum surrounding the outer surface of the tamper is a hollow sphere 460 mm in diameter composed of two halves fitted together by tongue and groove. One of the hemispheres has a hole for access to the interior, with a cover. The purpose of the aluminum shell is the uniform transformation of the explosive shock wave to the center of the device.

5. Explosive layer and lenses

The explosive layer outside the aluminum is made of 32 blocks of a special type. The blocks are placed around the outside of the Aluminum layer, pointing towards the center of the device. Special grooves on the outside of the blocks form 20 hexagonal and 12 pentagonal lenses.[8] Perpendicular to the axis of the sphere a layer of felt 1/16 "thick separates the surfaces of the lenses and the other high explosives, while along the radii there is blotting paper. Space between the surfaces does not exceed 1/32," to prevent the escape of air outward as the implosion proceeds. The lenses are cast in special forms prepared from acetate of cellulose. Each lens consists of two types of high explosive – one fast explosive and the other slow.[9] At the installation of the lenses in place, the fast explosive part adjoins the layer of high explosive. The explosive layer weighs about 2 tons total. There is one detonator for each lens, and two fuses for each detonator, to guarantee reliability. There are 64 wires in all, grouped into 4 quadrants of 16 each, each lens has two wires from different quadrants.

---

[8] Like a soccer ball.

[9] The Manhattan Project called the explosives Composition B and Baratol. "Fast and slow" refer to the explosive velocity, the speed with which the substance detonates. Two different speeds – a change in refractive index -help to focus the implosion wave. Baratol is a mixture of TNT and barium nitrate – about 30% TNT. Composition B is RDX/TNT in a 60/40 ratio. RDX is $C_3H_6N_6O_6$ in a hexagonal structure similar to TNT. See Appendix F for a description of TNT.

6. Duralumin shell

The layer of high explosives and lenses is covered by a duralumin shell which protects the blasting apparatus. Interior diameter is about 1400 mm, and total weight, including 180 kg for the blasting apparatus, is about 700 kg.

7. Bomb assembly

The uranium sphere is placed inside the aluminum shell so that the access holes match up. One of the high explosive blocks and lenses is missing from the explosive layer, and this gap also lines up with the other holes. Finally, there is a hole in the duralumin cover. In this form the bomb is ready to be transported to the launching or testing place. The final assembly is done there. The initiator is inserted through all of the holes into the center of the device, and the access channel is covered from inside out with the appropriate materials, plutonium, uranium, aluminum, explosive lens, duralumin, and steel. Natural radioactivity of plutonium in the active material and polonium in the initiator generates heat, and temperature can reach 90^0 C above ambient temperature, so the bomb is carried in a special refrigerated container.

This document is dated October, 1945, after the Hiroshima and Nagasaki bombs. It describes the implosion bomb, tested at Trinity and used at Nagasaki. The detail is striking. It is not known whether this description was accompanied by drawings, which would have helped considerably in following the text. The best description of the early Manhattan Project weapons is that of John Coster-Mullen, who presents a cutaway drawing of a plutonium implosion bomb on p 188 of his book.[10] The components follow the above account to the letter.

To compare the descriptions, we reproduce the list from the Soviet version, and append dimensions (converted to the metric system) from the drawing in Coster-Mullen's book. Going from the center out:

---

[10] John Coster-Mullen, "Atom Bombs," privately published, 2014, p 188.

Table H.2.

Component	Soviet outer dia	Manhattan Proj outer dia
Initiator	2 cm	2 cm
Plutonium	8-9 cm	9 cm
U tamper	23 cm	22 cm
Al shell	46 cm	47 cm
HE ring	127 cm[11]	140 cm

The space left for the explosive lenses, which occupied 96% of the total volume of the bomb, is underestimated, and it is not clear how to reproduce the sophisticated design of these lenses from the Soviet text, nor from the drawing in Coster-Mullen's book. The lenses, and the associated electronic firing mechanism, are probably the biggest gaps in the picture. It is an intricate three dimensional assembly, that has to work perfectly to uniformly compress the plutonium sphere. That aside, the information is all there, even to the use of felt, and the boron coating outside the uranium tamper. The detail on the initiator, where Los Alamos spent a lot of effort in design and manufacture, is impressive, but again a few drawings would help a lot in understanding it.

A scale drawing based on the data in the right hand column of the above table is shown in Figure H.1. Going in from the outside, the high explosive shell is blue, the aluminum pusher shell is blank, the uranium tamper is black, and the plutonium bomb is red. The small black circle at the center is the location of the initiator. The outer high explosive shell is surrounded by fuses that are simultaneously electrically fired.

---

[11] The Soviet intelligence report does not give this number, but cites the overall outer diameter, which is at least 13 cm too small.

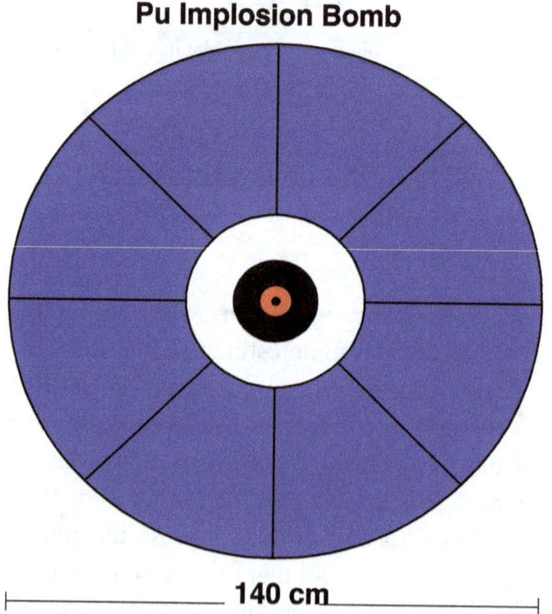

Figure H.1. Schematic of an early implosion bomb design based on the dimensions given in the table. The array of explosives in an eightfold pattern is illustrative rather than accurate. The large outer blue shell is high explosive chemical. Next is the aluminum pusher, then the black uranium tamper, and finally the red plutonium bomb with initiator in the center.

The active material is quoted as 'delta phase plutonium with specific gravity 15.8.' Plutonium metal has six allotropic forms.[12] At atmospheric pressure the metal changes phase as the temperature is increased. At room temperature it is in the alpha phase, which has the highest density (19.86 gm/cm^3), and is brittle. At atmospheric pressure the solid melts at 900 K. Like water, it increases in density when it melts. That is, solid plutonium floats on the liquid. At higher temperatures the solid metal changes phase, density, and physical properties. The delta phase normally exists between 583 and 725 K, but is stable at room

---

[12] en.wikipedia.org/wiki/Allotropes_of_plutonium. See Figure 14.2 in Chapter 14.

temperature (300 K) if alloyed with a small amount of gallium. The delta phase is preferred for machinability, welding, etc. The handbook density for delta phase plutonium is 15.92 gm/cm^3. The Manhattan Project metallurgists had to learn all of this from the small amounts of plutonium metal obtained from the Hanford reactors. While more information on plutonium may have come from other intelligence reports, the content of this Merkulov to Beria document leaves out a lot of detail, which has to be known in order to handle the components.

Nevertheless, this has to be chalked up as a high quality piece of work. The Soviet project still had plenty of work to do, but they were not starting out cold. The value of the information to the Soviet Atomic Project is obvious. Because the original stolen documents, in English, are not available, it is not possible to determine whether this description is the work of a single agent, or is a mosaic of several reports received in Moscow. The latter is more likely. In the event, an unknown person at NKVD headquarters did a remarkable job in putting together an accurate picture of a plutonium implosion bomb.

While the original documents in English are not available, some of the Russian translations are given in "Documents and Materials of the Atomic Project of the USSR".[13] Many documents were edited for accuracy by Yakov Terletsky, the physicist who visited Niels Bohr in Copenhagen in 1945. Dimensions remained in inches, feet, and miles and weight in pounds – convincing evidence that the original sources came from the USA. Descriptions of the metallurgy of plutonium, the design of the implosion bomb, and the results of the Trinity test from NKVD sources were delivered to Beria's committee – the PGU – in 1945 and 1946.

There are at least two versions of Sudoplatov's memoirs, "Special Tasks," in English, and "Razvedka I Kreml" in Russian. Both were first published in 1994, although the Russian version was revised and reprinted in 1996. The English version caused quite a stir when it first

---

[13] Atomnyy_proekt_SSSR._T.2.Kn.6.(2006).pdf, reports #330 through #340.

appeared. The Russian version is not as accessible, but also not as outrageous in its claims. So it is natural to assume that Russian was written, or at least modified, after English, in an attempt to tone down some of the language. However that may be, we will discuss some of the salient differences, and some statements in both versions that are not considered correct.

We are interested in Chapter 8 in both versions, entitled 'Atomic Spies.' "Special Tasks" starts right out with outrageous statements:

"Oppenheimer, Fermi, Szilard, and Szilard's secretary were often quoted in the NKVD files from 1942-1945 as sources for information of the development of the first American atomic bomb. It is in the record that on several occasions they agreed to share information on nuclear weapons with Soviet scientists."

The Russian version has no such paragraph. On page 211 Sudoplatov writes: "In the traditional sense Oppenheimer, Fermi, and Szilard were never our agents. This was confirmed by Kvasnikov, who headed the Soviet scientific-technical intelligence from 1947-1960: 'Scientist, working with our intelligence, could not be called agents." Subsequently, on pages 223-24:

"Oppenheimer and Fermi did not know that already by that time (1943) they were part of our operations, as sources of information under the code names 'Direktor reservatsii (Director of the reservation),' 'Veksel' (Money),' and 'Zayats (Rabbit).' The code name 'Money' was used at times for the source of general material from physicists participating in the American atomic project. I recall that both Oppenheimer and Fermi were referred to as 'Star (Elder).' I repeat one more time – none of them was ever our recruited intelligence agent."

Then on page 228 he writes: "It seems to me that between Bohr, Fermi, Oppenheimer, and Szilard there was an informal agreement to share secrets on atomic weapons with the group of left-leaning anti-fascist scientists." Sudoplatov subsequently says on page 260 "In May, 1995, the FBI refuted my version of the receipt of atomic weapons data by our intelligence. FBI stated that Fermi, Oppenheimer, Szilard, and

Bohr, according to their data, were never spies. But I have not confirmed this."

Kvasnikov's statement is ambiguous, and Sudoplatov is basically unapologetic, but the lives of the chief scientists, in particular the complicated story of J. Robert Oppenheimer, have never revealed any duplicity.

There are some other topics in the Russian version which raise eyebrows. He says that Klaus Fuchs was invited to be a member of the British delegation to Los Alamos by Oppenheimer (p 229). This claim is attributed to Aleksandr Feklisov, who wrote: "By the end of 1943 Robert Oppenheimer, the leader of the work on the creation of the American atomic bomb, who highly appreciated the theoretical works of Fuchs, asked to include Fuchs as part of the British scientific mission coming to the USA to assist the project."[14] This seems very implausible. Does Feklisov mean that Oppenheimer knew what Fuchs was doing on Tube Alloys? Rudolf Peierls, who recruited Fuchs for the project, and knew him best, included him as an essential member of the British delegation, and makes no mention of Oppenheimer. The delegation first went to Canada, then to New York City, and eventually to Los Alamos. Groves in his book "Now It Can be Told" states that he made no special effort to validate Fuchs' loyalty, assuming that was the responsibility of the British. Fuchs profited by Oppenheimer's policy of openness for workers at Los Alamos. All scientists could attend the weekly seminars, giving Fuchs access to information at the laboratory on topics with which he was not directly involved. On the other hand, he had very limited knowledge of what was going on at other sites, like Oak Ridge.

In addition, on page 247, Sudoplatov claims that Oppenheimer offered Fuchs an appointment at the Institute for Advanced Study in Princeton after the war. The dates don't match up. Oppenheimer became director of the IAS in August, 1947, but Fuchs had left the US for

---

[14] Voyennoi Istoricheski Zhurnal, Dec 1990, p 25. Quoted by Sudoplatov (E) p 193.

England already in 1946. There is no mention of such an offer in Oppenheimer's biography.[15]

Regarding the matter of Bruno Pontecorvo, Sudoplatov is clearly off base. Pontecorvo's interesting and sad story is told in Chapter 8, and does not need to be repeated here. But because of his defection, he appeared on Sudoplatov's radar screen, and Sudoplatov mistakenly associated Pontecorvo with many things. One example is on page 216, where Sudoplatov writes that at the end of January, 1943, the NKVD obtained information on Fermi's successful CP-1 graphite-uranium reactor, which went critical in December, 1942. Sudoplatov ascribes this information to Pontecorvo, and that he (Pontecorvo) sent the phrase "The Italian Navigator has reached the new world."

This quote was in fact from Arthur Compton to James Conant.[16] It had nothing to do with either Pontecorvo, who was in Montreal at that time,[17] or the NKVD. On the next page Sudoplatov writes that his deputy Vasilevsky – who later accompanied Terletsky on the mission to Copenhagen – contacted Pontecorvo in Canada. "Pontecorvo told Vasilevsky that Fermi was in favor of the idea of sharing information on atomic energy with scientists in anti-Nazi countries."

Further, on pages 236-237, Sudoplatov writes:

"Twelve days after the first atomic bomb test at Los Alamos (Trinity) we obtained descriptions of its construction from Washington and New York. Moscow Center received the first telegram on June 13, and the second on July 4, 1945. These telegrams were decoded five years later, and served to pressure Fuchs into confessing to espionage. I cannot fully confirm this, although I want to emphasize that the sources mentioned in the telegrams, 'Charles' and 'Mlad' were Fuchs and Pontecorvo."

---

[15] Kai Bird and Martin J. Sherwin, "American Prometheus" Alfred A. Knopf, New York, 2000. The authors emphasize that Oppenheimer had nothing to do with Fuchs' appointment to Los Alamos.

[16] Richard Rhodes, "Making of the Atomic Bomb," *ibid.*, p 442.

[17] Simone Turchetti, "The Pontecorvo Affair," University of Chicago Press, Chicago, 2012, p 48.

Fuchs was called 'Charles,' among other code names, but 'Mlad,' which means youngster, is now believed to have been Theodore Hall, not Bruno Pontecorvo. A possible code name for Pontecorvo was 'Guran,' for Lake Huron.[18] If Pontecorvo was indeed a spy for the Soviet Union during the Manhattan Project, he took that secret with him to the grave.

With all of its faults and outlandish claims, "Razvedka I Kreml" is a fascinating book. The Russian version was clearly edited after some criticism of the English version, but the two follow one another fairly closely. Discussions of Fuchs and Pontecorvo are pretty much the same. Pontecorvo is incorrectly identified as 'Mlad' in the English version as well. Pavel Sudoplatov was a capable individual. He was involved in some murders, including Trotsky in Mexico City. He carried out at least one himself (a Ukranian independence fighter named Evgeny Konovalets, on Stalin's personal orders). He directed an extensive spy network efficiently. As he has pointed out, the agents were never caught red handed. And as a reward, after the death of Stalin and the arrest of Beria, he served 15 years in prison.

H.3. Aleksandr Feklisov

Feklisov wrote at least two books. The one in Russian used here is "Kennedi I Sovietskaya Agentura," which is identical to the English version "Beyond the Ocean and on an Island." These books do not mention Julius Rosenberg. A later book (although published before "Kennedi") originally in French, but translated into English is "The Man Behind the Rosenbergs." This book does tell the Rosenberg story from Feklisov's point of view. Both books discuss in some detail Feklisov's work with Klaus Fuchs in England after the war.

In "The Man Behind the Rosenbergs" Feklisov explains that the earlier work contained some information regarding the Rosenberg spy

---

[18] John Earl Haynes and Harvey Klehr, "VENONA," Yale University Press, New Haven, 2000, Venona 259 Moscow to New York, 21 March 1945.

ring, "portrayed in such veiled form that it was barely recognizable."[19] The reason given for such obfuscation was that the Russian Federation even in the 1990's did not wish to be identified with the Rosenbergs. Apparently, the Fuchs link was OK, despite the denial in Pravda in 1950 that the USSR had any knowledge of him whatsoever.[20]

Some of the misinformation given by Feklisov is worth recording. The code name 'Stanley' for Julius Rosenberg was made up by Feklisov. 'Stanley' was indeed a genuine NKVD code name, but for Kim Philby – a British spy, one of the Cambridge five. Feklisov goes on to say that 'Stanley' was 33 years old (Rosenberg was born in 1918, so was only 26), that his parents were born in Sweden (they were born in Russia), that his wife was a practicing medical psychologist (Ethel worked as a clerk), that they had no children (one son at the time, eventually two sons), and that they lived in the Bronx (they lived in Knickerbocker Village in lower Manhattan). Feklisov clearly did not want you to know whom he was talking about. So if the unsuspecting reader is trying to figure out who 'Stanley' really was, he/she has a formidable task ahead. The second book "The Man behind the Rosenbergs" comes to the rescue – learning who 'Stanley' really was is hopeless without it.

---

[19] Aleksandr Feklisov, "The Man behind the Rosenbergs," *ibid.*, p 103.
[20] Pravda, March 8, 1950, p 5.

Appendix J

Critical Assemblies

The Arzamas-16 critical assembly rig was called 'fast neutron boiler', with the Russian acronym FIKOBYN. It consisted of two subcritical parts separated by a gap that could be varied. As the gap was decreased, the activity increased, giving an intense source of fission neutrons and fragment gamma rays.

Figure J.1 shows the setup. A hemispherical shell of natural uranium was fixed in space over a similar hemispherical shell that held a plutonium sphere.[1] A hydraulic lift moved the bottom uranium tamper and the 8 kg ball of δ-phase ^{239}Pu up and down together by remote control. The operators were in a separate room behind 1 meter of reinforced concrete shielding. The ball was also split into two hemispheres, and a neutron source was placed at the center. The two shells were separated at the distance of closest approach by steel spacer rings of variable thickness. Below the spacers was a 5 mm thick duralumin plate grooved to contain 24 80 mg pellets of U_3O_8 (90% enriched in ^{235}U). The integrated neutron flux from the assembly was measured by the resulting fission of the ^{235}U. The hydraulic lift stopped when the bottom spacer touched the top uranium tamper. The activity of the plutonium depended of course on the amount present, on the thickness of the spacer rings, and on the outer radius of the uranium shell. The accident described in Chapter 10, and in Sakharov's memoirs, occurred in April, 1953, when 5 mm rather than 10 mm steel spacer rings

---

[1] Thomas P. McLaughlin, et al., "A Review of Criticality Accidents," LA-13638, *ibid.*, p 78. Figure 46,

were inserted by an operator working alone over lunch hour – all in violation of safety procedures. A criticality accident generated enough heat to partly melt the plutonium sphere. Alarms went off, and the operator lowered the hydraulic lift, stopping the activity. The amount of ^{235}U that fissioned in the pellets indicated that there were about 1×10^{16} ^{239}Pu fissions in the sphere – a tiny amount of the 2×10^{25} nuclei available.

A variation of the setup in Figure J.1 is shown in Figure J.2, also at Arzamas-16. Here the top was moveable, rather than the bottom, and the thickness of the uranium tamper could be varied. The activity, measured by the neutron flux, was defined by the neutron multiplication factor, k. For criticality, $k \rightarrow \infty$, so 1/k was plotted as a function of the thickness d of the spacers between the uranium hemispheres, for various outer radii of the tamper. Figure J.3 shows the results, with the extrapolation to 1/k = 0 for the thickest tamper radius. The actual dimensions were not given. The intercept 1/k=0 was not actually reached – only inferred by extrapolation. This setup was a technique for studying the amount of material needed for a critical mass without compression, as well as being a useful neutron source for other experiments.

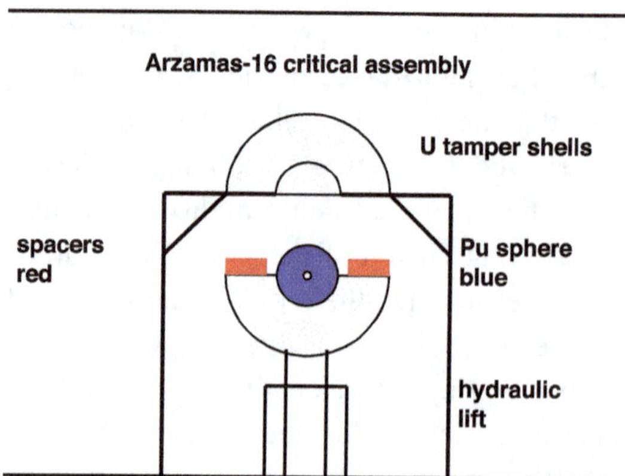

Figure J.1. Critical assembly FIKOBYN at Arzamas-16. Steel spacers prevented the upper uranium hemisphere from contact with the plutonium. The hydraulic lift raised the bottom part into the fixed upper tamper. This figure is adopted from Figure 46 in LA-13638.

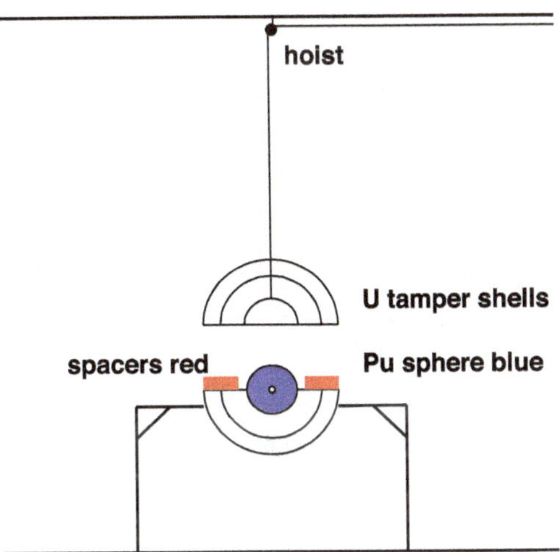

Figure J.2. A later variation of the apparatus in Figure 1, also at Arzamas-16. Here the top moved over a fixed bottom, and the thickness of the uranium shells could be varied as well as the spacers.

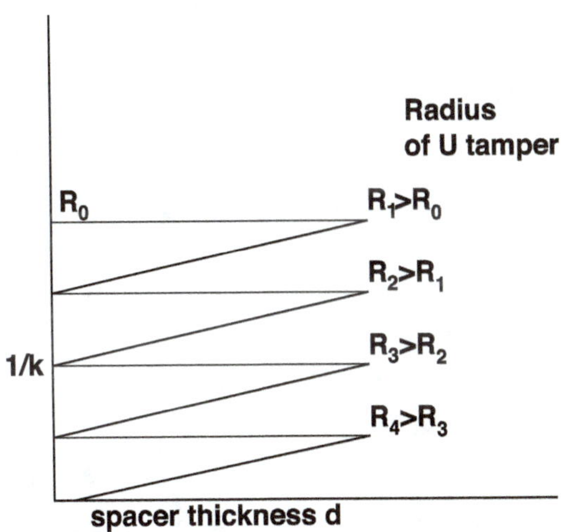

Figure J.3. Extrapolation to criticality by varying both the spacer thickness and the outer radius of the uranium tamper.

The plots shown in Figure J.3 require some explanation. No dimensions are given in the Figure, so one must guess how to proceed. The initial tamper radius R_0 in contact with the sample produces an activity $1/k_0$ at the top intercept. Then as the top tamper hemisphere is raised, the activity drops – $1/k$ rises – but when the assembly is fully opened to the spacer thickness endpoint, tamper is added to bring the activity up to $1/k_0$. Bringing the assembly in gives a new, lower intercept (higher activity). Taking it out again follows the same line, but then the thickness is increased to match the activity at the second intercept. The process is repeated with thicker and thicker tampers until criticality is reached by extrapolation. This procedure clearly illustrates the fact that the tamper for a critical assembly increases the neutron density by reflection, while in the bomb case it is the suppression of free expansion that is the dominant effect. The critical assembly is not supposed to melt. The bomb explodes so fast that reflected neutrons are not as useful.

Bibliography

Monographs and Memoirs

Allilueva, Svetlana, "Doch Stalina," Algoritm, Moskva, 2013. English version "Twenty Letters to a Friend," Avon Books, New York, 1967.

Applebaum, Anne, "Gulag," Anchor Books, New York, 2004.

Ardenne, Manfred von "Erinnerungen," Droste Verlag, Dusseldorf, 1997.

Barwich, Heinz and Elfi, "Das Rote Atom," Scherz Verlag, Munich, 1967.

Beria, Lavrenty Pavlovich "Tainy Dnevnik 1937-1953," Yauza – Press, Moskva, 2012.

Beria, Sergo, "Moi Otets Narkom Beria," Algoritm, Moskva, 2013.

Bernstein, Jeremy, "Hitler's Uranium Club," Copernicus Books, Springer Verlag, New York, 2001.

Bessarab, Maiya, "Lev Landau," Oktopus Press, Moskva, 2007.

Bird, Kai and Martin J. Sherwin, "American Prometheus," Alfred A. Knopf, New York, 2005.

Carr, Barnes, "Operation Whisper," University Press of New England, Lebanon, NH, 2016.

Chambers, James, "The Devil's Horsemen," Athenium, New York 1979. (History of Kazan).

Cochran, Bert, "Harry Truman and the Crisis Presidency," Funk and Wagnalls, New York, 1973.

Conant, Jennet, "109 East Palace," Simon and Schuster, New York, 2005.

Coster-Mullen, John, "Atom Bombs," copyright John Coster-Mullen, 2014.

Cray, Ed, "General of the Army George C. Marshall," WW Norton & Co, New York, 1990.

Dalitz, Richard H., and Sir Rudolf Peierls, editors, "Selected Scientific Papers of Sir Rudolf Peierls," World Scientific, Singapore, 1997.

Davidson, Eugene, "The Trial of the Germans," McMillan Co, New York, 1966.

Davis, Nuel Pharr, "Lawrence and Oppenheimer," Fawcett Premier Books, Greenwich, Conn, 1968.

Djilas, Milovan, "Conversations with Stalin," Harcourt Brace, Orlando, Florida, 1990.

Dornberger, Walter, "V-2," The Viking Press, New York, 1954.

Dyson, George, "Turing's Cathedral," Pantheon Books, New York, 2012.

Eisenhower, David, "Eisenhower at War 1943-1945," Random House, New York, 1986.

Esposito, S. and O. Pisanti, eds. "Neutron Physics for Nuclear Reactors," World Scientific, Singapore, 2010.

Farmelo, Graham, "Churchill's Bomb," Basic Books, New York, 2013.

Feklisov, Aleksandr, "Kennedy I Sovetskaya Agentura," Algoritm, Moskva, 2011.

Feklisov, Aleksandr, "The Man Behind the Rosenbergs," Enigma Books, New York, 2001.

Fermi, Laura, "Atoms in the Family," University of Chicago Press, Chicago, 1954.

Fermi, Enrico, "Collected Papers of Enrico Fermi," University of Chicago Press, Chicago, 1962.

Fischer, John, "Why they Behave like Russians," Harper & Brothers, New York, 1947.

Frayn, Michael "Copenhagen," London Methuen, New York, Anchor Books 1998 (a play).

Friedel, Frank, "Franklin D. Roosevelt," Little Brown and Company, Boston, 1990.

Gaister, Inna, "Deti Vragov Naroda," Vozvrashenie, Moskva, 2012.

Galler, Meyer and Harlan Marquess, "Soviet Prison Camp Speech," University of Wisconsin Press, Madison, WI. 1972.

Gamow, George, "My World Line, an Informal Autobiography," Viking Press, New York, 1970.

Golovin, Igor, "Igor Kurchatov," Atomizdat, Moskva, 1967.

Gorobets, B.S. "Sovietskie Fiziki Shutyat," Knizhny Dom Librokom, Moskva, 2012.

Griffiths, David J. "Introduction to Electrodynamics, 3rd Ed," Prentice Hall, New Jersey, 1999.

Gubarev, Vladimir S., "Atomnaya Bomba," Algoritm, Moskva, 2009.

Gubarev, Vladimir S., "Arzamas-16," Izdat, Moskva,1992.

Harriman, W. Averell and Elie Abel, "Special Envoy to Churchill and Stalin," Random House, New York, 1975.

Haynes, Earl and Harvey Klehr, "Venona," Yale University Press, New Haven, Conn, 2000.

Hirsch-Heisenberg, Anna Maria, "My Dear Li," Yale University Press, 2016. (Correspondence between Werner and Elisabeth Heisenberg).

Holloway, David, "Stalin and the Bomb," Yale University Press, New Haven, 1994.

Jacobsen, Annie, "Operation Paperclip," Little, Brown and Co, New York, 2014.

Jeffery, Inez C., "Inside Russia, the Life and Times of Zoya Zarubina," Eakin Press, Austin, Texas, 1999.

Jordan, George Racey, "From Major Jordan's Diaries," Harcourt Brace, New York, 1952.

Kahn, David, "The Codebreakers," Scribners, New York, 1996.

Khalatnikov, I. M. "Dau, Kentavr, I Drugie," Fizmatlit, Moskva, 2012.

Kiernan, Denise, "The Girls of Atomic City," Simon and Schuster, New York, 2013.

Khrushchev, Sergei, "Trilogia ob Otse, Reformator," Vremya, Moskva, 2010.

Kuznetsova, Raisa, "Kurchatov v Zhizni," Glavarkhiva, Moskva, 2007.

Knight, Amy, "Beria, Stalin's First Lieutenant," Princeton University Press, Princeton, N.J. 1993.

Kojevnikov, Alexei B., "Stalin's Great Science," Imperial College Press, London, 2004.

Lamarsh, John R. "Nuclear Reactor Theory," Addison Wesley, Reading, MA, 1966.

Lamphere, Robert J. and Tom Shachtman, "The FBI-KGB War," Berkley Books, New York, 1987.

Laurence, William L "Dawn over Zero," Alfred A. Knopf, New York, 1953.

Layton, Edwin T. "And I Was There," William Morrow, New York, 1985.

Lewin, Ronald, "Ultra goes to War," McGraw Hill, New York, 1978.

Libby, Leona Marshall, "The Uranium People," Crane Russak & Co, Charles Scribner's Sons, New York, 1979.

Livingston, M. Stanley, and John P. Blewett, "Particle Accelerators," McGraw Hill, New York, 1962.

Loeb, Leonard, "Kinetic Theory of Gases," McGraw Hill, New York, 1934.

Medvedev, Zhores A. "Nuclear Disaster in the Urals," W.W. Norton, New York, 1979.

Melvin, Mungo, "Manstein," Thomas Dunne Books, St Martin's Press, New York, 2010.

Mikoyan, Anastas, "Tak Bylo," Vagrius, Moskva, 1999.

Modin, Yuri, "My Five Cambridge Friends," Farrar Straus Giroux, New York, 1994.

Montefiore, Simon Sebag, "Stalin, the Court of the Red Tsar," Vintage Books, New York, 2003.

Morath, Inge, and Arthur Miller, "In Russia," Viking Press, New York, 1969.

Naimark, Norman, "The Russians in Germany," Belknap Press of Harvard University, Cambridge, MA, 1995.

Nelson, Anne, "Red Orchestra," Random House, New York, 2009.

Northrup, Doyle L. and Donald H. Rock, "Detection of Joe-1" Central Intelligence Agency Studies in Intelligence, September, 1966. Available on cia.gov.

Pais, Abraham, "Niels Bohr's Times," Oxford University Press, 1991.

Peierls, Rudolf, "Bird of Passage," Princeton University Press, Princeton, NJ, 1985.

Powers, Thomas, "Heisenberg's War," Alfred A. Knopf, New York, 1993.

Radzinsky, Edvard, "Stalin, Zhizn I Smert," AST, Moskva, 2010.

Reed, Bruce C. "The Physics of the Manhattan Project, 3rd Ed," Springer, Heidelberg, 2015.

Rhodes, Richard, "Making of the Atomic Bomb," Simon and Schuster, New York, 1986.

Rhodes, Richard, "Dark Sun," Simon and Schuster, New York, 1995.

Richards, Hugh T. "Through Los Alamos, 1945," The Arlington Place Press, Madison, WI, 1993.

Riehl, Nikolaus and Frederick Seitz, "Stalin's Captive," American Chemical Society and Chemical Heritage Foundation, New York, 1996.

Rokossovsky, Konstantin K. "Soldatsky Dolg," Olma Press, Moskva, 2002.

Shifman, M. "Under the Spell of Landau," World Scientific, Singapore, 2013.

Sinev, N.M. "Enriched Uranium for Atomic Weapons and Energy," TsNIIAtominform, Moskva, 1991.

Stalina, Galina Djugashvili, "Taina cemi Voshdya," AST Zebra, Moskva, 2007.

Starinov, Ilya G. "Over the Abyss," Mass Market Paperbacks New York, 1995.

Sudoplatov, Pavel, "Razvedka I Kreml," TOO Geya, Moskva,1996.

Sudoplatov, Pavel and Anatoly Sudoplatov, "Special Tasks," Little Brown & Co, New York, 1994.

Szanton, Andrew, "Recollections of Eugene P. Wigner," Plenum Press, New York, 1992.

Tipler, Paul and Ralph Llewellyn, "Modern Physics 5th Ed," W.H. Freeman, New York 2008.

Tuchman, Barbara, "The Zimmerman Telegram," Ballentine Books, New York, 1966.

Tucker, Robert C. "Stalin in Power," W.W. Norton and Company, New York, 1990.

Turchetti, Simone, "The Pontecorvo Affair," The University of Chicago Press, Chicago, 2012.

Walker, Mark, "German National Socialism and the Quest for Nuclear Power," Cambridge University Press, Cambridge, 1989.

Warner, Edith, "In the Shadow of Los Alamos," University of New Mexico Press, Albuquerque, NM, 2008.

Walker, Stephen, "Shockwave," Harper Collins, New York, 2005.

Walling, Michael G. "Forgotten Sacrifice," Osprey Publishing, Oxford, England, 2012.

Weinstein, Allen, and Alexander Vassiliev, "The Haunted Wood," Random House, New York, 1999.

Werth, Alexander, "Russia at War 1941-1945," E.P. Dutton & Co, New York, 1964.

Wolf, B. H. "Handbook of Ion Sources," CRC Press, 1995.

Zeldovich, Ya. B. "Theory of Shock Waves and Introduction to Gas Dynamics," Foreign Technology Division, Air Force Systems Command FTD-HT-66-258, 1967. (Translation from Russian.)

Zhukov, Georgy K. "Vospominaniya I Razmyshleniya," Olma Press, Moskva, 2002.

Ziegler, Charles A. and David Jacobson, "Spying without Spies," Praeger Publishers, Westport, Connecticut, 1995.

Official Histories

"Istorii Velikoi Otetchestvennoi Voiny Sovetskogo Soyuza," abbreviated IVOVSS, the official history of WWII from the perspective of the Soviet Union. Five Volumes.

Federal Atomic Energy Agency of the Russian Republic, "Atomny Proekt SSSR." A collection of transcriptions of declassified official documents in Russian pertaining to the Soviet Atomic Project, available on the website: work.atomlandonmars.com, built by Alex Wellerstein. They are referenced in the text as Atomnyy_proekt_SSSR. There are 12 separate pdf files in the directory. Entries in each pdf file are referenced by number in sequence.

Conference Proceedings

HISAP'96, International Symposium "Science and Society, History of the Soviet Atomic Project" Dubna, 14-18 May, 1996. Three volumes. Isdat, Moskva, 1997. The following articles in Volumes 1 and 2 are referred to in the text by author and page number.

Vol 1

Page 23. L.D. Ryabev, et al, Historical introduction.

Page 49 V.B. Barkovsky, Scientific-technical Intelligence.

Page 62 A.K. Kruglov, Participants in the Soviet Atomic Project.

Page 80 E.P. Ryazantsev, Research Reactors.

Page 101 Vetrov, V.I. and A.N. Eremeev, Sources of Uranium ore.

Page 116 V.I. Merkin, Industrial Reactors.

Page 135 Leya Sokhina, Obtaining high purity plutonium.

Page 146 F.G. Reshetnikov, Industrial production of uranium and transuranic elements.

Page 156 V.N. Prusakov and A.A. Sazykin, Uranium enrichment

Page 176 B.F. Gromov et al., Laboratory 'V'.

Page 184 L.V. Altshuler and K. K. Krupnikov, Experimental studies at Arzamas-16.

Page 192 A.A. Brish, Branch KB-11 (the same place as Arzamas-16).

Page 214 A.A. Samarsky, Calculation of the energy of the bomb.

Page 264 V.I. Mostovoy, Nuclear physics at Laboratory # 2.

Page 284 V.P. Dzhelepov, Early history of Dubna.

Page 291 V.V. Alekseev and B.V. Litvinov, Soviet Atomic Project and national economy.

Vol 2

Page 43 K.A. Petrzhak, A.A. Rimsky-Korsakov, and V.P. Eismont, Nuclear Physics at the Radium Institute.

Page 68 I.F. Zhezherun, Nuclear Reactor F-1.

Page 99 N.V. Knyazkaya, Igor Kurchatov and the beginnings of the Soviet Atomic Project.

Page 139 Yu. N. Smirnov, Georgy Flerov and the beginnings of the Soviet Atomic Project.

Page 172 A.A. Brish Sarov 1947-1955.

Page 179 I.S. Drovenikov and S. V. Romanov, Journey of Soviet physicists to Germany May, June 1945.

Page 189 G.A. Sosnin, Founders of atomic industry (Mostly Boris Vannikov).

Page 195 A.G. Plotkina and E.M. Voinov, Academician Isaak K. Kikoin.

Page 207 B.L. Ioffe, Why 'Truba' did not work. (This article has much valuable information.)

Page 245 S.S. Ilizarov, L.D. Landau and the Soviet Atomic Project.

Page 260 P.E. Rubinin, Kapitza, Beria, and the bomb.

Page 280 R.G. Pose, Recollections of Obninsk.

Page 300 Ya. P. Dokuchaev, Test of RDS-1.

Page 320 R.B. Kotelnikov and V.A. Tumbakov, Industrial resources of the USSR.

Page 331 A. Heinman-Gruder, Need for Uranium Ore.

Page 340 J. Janouch, Czechoslovak uranium.

Page 351 E.I. Ilenko and N.A. Abramova, Early radiochemical technology.

Page 360 G.N. Yakovlev, First work to obtain plutonium.

Page 382 V.I. Zemlyanukhin, Industrial Radiochemistry for plutonium separation.

Page 391 I.A. Reformatsky and G.N. Yakovlev, First Hot Lab at Laboratory #2.

Page 398 N.S. Burdakov, et al., First Reactor 'A' at Mayak.

Page 411 V.A. Khalkin, First Extraction of plutonium from solution in factory 'B.'

Page 431 N.I. Ivanov, First nuclear explosive.

Page 438 O.D. Simonenko, Development of Centrifuges for Isotope Separation.

Page 463 N.G. Makeev, Impulse x-ray technology.

Page 476 Y. Zu. S. Zamyatnin, First critical mass experiments at Arzamas-16.

Journal and Periodical Articles

Pravda, January 28, 1936 p 3 "Muddle instead of Music" critique of an opera by Shostokovich.

Pravda, June 25, 1945, WWII victory parade.

Pravda, November 7, 1945, Molotov's speech on atomic energy

Pravda, 25 September, 1949, TASS statement regarding Truman's announcement of a Soviet atomic bomb test.

Pravda, March 8, 1950, p 5 USSR had no knowledge of Klaus Fuchs

Pravda, March 1, 1955, p 3. Essay by Bruno Pontecorvo

Izvestia, 1941, #300, p 3, Story about setting up a factory moved from west to east.

Izvestia June 26, 1940, p 4. Use of Atomic Energy

Milwaukee Journal Sentinal, November 28, 2011, Obituary for Lana Peters (Stalin's daughter).

New York Times, March 23, 2004, "Slender and Elegant, it Fuels the Bomb" by William J. Broad. (Article about centrifuges)

Science Magazine Vol **348**, #6241 (19 June 2015) p 1320. Article on isotope separation by centrifuge.

Scientific American "In the World of Science," special edition on the 70th anniversary of the Kurchatov Institute. 2013.

Wall Street Journal October 30, 2013. Radiation cleanup problems.

Washington Post, December 1, 2016. Exoneration request by Rosenberg sons.

The journal Voprosy Istorii Estestvoznaniya I Technikhi (Questions in history of natural science and technology) is abbreviated VIET.

Akhiezer, Aleksandr, Physicalishe Zeitschrift der Sowietunion **11**, 263 (1937).

Albright, Joseph and Marcia Kunstel, "Mlad and Star," Bulletin of the Atomic Scientists, Jan/Feb 1998, p 48ff.

Allison, S.K., Lester Skaggs, and Nicholas Smith, Jr. Phys Rev **54**, 171, (1938). Early nuclear reaction experiment.

Altshuler, Lev, Oral Interview, Literaturnaya Gazeta June 6, 1990, p 13.

Anderson, Herbert L. "Scientific Uses of the MANIAC," Journal of Statistical Physics, **43**, 731 (1986).

Barkovsky, V.B., VIET, 1994, #4, p 122.

Bartlett, Neil, Proceedings of the Chemical Society of London (6), 218, (1962).

Beams, J.W. and F.B. Haynes, Phys Rev **50**, 491, (1936). Mass separation by centrifuge.

Bethe, Hans, Kurt Gottfried, and Roald Sagdeev, "Did Bohr Share Nuclear Secrets?," Scientific American, May, 1995, p 85.

Bohr, Niels, Philosophical Magazine (**6**), 26, 1 (1913).

Bohr, Niels, and John A. Wheeler, Phys Rev **56**, 426 (1939).

Brill, T. and H.V. Lichtenberger, Phys Rev **72**, 585 (1947). Neutron velocity selector.

Condon, Edward U, Memoirs of Members of the National Academy of Sciences.

Cherenkov, Pavel A. Nobel Address, 1958.

Clusius, K. and G. Dickel, Naturwiss. **26**, 546 (1938). Thermal diffusion.

Cockcroft, J.D. and E.T. Walton, Proc Roy Soc (London) **A136**, 229 (1932), and **A144**, 704, (1934). Electrostatic accelerator.

Curie, Marie, Nobel Address, 1911.

Dempster, Arthur J. Biographies of members of the National Academy of Sciences.1952.

Feder, Toni, "Building a cyclotron on a shoestring," Physics Today **57**, 11, 30 (2004).

Fermi, Enrico, Nobel Address, 1938.

Fermi, Enrico, J. Marshall, and L. Marshall, Phys Rev **72**, 193(1947). Neutron velocity selector.

Feynman, Richard, "Los Alamos from Below," Engineering and Science 39, #2 (Jan-Feb 1976) p 12.

Friedman, Herbert, Luther B. Lockhart, and Irving H. Blifford, Physics Today, November, 1996, p 38. Observation of 'Joe-1' in rainwater. This issue of Physics Today is devoted to a review of the international conference at Dubna, HISAP'96.

Frisch, Otto, Nature **143**, 276, (1939). Observing nuclear fission fragments in an ion chamber.

Furry, W.H., R.C. Jones and L: Onsager, Phys Rev **55**, 1083 (1939). Thermal diffusion

Gamow, George, Zeitschrift für Physik **51**, 204 (1928). Alpha decay.

Goleusova, L.P. "Arzamas-16," VIET 1994, #4, p 89.

Goloborodko and Leipunsky, Phys Rev **56**, 891 (1939).

Gorelik, G.E. "Those who did not become academicians," Priroda, 1990, #1 p 123.

Gorelik, G.E. "Natural philosophy physics problems in 1937," Priroda, 1990, #2, p 93.

Greinacher, H. Zeitschrift für Physik **4**, 195 (1921). Voltage multiplier.

Gurney, R.W. and E.U. Condon, Nature **122**, 439 (1928). Alpha decay.

Hahn, Otto, and Fritz Strassmann, Naturwissenschaften **27**, 11 (1939).

Halban, H. von, F. Joliot, and L. Kowarski, Nature **143**, 680 (1939).

Herb, Raymond G. Biographical Memoirs of members of the National Academy of Sciences, 1996.

Hinterberger, F. "Electrostatic Accelerators," available online at cds.cern.ch.

Hogerton, John F. Chemical and Metallurgical Engineering, December, 1945, p 98.

Hogerton, John F. and Ellsworth Raymond, "When will Russia have the Atomic Bomb?" Look Magazine, March 16, 1948, p 27.

Hornig, Donald Obituary New York Times January 27, 2013.

Kaftanov, S.V. "Po Trevoge," Khimia I Zhizn, #3, 1985, p 6ff.

Keith, P.C. Chemical and Metallurgical Engineering, February, 1946, p 112.

Kramer, David, "Cleanup of Cold War Nuclear Waste Drags On," Physics Today, July, 2017 p 28.

Kurchatov, Igor, et al. Comptes Rendus Acad des Sciences, Vol **200**, p 1201 (1935).

Khriplovich, Iosif, "The Eventful Life of Fritz Houtermans," Physics Today Vol **45**, p 29 (July 1992).

Landau, L.D. and R. Peierls, Zeitschrift für Physik, **62**, 188 (1930).

Landau, L.D. Zeitschrift für Physik **64**, 629 (1930).

Landau, L.D. and R. Peierls, Zeitschrift für Physik **69**, 56 (1931).

Larin, Vladislav, "Unknown Accidents at Mayak," Energia #1, 2000.

Libby, Willard, Phys Rev **55**, 1269 (1939). Search for spontaneous fission of uranium

Lifshitz, E.M. "Life of L.D. Landau," Uspekhi fizicheskikh nauk, **97**, 169 (1969).

Lunghi, T. et al., Phys Rev Lett **114**, 150501, (2015), Quantum random number generator.

Malik, John, "Yields of Hiroshima and Nagasaki Nuclear Explosions," LA-8819, Los Alamos National Laboratory, 1985.

McLaughlin, Thomas P. et al. "A Review of Criticality Accidents" Los Alamos National Laboratory LA-13638, 2000.

McMillan, Edward M. Nobel Address, 1951.

Meitner, Lise, and O.R. Frisch, Nature **143**, 239 (1939).

Nier, A.O., E.T. Booth, J.R. Dunning, and A.V. Grosse, Phys Rev **57**, 546 (1940). Uranium fission cross sections.

Nemenov, Leonid, Tekhnika Molodezhi, June 1975, p18. "Poslezavtra Nachnem Obluchenia."

Oleinikov, Pavel, "German Scientists in the Soviet Atomic Project," The Nonproliferation Review, Vol **7**, #2 (2000).

Pagano, Owen N., "The Spy who stole the Urchin," Senior Thesis, History Department, Washington University, class of 2014.

Parkins, W.E. Physics Today **58**, 45 (2005). Electromagnet isotope separation at Oak Ridge

Penney, Lord William, D.E.J. Samuels, and G.C. Scorgie, "Nuclear Explosive Yields at Hiroshima and Nagasaki," Philosophical Transactions of the Royal Society of London, A266(1177) 357(1970). This journal first published in 1666.

Pestov, Stanislav, "Stranitsy Istorii Tainy Atomnoi Bomby," Argumenty I Fakty, #41, October 14, 1989.

Petrzhak, K.A. and G.N. Flerov, "Spontaneous Fission of Uranium" Zhurnal Eksperimentalnoi I Teoreticheskoi Fiziki (JETP) 9, 143 (1939).

Reines, Frederick, Nobel Address, 1995.

Rossi, Bruno, Nature 125, 636 (1930). Early coincidence circuit.

Rukavichnikov, V.N. Phys Rev 52, 1077, (1937). Leningrad cyclotron.

Schoenberg, D. "Piotr Leonidovich Kapitza," Biographical Memoirs of Fellows of the Royal Society, 1984.

Seaborg, Glenn T. Nobel Address, 1951.

Segre, Emilio, "The Discovery of Nuclear Fission," Physics Today 42, 38 (1989).

Shpolsky, E.V., Uspekhi Fizicheskikh Nauk, 21, 253 (1939).

Smith, L.P., W.E. Parkins, and A.T. Forrester, Phys Rev 72, 989 (1947). Control of space charge in electromagnetic mass separators.

Sonin, A.S. "The Meeting that did not take place," Priroda, 1990 #3, #4, and #5.

Soran, Diane M., and Danny B. Stillman, "An Analysis of the Alleged Kyshtym Disaster," Los Alamos National Laboratory LA-9217 MS.

Steury, Donald P. Studies in Intelligence 49, no 1 (2005), p 19-26.

Stimson, Henry L. Obituary New York Times, October 21, 1950.

Sumerville, I.C. "A Simple Compact 1 MV 4kJ Marx," Proceedings of the Pulsed Power Conference, Monterey, CZ, June 11, 1989, p 744.

Tamm, Igor, Uspekhi Fizicheskikh Nauk, 16, 922 (1936).

Terletsky, Yakov P. Uspekhi Fizicheskikh Nauk, 164, #2 1994. Memorial biography.

Terletsky, Yakov P, "Copenhagen Operation of Soviet Intelligence," VIET #2, p 18-44, 1994.

Truman Library, Public papers of Harry S. Truman.

Urey, Harold, F. Brickwedde, and G. Murphy, Phys Rev 39, 164 (1932). Discovery of deuterium.

Van de Graaff, R.J. Phys Rev 38, 1919A, (1931). Electrostatic accelerator.

Van Delft, Dirk, "Paul Ehrenfest's final years," Physics Today 67, 1 (2014) p 41.

Way, K. and Eugene Wigner, Phys Rev 73, 1318 (1948), Radiation from fission products.

Wigner, Eugene, "Unreasonable Effectiveness of Mathematics in the Natural Sciences," Comm. In Pure and Applied Mathematics, Vol 13, no. 1 (Feb. 1960) John Wiley & Sons, Inc., NY.

Wood, H.G. Alexander Glaser, and R. Scott Kemp, Physics Today 61, #9, p 40 2008.

Yatskov, A.A. VIET 1992, #3, p 103.

Zeldovich, Yakov B. and Yu. B. Khariton, Uspekhi Fizicheskikh Nauk 25, 381 (1941).

Zorich, I. Priroda 1990, #9, p 106. Stalin's quote on the physicists.

WEB Sites

If a web address does not work, Google the subject and follow its lead.

Oral histories can be found at www.aip.org (American Institute of Physics).

Work.atomlandonmars.com has the directories for 12 pdf files of Atomnyy_proekt_SSSR

ru.wikipedia.org/wiki/Fok_Vladimir_Alexandrovich

ru.wikipedia.org/wiki/Pulkovskoe_delo.

https://archive.org/details/sovietatomicespi1951unit/ transcript of JCAE hearings

users.telenet.be/d_rijmenants/en/onetimepad.htm; description of coding from a one time pad.

www.atomicheritage.org/profile/george-koval

www.world-nuclear.org contians much valuable information

www.nrc.gov information for radiation workers.

www.wikipedia.org/wiki.Separative_work_units. For the separative work formulas used to compare different methods of isotope separation,

www.fas.org/man/dod-101/navy Data for chemical explosives

www.kayelaby.npl.co.uk fission cross section data

www.nuclearweaponarchive.org useful current nuclear data for fissile materials

Index